Lectures on Coulomb and Riesz Gases

American Mathematical Society

Colloquium Publications
Volume 70

Lectures on Coulomb and Riesz Gases

Sylvia Serfaty

AMS AMERICAN MATHEMATICAL SOCIETY

2020 *Mathematics Subject Classification.* Primary 60K35, 82C22, 35Q82, 82B05, 60F05, 60B20, 82B21, 82B26.

For additional information and updates on this book, visit
www.ams.org/bookpages/coll-70

Library of Congress Cataloging-in-Publication Data

Names: Serfaty, Sylvia author
Title: Lectures on Coulomb and Riesz gases / Sylvia Serfaty.
Description: Providence, Rhode Island : American Mathematical Society, [2026] | Series: Colloquium publications, 0065-9258 ; volume 70 | Includes bibliographical references and index.
Identifiers: LCCN 2026015824 | ISBN 9781470486044 hardcover | ISBN 9781470486211 ebook
Subjects: LCSH: Statistical mechanics–Mathematics | Coulomb functions | Random matrices | Limit theorems (Probability theory) | AMS: Probability theory and stochastic processes – Special processes – Interacting random processes; statistical mechanics type models; percolation theory | Statistical mechanics, structure of matter – Time-dependent statistical mechanics (dynamic and nonequilibrium) – Interacting particle systems | Partial differential equations – Equations of mathematical physics and other areas of application – PDEs in connection with statistical mechanics | Statistical mechanics, structure of matter – Equilibrium statistical mechanics – Classical equilibrium statistical mechanics (general) | Probability theory and stochastic processes – Limit theorems – Central limit and other weak theorems | Probability theory and stochastic processes – Probability theory on algebraic and topological structures – Random matrices (probabilistic aspects) | Statistical mechanics, structure of matter – Equilibrium statistical mechanics – Continuum models (systems of particles, etc.) | Statistical mechanics, structure of matter – Equilibrium statistical mechanics – Phase transitions (general)
Classification: LCC QC174.82 .S474 2026
LC record available at https://lccn.loc.gov/2026015824

Colloquium Publications ISSN: 0065-9258 (print); 2473-3946 (online)
DOI: https://doi.org/10.1090/coll/70

10 9 8 7 6 5 4 3 2 1 31 30 29 28 27 26

Contents

Preface

This book grew out of notes written for graduate courses that I taught at the Courant Institute, the Ecole Normale Supérieure at the invitation of the Fondation Sciences Mathématiques de Paris, and at the 2024 Saint-Flour Probability summer school. They were meant to reflect the advances made since the previous lecture notes [**Ser15**].

The goal of this book is not to provide an exhaustive view of the topic but rather a necessarily biased but self-contained presentation of the approach to Coulomb gases that has emerged from a body of work initiated in collaboration with Etienne Sandier and continued with Nicolas Rougerie, Mircea Petrache, Thomas Leblé, Simona Rota Nodari, Scott Armstrong, Matthew Rosenzweig, and Luke Peilen; and which one may characterize as an electric-formulation-based analysis of the statistical mechanics of Coulomb and Riesz gases.

The book assumes some standard knowledge of analysis, measure theory, PDE, and basic probability at the level of a first- or second-year graduate text. It starts by reviewing some standard notions and facts, particularly on equilibrium measures, before moving on to more recent research. It is meant to serve both as a text for researchers interested in learning about the topic and as a point of reference collecting the various results in one place.

It introduces the main concepts used in our analytic, electrostatic-based approach: the electric formulation, the modulated energy, the screening procedure, the renormalized jellium energy, the transport method, with streamlined and updated definitions and results. For instance, all cases of $\mathsf{s} \in [\mathsf{d}-2, \mathsf{d})$ are treated, including when $\mathsf{s} \leq 0$ and the Coulomb case in dimension 1. I tried to minimize the assumptions as much as possible and to allow for the broadest temperature regimes possible. The text focuses mostly on the analysis of the canonical Gibbs measure of Coulomb and Riesz gases in an external potential, but takes a detour to discuss the application of the tools to the modulated energy method for mean-field limits of the dynamics of such gases. I have chosen to present results for both the Coulomb and Riesz cases whenever treating the Riesz case did not add too much complexity, and to restrict discussion to the Coulomb case and refer the reader to the relevant papers when it did, that is, in all instances where the screening procedure needs to be used.

I would like to thank the students who followed my courses for their feedback. I owe much gratitude to Thomas Leblé for his invaluable help all along this project. Many thanks to him, Octave Mischler, and Eric Thoma for their help with some proofs. Thanks also to Sungsoo Byun, Antonin Chodron de Courcel, Luke Peilen, Matt Rosenzweig, and Ofer Zeitouni for their careful reading and feedback. Let me also thank Peter Forrester, Djalil Chafaï, Yacin Ameur, and Paul Bourgade for help with references.

This project was supported by the Simons Foundation through the Simons Investigator program, by NSF grants DMS-2000205 and DMS-2247846 and by the Fondation Sciences Mathématiques de Paris.

Sylvia Serfaty
New York, July 2024
Paris, December 2025

CHAPTER 1

Introduction

1.1. Setting: Coulomb, logarithmic, and Riesz gases

Let $N \geq 1$ and let X_N denote an N-tuple of points $(x_1, \ldots, x_N)$ in $(\mathbb{R}^\mathsf{d})^N$, where $\mathsf{d} \geq 1$ is the dimension. We are interested in energy functionals $\mathcal{H}_N : (\mathbb{R}^\mathsf{d})^N \to (-\infty, +\infty]$ of the form:

$$\mathcal{H}_N(X_N) := \frac{1}{2} \sum_{1 \leq i \neq j \leq N} \mathsf{g}(x_i - x_j) + N \sum_{i=1}^N V(x_i), \tag{1.1.1}$$

where $\mathsf{g} : \mathbb{R}^\mathsf{d} \to (-\infty, +\infty]$ is called the *pair interaction potential*, and $V : \mathbb{R}^\mathsf{d} \to (-\infty, +\infty)$ is called the *external field* or *confining potential.*

In the cases that we study, g is given by

$$\mathsf{g}(x) = \begin{cases} \frac{1}{\mathsf{s}} |x|^{-\mathsf{s}} & \mathsf{s} \neq 0 \\ -\log |x| & \mathsf{s} = 0, \end{cases} \tag{1.1.2}$$

and we call the first case the Riesz case and the second the logarithmic case. The latter can be obtained as the formal $\mathsf{s} \to 0$ limit of the general Riesz case by noting that $-\log|x| = \lim_{\mathsf{s} \to 0} \left(\frac{1}{\mathsf{s}} |x|^{-\mathsf{s}} - 1\right)$. Whenever the parameter s appears, according to the definition (1.1.2) it will be with the convention that $\mathsf{s} = 0$ in the logarithmic case.

We are particularly interested in the case $\mathsf{s} \geq 0$ where g is singular, but can handle some instances of $\mathsf{s} \leq 0$ as well. The text will focus on the more specific regime

$$\mathsf{d} - 2 \leq \mathsf{s} < \mathsf{d} \tag{1.1.3}$$

in all dimensions, which includes the important Coulomb case

$$\mathsf{s} = \mathsf{d} - 2. \tag{1.1.4}$$

When $\mathsf{s} < \mathsf{d}$, the interaction kernel g is integrable near 0, making this the *potential case*, for which mean-field theory and potential theory can be applied (see Chapters 2 and 3). The case $\mathsf{s} \geq \mathsf{d}$, where g is not integrable near the origin, is called the *hypersingular case.* It is not amenable to potential theory, and behaves much more like a short-range interaction problem. We will not discuss that case but refer instead to [**BHS19**, **HLSS18**] for instance.

Let us underline that the interaction is in all cases *purely repulsive*, which explains the need for the confining potential V, on which we will make precise assumptions in the next chapters. In short, we take it fairly regular and growing sufficiently fast at infinity. Alternatively, one could work on the torus $\mathbb{T}^\mathsf{d}$. The techniques developed here would apply as well and would even suppress certain boundary difficulties.

Let β be a positive real number called the *inverse temperature*, which may depend on N. We let $\mathbb{P}_{N,\beta}$ be the probability measure on $(\mathbb{R}^{\mathsf{d}})^N$ whose density with respect to the standard Lebesgue measure $dX_N := dx_1 \dots dx_N$ is given by:

$$d\mathbb{P}_{N,\beta}(X_N) := \frac{1}{Z_{N,\beta}} \exp\left(-\beta N^{-\frac{\mathsf{s}}{\mathsf{d}}} \mathcal{H}_N(X_N)\right) dX_N, \tag{1.1.5}$$

where $Z_{N,\beta}$ is a normalizing constant called the *partition function*:

$$Z_{N,\beta} := \int_{(\mathbb{R}^{\mathsf{d}})^N} \exp\left(-\beta N^{-\frac{\mathsf{s}}{\mathsf{d}}} \mathcal{H}_N(X_N)\right) dX_N. \tag{1.1.6}$$

The measure $\mathbb{P}_{N,\beta}$ is called the *canonical Gibbs measure* associated to the energy $\mathcal{H}_N$ at inverse temperature β. We call this ensemble a Riesz gas (or if applicable, a Coulomb gas or log gas). The factor $N^{-\frac{\mathsf{s}}{\mathsf{d}}}$ is a convenient scaling choice, the reason for which will appear later, but it does not reduce generality, since β may itself depend on N. Note that in the case $\mathsf{s} \le 0$, if two particles x, y are sent to infinity in opposite directions, their pair interaction $\mathsf{g}(x-y)$ tends to $-\infty$. Thus being far away from the origin needs to be penalized by the *confining* potential V, otherwise the integral (1.1.6) defining $Z_{N,\beta}$ may not converge.

To illustrate and fix ideas, the reader may keep in mind the important particular case of $\mathsf{s} = 0$ and quadratic confinement, say $V(x) = \frac{1}{4}|x|^2$, situation in which

$$\mathcal{H}_N(X_N) = -\frac{1}{2} \sum_{i \neq j} \log|x_i - x_j| + \frac{N}{4} \sum_{i=1}^N |x_i|^2$$

and

$$d\mathbb{P}_{N,\beta}(X_N) = \frac{1}{Z_{N,\beta}} \prod_{i<j} |x_i - x_j|^\beta e^{-\frac{\beta}{4} N \sum_{i=1}^N |x_i|^2} dx_1 \dots dx_N, \tag{1.1.7}$$

which one may view as a Gaussian law with pairwise repulsion added.

Most of our study is focused on understanding the typical and atypical behavior of particles X_N when randomly distributed according to the Riesz gas law $\mathbb{P}_{N,\beta}$, or on the deterministic behavior of X_N minimizing $\mathcal{H}_N$ which formally corresponds to taking $\beta = +\infty$.

It is also of interest to study evolutions, in particular the stochastic differential equations (SDE) system

$$dx_i = -\frac{1}{N} \nabla_i \mathcal{H}_N(x_1, \dots, x_N) dt + \sqrt{\frac{2}{\beta N^{1-\frac{\mathsf{s}}{\mathsf{d}}}}} dW_i^t \tag{1.1.8}$$

with W_i^t independent Brownian motions, which is the overdamped Langevin/Glauber dynamics for (1.1.5). The measure (1.1.5) can be seen as the invariant measure for this dynamics. Although this is largely open, understanding the features of the evolution (1.1.8) and its convergence to the equilibrium state (1.1.5) is an important statistical mechanics problem of interest in its own right, and can also provide information on the Riesz gas itself.

Other dynamics are possible and also physically very interesting. A first one is the class of conservative dynamics of the form

$$dx_i = \frac{1}{N} \mathbb{J} \nabla_i \mathcal{H}_N(x_1, \dots, x_N) dt + \sqrt{\frac{2}{\beta N^{1-\frac{\mathsf{s}}{\mathsf{d}}}}} dW_i^t, \tag{1.1.9}$$

where $\mathbb{J}$ is an antisymmetric matrix, and another is the class of second-order evolution according to Newton's law

$$dx_i = v_i dt, \qquad dv_i = -\frac{1}{N}\nabla_i \mathcal{H}_N(x_1, \dots, x_N)dt + \sqrt{\frac{2}{\beta N^{1-\frac{s}{d}}}}dW_i^t. \tag{1.1.10}$$

Dynamics will be discussed in Chapter 6.

Coulomb case. The choice (1.1.4) for the pair interaction potential g is called the *Coulomb case*, because g is then (up to a multiplicative constant) the Coulomb kernel, i.e., the fundamental solution to the Laplace operator, solving

$$-\Delta \mathsf{g} = \mathsf{c_d}\delta_0, \tag{1.1.11}$$

where δ_0 is the Dirac mass at the origin and $\mathsf{c_d}$ is an explicit constant depending on the dimension, given by:[1]

$$\mathsf{c_d} = 2\pi \quad \text{if } \mathsf{d} = 2 \qquad \mathsf{c_d} = |\mathbb{S}^{\mathsf{d}-1}| \quad \text{for} \quad \mathsf{d} \geq 3. \tag{1.1.12}$$

Coulomb interactions are ubiquitous in physics, most notably as the classical electrostatic interaction potential between charged particles. The one-dimensional Coulomb interaction with kernel $\mathsf{g}(x) = -|x|$ is the most "explicitly solvable", hence the best understood (see [**Len61, Len63, Bax63, Kun74, BL75, AM81**]), however we will still provide new results for this case.

Logarithmic case. The choice $\mathsf{s} = 0$ or $\mathsf{g}(x) = -\log|x|$ for the pair interaction potential is called the *logarithmic case*. It is important in random matrix theory and several physics models, as we will see below. A one-dimensional log gas for arbitrary $\beta > 0$ and general confinement V is also called a *β-ensemble*. Among all the Coulomb and Riesz gases, the β-ensembles have probably been the most extensively studied in the mathematical literature; see (1.1.7), the principal reference case. They are also the most ubiquitous, as one encounters them in random matrix theory, quantum mechanics, self-avoiding random walks, random tilings, and even proofs of functional inequalities (see [**DFGZ23**] for an example)!

Riesz cases. The general choices $\mathsf{d} - 2 \leq \mathsf{s} < \mathsf{d}$ in (1.1.2) are called the *Riesz cases* and the Coulomb case can be considered as a special instance of them. As seen above, the logarithmic cases $\mathsf{s} = 0$ can be thought of as the $\mathsf{s} \to 0$ limit of Riesz cases, so by extension we will consider them included in Riesz cases. In the Riesz case with $\mathsf{d} - 2 < \mathsf{s} < \mathsf{d}$, instead of (1.1.11), g is known to be the kernel of a *fractional Laplacian* operator, in the sense that

$$(-\Delta)^{\frac{\mathsf{d}-\mathsf{s}}{2}}\mathsf{g} = \mathsf{c_{d,s}}\delta_0 \tag{1.1.13}$$

for some normalization constant $\mathsf{c_{d,s}}$ given by

$$\mathsf{c_{d,s}} = \begin{cases} \frac{2^{\mathsf{d}-\mathsf{s}}\pi^{\mathsf{d}/2}\Gamma(\frac{\mathsf{d}-\mathsf{s}}{2})}{\Gamma(\frac{\mathsf{s}}{2})} & \text{for } \mathsf{s} > \max(0, \mathsf{d}-2) \\ \frac{2\pi^{\mathsf{d}/2}}{\Gamma(\frac{\mathsf{d}}{2})} = |\mathbb{S}^{\mathsf{d}-1}| & \text{if } \mathsf{s} = \mathsf{d}-2 > 0 \\ 2\pi & \text{if } \mathsf{s} = 0, \mathsf{d} = 1 \text{ or } \mathsf{d} = 2; \end{cases} \tag{1.1.14}$$

see [**GS16**] for the basis of the computation.

[1]Here $\mathbb{S}^{\mathsf{d}-1}$ is the unit sphere in $\mathbb{R}^{\mathsf{d}}$ and $|\mathbb{S}^{\mathsf{d}-1}|$ denotes its volume for the standard $\mathcal{H}^{\mathsf{d}-1}$ Hausdorff measure.

The fractional Laplacian is a nonlocal operator, it can be defined via Fourier multipliers or in real space by

$$(-\Delta)^\alpha f(x) = C_{\mathsf{d},\alpha} \int_{\mathbb{R}^\mathsf{d}} \frac{f(x)-f(y)}{|x-y|^{\mathsf{d}+2\alpha}} dy, \quad \alpha \in (0,1); \tag{1.1.15}$$

see for instance [**Kwa17**]. This property of g will play an important role for us. Note that it is only true in the super-Coulomb range $\mathsf{s} \in [\mathsf{d}-2, \mathsf{d})$, which is the main reason we focus on this regime.

1.2. Motivation

1.2.1. Statistical and quantum mechanics. Classical statistical mechanics views large physical systems of interacting deterministic particles as a random object. If $X \mapsto \mathcal{H}(X)$ is the interaction energy of the system in a state X and if the ambient temperature is β^{-1} then at equilibrium the probability of observing a given state X is proportional to the Boltzmann factor $\exp(-\beta\mathcal{H}(X))$. This justifies the introduction of the Gibbs measure (1.1.5) from a statistical physics point of view, in order to understand the statistical properties of a system of particles in $\mathbb{R}^\mathsf{d}$ with interaction energy given by (1.1.1). Grand canonical ensembles, i.e., where the number of points N is not fixed but also part of the variables, are also considered both in general and in the particular instance of Coulomb and Riesz gases, we refer to the survey [**Lew22**]. The canonical ensemble is usually considered more difficult to study than the grand canonical one.

Most of the interactions considered here are *singular* at the origin, and particles live in $\mathbb{R}^\mathsf{d}$ and not only on a lattice, which would ensure a minimal particle separation. Moreover, in the Riesz cases with $\mathsf{s} < \mathsf{d}$ that we consider, the pair interaction is considered *long-range* because it decays slower than $|x|^{-\mathsf{d}}$ at infinity, which gives rise to specific phenomena; see, e.g., [**BBDR05**] and references therein. These are the main difficulties of these models.

Coulomb gas. The most physical case is, of course, the Coulomb case, since Coulomb is the fundamental electrostatic interaction, with the three-dimensional situation being the most natural. The ensemble given by (1.1.5) in the case (1.1.4) is called a *Coulomb gas* or *one-component plasma* (which refers to the fact that there is only one type of charges, i.e., positive charges). With a neutralizing background, it is called a *jellium*. Closely related to the jellium is the Uniform Electron Gas (UEG) model which is defined rather via a density constraint, or imposing the one-point marginal. Relations between the jellium and the UEG have been explored in particular recently in [**LLS18, Lau21**].

The Coulomb gas can be seen as a toy model for classical matter, ignoring quantum effects. For instance, Gamov's "liquid drop model" for the atomic nucleus (see [**CMT17**]) is also a simplified model for electrons and atoms, and in some regime where one phase is in large majority can be reduced to a system of points interacting like (1.1.1) (see [**ACO09, GMS13, GMS14**] and references therein).

The Coulomb gas is thus a classical ensemble of statistical mechanics and has been well studied since the 1970s; see, e.g., [**AJ81, JLM93, Jan95, Ala86, SM76, PS72, JLM93, LL69, LN75, Frö76, FS81a, FS81b, Kie93, KS99**]. The quantum Coulomb gas has also been the object of much attention, and some of the techniques developed there also apply to the classical setting [**GS95, HLS08, LLS18**]. Density functional theory also involves the Coulomb interaction in a crucial way

[**EL81, LS10, LLS18, LLS22, CFK13, CP19**], in particular through the "indirect Coulomb energy" and its bounds via Lieb-Oxford's inequality and N-marginal optimal transport with Coulomb costs.

An important instance of the Coulomb case is the two-dimensional Coulomb gas, which coincides with the two-dimensional logarithmic case, see, e.g., [**Dys62, Meh04, Mar88, MY80**]. It is also called in the physics literature *log gas*, *two-dimensional one-component plasma* (which gets abbreviated as 2DOCP), *two-dimensional jellium*, or *Dyson gas*. The dynamics (1.1.8) (with proper scaling of β) is in this two-dimensional logarithmic case called *Dyson Brownian motion*. The 2DOCP is deemed interesting as a natural toy model to do statistical physics in two dimensions in a singular, long-range setting, as a reasonable model for plasmas in astrophysics (in particular (1.1.10)) and due to its connection with quantum mechanics, the fractional quantum Hall effect and random matrices, as we will see below.

We also refer the reader to the very recent physics book [**AM25**], which addresses all statistical mechanics aspects (exact results, correlations, phase transitions) of classical and quantum Coulomb gases.

Riesz gases. The Riesz case can be seen as a generalization of the Coulomb case. Motivations for studying Riesz gases are numerous in the physics literature (in solid state physics, ferrofluids, elasticity); see for instance [**Maz11, BBDR05, CDR09, CDFR14, Tor18**]. For integer $\mathsf{s} < \mathsf{d}$, the pair interaction Riesz systems can also be seen as systems with Coulomb interaction constrained to a lower-dimensional subspace, take for example $\mathsf{d} = 2$ and $\mathsf{s} = 1$: one gets the usual three-dimensional Coulomb interaction, only restricted to the two-dimensional plane. The jellium and uniform electron gas have also been studied for Riesz interactions [**LLS18, CP19**]. We note that even one-dimensional Riesz systems, possibly with nonsingular repulsion $\mathsf{s} \leq 0$, are currently attracting attention [**KKK$^+$22, KKK$^+$21, ADK$^+$19, FMS22**]. We also refer to the very nice recent survey [**Lew22**] with many open questions.

Link with wave functions. From the physics point of view, another motivation for studying the probability measure $\mathbb{P}_{N,\beta}$ is that in several cases with logarithmic interaction, it happens to be the square of the wave function of certain quantum systems. Examples corresponding to the one-dimensional logarithmic case are the Tonks-Girardeau model of impenetrable bosons [**GWT01, FFGW03**], the Calogero-Sutherland quantum many-body Hamiltonian [**FJM01, For10**], and finally the density of the many-body wave function of noninteracting fermions in a harmonic trap [**DLDMS19, KLDMS23**].

Examples in the two-dimensional logarithmic case are free fermions in a magnetic field in the lowest Landau level, and the Laughlin wave function for the fractional quantum Hall effect [**Gir05, STG99**], which is the Gibbs measure (1.1.5), or more specifically (1.1.7), for a 2DOCP. This is called the "plasma analogy" in the physics literature; see [**Lau83, Lau, Lau99**], and [**Rou22b, Rou22a**] for an introduction. For recent mathematical progress using this correspondence, see [**RSY14, RY15, LRY18**]. Moreover, for a certain choice of the inverse temperature ($\beta = 2$), the same measure also arises as the (square of the) wave function of the ground state for N noninteracting fermions confined to a plane with a perpendicular magnetic field; see [**For10**, Chap. 15].

1.2.2. Random matrix theory and related questions. The origins of Random Matrix Theory (RMT) trace back to Hurwitz, with later foundational works by statisticians (Wishart) to understand sample covariance matrices, and physicists (Wigner, Dyson) to understand the spectrum of large atoms and for mathematical curiosity (Ginibre). We refer to [**DF17**] for a historical perspective and extensive references, and to [**Meh04, AGZ10**] for a mathematical introduction. In short, the aim of RMT is to understand the eigenvalues and eigenvectors of (large) matrices drawn at random for certain distributions called *ensembles* or *models*, and it has been the object of a vast mathematical literature.

As noticed early on, for certain natural random matrix ensembles the joint law of the eigenvalues can be computed explicitly and happens to coincide with a Gibbs measure of the form (1.1.5), where g is chosen to be the logarithmic interaction $\mathsf{s} = 0$. In particular, choosing $V(x) = \frac{1}{4}|x|^2$ as the external field or confinement, the measure (1.1.5) on $\mathbb{R}^N$ or $(\mathbb{R}^2)^N$ becomes (1.1.7) and coincides in that case with the joint law of eigenvalues of several important random matrix ensembles:

GUE: With $\beta = 2$, one gets the law of eigenvalues of an $N \times N$ *Hermitian* matrix with i.i.d. complex Gaussian entries. This distribution of random matrices is called the Gaussian Unitary Ensemble or GUE.

GOE: With $\beta = 1$, one gets the law of eigenvalues of an $N \times N$ real *symmetric* matrix with i.i.d. real Gaussian entries. This distribution of random matrices is called the Gaussian Orthogonal Ensemble or GOE.

GSE: With $\beta = 4$, one gets the law of eigenvalues of an $N \times N$ self-dual matrix with i.i.d. quaternionic Gaussian entries. This distribution of random matrices is called the Gaussian Symplectic Ensemble or GSE.

Tridiagonal models for arbitrary β: In fact, as was realized much later, for every choice $\beta > 0$ (and still $V(x) = \frac{1}{4}x^2$) there exists a random matrix ensemble whose eigenvalues are distributed on the real line according to the law $\mathbb{P}_{N,\beta}$ for $\mathsf{s} = 0$. Dumitriu-Edelman and Killip-Nenciu [**DE02, KN04**] construct such models, that are sometimes referred to as the *tridiagonal* (or *pentadiagonal* in the case of [**KN04**]) models.

Ginibre ensemble: An important addition coming from non-Hermitian random matrix theory is the law of complex eigenvalues of an $N \times N$ matrix where the entries are chosen to be i.i.d. complex Gaussian variables with no symmetry imposed. This law can again be explicitly computed and coincides with (1.1.5) in the two-dimensional logarithmic case $\mathsf{s} = 0$, changing V into $V(x) = \frac{1}{2}|x|^2$ and taking $\beta = 2$. This is called the Ginibre ensemble, after Jean Ginibre; see [**Gin65**]. Its eigenvalue statistics arise in quantum chaos [**BF23**]. Quaternionic and real variants are also discussed in [**BF23**].

We refer to [**For10**] for an extensive discussion of this connection between random eigenvalues and random particles with logarithmic interaction, which is particularly strong in the one-dimensional case.

This connection is among the reasons why the logarithmic case $\mathsf{s} = 0$, in particular the one-dimensional one, are by far the most intensively studied among Riesz gases. The random matrix correspondence provides a physical intuition as well as tools and methods from mathematical physics in order to study the eigenvalues drawn from certain classical ensembles of random matrices. For example, one may readily give a physical interpretation for a well-known phenomenon called "repulsion of eigenvalues": eigenvalues are less likely to be close to each other than if they were drawn independently at random, in fact one may say that they repel each other logarithmically. Conversely, the random matrix models provide access to computing explicitly certain quantities for the log gas in one and two dimensions; see below.

Determinantal case. In both the one- and two-dimensional logarithmic cases at the specific temperature[2] $\beta = 2$, the Gibbs measure $\mathbb{P}_{N,\beta}$ acquires a special algebraic feature, which can be seen by rewriting it as

$$d\mathbb{P}_{N,2}(X_N) = \frac{1}{Z_{N,2}} \left(\prod_{i<j} |x_i - x_j| \right)^2 e^{-2N \sum_{i=1}^N V(x_i)} dX_N \tag{1.2.1}$$

with $\prod_{i<j} |x_i - x_j|$ equal to the Vandermonde determinant of the points $x_1, \ldots, x_N$. This makes the log gas ensemble in that particular instance belong to the class of *determinantal* point processes. An important consequence is that all of its correlation functions can be obtained explicitly by computing certain determinants, which allows to give very precise answers to many questions through algebraic computations; see [**For10, BF23**]. Again, this is the reason why the log gas with $\beta = 2$ is the best understood of all, as we will see with many examples of results. The rewriting (1.2.1) also makes the $\beta = 2$ log gases a particular instance of orthogonal polynomial ensembles, which are Gibbs measures on $\mathbb{R}^N$ of the form

$$\frac{1}{Z_N} \left(\prod_{i<j} |x_i - x_j| \right)^2 \prod_{i=1}^N d\mu(x_i),$$

and form another well-developed field of study [**Kö5**]. We refer to [**HKPV09, Bor11**] for more on determinantal point processes. In the physics literature, one sometimes speaks of systems that are *exactly solvable.*

Understanding the behavior of eigenvalues in certain ensembles can, for some questions, be sufficient because one expects many properties to be *universal* across a broad family of random matrix models (see for instance [**TV11, EPR$^+$10**]). Universality has been well understood in determinantal cases $\beta = 2$, see [**DKM$^+$99, Joh98, DG07, AHM11**] and the survey [**Kui11**].

Self-avoiding walks in probability and random tilings. Motivated by statistical physics, the analysis of self-avoiding walks and random tilings has been a very active field of probability and integrable probability. At their heart, one encounters again points with logarithmic-like repulsion and a discretized log gas on the real line at general temperature β [**BGG17, Gor21**], leading to similar limit point processes, questions and results as those encountered for the log gas. For reference on these topics as well as a general introduction to the field of integrable probability, see [**BG16b**].

For more on the aspects mentioned in Sections 1.2.1 and 1.2.2, we also refer to the very nice recent survey [**Cha23a**].

[2]For $\mathsf{s} = 0$ at $\beta = 1$ and 4, which correspond to the GOE and GSE models mentioned above, there is another specific algebraic structure called *pfaffian process*, for which exact computations are still possible in principle, but less tractable.

1.2.3. Vortices in condensed matter physics and fluids. In superconductors with applied magnetic fields, in rotating superfluids and in Bose-Einstein condensates, one observes the occurrence of quantized *vortices*, which are local point defects of superconductivity or superfluidity, surrounded by a current loop. The vortices repel each other, while being confined together by the effect of the magnetic field or rotation, and the result of the competition between these two effects is that, as predicted by Abrikosov [**Abr57**], they arrange themselves in a particular *triangular lattice* pattern, called *Abrikosov lattice*, cf. Figure 1.2.1 (for more pictures, see `www.fys.uio.no/super/vortex/`). Superconductors and super-

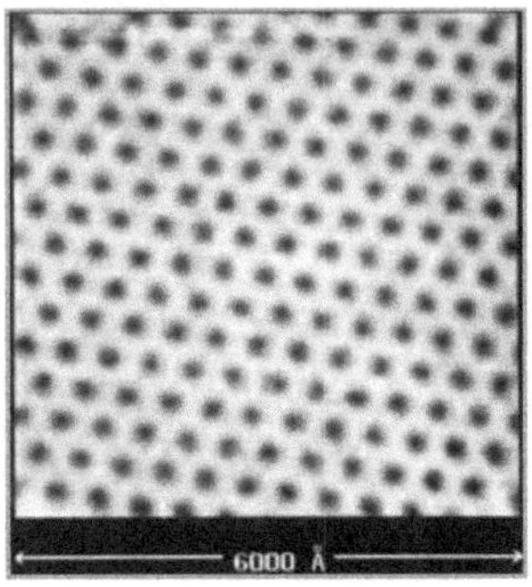

FIGURE 1.2.1. Abrikosov lattice. Reprinted figure with permission from H. F. Hess et al., *Scanning-Tunneling-Microscope Observation of the Abrikosov Flux Lattice and the Density of States near and inside a Fluxoid*, Phys. Rev. Lett., vol. 62, pp. 214–216, 1989. Copyright 1989 by the American Physical Society. `https://doi.org/10.1103/PhysRevLett.62.214`

fluids are modeled by the celebrated Ginzburg-Landau energy [**GL65**], which in simplified form[3] can be written

$$\int |\nabla\psi|^2 + \frac{(1-|\psi|^2)^2}{2\varepsilon^2}, \tag{1.2.2}$$

where ψ is a complex-valued unknown function (the "order parameter" in physics) and ε is a small parameter, and gives rise to the associated Ginzburg-Landau equation

$$\Delta\psi + \frac{1}{\varepsilon^2}\psi(1-|\psi|^2) = 0 \tag{1.2.3}$$

and its dynamical versions, the heat flow

$$\partial_t\psi = \Delta\psi + \frac{1}{\varepsilon^2}\psi(1-|\psi|^2) \tag{1.2.4}$$

and Schrödinger-type flow (called the Gross-Pitaevskii equation in the physics literature)

$$i\partial_t\psi = \Delta\psi + \frac{1}{\varepsilon^2}\psi(1-|\psi|^2). \tag{1.2.5}$$

[3]The complete form for superconductivity contains a gauge field, but we omit it here for simplicity.

When restricting to a two-dimensional situation, it can be shown rigorously (this was pioneered by [**BBH94**] for (1.2.2) and extended to the full gauged model [**BR95, SS07, SS12**]) that the minimization of (1.2.2) can be reduced, in terms of the vortices and as $\varepsilon \to 0$, to the minimization of an energy of the form (1.1.1) in the case $\mathsf{d} = 2, \mathsf{s} = 0$ (for a formal derivation, see also [**Ser15**]) and this naturally leads to the question of understanding the connection between minimizers of (1.1.1) and the Abrikosov triangular lattice. Similarly, the dynamics of vortices under (1.2.4) can be formally reduced to the gradient flow of (1.1.1) which is (1.1.8) with $\beta = \infty$, respectively under (1.2.5) to the Hamiltonian flow associated to (1.1.1), (1.1.9) with $\beta = \infty$ and $\mathbb{J}$ equal to the matrix of rotation by $\pi/2$. This was established formally for instance in [**PR93, CRS96, E94**] and proven for a fixed number of vortices N and in the limit $\varepsilon \to 0$ in [**Lin96, JS98, CJ98, LX99a, LX99b, BJS08**] until the first collision time and in [**BOS05, BOS07, SBO07, Ser07**] including after collision.

Vortices also arise in classical fluids, where in contrast with the situation of superconductors and superfluids, their charge is not quantized. In that context the energy (1.1.1) with $\mathsf{d} = 2, \mathsf{s} = 0$, is sometimes called the Kirchhoff energy and the corresponding Hamiltonian system (1.1.9) with $\mathbb{J}$ taken to be a rotation by $\pi/2$, known as the point-vortex system, corresponds to the dynamics of idealized vortices in an incompressible fluid whose statistical mechanics analysis was initiated by Onsager; see [**ES06**]. One of the motivations for studying the Hamiltonian flow with additive noise, as in (1.1.9), is precisely to understand fluid turbulence as he conceived. It has thus been quite studied as such, see [**MP84**] for further reference. The study of Newton's law (1.1.10) with interaction (1.1.1) is also motivated by plasma physics in which the interaction between ions is Coulombic; see the review [**Jab14**].

1.2.4. Fekete points, approximation theory, statistics and sampling.

Best packings and minimal energy configurations. Finding point configurations that are optimal in some respect is an old, recurrent question in mathematics. One may think of the famous *optimal (sphere) packing problem*: among all possible arrangements of disks, balls, etc. of fixed radius, which one is the most compact, i.e., has the highest density? Since the radius is fixed, only the centers can be chosen, so it is really a problem about point configurations, that can be seen as the $\mathsf{s} \to \infty$ limit of the minimization of the s-Riesz energy $\mathcal{H}_N$. In the more general problem of *energy minimization*, one fixes a certain pair potential and asks: among all point configurations of fixed density, which one has the minimal interaction energy? Perhaps surprisingly, these questions are extremely difficult to answer in general, except in dimension 1. It is believed that in low enough dimension, many such problems are minimized by *lattice* point configurations. This is the so-called *crystallization conjecture*, we refer to [**BL15**] for a recent survey.

The solution to the optimal sphere packing problem is for instance only known for a handful of dimensions: $\mathsf{d} = 1, 2, 3$, and thanks to very recent progress $\mathsf{d} = 8, 24$. This is a special case of the Cohn-Kumar conjecture [**CK07**] (relying on linear programming bounds at the level of Fourier transforms), which asserts that there are some universally minimizing lattices in dimensions $\mathsf{d} = 2, 8, 24$, more precisely the triangular lattice A_2 in dimension 2, the E_8 lattice in dimension 8 and the Leech Λ_{24} lattice in dimension 24, which minimize not only the sphere packing problem but also all interaction energies which are of the form

$$\sum_{i,j} f(|x_i - x_j|^2)$$

with f a *completely monotonic*[4] function. This conjecture was recently proven for $\mathsf{d} = 8, 24$, first for the sphere packing problem in [**Via17, CKM+17**], and then in [**CKM+22**] for all completely monotonic interactions, and it implies the same result for Coulomb and Riesz interactions in the same dimensions, as shown in [**PS20**]. The conjecture remains open for A_2 in $\mathsf{d} = 2$. If proven true, it implies in view of [**SS12**] combined with [**PS20**] that points that minimize $\mathcal{H}_N$ in the two-dimensional logarithmic or Riesz cases $\mathsf{d} - 2 \le \mathsf{s} < \mathsf{d}$ arrange themselves along a triangular lattice, the same as the Abrikosov lattice in superconductors.

In high dimension, where the problem is important for error-correcting codes, it is expected that the solution is *not* a lattice (in dimension 10 already, the so-called "best lattice", a non-lattice competitor, is known to beat the lattices); see [**CS99**] for these aspects. We refer to [**Coh17**] for an introduction to this topic, which we will discuss further in Chapter 11.

Fekete points. Fekete points arise in approximation theory as the points minimizing interpolation errors for numerical integration [**ST97**]. More precisely, if one is looking for N interpolation points $\{x_1, \dots, x_N\}$ in some nice compact subset $K \subset \mathbb{R}^{\mathsf{d}}$ such that the relation

$$\int_K f(x)dx = \sum_{j=1}^N w_j f(x_j)$$

holds when f is an arbitrary polynomial of degree $\le N-1$, one needs to compute the coefficients w_j such that $\int_K x^k = \sum_{j=1}^N w_j x_j^k$ for $0 \le k \le N-1$. This computation turns out to be easy if one knows how to invert the Vandermonde matrix of the $\{x_j\}_{j=1\dots N}$. The numerical stability of this operation is as large as the *condition number* of the matrix, i.e., as the Vandermonde determinant of the x_j's. In fact, the N-tuple of points that minimize the maximal interpolation error for general functions can be shown to be the Fekete points, defined as those that maximize the Vandermonde determinant $\prod_{1\le i<j\le N} |x_i - x_j|$, or equivalently minimize the energy $-\frac{1}{2}\sum_{1\le i\ne j\le N} \log|x_i - x_j|$ among all configurations in K. Such problems are often studied on compact manifolds, such as the d-dimensional sphere [**Bra08, BH19, BS18**].

In Euclidean space, one also considers *weighted Fekete points*, by introducing a weight V and asking which configuration maximizes the weighted Vandermonde determinant

$$\prod_{1\le i<j\le N} |x_i - x_j| e^{-N\sum_{i=1}^N V(x_i)}$$

or equivalently minimizes the logarithmic energy functional

$$-\frac{1}{2}\sum_{1\le i\ne j\le N} \log|x_i - x_j| + N\sum_{i=1}^N V(x_i),$$

[4]A function f is said to be completely monotonic when $(-1)^k f^{(k)} \ge 0$ for each $k \ge 0$, an important example is e^{-ct}, leading to Gaussian interactions.

which corresponds exactly to the minimization of $\mathcal{H}_N$ with $\mathsf{s} = 0$. Fekete points can also be characterized as the zeroes of a family of orthogonal polynomials; see [**Sim08**].

Finally there is also interest in the approximation theory literature in studying Riesz "s-energies," i.e., the minimization of $\sum_{1\le i\neq j\le N} \frac{1}{|x_i-x_j|^{\mathsf{s}}}$ for all possible s, which provides a motivation for studying the Riesz case (1.1.2). For these aspects, we refer to the review papers [**SK97, BHS14**] and the book [**BHS19**].

Let us note that varying s from 0 to $+\infty$ connects Fekete points to the sphere packing problem (which as mentioned above formally corresponds to $\mathsf{s} = \infty$).

Statistics, sampling, and maximum mean discrepancy. The analysis of (1.1.1) will quickly lead us to consider quantities of the form (see in particular (1.4.1) and (1.4.2) below)

$$\iint_{\mathbb{R}^{\mathsf{d}}\times\mathbb{R}^{\mathsf{d}}} \mathsf{g}(x-y)d\Big(\frac{1}{N}\sum_{i=1}^N \delta_{x_i} - \mu\Big)(x)d\Big(\frac{1}{N}\sum_{i=1}^N \delta_{x_i} - \mu\Big)(y). \tag{1.2.6}$$

For good choices of g, such a quantity can be seen as quantifying the distance between the empirical measure $\frac{1}{N}\sum_{i=1}^N \delta_{x_i}$ and the target probability measure μ, and it goes by the name of maximum mean discrepancy in statistics. It is also widely used in machine learning [**GBR$^+$12**]. Sampling from a target distribution is a fundamental question, and in order to provide sampling or numerical integration guarantees, algorithms of the family of kernel herding greedily minimize quantities of the form (1.2.6), see [**CWS10, BLJO12, CMG$^+$18, PZ20**] also relying on dynamics similar to the gradient flows / overdamped Langevin dynamics of the type (1.1.8) [**KAFMA21, RBM25**]. The situation is similar in the Stein variational gradient descent [**LW16**]. In all these instances, the singular Coulomb/Riesz interaction is not the typical choice, but regular powers of the distance (i.e., the case $\mathsf{s} < 0$ here in (1.1.2)) are a very relevant choice and they can be studied with some of our tools [**HCRS**]. On the other hand, mollifications of singular Riesz interactions have also been used [**LLK$^+$23**].

1.2.5. Other systems and further motivations.

Two-component plasma. The *two-dimensional, two-component plasma* (2DTCP) is a counterpart to the 2DOCP introduced above, which consists in N particles $X_N = (x_1, \dots, x_N)$ of charge $+1$ and N particles $Y_N = (y_1, \dots, y_N)$ of charge -1 with a logarithmic interaction, which is now attractive for particles with opposite charges. The interaction energy is given by:

$$\begin{aligned}\mathcal{H}_N(X_N, Y_N) := &-\frac{1}{2}\sum_{1\le i\neq j\le N} \log|x_i - x_j| \\ &-\frac{1}{2}\sum_{1\le i\neq j\le N} \log|y_i - y_j| + \sum_{1\le i,j\le N} \log|x_i - y_j|,\end{aligned}$$

and the particles are for instance constrained to a square in $\mathbb{R}^2$. The canonical Gibbs measure associated to the system is again defined as:

$$\frac{1}{Z_{N,\beta}} \exp\left(-\beta\mathcal{H}_N(X_N, Y_N)\right) dX_N dY_N,$$

with obvious notation.

The energy is no longer bounded below. In fact $\mathcal{H}_N(X_N, Y_N) = -\infty$ if two particles with opposite charges happen to have the same position, and more generally, $\mathcal{H}_N$ can be very negative if two such particles are very close. However, since such configurations are rare for the Lebesgue measure $dX_N dY_N$, the Gibbs measure is still well defined for high temperatures, i.e., small values of β. More precisely, the partition function converges for $\beta < 2$, the threshold for convergence of $\int e^{-\beta \log|x-y|} dx dy$. Note that the convergence at high temperature is only true for attractive interactions that are less singular than the logarithm.

The 2DTCP is interesting due to its close relation to two important theoretical physics models: the XY model and the sine-Gordon model, which exhibit a Berezinski-Kosterlitz-Thouless phase transition [**BG16a**]. We refer to [**Spe97**] for a presentation of this connection, and [**Frö76, DL74, FS81a, FS81b, GP77**] for studies of the 2DTCP in the physics literature. We will not discuss the 2DTCP further in this book, but some of the techniques developed here can be adapted to a mathematical investigation of its properties; see [**LSZ17, BS24, BS25**].

Multicomponent Coulomb gas. Even further, one can consider *multicomponent* Coulomb systems, including a mixture of charges of arbitrary integer values with a neutralizing background [**Mar88**].

Other boundary conditions. There has also been interest in considering a Coulomb gas in a domain with boundary conditions, either Dirichlet boundary conditions (corresponding to a "metal wall") [**JT96, AM25**], or Neumann boundary conditions [**KS99, For16, BKS23**]. Predictions are made for the decay of the correlations along the boundary (exponentially fast rather than the algebraic decay with free boundary conditions), and of the coefficient D_β in (1.3.5) below. We also refer to the related discussion in Chapter 7.

Complex geometry and theoretical physics. Coulomb systems and higher-dimensional analoguess involving powers of determinantal densities are also of interest to geometers as a way to construct Kähler-Einstein metrics with negative Ricci curvature on complex manifolds [**Ber14, BBWN11**]. Another motivation is the construction of Laughlin states for the fractional quantum Hall effect on complex manifolds, which effectively reduces to the study of a two-dimensional Coulomb gas on a manifold, possibly with nontrivial topology. The coefficients in the expansion of the (logarithm of the) partition function have interpretations as geometric invariants and conformal field theories; see for instance [**Kle16, Kle19, Kle14, KMMW17**], and the end of Section 9.4. Finally, recent work studying the Coulomb gas on a Jordan curve or a Jordan domain in the determinantal case $\beta = 2$ [**JV23, CJ24**] highlights the connection with geometry via Grunsky operators and the Loewner energy of Weil-Peterson curves.

1.3. Questions

As usual in statistical mechanics [**Hua63, Rue99, AM25**], one would like to understand the typical behavior of the system under (1.1.5) in the *thermodynamic limit* $N \to \infty$, and investigate hypothetical phase transitions and critical phenomena as the inverse temperature β varies. Compared to many of the topics listed above where exact formulae and determinantal structures play an important role, we will be looking for methods and results that apply to *all inverse temperatures*

β even possibly depending on N, and to the whole class of Coulomb and Riesz interactions constrained by (1.1.3).

Let us now list a set of questions that naturally arise, noting that we are still far from able to address them all.

We note that the scaling of the problem is set up so that most particles will be confined to a region of size $O(1)$ of the space, which we call the *macroscale*, while the scale at which we see a finite number of particles in a box is the *microscale* $N^{-1/\mathsf{d}}$. Intermediate scales are called *mesoscales*.

Universality, phase portrait, phase transitions. Looking back to the very definition of our systems in (1.1.1) and (1.1.5), we see that there are three natural parameters: the pair interaction g, the confining potential V, and the inverse temperature β. Every time one studies the behavior of a given observable, one may ask:

- Does it depend on V? A negative answer is usually called a form of *universality*. For example, the global distribution of the particles depends on V but their local arrangement is expected to be mostly universal, i.e., independent of V up to rescalings. This has been proven only in the one-dimensional logarithmic case of β-ensembles in [**BEY14**, **BEY12**], and in the two-dimensional Coulomb case only for $\beta = 2$ [**AHM11**, **HW21**].
- How much does it depend on g? This kind of universality *with respect to the interaction* is, in fact, not much explored, besides extensions to log-like cases in dimension 1 [**BGG17**, **BGK15**, **Ven13**]. In the present text, we rely crucially on the fact that g has a very specific form, but it would be very interesting to understand how much of the qualitative properties of Riesz gases are preserved under perturbation and depend only on the singularity of g at the origin and at infinity.
- Does it depend on β? If so, is there any *critical point*, i.e., value of β at which the dependence as a function of the inverse temperature ceases to be smooth? Usually, this never happens at finite N, but may arise in the thermodynamic limit $N \to \infty$. If one may pinpoint a significant *order parameter* (an observable that encodes the order/disorder of the system) which, in the thermodynamic limit, does not depend smoothly on the inverse temperature, then there is a *phase transition*. Drawing the *phase portrait* of a system consists in listing such critical phenomena. It is a very delicate problem and, in some sense, the ultimate goal.

Observables. Many questions about such systems can be formulated as follow. Given a map $\mathcal{O}_N$ from $(\mathbb{R}^\mathsf{d})^N$ to some space, we call observable the random variable $\mathcal{O}_N(X_N)$, whose law is the push-forward of $\mathbb{P}_{N,\beta}$ by $\mathcal{O}_N$. Natural questions include:

- Does the observable have a limit in law as $N \to \infty$? Can we characterize this limit (besides its mere existence)? How does it depend on the scale at which the observable lives?
- Does the observable concentrate around certain values as $N \to \infty$? If yes, at which speed: can we state concentration inequalities? Can we find a physical or mathematical characterization of those values?
- If the observable has a typical value in the limit, can we study the fluctuations? Are they Gaussian?

By extension, one can ask exactly the same questions about energy minimizers, for which observables and their limits are deterministic instead of random.

A first natural observable is the *empirical measure* $\widehat{\mu_N}$ with values in $\mathcal{P}(\mathbb{R}^{\mathsf{d}})$, defined by

$$\widehat{\mu_N}(X_N) := \frac{1}{N}\sum_{i=1}^{N}\delta_{x_i}.$$

Understanding its limit as $N \to \infty$ corresponds to a Law of Large Numbers and provides a description of the particles density at the global or macroscale. This is well understood; see Chapters 2 and 3.

Another important class of observables consists in counting the number of points in balls (or other more general sets), or rather its difference with the expected limit,

$$D(B_R) := \int_{B_R} d\left(\sum_{i=1}^{N}\delta_{x_i} - N\mu\right), \tag{1.3.1}$$

which we call the *discrepancy* in the ball B_R. We can let R depend on N, allowing to study point discrepancies at the macroscale ($R = 1$) down to the microscale $R = N^{-1/\mathsf{d}}$. The discrepancy in the number of points measures how regular a point distribution is, and, together with its variance, is a very important quantity from the point of view of the analysis of point processes; see, e.g., [**Tor18, TS03**]. In approximation theory, the discrepancy is exactly the measure of the accuracy (or error) in the approximation, see for instance the surveys [**Bil20, Cos**]. We refer to Section 10.3 for a discussion of conjectures and results on the discrepancy in the two-dimensional Coulomb case.

A variant of the discrepancy consists in testing $\sum_{i=1}^N \delta_{x_i} - N\mu$, where $\mu = \lim_{N\to\infty}\widehat{\mu_N}$, not against indicator functions as in (1.3.1), but against more regular (say smooth) functions φ, and study the size and fluctuations of regular linear statistics $\sum_{i=1}^N \varphi(x_i)$ in the form

$$\int_{\mathbb{R}^{\mathsf{d}}} \varphi\, d\left(\sum_{i=1}^{N}\delta_{x_i} - N\mu\right) = N\int_{\mathbb{R}^{\mathsf{d}}} \varphi\, d(\widehat{\mu_N} - \mu). \tag{1.3.2}$$

Again, φ can be taken to depend on N and be supported at macro, meso or microscales. Precise results are known for one- and two-dimensional logarithmic cases; see Chapter 10.

For other relevant observables, one may think of something that encodes the arrangement of the particles at a certain scale. At the local or microscale, a natural choice would be the local configuration observable $\mathcal{C}^{\mathrm{loc}}_{N,\bar{x}}$ defined by fixing some point $\bar{x}$ in $\mathbb{R}^{\mathsf{d}}$ and looking at:

$$\mathcal{C}^{\mathrm{loc}}_{N,\bar{x}}(X_N) := \sum_{i=1}^{N}\delta_{N^{1/\mathsf{d}}(x_i-\bar{x})}, \tag{1.3.3}$$

with values in the space of point configurations. The law on the limit of $\mathcal{C}^{\mathrm{loc}}_{N,\bar{x}}$ as $N \to \infty$ is called a limiting point process, or local limit, for the ensemble. Even for energy minimizers, understanding the patterns formed by the points in the limit is very hard, except in dimension 1 which is the only one for which the interaction is convex. In contrast to the macroscopic behavior, several observations (e.g., by numerical simulations, see Figure 1.3.2) indicate that the behavior of the system at

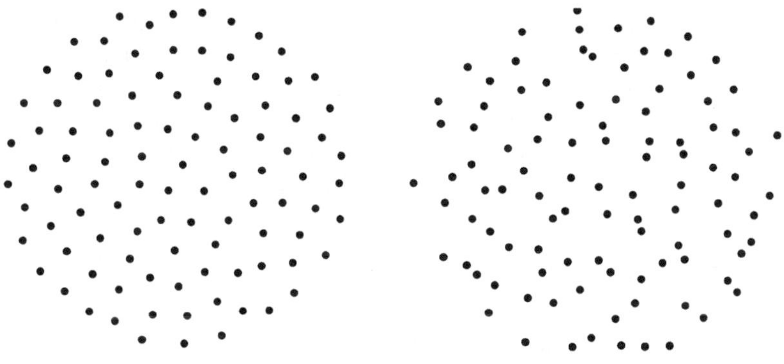

FIGURE 1.3.2. Case $\mathsf{d} = 2$, $\mathsf{s} = 0$, with $N = 100$ and $V(x) = |x|^2$, for $\beta = 400$ (left) and $\beta = 5$ (right). Pictures by Thomas Leblé.

the microscopic scale depends on β in a clear, nontrivial way. It is one of our main goals to analyze and characterize that β-dependence and the microscopic behavior.

Limit point processes. There are only a few instances where the existence and nature of a limiting point process is known. The central ones are the case of the logarithmic interaction in dimension 1 with $\beta = 2$ for which the process is the sine-process [**Meh04, For10, Bor11**], and the case of the logarithmic interaction in dimension 2 with $\beta = 2$, or Ginibre ensemble, for which the limit is the *Ginibre point process* [**Meh04, For10**]; see Figure 1.3.3. These are all instances of determinantal

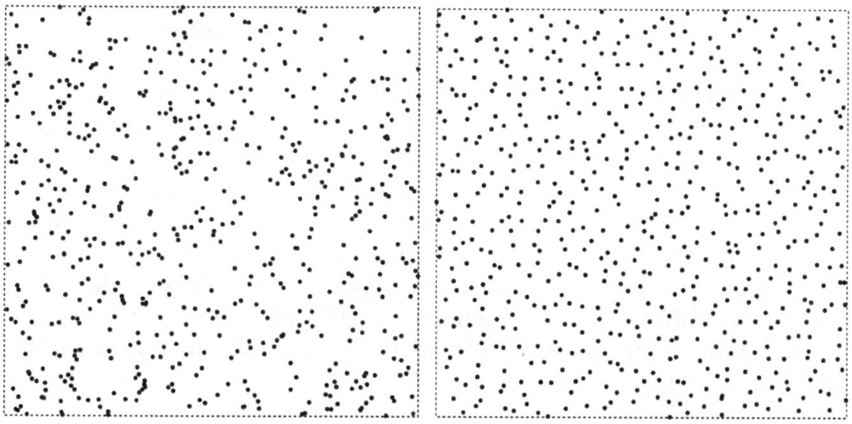

FIGURE 1.3.3. A realization of the Poisson point process (left) and the Ginibre point process (right). Pictures by Alon Nishry.

point processes. The special cases $\beta = 1$ and $\beta = 4$ also allow a similar treatment as pfaffian point processes. In the case of the logarithmic interaction in dimension 1 with general β, the existence of a limiting point process, called the sine-β process, was established in [**VV09, KS09**]. Recently [**GK24**] proposed an approach to defining a $G_\infty E$ ensemble corresponding to $\beta = \infty$, and which also appears as a universal limit. Also recently, [**Boua**] established the existence of a limit point process in the case of the one-dimensional circular Riesz gas (i.e., set on a torus).

We will discuss the existence of subsequential limit points, and other ways to characterize the limits of (1.3.3) in Chapters 12 and 13. Once limiting point processes are obtained, one can ask for instance whether they satisfy Dobrushin-Lanford-Ruelle (DLR) equations, whether these point processes are *number rigid* in the sense of [**GP17**], and whether they are *hyperuniform* in the sense of [**Tor18, TS03**]. We refer to [**OS08, DHLM21, DV23, Tho23, Leb, HL24, Leb24**] for some results and to [**Cos**] for a survey of questions and results.

Correlations and statistics. With a slightly different mindset, one can try to understand the law of X_N itself. The k-point correlation function, or k-marginal (for $k \geq 1$) is the function $\rho_N^{(k)} : (\mathbb{R}^{\mathsf{d}})^k \to [0, +\infty)$ defined by:

$$\rho_N^{(k)}(x_1, \dots, x_k) = \frac{1}{Z_{N,\beta}} \int_{(\mathbb{R}^{\mathsf{d}})^{N-k}} \exp\left(-\beta N^{-\frac{\mathsf{s}}{\mathsf{d}}} \mathcal{H}_N(X_N)\right) dx_{k+1} \dots dx_N. \tag{1.3.4}$$

It can be thought of as giving the probability density of observing simultaneously a particle at x_1, a particle at x_2, ..., and a particle at x_k. Correlation functions are a very powerful tool, and obtaining knowledge of the $\rho_N^{(k)}$'s allows one to answer many of the relevant questions about the system through (possibly challenging) computations. In particular, the decay rate as $|x_1 - x_2| \to \infty$ of $\rho^{(2)}(x_1, x_2)$, defined as the scaled $N \to \infty$-limit of $\rho_N^{(2)}(x_1, x_2)$ indicates the state of the system (solid if no decay, liquid for algebraic decay, and gas for exponential decay), and transitions in that decay rate depending on β correspond to phase-transitions.

In the special cases $\beta = 2$ and $\mathsf{s} = 0$ mentioned above where the particles form a determinantal point process, the $N \to \infty$ limits of the $\rho_N^{(k)}$ are identified as determinants of the form $\rho^{(k)} = \det[K(x_i, x_j)]_{1 \leq i,j \leq k}$ for an explicit kernel $K(x, y)$: the sine-kernel for the case of the sine process, and an exponential kernel for the Ginibre point process. Aside from these special cases and other cases where correlation functions can be accessed via kernels expressed via orthogonal polynomials, no explicit formulae exist in general and estimating the correlation functions (let alone computing them exactly) or getting a handle on their decay rate is extremely difficult, with the exception of the one-dimensional situations. Note that a lot of the statistical physics literature [**Mar88, MY80**] establishes "sum rules" and charge fluctuations estimates but implicitly assuming properties of $\rho^{(2)}$ at large distances and bootstrapping via BBGKY hierarchies satisfied by the higher order correlation functions, see [**AM25**] and also comments in Section 5.3.2.

A weaker form of the question consists in fixing a test function $\varphi : (\mathbb{R}^{\mathsf{d}})^k \to \mathbb{R}$ and considering the k-point *statistics*

$$\sum_{1 \leq i_1, \dots, i_k \leq N} \varphi(x_{i_1}, \dots, x_{i_k})$$

as a random variable whose law we try to understand as $N \to \infty$, which is the generalization of the linear statistics case $k = 1$ of (1.3.2).

For the one-dimensional Coulomb gas, the behavior of $\rho^{(1)}$ and $\rho^{(2)}$ was elucidated, proving crystallisation [**Kun74, BL75, AM81**]; see also the discussion in [**Lew22**, p. 54]. For the one-dimensional circular Riesz case with $\mathsf{s} > 0$, [**Boua**] proves a rate of decay of correlations of the particles gaps.

For the two- and three-dimensional Coulomb gases, a phase transition at finite, nonzero temperature has been conjectured in the physics literature based on numerical observations: in dimension 2 it happens around $\beta = 140$ [**CLWH82, CC83**] and in dimension 3 around $\beta = 175$ [**BST66, JC96**]; see also the review [**KK16**]. This transition should correspond to a change of decay of $\rho^{(2)}$ from exponential to algebraic as β crosses the hypothetical transition temperature. Its existence and precise nature is still disputed in the physics literature; see [**CSA21**], and getting any mathematical understanding about it remains a fascinating challenge.

Free/minimal energy expansions. For energy minimizers, it is natural to understand the value of $\min \mathcal{H}_N$ and its asymptotic expansion as $N \to \infty$, if it exists.

In the statistical mechanics setting, a similar role is played by the partition function defined in (1.1.6), and more precisely by the *free energy* $-\frac{1}{\beta}\log Z_{N,\beta}$. Knowing it is important because it provides access to many physical quantities associated to the system (e.g., for differentiating $Z_{N,\beta}$ with respect to β yields the average energy, etc.); see statistical mechanics textbooks such as [**Hua63**]. We will also see that evaluating the free energy is directly connected with (hence key to understanding) the fluctuations of linear statistics.

For the one-dimensional Coulomb gas with quadratic confinement, the $N \to \infty$ expansion of $\log Z_{N,\beta}$ is given in [**Kun74**], and the coefficients are shown to be analytic in β. For the one-dimensional log gas, the value of $Z_{N,\beta}$ is known explicitly for all $\beta > 0$ when $V(x) = x^2$ via the exact computation of the integral in (1.1.6), which uses so-called *Selberg integrals*; see, e.g., [**Meh04**]. For more general V's an expansion in N to any order is also known [**BG13, Shc13**]. For the two-dimensional Coulomb gas however, no equivalent of the Selberg integral exists and the exact value of $Z_{N,\beta}$ is only known for the Ginibre case $\beta = 2$ and $V(x) = |x|^2$ [**Meh04**], and a few other determinantal cases mentioned at the end of Chapter 9.

The works [**Imb83, BF80b, BF80a**] studied the grand canonical case for small β in dimensions 2 or 3, mapping the model to a quantum field theory by sine-Gordon transformation. This indicates that the coefficients of $\log Z_{N,\beta}$ should be analytic in β when β is close enough to 0.

There are conjectures on the expansion of $\min \mathcal{H}_N$ for Fekete points (energy minimizers for $\mathsf{s} = 0$) on the 2-sphere [**RSZ94**]; see also [**BHS12**], as well as conjectures for the expansion of the free energy in the two-dimensional logarithmic case [**JGP94, TF99, ZW06, CFTW15, Kle16**]. In both cases the (free) energy expansion is conjectured (and this is often called the Zabrodin-Wiegmann conjecture) to be of the form

$$\log Z_{N,\beta} = A_\beta N^2 + \frac{\beta}{4} N \log N + B_\beta N + C_\beta \sqrt{N} + D_\beta \log N + E_\beta + o(1), \tag{1.3.5}$$

with the coefficients having physical and geometric interpretations (we refer to the end of Section 9.4 for the precise formulas). The existence of the $B_\beta N$ term corresponds to a thermodynamic limit (or limit of free energy per particle) in the statistical mechanics language. In particular, if these coefficients are found to depend on the inverse temperature in a nonsmooth way, it is again an indication of a phase transition—in fact a phase transition can be expected to be manifested by a joint change of smoothness of the free energy and change of decay rate of the two-point correlation function as described in the previous paragraph.

Other observables. One can think of other natural observables, such as the minimal and maximal distance between two particles [**BAB13, BLZ25, FW21, FTW19, Ame18, AR23, Tho24**], or the maximal distance to the support of μ

[**Ame21**]. We will discuss them a bit in Section 5.3. The electrostatic field generated by the particles, and its maximal value, are also observables of interest; see [**Lew22**]. Closely related is the characteristic polynomial for β-ensembles and the two-dimensional log gas, together with its maximum, which have been the object of many studies [**CMN18, BMP22, LP23, ABZ23, LLZ24**].

1.4. Plan of the book

The book starts with the analysis of the macroscopic behavior of the Coulomb or Riesz gas, which is governed by the Frostman *equilibrium measure* μ_V, unique minimizer among probability densities of

$$\mathcal{E}(\mu) = \frac{1}{2}\iint_{\mathbb{R}^d \times \mathbb{R}^d} \mathsf{g}(x-y)d\mu(x)d\mu(y) + \int_{\mathbb{R}^d} V(x)d\mu(x). \tag{1.4.1}$$

Chapter 2 provides the characterization and description of the equilibrium measure, which are basics of potential theory. We then describe the less well-known connection of this minimization problem with the *classical obstacle problem* of calculus of variations in the Coulomb case, and the *fractional obstacle problem* in the general Riesz case. We then examine and describe the *thermal equilibrium measure*, a version of the minimization problem (1.4.1) with an added entropy cost $\int \mu \log \mu$, for which we present a general existence theorem. Even though the equilibrium measure is what is generally used, the thermal measure provides a more accurate description of the particles density, particularly when the inverse temperature gets small.

Chapter 3 connects the Coulomb/Riesz gas to the (thermal) equilibrium measure, via the framework of Γ-convergence and large deviations, which we start by reviewing. The first main result is that, for minimizers of $\mathcal{H}_N$, the large-N limit of the empirical measure $\frac{1}{N}\sum_{i=1}^N \delta_{x_i}$ converges to the equilibrium measure μ_V. For general configurations the energy $\mathcal{H}_N$ can be described in terms of the limit energy (1.4.1) for the limit of the empirical measure. These facts are standard for logarithmic interactions but here we give a presentation that applies to fairly general interaction g, often left out in the literature. In particular we present a streamlined generalization to all Riesz cases and all temperature regimes of the Large Deviations Principle (LDP) of [**HP00, BAZ98, BAG97**], asserting roughly that

$$\mathbb{P}_{N,\beta}(\mu_N \simeq \mu) \approx e^{-\beta N^{1-\frac{s}{d}}(\mathcal{E}(\mu)-\min \mathcal{E})},$$

i.e., the probability that the empirical measures limits to something else than the equilibrium measure is exponentially small as $N \to \infty$. This elucidates the macroscopic behavior of the Coulomb/Riesz gas.

The second part of the book is focused on the next-order electric (called *modulated* in the dynamical context) energy. This quantity, defined for any configuration X_N and any reference probability density μ by

$$\begin{aligned} \mathsf{F}_N(X_N,\mu) &= \frac{1}{2}\iint_{\{x\neq y\}} \mathsf{g}(x-y)d\Big(\sum_{i=1}^N \delta_{x_i} - N\mu\Big)(x)d\Big(\sum_{i=1}^N \delta_{x_i} - N\mu\Big)(x) \\ &= \frac{N^2}{2}\iint_{\{x\neq y\}} \mathsf{g}(x-y)d(\widehat{\mu_N}-\mu)^{\otimes 2}(x,y), \end{aligned} \tag{1.4.2}$$

corresponds to a Coulomb/Riesz interaction of a system of discrete point charges and a negative background charge $-\mu$. It can be used as a "metric" to quantify the convergence of the empirical measure to the reference measure μ. In Chapter 4, the energy is studied in detail, in particular we present its *electric reformulation* as the (weighted) Dirichlet energy of the electric potential

$$h_N = \mathsf{g} * \left(\sum_{i=1}^{N} \delta_{x_i} - N\mu \right), \tag{1.4.3}$$

or square of the electric field $\int |\nabla h_N|^2$, which in particular allows to *localize* the energy. The electric reformulation involves a *renormalization* of the integral, performed via a *truncation* procedure by smearing of the charges on spheres with *point-dependent* radii. We describe a crucial *monotonicity property*, with respect to the truncation radii, of this renormalization, which allows to deduce the coercivity of the modulated energy and the fact that $\frac{1}{N^2}\mathsf{F}_N$ is essentially a metric $\|\widehat{\mu_N} - \mu\|^2$ in a weak Sobolev norm, but computed in a renormalized fashion. This way, it controls charge discrepancies as in (1.3.1) or the difference between the empirical measure and the reference measure μ.

In Chapter 5, we show how the quantity F_N appears at the next order in an exact *splitting formula* for $\mathcal{H}_N$, after subtracting off the energy of either the equilibrium measure or the thermal equilibrium measure. More precisely, one checks that for any configuration

$$\mathcal{H}_N(X_N) = N^2\mathcal{E}(\mu_V) + \mathsf{F}_N(X_N, \mu_V) + \text{effective confinement energy},$$

and a convenient variant with respect to the thermal equilibrium measure. This allows to effectively reduce the Riesz gas energy to the modulated energy F_N. Inserting the general lower bound on F_N found in Chapter 4, one then obtains an easy lower bound on the partition function in terms of N. It can be complemented with an upper bound, and together these very directly imply (in Chapter 5) *concentration bounds* quantifying the closeness of the empirical measure to μ_V thanks to the fact that $\mathsf{F}_N(X_N, \mu_V)$ is like a (square) distance between them. At the end of Chapter 5 we discuss various questions of localization (near the support of μ_V), separation, charge excess and discrepancy; on these topics we describe some results of Ameur et al. and the isotropic averaging method of Thoma.

Chapter 6 makes a detour through the question of dynamics for systems with Coulomb/Riesz interactions, i.e., analyzing the large N limit of gradient flows (1.1.8) and conservative flows (1.1.9). We first present a functional inequality on F_N that has been termed *commutator estimate.* This functional inequality allows to control *derivatives of* $\mathsf{F}_N(X_N, \mu)$ *along a generic transport* applied to both the points of the configuration X_N and μ. These derivatives are bounded in terms of F_N and the Lipschitz norm of the transport map. This commutator estimate is the key to proving quantitative mean-field convergence for dynamics of systems of points with Coulomb/Riesz interactions, either (1.1.9) or (1.1.8), via the *modulated energy method*, or *modulated free energy method* in the case with noise. These convergence results are presented in Chapter 1 and are of independent interest. The commutator estimate is also important later for the transport calculus approach to fluctuations of Chapter 9.

The third part of the book concerns the mesoscopic behavior of Coulomb gases (in that part we focus exclusively on the Coulomb case for simplicity) and in particular in proving *local laws*, i.e., exponential moment bounds on the localized version of F_N introduced in Chapter 4. The idea to do so, presented in Chapter 7, is to use a *bootstrap on scales* and compare two energy quantities, one defined with Dirichlet boundary conditions and one with Neumann boundary conditions, and show that they are close. The crucial technical tool to show this is the *screening procedure*. In Chapter 8, we explain how to perform the bootstrap on scales to derive local laws: local laws at the macroscale are easy consequences of the free energy bounds of Chapter 5, then the screening procedure allows to deduce almost additivity of the free energy and of the system along slightly smaller than macroscopic boxes, up to surface errors. The local laws then hold down to this slightly smaller scale. The reasoning can then be iterated to get the local law on smaller and smaller scales, until the estimates saturate at a new (temperature-dependent) minimal scale, which is the microscale $N^{-\frac{1}{\mathsf{d}}}$ if β is bounded below, $\beta^{-\frac{1}{2}}N^{-\frac{1}{\mathsf{d}}}$ otherwise. The local laws being a byproduct of the almost additivity of the free energy, they naturally come together with free energy expansions with explicit error rates in $N^{1-\frac{1}{\mathsf{d}}}$ corresponding to surface errors, for uniform equilibrium measures.

In Chapter 9, we introduce the *transport method* and transport calculus with a view on studying fluctuations. This method consists in viewing the comparison of the free energy of a Coulomb gas with an external potential V and that of a Coulomb gas with perturbed potential $V + t\varphi$ in terms of a transport map between the corresponding equilibrium measures. This allows in particular to obtain free energy expansions with explicit error rates for varying equilibrium measures, corresponding to the expansion down to the $B_\beta N$ term in (1.3.5), with explicit $o(N)$ error. In Chapter 10, we combine the transport method of Chapter 9 and the commutator estimates of Chapter 6 to obtain bounds on fluctuations of linear statistics, again for Coulomb gases in arbitrary dimension. When specializing to one- and two-dimensional Coulomb gases, we prove a complete central limit theorem (CLT), which we now state for dimension 2.

Sample Theorem. *If* $\mathsf{d} = 2$ *and* $\mathsf{s} = 0$*, for* φ *regular enough,*

$$\int_{\mathbb{R}^{\mathsf{d}}} \varphi \, d\left(\sum_{i=1}^N \delta_{x_i} - N\mu_V\right)$$

converges as $N \to \infty$ *to a Gaussian with explicit mean and variance* $\frac{1}{2\pi\beta}\int |\nabla\varphi|^2$. *In particular, for* $\beta = \infty$*, i.e., for minimizers of* $\mathcal{H}_N$*,* $\int \varphi \, d\left(\sum_{i=1}^N \delta_{x_i} - N\mu_V\right)$ *converges as* $N \to \infty$ *to an explicit constant.*

This result can be rephrased as the convergence of the electric potential (1.4.3) to the two-dimensional Gaussian free field. It is also valid for test functions φ that are localized on mesoscales, all the way down to large multiples of the microscale. Chapters 9–10 are rather independent from what follows.

The last part of the book focuses on the microscopic description of the configurations via the *jellium renormalized energy* $\mathbb{W}$, an infinite volume energy version of the Coulomb / Riesz interaction of an infinite system of point charges neutralized by a uniform background charge (i.e., a jellium), defined via the electric formulation of the energy. In this part we return to the general setting of Riesz interactions. This jellium energy is introduced in Chapter 11 and its properties are described. The question of its minimization is connected with crystallization questions and the Cohn-Kumar conjecture alluded to in Section 1.2.4.

Chapter 12 derives this energy $\mathbb{W}$ as the $N \to \infty$ limit energy of the next order energy F_N. This is expressed in terms of *empirical fields* or limit point processes, i.e., probability measures P on point configurations that are obtained as limits of quantities like (1.3.3). The connection between F_N and $\mathbb{W}$, again obtained via the electric formulation of the energy, is first expressed via a general lower bound on F_N by the average of $\mathbb{W}$ with respect to the probability P. We then obtain the main theorem stating that the empirical fields (i.e., local limits at the microscale) of minimizers of $\mathcal{H}_N$ must converge to minimizers of $\mathbb{W}$, thus connecting to the crystallization questions of the prior chapter. This is proven by combining the general lower bound with an upper bound construction relying on the screening procedure.

In Chapter 13, we adapt this to the probabilistic situation of the Coulomb/Riesz gas ensemble. This requires combining the energetic effects (the derivation of $\mathbb{W}$) with the entropic effects, which are accounted for via a Sanov-type theorem at the level of empirical fields. The combination of the two allows to derive a large deviations principle (LDP) for the empirical field, with a rate function that takes the form of a free energy, as a sum of β times the energy $\mathbb{W}$ and the specific relative entropy of the limit point process with respect to the Poisson point process. This provides a variational interpretation of the limiting point processes, where the energetic (favoring ordered configurations) and entropic (favoring disorder) effects compete. The LDP, which is for macroscopic averages of the empirical field, is complemented in the Coulomb case by a local version, for mesoscopic and down to microscopic averages.

Part 1

Macroscopic behavior

CHAPTER 2

The equilibrium measure(s)

As mentioned in Chapter 1, our setting in this text is that of Riesz interaction potentials

$$\mathsf{g}(x) = \begin{cases} \frac{1}{\mathsf{s}}|x|^{-\mathsf{s}} & \mathsf{s} \neq 0 \\ -\log|x| & \mathsf{s} = 0 \\ \mathsf{d}-2 \leq \mathsf{s} < \mathsf{d}. \end{cases} \tag{2.0.1}$$

We note that either $\mathsf{s} \geq 0$ and g is singular at the origin, or $\mathsf{d} = 1$ and $-1 \leq \mathsf{s} < 0$. In both cases, g is radial decreasing. We will abuse notation by also denoting for $t \in \mathbb{R}_+$,

$$\mathsf{g}(t) = \begin{cases} \frac{1}{\mathsf{s}}t^{-\mathsf{s}} & \mathsf{s} \neq 0 \\ -\log t & \mathsf{s} = 0. \end{cases}$$

We recall the definition of the energy

$$\mathcal{H}_N(X_N) := \frac{1}{2} \sum_{1 \leq i \neq j \leq N} \mathsf{g}(x_i - x_j) + N \sum_{i=1}^N V(x_i). \tag{2.0.2}$$

As mentioned previously, the leading order behavior of the energy $\mathcal{H}_N$ is governed by the functional

$$\mathcal{E}(\mu) := \frac{1}{2} \iint_{\mathbb{R}^{\mathsf{d}} \times \mathbb{R}^{\mathsf{d}}} \mathsf{g}(x-y)d\mu(x)d\mu(y) + \int_{\mathbb{R}^{\mathsf{d}}} V(x)d\mu(x) \tag{2.0.3}$$

defined over the space $\mathcal{P}(\mathbb{R}^{\mathsf{d}})$ of probability measures on $\mathbb{R}^{\mathsf{d}}$.

Note that $\mathcal{E}(\mu)$ is simply a continuum version of the discrete Hamiltonian $\mathcal{H}_N$ defined over all $\mathcal{P}(\mathbb{R}^{\mathsf{d}})$, which may also take the value $+\infty$. From the point of view of statistical mechanics, $\mathcal{E}$ is the "mean-field" limit energy of $\frac{1}{N^2}\mathcal{H}_N$, while we will see in the next chapter that from the point of view of probability, $\mathcal{E}$ also plays the role of a *rate function.*

Its minimization turns out to be a classical problem of electrostatics, that of finding the equilibrium distribution of charges in a capacitor with an external potential also called the "capacitance problem". It was historically studied by Gauss and settled by Frostman [**Fro35**]. It is thus a fundamental question in *potential theory*, a topic which itself grew out of the study of the electrostatic or gravitational potential; see, e.g., [**Lan72, AH96, Doo01, ST97**] and references therein. The case of $\mathsf{d} = 2$ and $\mathsf{g}(x) = -\log|x|$ is precisely treated in [**ST97**, Chap. 1]. Higher-dimensional and more general singular interaction potentials are treated for instance in [**CGZ14**]. The general case, that we will treat, is not more difficult.

We will see that, under appropriate assumptions, $\mathcal{E}$ has a unique minimizer μ_V, called the *equilibrium measure*, or the Frostman equilibrium measure, or sometimes the extremal measure. The equilibrium measure arises as the law of large numbers limit of the Gibbs measure (1.1.5), we will show in the next chapter a stronger version of this result in the form of a large deviations principle.

We will also discuss the *thermal equilibrium measure* μ_θ defined for any given inverse temperature θ as the minimizer in $\mathcal{P}(\mathbb{R}^\mathsf{d})$ of the functional $\mathcal{E}$ with an added entropy term, i.e.,

$$\mathcal{E}_\theta(\mu) := \mathcal{E}(\mu) + \frac{1}{\theta}\int_{\mathbb{R}^\mathsf{d}} \mu \log \mu. \tag{2.0.4}$$

This energy functional arises when taking into account temperature effects in the study of (1.1.5), this was previously used for instance in [**CLMP92, Kie93, RSY14, ABG12**]. As we will see throughout the text, when temperature is present, the thermal equilibrium measure, with the choice $\theta = \beta N^{1-\frac{\mathsf{s}}{\mathsf{d}}} \gg 1$ in our scaling, always provides a more precise description of the point distribution (see Remark 2.24 below), and it will help in our study of regimes of β very small when $N \to \infty$. The discussion of the thermal equilibrium measure is new in this level of generality and we have reduced to rather minimal assumptions for its existence.

2.1. Existence, uniqueness, and characterization of the equilibrium measure

As previously we consider g given by (1.1.2) with $\mathsf{d} - 2 \le \mathsf{s} < \mathsf{d}$.

Lemma 2.1 (Strict convexity). *Assume $\mu_\pm \in \mathcal{P}(\mathbb{R}^\mathsf{d})$ and*

$$\iint_{\mathbb{R}^\mathsf{d}\times\mathbb{R}^\mathsf{d}} |\mathsf{g}(x-y)|d\mu_\pm(x)d\mu_\pm(y) < +\infty.$$

Letting $\mu = \mu_+ - \mu_-$, we have

$$\iint_{\mathbb{R}^\mathsf{d}\times\mathbb{R}^\mathsf{d}} \mathsf{g}(x-y)d\mu(x)d\mu(y) \ge 0,$$

and the functional $\mathcal{E}$ is strictly convex.

PROOF. The map $\mu \mapsto \int V\,d\mu$ is linear, so proving the convexity of $\mathcal{E}$ reduces to showing that the quadratic function $Q(f) = \iint \mathsf{g}(x-y)\,df(x)\,df(y)$, is strictly convex over probability measures. Since

$$Q(tf_1 + (1-t)f_2) = t^2 Q(f_1) + (1-t)^2 Q(f_2) + 2t(1-t)\iint \mathsf{g}(x-y)df_1(x)df_2(y),$$

we have

$$Q(tf_1 + (1-t)f_2) - tQ(f_1) - (1-t)Q(f_2) = -t(1-t)Q(f_1 - f_2),$$

and thus to prove the claimed convexity it suffices to show that $Q(f_1 - f_2) \ge 0$ for f_1, f_2 two probability measures, for which it suffices to prove that $Q(f) \ge 0$ if f is a signed Radon measure with $\int f = 0$ and $\int |f| < +\infty$, with equality if and only if $f = 0$.

Let us first consider f to be a Schwarz function with integral 0, and recall that $\hat{\mathsf{g}}(\xi) = C_{\mathsf{d},\mathsf{s}}|\xi|^{\mathsf{s}-\mathsf{d}} \ge 0$ for our choice (1.1.2) with $\mathsf{d}-2 \le \mathsf{s} \le \mathsf{d}$ (see [**Gra14**, Theorem 2.4.6]) where $\hat{\cdot}$ denotes the Fourier transform.[1] Also $|\hat{f}(\xi)| \le C\min(|\xi|, 1)$ since

[1] When $\mathsf{s} \le 0$, $\hat{\mathsf{g}}$ has an additional Dirac at the origin, but it doesn't affect the argument since we test against $|\hat{f}|^2$ which vanishes at the origin.

$\int f = 0$ and $\int |f| < +\infty$, while $\hat{f}$ decays rapidly at infinity, thus $\hat{\mathsf{g}}(\xi)|\hat{f}|^2(\xi) \in L^1(\mathbb{R}^\mathsf{d})$ and

$$Q(f) = \iint_{\mathbb{R}^\mathsf{d}\times\mathbb{R}^\mathsf{d}} \hat{\mathsf{g}}(\xi)|\hat{f}|^2(\xi)d\xi.$$

We thus find $Q(f) \geq 0$ in that case. For general f, we mollify it by convolution with a Gaussian approximation of the Dirac mass δ_0, call it ϕ_ε, and let $f_\varepsilon = \phi_\varepsilon * f$. Then $\int_{\mathbb{R}^\mathsf{d}} f_\varepsilon = \int_{\mathbb{R}^\mathsf{d}} f = 0$ and $\hat{f}_\varepsilon = \hat{\phi}_\varepsilon \hat{f}$ decays like a Gaussian, so the previous reasoning applies and we find $Q(f_\varepsilon) \geq 0$. To conclude we then argue that $Q(f_\varepsilon) \to Q(f)$ by applying the dominated convergence theorem in physical space, and using the assumption. Also $Q(f) = 0$ implies $\hat{f} = 0$ by the same steps, hence $f = 0$ and Q is strictly convex over probability measures. □

Remark 2.2. Less restrictive assumptions on g, for instance $\hat{\mathsf{g}} \geq 0$, where $\hat{\mathsf{g}}$ stands for the Fourier transform, suffice for the above result.

As a consequence of the lemma, there is a unique (if any) minimizer to (2.0.3) among probability measures.

The existence of a minimizer is a bit more delicate. To obtain it we make the following assumptions on the potential V:

(A1) V is lower semicontinuous (l.s.c.) and bounded below.
(A2) (growth assumption)

$$\lim_{|x|\to+\infty} (V(x) + \mathsf{g}(x)) = +\infty.$$

The first condition is there to ensure the lower semicontinuity of $\mathcal{E}$ and that $\inf \mathcal{E} > -\infty$, the second is made to ensure that $\mathcal{E}$ is coercive. Of course, if $\mathsf{s} > 0$, condition (A2) is equivalent to the condition that V tends to $+\infty$ at infinity.

We will repeatedly use the following

Remark 2.3. If g is as in (2.0.1), we have

$$\mathsf{g}(x-y) \geq \mathsf{g}(x) \wedge \mathsf{g}(y)\mathbf{1}_{\mathsf{s}=0} + (\mathsf{g}(x) + \mathsf{g}(y))\mathbf{1}_{\mathsf{s}<0} - C, \tag{2.1.1}$$

for some $C > 0$ depending only on s, where $\wedge$ denotes the minimum of two numbers and $\mathbf{1}_A$ means 1 if A holds and 0 otherwise. Thus, under assumptions (A1), (A2), the function $\mathsf{g}(x-y) + V(x) + V(y)$ is bounded below and tends to $+\infty$ as $|x|$ or $|y| \to +\infty$.

PROOF. For (2.1.1) we observe that when $\mathsf{s} = 0$, $\mathsf{g}(x-y) = -\log|x-y| \geq -\log 2 - \log\max(|x|, |y|)$, and when $-1 \leq \mathsf{s} < 0$, we can argue that $\mathsf{g}(x-y) \geq \mathsf{g}(x) + \mathsf{g}(y)$. The case $\mathsf{s} > 0$ is obvious. The second assertion then follows from (2.1.1) and assumptions (A1), (A2), after distinguishing the cases $\mathsf{s} > 0$, $\mathsf{s} = 0$ and $-1 \leq \mathsf{s} < 0$. □

Lemma 2.4 (Coercivity of $\mathcal{E}$). *Assume* (A1) *and* (A2) *are satisfied, and let* $\{\mu_n\}_n$ *be a sequence in* $\mathcal{P}(\mathbb{R}^\mathsf{d})$ *such that* $\{\mathcal{E}(\mu_n)\}_n$ *is bounded. Then, up to extraction of a subsequence,* μ_n *converges to some* $\mu \in \mathcal{P}(\mathbb{R}^\mathsf{d})$ *in the weak sense of probabilities, and*

$$\liminf_{n\to\infty} \mathcal{E}(\mu_n) \geq \mathcal{E}(\mu), \tag{2.1.2}$$

and $\inf \mathcal{E} > -\infty$. *In other words,* $\mathcal{E}$ *is lower semicontinuous, bounded below, and its sublevel sets are compact.*

PROOF. Assume that $\mathcal{E}(\mu_n) \le C_1$ for each n. By Remark 2.3, given any constant $C_2 > 0$ there exists a compact set $K \subset \mathbb{R}^\mathsf{d}$ such that

$$\min_{(K\times K)^c} (\mathsf{g}(x-y) + V(x) + V(y)) > C_2. \tag{2.1.3}$$

Also, by the same remark, we may write

$$\mathsf{g}(x-y) + V(x) + V(y) \ge -C_3 \tag{2.1.4}$$

with $C_3 > 0$. Rewriting then $\mathcal{E}$ as

$$\mathcal{E}(\mu) = \frac{1}{2}\iint_{\mathbb{R}^\mathsf{d}} (\mathsf{g}(x-y) + V(x) + V(y))\, d\mu(x) d\mu(y), \tag{2.1.5}$$

we deduce from (2.1.3) that $\mathcal{E}(\mu) \ge \frac{C_2}{2}$ for all μ, hence $\inf \mathcal{E} > -\infty$, and the relations (2.1.3)–(2.1.4) and our assumption on μ_n imply that

$$C_1 \ge \mathcal{E}(\mu_n) \ge -C_3 + C_2(\mu_n \otimes \mu_n)((K\times K)^c) \ge -C_3 + C_2\mu_n(K^c)^2.$$

Since C_2 can be made arbitrarily large, $\mu_n(K^c)$ can be made arbitrarily small, which means precisely that $\{\mu_n\}_n$ is a tight sequence of probability measures. By Prokhorov's theorem, it thus has a convergent subsequence (still denoted $\{\mu_n\}_n$), which converges to some probability μ. For any n and any $M > 0$, we may then write

$$\iint \mathsf{g}(x-y) d\mu_n(x) d\mu_n(y) \ge \iint (\mathsf{g}(x-y) \wedge M) d\mu_n(x) d\mu_n(y), \tag{2.1.6}$$

where $\wedge$ denotes the minimum of two numbers. For each given M, $\mathsf{g}(x-y) \wedge M$ is continuous hence $\mathsf{g}(x-y) \wedge M + V(x) + V(y)$ is l.s.c and bounded below (by Remark 2.3), and thus the weak convergence of μ_n to μ, which implies the weak convergence of $\mu_n \otimes \mu_n$ to $\mu \otimes \mu$, yields

$$\begin{aligned}
\liminf_{n\to+\infty} \mathcal{E}(\mu_n) &= \liminf_{n\to+\infty} \frac{1}{2}\iint (\mathsf{g}(x-y) + V(x) + V(y)) d\mu_n(x) d\mu_n(y)\\
&\ge \liminf_{n\to+\infty} \frac{1}{2}\iint (\mathsf{g}(x-y) \wedge M + V(x) + V(y)) d\mu_n(x) d\mu_n(y)\\
&\ge \frac{1}{2}\iint (\mathsf{g}(x-y) \wedge M + V(x) + V(y)) d\mu(x) d\mu(y).
\end{aligned}$$

The monotone convergence theorem then allows to let $M \to +\infty$, and conclude that (2.1.2) holds. □

We have seen above that $\inf \mathcal{E} > -\infty$ thanks to (A1). The next question is to see whether $\inf \mathcal{E} < +\infty$, i.e., that there exist probabilities with finite $\mathcal{E}$'s. For $\mathsf{s} < 0$ this is clear. For $\mathsf{s} \ge 0$, this is directly related to the notion of (electrostatic, Bessel, or logarithmic) capacity, whose definition we now give. One may find it in [**Lan72, ST97, EG15, AH96**] or [**LL01**, Sec. 11.15], the formulations differ a bit but are essentially equivalent. It is usually not formulated this way for $\mathsf{s} = 0$ but it can be extended to that case without trouble.

Definition 2.5 (Capacity of a set). We define the capacity of a compact set $K \subset \mathbb{R}^\mathsf{d}$ by

$$\mathrm{cap}(K) := \Phi\left(\inf_{\mu\in\mathcal{P}(K)} \iint_{\mathbb{R}^\mathsf{d}} \mathsf{g}(x-y) d\mu(x) d\mu(y)\right), \tag{2.1.7}$$

with $\Phi(t) = e^{-t}$ if $\mathsf{s} = 0$ and $\Phi(t) = t^{-1}$ if $\mathsf{s} > 0$, and where $\mathcal{P}(K)$ denotes the set of probability measures supported in K. Here the inf can be $+\infty$ if there exists no probability measure $\mu \in \mathcal{P}(K)$ such that $\iint_{\mathbb{R}^d} \mathsf{g}(x-y)d\mu(x)d\mu(y) < +\infty$. For a general set E, we define $\mathrm{cap}(E)$ as the supremum of $\mathrm{cap}(K)$ over the compact sets K included in E.

It is easy to check that capacity is increasing with respect to the inclusion of sets.

For $\mathsf{s} < 0$, in order to have a unified presentation, we use the word "capacity" to mean "cardinality".

A basic fact is that a set of zero capacity also has zero Lebesgue measure (see the references above). In fact $\mathrm{cap}(E) = 0$ is stronger than $|E| = 0$, it implies for example that the perimeter of E (whenever well defined) is also 0. A property is said to hold "quasi-everywhere" (q.e.) if it holds everywhere except on a set of capacity zero. By the preceding lemma a property that holds q.e. also holds Lebesgue-almost everywhere (a.e.), whereas the converse is in general not true. In the case $\mathsf{s} < 0$, since by capacity we mean cardinality, saying that a property holds q.e. just means that it holds everywhere.

For the sake of generality, it is interesting to consider potential V's which can take the value $+\infty$ (this is the same as imposing the constraint that the probability measures only charge a specific set, the set where V is finite). We then need to place a third assumption

(A3) $$\{x \in \mathbb{R}^{\mathsf{d}}, V(x) < +\infty\} \text{ has positive capacity.}$$

Note that with our notion of capacity, if $\mathsf{s} < 0$ this just means that $\{x \in \mathbb{R}^{\mathsf{d}}, V(x) < +\infty\}$ is nonempty.

Lemma 2.6. *Under assumptions* (A1)–(A3), *we have* $\inf \mathcal{E} < +\infty$.

PROOF. The case $\mathsf{s} < 0$ is obvious: it suffices to take μ equal to a Dirac mass at a point where V is finite. Let us turn to the case $\mathsf{s} \geq 0$. Let us define for any $\varepsilon > 0$ the set $\Sigma_\varepsilon = \{x \mid V(x) \leq \frac{1}{\varepsilon}\}$. Since V is l.s.c. the sets Σ_ε are closed, and it is easy to see that assumption (A2) implies that they are also bounded, since it implies in all cases that $V(x)$ tends to $+\infty$ when $|x| \to +\infty$.

The capacity of $\Sigma_0 = \{x \in \mathbb{R}^{\mathsf{d}}, V(x) < +\infty\}$ is positive by assumption. It is easily seen that the sets $\{\Sigma_\varepsilon\}_{\varepsilon>0}$ form a decreasing family of compact sets with $\bigcup_{\varepsilon>0} \Sigma_\varepsilon = \Sigma_0$. By standard results on capacity, we then have $\mathrm{cap}(\Sigma) = \lim_{n\to\infty} \mathrm{cap}(\Sigma_{1/n})$. Hence, since $\mathrm{cap}(\Sigma) > 0$, we have that $\mathrm{cap}(\Sigma_\varepsilon)$ is positive for ε small enough. Then by definition there exists a probability measure μ_ε supported in Σ_ε such that

(2.1.8) $$\iint \mathsf{g}(x-y)d\mu_\varepsilon(x)d\mu_\varepsilon(y) < +\infty.$$

Of course, we also have $\int V d\mu_\varepsilon < +\infty$ by definition of Σ_ε. Hence $\mathcal{E}(\mu_\varepsilon) < +\infty$, in particular $\inf \mathcal{E} < +\infty$. □

We may now give the main existence result, together with the characterization of the minimizer.

Remark 2.7. In the following we will not really use the particulars of the logarithmic and Riesz kernels. While uniqueness essentially requires g to have positive

Fourier transform, the rest of the theorem still holds for a much more general class of g's, say g positive, monotone radial and satisfying $\iint \mathsf{g}(x-y)\,dx\,dy < \infty$.

Theorem 2.8 (Frostman [**Fro35**], existence and characterization of the equilibrium measure). *Under the assumptions* (A1)–(A3), *the minimum of* $\mathcal{E}$ *over* $\mathcal{P}(\mathbb{R}^{\mathsf{d}})$ *exists, is finite and is achieved by a unique* μ_V, *which has a compact support of positive capacity. In addition* μ_V *is uniquely characterized by the fact that there exists a constant* c *such that*

$$\begin{cases} h^{\mu_V} + V \geq c & q.e.\ in\ \mathbb{R}^{\mathsf{d}} \\ h^{\mu_V} + V = c & q.e.\ in\ the\ support\ of\ \mu_V \end{cases} \tag{2.1.9}$$

where

$$h^{\mu_V}(x) := \int_{\mathbb{R}^{\mathsf{d}}} \mathsf{g}(x-y)d\mu_V(y) \tag{2.1.10}$$

is the electrostatic potential generated by μ_V. *Moreover, the constant* c *must be*

$$c = 2\mathcal{E}(\mu_V) - \int_{\mathbb{R}^{\mathsf{d}}} V(x)d\mu_V(x). \tag{2.1.11}$$

Definition 2.9. From now on, we denote by ζ the function

$$\zeta = h^{\mu_V} + V - c. \tag{2.1.12}$$

We also denote $\omega := \{x \in \mathbb{R}^{\mathsf{d}}, \zeta(x) = 0, \text{q.e.}\}$ and $\Sigma := \operatorname{supp} \mu_V$.

We note that in view of (2.1.9), $\zeta \geq 0$ a.e. and $\zeta = 0$, μ_V-a.e. Also,

$$\Sigma \subset \{\zeta = 0, q.e.\} = \omega. \tag{2.1.13}$$

The set ω corresponds to the *contact set* or *coincidence set* of the obstacle problem, while Σ is the set where the obstacle is *active*, sometimes called the *droplet*. We may place assumptions so that they coincide.

Here h^{μ_V} is the self-generated (electrostatic) potential, while V is the external potential, so $h^{\mu_V} + V$ corresponds to the total potential; ζ is thus an effective potential. The theorem says that the equilibrium distribution μ_V is one for which the total potential is minimized, and constant, in the support of the distribution.

Moreover, we claim that for every $x \in \mathbb{R}^{\mathsf{d}}$

$$\zeta(x) \geq \mathsf{g}(x) + V(x) - C \tag{2.1.14}$$

where C depends only on d, s and V. Indeed,

$$h^{\mu_V}(x) - \mathsf{g}(x) = \int_{\mathbb{R}^{\mathsf{d}}} (\mathsf{g}(x-y) - \mathsf{g}(x))d\mu_V(y), \tag{2.1.15}$$

and since μ_V is compactly supported the integral reduces to a compact set of y, thus using (2.1.1) in the case $\mathsf{s} \leq 0$, or direct estimate if $\mathsf{s} > 0$ we easily deduce that (2.1.15) is bounded below. The relation (2.1.14) then follows from (2.1.12).

The effective potential ζ thus grows like $V + \mathsf{g}$ at infinity.

Remark 2.10. It is sometimes of interest to consider "weakly confining potentials" for which (A2) barely fails, and the equilibrium measure still exists but will fail to be compactly supported; see for instance [**Har12**].

Example 2.11 (Capacity of a compact set)**.** Let K be a compact set of positive capacity, and let $V = 0$ in K and $V = +\infty$ in K^c. In that case the minimization of $\mathcal{E}$ is the same as the computation of the capacity of K as in (2.1.7). The support of the equilibrium measure μ_V is contained in K, and the associated Euler-Lagrange equation (2.1.9) states that the potential h^{μ_V} is constant q.e. on the support of μ_V, a well-known result in physics.

Remark 2.12. Note that by (1.1.11), if $\mathsf{s} = \mathsf{d} - 2$, the function h^{μ_V} solves:

$$-\Delta h^{\mu_V} = \mathsf{c_d}\mu_V, \tag{2.1.16}$$

where $\mathsf{c_d}$ is the constant defined in (1.1.12). In particular in the example above, if K is sufficiently regular and $\mathsf{s} = \mathsf{d} - 2$, one finds that $\mu_V = 0$ q.e. in K, which indicates that μ_V is supported on ∂K.

Example 2.13 ($C^{1,1}$ potentials and RMT examples)**.** In the Coulomb case $\mathsf{s} = \mathsf{d} - 2$, using again Remark 2.12 yields that if $V \in C^{1,1}$,

$$\mu_V = \frac{\Delta V}{\mathsf{c_d}} \quad \text{in } \overset{\circ}{\Sigma}, \tag{2.1.17}$$

where $\overset{\circ}{\Sigma}$ is the interior of Σ, i.e., the density of the measure on the interior of its support is given by $\frac{\Delta V}{\mathsf{c_d}}$. This will be proven in Proposition 2.17. For example if V is quadratic, then the measure μ_V has constant density in the interior of its support.

This corresponds to the important examples of the energies that arise in random matrix theory, more precisely:

- in dimension $\mathsf{d} = 2$, for $V(x) = |x|^2$, one may check that $\mu_V = \frac{1}{\pi}\mathbf{1}_{B_1}$, where $\mathbf{1}$ denotes a characteristic function and B_1 is the unit ball, i.e., the equilibrium measure is the normalized Lebesgue measure on the unit disk (by uniqueness, μ_V should be radially symmetric, and the combination of (2.1.17) with the constraint of being a probability measure imposes the support to be B_1). This is known as the *circle law* for the Ginibre ensemble in the context of Random Matrix Theory (RMT). Its derivation (which can be seen as a consequence of Section 3.2 below) is attributed to Ginibre, Mehta, an unpublished paper of Silverstein and Girko [**Gir84**] (the latter includes the non-Gaussian i.i.d. case).
- in all Coulomb cases $\mathsf{s} = \mathsf{d} - 2$, the same holds, i.e., for $V(x) = |x|^2$ we have $\mu_V = \frac{\mathsf{d}}{\mathsf{c_d}}\mathbf{1}_{B_{(2/\mathsf{d})^{1/\mathsf{d}}}}$ by the same reasoning.
- in dimension $\mathsf{d} = 1$, with $\mathsf{s} = 0$ and $V(x) = x^2$, the equilibrium measure is $\mu_V(x) = \frac{1}{2\pi}\sqrt{4 - x^2}\mathbf{1}_{|x|\le 2}$, which corresponds in the context of RMT (GUE and GOE ensembles) to *Wigner's semicircle law*, cf. [**Wig55, Meh04**].

Example 2.14 (Radial example)**.** When $V(x) = |x|^2$, in the general Riesz case (2.0.1), then μ_V is given by the so-called Barenblatt profile of the form $(a - bx^2)_+^{1-\frac{\mathsf{d}-\mathsf{s}}{2}}$, see [**BIK15**, Theorem 2.2] and [**CV11**, Theorem 3.1], which generalizes the classical Barenblatt profile for self-similar solutions to the porous medium equation.

More examples of exact computations of equilibrium measures, with some surprising phenomena, are given in [**CSW22, CSW23, Byu24**]. Note that given any measure μ with finite energy, one can choose the confinement

$$V := -\mathsf{g} * \mu + \frac{1}{2} \int_{(\mathbb{R}^{\mathsf{d}})^2} \mathsf{g}(x-y) d\mu(x) d\mu(y)$$

so that the associated equilibrium measure μ_V equals μ.

Remark 2.15. In the Coulomb case $\mathsf{s} = \mathsf{d} - 2$, in view of (2.1.17) the density $\mu_V(x)$ typically has a discontinuity on $\partial\Sigma$. In contrast, in Riesz cases $\mathsf{s} \in (\mathsf{d}-2, \mathsf{d})$, the density $\mu_V(x)$ is typically continuous and vanishing on $\partial\Sigma$, as the semicircle law provides an example of. In fact, as shown in [**CF25**], the generic vanishing rate in Riesz cases is in $\mathrm{dist}(x, \partial\Sigma)^{1-\frac{\mathsf{d}-\mathsf{s}}{2}}$, which is deduced from the connection with the obstacle problem described in Section 2.4 below. We also refer to [**RO**] and [**RS25a**].

We now turn to the proof of Theorem 2.8, adapted from [**ST97**, Chap. 1].

Proof of Theorem 2.8. **Step 1: Existence and uniqueness modulo compact support.**

The existence of a minimizer μ_V follows directly from Lemmas 2.4 and 2.6, its uniqueness from Lemma 2.1 once compactness of the support is proven. Indeed, if μ_+ and μ_- are two probabilities with compact support, to satisfy $\iint |\mathsf{g}(x-y)| d\mu_\pm(x) d\mu_\pm(y) < +\infty$ and be able to apply Lemma 2.1, it suffices to know that $\left|\iint \mathsf{g}(x-y) d\mu_\pm(x) d\mu_\pm(y)\right| < +\infty$. When $\mathsf{s} > 0$ or $\mathsf{s} < 0$ it is clear since g has a sign, and when $\mathsf{s} = 0$ it follows from the fact that $\mu_\pm$ are compactly supported that the negative part of the integral is bounded. Let us now justify that $-\infty < \iint \mathsf{g}(x-y) d\mu_V(x) d\mu_V(y) < +\infty$. First, since by (A1) V is bounded below and since $\mathcal{E}(\mu_V) < +\infty$ by Lemma 2.6, we find that $\iint \mathsf{g}(x-y) d\mu_V(x) d\mu_V(y) < +\infty$. Secondly, $\iint \mathsf{g}(x-y) d\mu_V(x) d\mu_V(y) > -\infty$ since by definition g is bounded below on compact sets and μ_V has compact support.

It remains to show that μ_V has compact support of finite capacity and that (2.1.9) holds.

Step 2: Compact support. Using Remark 2.3, we may find a compact set K such that $\mathsf{g}(x-y) + V(x) + V(y) \geq 2(\mathcal{E}(\mu_V) + 1)$ outside of $K \times K$.

Assume that μ_V has mass outside K, i.e., assume $\mu_V(K) < 1$, and define the new probability measure

$$\tilde{\mu} := \frac{\mu_V|_K}{\mu_V(K)}. \tag{2.1.18}$$

We want to show that $\tilde{\mu}$ has less or equal energy $\mathcal{E}(\tilde{\mu})$ than μ_V, in order to get a contradiction. One may compute $\mathcal{E}(\mu_V)$ in the following way:

$$\begin{aligned}
\mathcal{E}(\mu_V) &= \frac{1}{2} \iint_{K \times K} (\mathsf{g}(x-y) + V(x) + V(y)) \, d\mu_V(x) d\mu_V(y) \\
&\quad + \frac{1}{2} \iint_{(K \times K)^c} (\mathsf{g}(x-y) + V(x) + V(y)) \, d\mu_V(x) d\mu_V(y) \\
&\geq \mu_V(K)^2 \mathcal{E}(\tilde{\mu}) \\
&\quad + \frac{1}{2} \min_{(K \times K)^c} (\mathsf{g}(x-y) + V(x) + V(y)) \left(1 - \iint_{K \times K} d\mu_V d\mu_V\right).
\end{aligned}$$

By choice of K, and since we assumed $\mu_V(K) < 1$, this implies that

$$\mathcal{E}(\mu_V) \geq \mu_V(K)^2\mathcal{E}(\tilde{\mu}) + (1 - \mu_V(K)^2)(\mathcal{E}(\mu_V) + 1) \tag{2.1.19}$$

and thus

$$\mathcal{E}(\tilde{\mu}) \leq \frac{\mathcal{E}(\mu_V)}{\mu_V(K)^2} + \frac{\mu_V(K)^2 - 1}{\mu_V(K)^2}(\mathcal{E}(\mu_V) + 1) < \mathcal{E}(\mu_V),$$

a contradiction with the minimality of μ_V. We thus conclude that μ_V has compact support. The fact that the support of μ_V has positive capacity is an immediate consequence of the fact that $\mathcal{E}(\mu_V) < \infty$ and the definition of capacity.

Step 3. We turn to the proof of the Euler-Lagrange equations (2.1.9). For this, we use the "method of variations" which consists in continuously deforming μ_V into other admissible probability measures.
Let ν in $\mathcal{P}(\mathbb{R}^\mathsf{d})$ such that $\mathcal{E}(\nu) < +\infty$, and consider the probability measure $(1-t)\mu_V + t\nu$ for t in $[0,1]$. Since μ_V minimizes $\mathcal{E}_V$, we have

$$\mathcal{E}\left((1-t)\mu_V + t\nu\right) \geq \mathcal{E}(\mu_V), \text{ for all } t \in [0,1]. \tag{2.1.20}$$

By letting $t \to 0^+$ and keeping only the first order terms in t, one obtains the "functional derivative" of $\mathcal{E}$ at μ_V. More precisely, writing

$$\iint \frac{1}{2}\mathsf{g}(x-y)d((1-t)\mu_V + t\nu)(x)d((1-t)\mu_V + t\nu)(y) + \int V(x)d((1-t)\mu_V + t\nu)(x) \geq \mathcal{E}(\mu_V), \tag{2.1.21}$$

one easily gets that

$$\mathcal{E}(\mu_V) + t\left[\frac{1}{2}\iint \mathsf{g}(x-y)d(\nu - \mu_V)(x)d\mu_V(y) + \frac{1}{2}\iint \mathsf{g}(x-y)d(\nu - \mu_V)(y)d\mu_V(x) + \int V(x)d(\nu - \mu_V)(x)\right] + O(t^2) \geq \mathcal{E}(\mu_V). \tag{2.1.22}$$

Here, we may cancel the identical order 0 terms $\mathcal{E}(\mu_V)$ on both sides, and note that in view of (2.1.10) the expression between brackets can be rewritten as $\int h^{\mu_V}(x)d(\nu - \mu_V)(x) + \int V(x)d(\nu - \mu_V)(x)$. Next, dividing the inequality by $t > 0$, and letting $t \to 0^+$, we obtain that for all ν in $\mathcal{P}(\mathbb{R}^\mathsf{d})$ such that $\mathcal{E}(\nu) < +\infty$,

$$\int h^{\mu_V}(x)d(\nu - \mu_V)(x) + \int V(x)d(\nu - \mu_V)(x) \geq 0, \tag{2.1.23}$$

or equivalently

$$\int \left(h^{\mu_V} + V\right)(x)d\nu(x) \geq \int \left(h^{\mu_V} + V\right)(x)d\mu_V(x). \tag{2.1.24}$$

Defining the constant c by
(2.1.25)

$$c := 2\mathcal{E}(\mu_V) - \int V d\mu_V = \iint \mathsf{g}(x-y)d\mu_V(x)d\mu_V(y) + \int V d\mu_V = \int \left(h^{\mu_V} + V\right)d\mu_V,$$

(2.1.24) asserts that

$$\int \left(h^{\mu_V} + V\right)d\nu \geq c \tag{2.1.26}$$

for all probability measures ν on $\mathbb{R}^{\mathsf{d}}$ such that $\mathcal{E}(\nu) < +\infty$. Note that at this point, if we relax the condition $\mathcal{E}(\nu) < +\infty$, then choosing ν to be a Dirac mass when applying (2.1.26) would yield

$$h^{\mu_V} + V \geq c \tag{2.1.27}$$

pointwise. However, Dirac masses have infinite energy $\mathcal{E}$, and we can only prove that (2.1.27) holds quasi-everywhere, which we do now.

Do not assume that there exists a set K of positive capacity such that (2.1.27) is false on K. By definition of the capacity of K as supremum of capacities over compact sets included in K, we may in fact suppose that K is compact. By definition, this means that there is a probability measure ν supported in K such that

$$\iint \mathsf{g}(x-y)d\nu(x)d\nu(y) < +\infty. \tag{2.1.28}$$

Let us observe that $-h^{\mu_V}$ is bounded above on any compact set (this is clear if $\mathsf{s} > 0$ because then g is positive and so is h^{μ_V}, and can be easily checked for $\mathsf{s} \leq 0$ because $-\mathsf{g}$ is then bounded above on any compact set and μ_V has compact support). By assumption, equation (2.1.27) is false on K, that is $V < c - h^{\mu_V}$ on K. Integrating this inequality against ν yields

$$\int V d\nu = \int_K V d\nu < \int_K \left(c - h^{\mu_V}\right) d\nu < +\infty \tag{2.1.29}$$

which, combined with (2.1.28) ensures that $\mathcal{E}(\nu)$ is finite. But (2.1.26) must hold, which contradicts (2.1.29). We have thus shown that

$$h^{\mu_V} + V \geq c \text{ q.e.}, \tag{2.1.30}$$

which is the first of the relations (2.1.9).

For the second one, let us denote by E the set where the previous inequality (2.1.30) fails. We know that E has zero capacity, but since μ_V satisfies $\mathcal{E}(\mu_V) < +\infty$, it does not charge sets of zero capacity (otherwise one could restrict μ_V to such a set, normalize its mass to 1 and get a contradiction with the definition of a zero capacity set). Hence we have

$$h^{\mu_V} + V \geq c \quad \mu_V\text{-a.e.} \tag{2.1.31}$$

Integrating this relation against μ_V yields

$$\int \left(h^{\mu_V} + V\right) d\mu_V \geq c, \tag{2.1.32}$$

but in view of (2.1.25) this implies that equality must hold in (2.1.31) μ_V-almost everywhere. This establishes the second Euler-Lagrange equation in (2.1.9) and (2.1.11) holds by (2.1.25).

Step 4. We show that the relations (2.1.9) uniquely characterize the minimizer of $\mathcal{E}_V$. Assume that μ is another probability solving (2.1.9) with some constant c', and set, for $t \in [0,1]$, $\mu_t := t\mu + (1-t)\mu_V$, hence $h^{\mu_t} = th^{\mu} + (1-t)h^{\mu_V}$. We have

$$\begin{aligned}
\mathcal{E}(\mu_t) &= \frac{1}{2}\int \left(th^{\mu}(x) + (1-t)h^{\mu_V}(x) + 2V(x)\right) d\mu_t(x) \\
&= \frac{t}{2}\int \left(h^{\mu}(x) + V(x)\right) d\mu_t(x) + \frac{(1-t)}{2}\int \left(h^{\mu_V}(x) + V(x)\right) d\mu_t(x) \\
&\quad + \frac{1}{2}\int V(x) d\mu_t(x).
\end{aligned}$$

By assumption, $h^\mu + V \geq c'$ and $h^{\mu_V} + V \geq c$ almost everywhere. We thus get that

$$\begin{aligned}\mathcal{E}(\mu_t) &\geq \frac{1}{2}(tc' + (1-t)c) + \frac{1}{2}\int V(x)\,(td\mu(x) + (1-t)d\mu_V(x)) \\ &= \frac{t}{2}\left(c' + \int V d\mu\right) + \frac{(1-t)}{2}\left(c + \int V d\mu_V\right).\end{aligned} \tag{2.1.33}$$

On the other hand, by (2.1.11) we have

$$\mathcal{E}(\mu) = \frac{1}{2}\left(c' + \int V d\mu\right) \text{ and } \mathcal{E}(\mu_V) = \frac{1}{2}\left(c + \int V d\mu_V\right).$$

Hence (2.1.33) asserts that $\mathcal{E}(\mu_t) \geq t\mathcal{E}(\mu) + (1-t)\mathcal{E}(\mu_V)$, which is impossible by strict convexity of $\mathcal{E}$ unless $\mu = \mu_V$. This proves that the two measures μ and μ_V must coincide. □

Henceforth, we will assume that μ_V is a measure that is absolutely continuous with respect to the Lebesgue measure and we will use the same notation for a measure $d\mu$ and its density $\mu(x)$.

2.2. A first electric rewriting

We now reformulate the energy $\mathcal{E}$ in terms of the "electric" potential of (2.1.10). This electric rewriting is easy in the Coulomb case and well-known in physics (see for instance [**JGP94**]). It consists in using the fact that g is the Coulomb kernel and thus satisfies (1.1.11), to deduce that the Coulomb potential defined by

$$h^\mu = \mathsf{g} * \mu = \int \mathsf{g}(\cdot - y)d\mu(y)$$

(whenever the integral is convergent) satisfies

$$-\Delta h^\mu = \mathsf{c_d}\mu,$$

which allows to rewrite

$$\iint_{\mathbb{R}^\mathsf{d}\times\mathbb{R}^\mathsf{d}} \mathsf{g}(x-y)d\mu(x)d\mu(y) = \int_{\mathbb{R}^\mathsf{d}} h^\mu(x)d\mu(x) = -\frac{1}{\mathsf{c_d}}\int_{\mathbb{R}^\mathsf{d}} h^\mu \Delta h^\mu = \frac{1}{\mathsf{c_d}}\int_{\mathbb{R}^\mathsf{d}} |\nabla h^\mu|^2. \tag{2.2.1}$$

This rewriting has used Green's formula and assumed that the boundary terms at infinity that it generates vanish. This is not always the case: it is true if $\mathsf{s} > 0$ (or $\mathsf{d} \geq 3$) but it is not true for $\mathsf{s} \leq 0$, i.e., for the logarithmic case in two dimensions or for the one-dimensional Coulomb kernel, because these do not vanish at infinity. In these cases, (2.2.1) will instead only hold for the differences of two probability measures $\mu = \mu_+ - \mu_-$ which are such that $\int_{\mathbb{R}^\mathsf{d}} \mu = 0$ and h^μ decays at higher order at infinity. This is closely related to what was done in Lemma 2.1.

In order to present results valid for the general Riesz case, it is useful to first describe the *dimension extension representation* of fractional Laplacians, that will be used throughout the text.

2.2.1. Extension representation of the fractional Laplacian. The extension method is found in work of Molchanov and Ostrovskii [**MO69**] in their studies of symmetric stable processes, and was made systematic by Caffarelli-Silvestre [**CS07**]. The difficulty with the fractional Laplacian is that it is a *nonlocal operator*. The extension representation allows to revert to a local operator. It consists in seeing $\mathbb{R}^{\mathsf{d}}$ as embedded into $\mathbb{R}^{\mathsf{d}+1} = \{(x,y), x \in \mathbb{R}^{\mathsf{d}}, y \in \mathbb{R}\}$, by identifying it with $\mathbb{R}^{\mathsf{d}} \times \{0\}$. Let us denote by $\delta_{\mathbb{R}^{\mathsf{d}}}$ the uniform measure on $\mathbb{R}^{\mathsf{d}} \times \{0\}$, i.e., the distribution such that for any smooth $\varphi(x,y)$ (with $x \in \mathbb{R}^{\mathsf{d}}, y \in \mathbb{R}$) we have

$$\int_{\mathbb{R}^{\mathsf{d}+1}} \varphi \delta_{\mathbb{R}^{\mathsf{d}}} = \int_{\mathbb{R}^{\mathsf{d}}} \varphi(x,0)\, dx. \tag{2.2.2}$$

Then, as observed in [**CS07**], g can be seen as the restriction to the $\mathbb{R}^{\mathsf{d}}$ hyperplane of the kernel of the weighted divergence form operator

$$-\mathrm{div}\,(|y|^\gamma \nabla \cdot) \tag{2.2.3}$$

where $y \in \mathbb{R}$ stands for the last (or $x_{\mathsf{d}+1}$) coordinate in $\mathbb{R}^{\mathsf{d}+1}$ and γ is given by

$$\gamma = \mathsf{s} - \mathsf{d} + 1. \tag{2.2.4}$$

For instance, in the one-dimensional log case $\mathsf{s} = 0$, $\mathsf{g}(x) = -\log|x|$ is the kernel of the half-Laplacian, and it is known that the half-Laplacian can be made to correspond to the Laplacian by adding one extra space dimension. In general dimension, the half-Laplacian is associated to $\mathsf{d} - \mathsf{s} = 1$, hence $\gamma = 0$, and it can be viewed as the kernel of a full Laplacian: if one wants to solve

$$(-\Delta)^{\frac{1}{2}} u = f \quad \text{in } \mathbb{R}^{\mathsf{d}}$$

it suffices to solve

$$-\Delta u = f \delta_{\mathbb{R}^{\mathsf{d}}} \quad \text{in } \mathbb{R}^{\mathsf{d}+1},$$

where $\delta_{\mathbb{R}^{\mathsf{d}}}$ is as above, and consider the $\mathbb{R}^{\mathsf{d}} \times \{0\}$ trace of the solution. Note that u is automatically harmonic away from $\mathbb{R}^{\mathsf{d}} \times \{0\}$, and that its Laplacian is, in the distributional sense, equal to (half) the jump of the normal derivative across $\mathbb{R}^{\mathsf{d}} \times \{0\}$.

We will use this extension representation in all the book whenever we deal with Riesz interaction. It allows to view g as the kernel of a *local* operator, $-\mathrm{div}\,(|y|^\gamma \nabla)$. The weight $|y|^\gamma$ does not introduce much change in the computations, and up to adding one dimension, this will allow us to treat the Riesz case similarly to the Coulomb case in many instances.

Indeed, the cases with extension can be treated in a unified way with the Coulomb cases by working in $\mathbb{R}^{\mathsf{d}+\mathsf{k}}$ where $\mathsf{k} = 0$ in the Coulomb cases, $\mathsf{k} = 1$ otherwise, and setting

$$\gamma = \mathsf{s} + 2 - \mathsf{k} - \mathsf{d}, \tag{2.2.5}$$

which is 0 in the Coulomb case. From now on, we use this unified formulation. The reader only interested in the Coulomb case can just set $\mathsf{k} = 0$ (i.e., remove the dimension extension) and $\gamma = 0$, i.e., remove the $|y|^\gamma$ weight and replace $\mathrm{div}\,(|y|^\gamma \nabla \cdot)$ by the Laplacian.

Again, the main feature we will use is that the extended g is a fundamental solution for the degenerate elliptic operator $-\frac{1}{\mathsf{c}_{\mathsf{d},\mathsf{s}}} \mathrm{div}\,(|y|^\gamma \nabla \cdot)$ in $\mathbb{R}^{\mathsf{d}+\mathsf{k}}$, where $\mathsf{c}_{\mathsf{d},\mathsf{s}}$ is as in (1.1.13), i.e.,

$$-\mathrm{div}\left(|y|^\gamma \nabla \mathsf{g}\right) = \mathsf{c}_{\mathsf{d},\mathsf{s}} \delta_0 \quad \text{in } \mathbb{R}^{\mathsf{d}+\mathsf{k}} \tag{2.2.6}$$

with equality in the sense of distributions. This way, for any distribution f on $\mathbb{R}^{\mathsf{d}}$, the potential

$$\mathsf{g} * f = \int_{\mathbb{R}^{\mathsf{d}}} \mathsf{g}(x - x')f(x')dx',$$

can naturally be extended into a potential on $\mathbb{R}^{\mathsf{d}+\mathsf{k}}$

$$h^f(x,y) := \int_{\mathbb{R}^{\mathsf{d}}} \mathsf{g}((x,y) - (x',0))f(x')dx'$$

satisfying

$$-\mathrm{div}\,(|y|^{\gamma}\nabla h^f) = \mathsf{c}_{\mathsf{d},\mathsf{s}} f\delta_{\mathbb{R}^{\mathsf{d}}} \quad \text{in } \mathbb{R}^{\mathsf{d}+\mathsf{k}}. \tag{2.2.7}$$

2.2.2. First electric rewriting in Riesz cases. The electric rewriting is another point of view on the fact that by Plancherel's theorem, since $\hat{\mathsf{g}}(\xi) = C_{\mathsf{d},\mathsf{s}}|\xi|^{\mathsf{s}-\mathsf{d}}$, we may rewrite formally

$$\begin{aligned}\iint_{\mathbb{R}^{\mathsf{d}}\times\mathbb{R}^{\mathsf{d}}} \mathsf{g}(x-y)d\mu(x)d\mu(y) &= \int_{\mathbb{R}^{\mathsf{d}}} \hat{\mathsf{g}}(\xi)|\hat{\mu}(\xi)|^2 d\xi \\ &= C_{\mathsf{d},\mathsf{s}} \int_{\mathbb{R}^{\mathsf{d}}} |\xi|^{\mathsf{s}-\mathsf{d}}|\hat{\mu}(\xi)|^2 d\xi \\ &= C_{\mathsf{d},\mathsf{s}} \|\mu\|^2_{\dot{H}^{\frac{\mathsf{s}-\mathsf{d}}{2}}(\mathbb{R}^{\mathsf{d}})},\end{aligned} \tag{2.2.8}$$

where $\dot{H}^m(\mathbb{R}^{\mathsf{d}})$ denotes the homogeneous Sobolev space defined as the completion of Schwartz functions for the seminorm

$$\|f\|^2_{\dot{H}^m(\mathbb{R}^{\mathsf{d}})} = \int_{\mathbb{R}^{\mathsf{d}}} |\xi|^{2m}|\hat{f}(\xi)|^2 d\xi. \tag{2.2.9}$$

Proposition 2.16 (Electric reformulation for Riesz interactions). *Assume* (2.0.1). *If* $\mathsf{s} > 0$, *for any bounded Radon measure* μ *of finite mass in* $\mathbb{R}^{\mathsf{d}}$ *such that* $\iint \mathsf{g}(x-y)d\mu(x)d\mu(y) < +\infty$, *letting* $h^\mu = \mathsf{g} * (\mu\delta_{\mathbb{R}^{\mathsf{d}}})$ *in the sense of distributions, we have* $h^\mu \in L^1_{\mathrm{loc}}(\mathbb{R}^{\mathsf{d}+\mathsf{k}})$ *and*

$$\iint_{\mathbb{R}^{\mathsf{d}}\times\mathbb{R}^{\mathsf{d}}} \mathsf{g}(x-y)d\mu(x)d\mu(y) = \frac{1}{\mathsf{c}_{\mathsf{d},\mathsf{s}}} \int_{\mathbb{R}^{\mathsf{d}+\mathsf{k}}} |y|^{\gamma}|\nabla h^\mu|^2. \tag{2.2.10}$$

If $\mathsf{s} \le 0$, *then for* μ_+ *and* μ_- *probability measures on* $\mathbb{R}^{\mathsf{d}}$, *each satisfying*

$$\iint_{\mathbb{R}^{\mathsf{d}}\times\mathbb{R}^{\mathsf{d}}} |\mathsf{g}(x-y)|d\mu_\pm(x)d\mu_\pm(y) < +\infty, \tag{2.2.11}$$

letting $\mu = \mu_+ - \mu_-$, *the potential* $h^\mu = \mathsf{g} * (\mu\delta_{\mathbb{R}^{\mathsf{d}}})$ *is in* $L^1_{\mathrm{loc}}(\mathbb{R}^{\mathsf{d}+\mathsf{k}})$ *and we have*

$$\iint_{\mathbb{R}^{\mathsf{d}}\times\mathbb{R}^{\mathsf{d}}} \mathsf{g}(x-y)d\mu(x)d\mu(y) = \frac{1}{\mathsf{c}_{\mathsf{d},\mathsf{s}}} \int_{\mathbb{R}^{\mathsf{d}+\mathsf{k}}} |y|^{\gamma}|\nabla h^\mu|^2 < +\infty. \tag{2.2.12}$$

This also coincides with the fractional Sobolev norm $\|h^\mu\|^2_{\dot{H}^{\frac{\mathsf{d}-\mathsf{s}}{2}}}$ or $\|\mu\|^2_{\dot{H}^{\frac{\mathsf{s}-\mathsf{d}}{2}}}$.

PROOF. **Step 1: Definition of the potential.** We first check that $h^\mu \in L^1_{\mathrm{loc}}(\mathbb{R}^{\mathsf{d}+\mathsf{k}})$. Let U be a bounded set in $\mathbb{R}^{\mathsf{d}+\mathsf{k}}$, we have $\int_U |\mathsf{g}(x-y)|dx < +\infty$ since $\mathsf{s} < \mathsf{d}$, and thus by Fubini-Tonelli, we may write

$$\int_U h^\mu = \int_U \int_{\mathbb{R}^{\mathsf{d}}} \mathsf{g}(x-y)d\mu(y)dx = \int_{\mathbb{R}^{\mathsf{d}}} \int_U \mathsf{g}(x-y)dxd\mu(y) < +\infty, \tag{2.2.13}$$

hence the claim.

Step 2: The case $\mathsf{s} \le 0$, μ smooth and compactly supported. Let us first assume that $\mu_\pm$ are both bounded, Lipschitz, and compactly supported densities. This guarantees that h^μ is well-defined as $\int \mathsf{g}(x-y)\mu(y)dy$, and the same for ∇h^μ as $\int \mathsf{g}(x-y)\nabla\mu(y)dy$. The fact that $\int \mu_+ = \int \mu_-$ implies that h^μ decays like $\frac{1}{|x|^{\mathsf{s}+1}}$ while ∇h^μ decays like $\frac{1}{|x|^{\mathsf{s}+2}}$. To see this it suffices to write

$$h^\mu(x) = \int_{\mathbb{R}^\mathsf{d}} (\mathsf{g}(x-y) - \mathsf{g}(x))d\mu_+(y) - (\mathsf{g}(x-y) - \mathsf{g}(x))d\mu_-(y)$$

and argue that $|\mathsf{g}(x-y) - \mathsf{g}(x)| \le \frac{C}{|x|^{\mathsf{s}+1}}$ for all x large enough when y remains in the compact support of $\mu_\pm$, and similarly for ∇h^μ.

Using Green's formula, we have for any R, B_R being the ball centered at 0 and of radius R in $\mathbb{R}^{\mathsf{d}+\mathsf{k}}$ and ν the outer unit normal,

$$\int_{B_R} |y|^\gamma |\nabla h^\mu|^2 = \int_{\partial B_R} |y|^\gamma \frac{\partial h^\mu}{\partial \nu} h^\mu - \int_{B_R} h^\mu \mathrm{div}\,(|y|^\gamma \nabla h^\mu). \tag{2.2.14}$$

The boundary term is bounded by $R^{-2\mathsf{s}-3} \int_{\partial B_R} |y|^\gamma = R^{-2\mathsf{s}-3+\mathsf{d}+\mathsf{k}-1+\gamma}$, and using (2.2.5) this is $R^{-\mathsf{s}-2}$ which tends to 0 as $R \to \infty$ since $\mathsf{s} \ge -1$ in the cases we consider. Taking the limit and using (2.2.7), we obtain that

$$\begin{aligned}\int_{\mathbb{R}^{\mathsf{d}+\mathsf{k}}} |y|^\gamma |\nabla h^\mu|^2 &= -\int_{\mathbb{R}^{\mathsf{d}+\mathsf{k}}} h^\mu \mathrm{div}\,(|y|^\gamma \nabla h^\mu) \\ &= \mathsf{c}_{\mathsf{d},\mathsf{s}} \int_{\mathbb{R}^\mathsf{d}} h^\mu \mu = \iint_{\mathbb{R}^\mathsf{d}\times\mathbb{R}^\mathsf{d}} \mathsf{g}(x-y)d\mu(x)d\mu(y).\end{aligned}$$

Step 3: The general case $\mathsf{s} \le 0$. Let us first consider, as in the proof of Lemma 2.1, a Gaussian mollification $\phi_\varepsilon * \mu_\pm$ of $\mu_\pm$, where ϕ_ε is a Gaussian approximation of the identity. Letting $\mu^\varepsilon = \phi_\varepsilon * \mu$, we have $\int \mu^\varepsilon = 0$, and $\mu^\varepsilon \to \mu$ weakly and in $\dot H^{\frac{\mathsf{s}-\mathsf{d}}{2}}$. Finally, by standard properties of mollifiers,

$$\lim_{\varepsilon\to 0} \iint_{\mathbb{R}^\mathsf{d}\times\mathbb{R}^\mathsf{d}} \mathsf{g}(x-y)d\mu^\varepsilon(x)d\mu^\varepsilon(y) = \iint_{\mathbb{R}^\mathsf{d}\times\mathbb{R}^\mathsf{d}} \mathsf{g}(x-y)d\mu(x)d\mu(y). \tag{2.2.15}$$

Moreover, (2.2.8) is correct for Gaussian decaying functions such as μ^ε, thus by finiteness of $\iint_{\mathbb{R}^\mathsf{d}\times\mathbb{R}^\mathsf{d}} \mathsf{g}(x-y)d\mu(x)d\mu(y)$, the sequence $\{\mu^\varepsilon\}_\varepsilon$ is Cauchy in $\dot H^{\frac{\mathsf{s}-\mathsf{d}}{2}}$ and converges in $\dot H^{\frac{\mathsf{s}-\mathsf{d}}{2}}$ to μ.

Next, we may consider truncations $\mu_\pm^{\varepsilon,n}$ which are bounded, Lipschitz and compactly supported, such that $\int \mu_+^{\varepsilon,n} = \int \mu_-^{\varepsilon,n}$, and $\mu_\pm^{\varepsilon,n} \to \mu_\pm^\varepsilon$ as $n \to \infty$, pointwise and with domination. For instance, it suffices to restrict the $\mu_\pm^\varepsilon$ to the compact level sets $\{\mu_\pm^\varepsilon > \frac{1}{n}\}$, subtract off $1/n$, and then normalize by a factor that makes the restricted measures probabilities. The result of Step 2 applies to $\mu^{\varepsilon,n} := \mu_+^{\varepsilon,n} - \mu_-^{\varepsilon,n}$ and yields that

$$\iint_{\mathbb{R}^\mathsf{d}\times\mathbb{R}^\mathsf{d}} \mathsf{g}(x-y)d\mu^{\varepsilon,n}(x)d\mu^{\varepsilon,n}(y) = \int_{\mathbb{R}^{\mathsf{d}+\mathsf{k}}} |y|^\gamma |\nabla h^{\mu^{\varepsilon,n}}|^2 = \|\mu^{\varepsilon,n}\|^2_{\dot H^{\frac{\mathsf{s}-\mathsf{d}}{2}}}. \tag{2.2.16}$$

We then need to take the limits $n \to \infty$, then $\varepsilon \to 0$ in this relation. For the left-hand side, we find the desired $n \to \infty$ limit by dominated convergence theorem applied to the positive and negative parts of the integrand, and using the assumptions (2.2.11).

Next, with the same argument, we have for any n, m,

$$\iint_{\mathbb{R}^d\times\mathbb{R}^d} \mathsf{g}(x-y)d(\mu^{\varepsilon,n}-\mu^{\varepsilon,m})(x)d(\mu^{\varepsilon,n}-\mu^{\varepsilon,m})(y) = \int_{\mathbb{R}^{d+k}} |y|^\gamma |\nabla h^{\mu^{\varepsilon,n}-\mu^{\varepsilon,m}}|^2$$
$$= \|\mu^{\varepsilon,n}-\mu^{\varepsilon,m}\|^2_{\dot H^{\frac{s-d}{2}}}.$$

Thus for fixed ε, the sequence $\{\mu^{\varepsilon,n}\}_n$ is a Cauchy sequence in $\dot H^{\frac{s-d}{2}}$, hence

$$\{\nabla h^{\mu^{\varepsilon,n}}\}_n$$

is also a Cauchy sequence in the weighted space $L^2_{|y|^\gamma}(\mathbb{R}^{d+k})$ thus $\nabla h^{\mu^{\varepsilon,n}}$ converges to some vector field E. Taking the weak limit in $h^{\mu^{\varepsilon,n}} = \mathsf{g} * \mu^{\varepsilon,n}$ we deduce we must have $E = \nabla h^{\mu^\varepsilon}$ and taking the limit in (2.2.16) we deduce that

$$\int_{\mathbb{R}^{d+k}} |y|^\gamma |\nabla h^{\mu^\varepsilon}|^2 = \|\mu^\varepsilon\|^2_{\dot H^{\frac{s-d}{2}}} = \iint_{\mathbb{R}^d\times\mathbb{R}^d} \mathsf{g}(x-y)d\mu^\varepsilon(x)d\mu^\varepsilon(y). \tag{2.2.17}$$

As mentioned above, the sequence $\{\mu^\varepsilon\}_\varepsilon$ converges in $\dot H^{\frac{s-d}{2}}$ to μ. Thus from (2.2.17) applied to $\mu^\varepsilon - \mu^{\varepsilon'}$ we find that $\nabla h^{\mu^\varepsilon}$ is Cauchy $L^2_{|y|^\gamma}$, hence converges in that space to a limit, which also coincides with ∇h^μ. Finally, taking the limit $\varepsilon \to 0$ in (2.2.17), in view of (2.2.15) we obtain the result.

Step 4: The case $\mathsf{s} > 0$. Assume first that μ has a bounded density and is compactly supported. Then $h^\mu = \int \mathsf{g}(\cdot - y)d\mu(y)$ is finite μ-a.e., moreover, $|h^\mu(x)| \le C|x|^{-\mathsf{s}}$ and $|\nabla h^\mu(x)| \le C|x|^{-\mathsf{s}-1}$ for $|x|$ large enough, and

$$\iint_{\mathbb{R}^d\times\mathbb{R}^d} \mathsf{g}(x-y)d\mu(x)d\mu(y) = \int_{\mathbb{R}^{d+k}} h^\mu d\mu.$$

Using (2.2.7) we may then write

$$\iint_{\mathbb{R}^d\times\mathbb{R}^d} \mathsf{g}(x-y)d\mu(x)d\mu(y) = -\frac{1}{\mathsf{c}_{d,s}}\int_{\mathbb{R}^{d+k}} h^\mu \mathrm{div}\,(|y|^\gamma \nabla h^\mu).$$

For any $R > 0$, as in (2.2.14), we have

$$\int_{B_R} h^\mu \mathrm{div}\,(|y|^\gamma \nabla h^\mu) = \int_{\partial B_R} |y|^\gamma \frac{\partial h^\mu}{\partial \nu} h^\mu - \int_{B_R} |y|^\gamma |\nabla h^\mu|^2.$$

Since h^μ decays like $|x|^{-\mathsf{s}}$ and ∇h^μ like $|x|^{-\mathsf{s}-1}$, the boundary term is bounded by $R^{\gamma+\mathsf{d}+\mathsf{k}-1-2\mathsf{s}-1}$ which in view of (2.2.5) is $R^{-\mathsf{s}}$ and tends to 0 as $R \to +\infty$. Thus we conclude that

$$\int_{\mathbb{R}^d} h^\mu d\mu = -\frac{1}{\mathsf{c}_{d,s}}\int_{\mathbb{R}^{d+k}} h^\mu \mathrm{div}\,(|y|^\gamma \nabla h^\mu) = \frac{1}{\mathsf{c}_{d,s}}\int_{\mathbb{R}^{d+k}} |y|^\gamma |\nabla h^\mu|^2.$$

The case of general μ then follows by approximation as in Step 3. □

2.3. Linking the equilibrium measure with the obstacle problem in the Coulomb case

In Section 2.1 we described the characterization of the equilibrium measure minimizing $\mathcal{E}$ via tools of potential theory. In this section, we return to this question and connect it instead to a well-studied problem in the calculus of variations called the *obstacle problem* in the Coulomb case and the *fractional obstacle problem* in the Riesz case. The connection with the classical obstacle problem is mentioned in passing in [**ST97**]. It allows us to use the rich PDE theory developed for this problem, such as methods based on the maximum principle and regularity theory to obtain additional information on μ_V, valid for the Coulomb case in any dimension. In two dimensions complex analytic methods provide a replacement via Sakai's regularity theory [**Sak91**], providing information on regularity and connectivity [**HM13, LM16**]. Finally, the connection of the equilibrium measure in the Riesz cases with the fractional obstacle problem is less well-known and we will describe it below.

2.3.1. Short presentation of the obstacle problem. The obstacle problem is generally formulated over a bounded domain $\Omega \subset \mathbb{R}^d$: given an $H^1(\Omega)$ function $\psi : \Omega \to \mathbb{R}$ (called the *obstacle*), which is nonpositive on $\partial\Omega$, find the function that achieves

$$\min\left\{\int_\Omega |\nabla h|^2,\ h \in H_0^1(\Omega),\ h \geq \psi\right\}. \tag{2.3.1}$$

In PDE formulation, the solution is characterized by

$$\min(-\Delta h, h - \psi) = 0. \tag{2.3.2}$$

For general background and motivation for this problem, see, e.g., [**KS00, Fri88, CK80**].

Here the space $H_0^1(\Omega)$ is the Sobolev space of trace-zero functions (informally, functions vanishing on the boundary in the sense of boundary trace), which is the completion of $C_c^1(\Omega)$ (C^1 functions with compact support in Ω) under the H^1-Sobolev norm $\|h\|_{H^1} = \|h\|_{L^2} + \|\nabla h\|_{L^2}$. The zero trace condition $h \in H_0^1(\Omega)$ may be replaced by different boundary conditions, e.g., a translation $h \in H_0^1(\Omega) + f$, where f is a given function. Note that the minimization problem (2.3.1) is a convex minimization problem under a convex constraint, hence it has at most one minimizer (it is not too hard to show that the minimum is achieved, hence there actually is a unique minimizer).

An admissible function for (2.3.1) has two options at each point: to touch the obstacle or not (and typically uses both possibilities over the domain). If h is the optimizer, the set $\{x \in \Omega, h(x) = \psi(x)\ q.e.\}$ is closed and called the *coincidence set* or the *contact set*. It is unknown (part of the problem), and the obstacle problem thus belongs to the class of so-called *free-boundary problems*, see [**Fri88**].

Trying to compute the Euler-Lagrange equation associated to this problem by perturbing h by a small function, one is led to two possibilities depending on whether $h = \psi$ or $h > \psi$. In a region where $h > \psi$, one can perform infinitesimal variations of h of the form $(1-t)h + tv$ with $t \in [0,1]$ and v, say, smooth (this still gives an admissible function, i.e., lying above the obstacle, as soon as t is small enough) which shows that $\Delta h = 0$ there (since the "functional derivative" of the Dirichlet energy is the Laplacian). In the set where $h = \psi$, only variations of the same form $(1-t)h+tv$ but with $v \geq \psi$ (equivalent to $v \geq h$ there) and $t \geq 0$ provide admissible functions, and this only leads to an inequality $-\Delta h \geq 0$ there. These two pieces of information can be grouped in the following more compact form:

$$\text{for all } v \text{ in } H_0^1 \text{ such that } v \geq \psi \text{ q.e.,} \quad \int_\Omega \nabla h \cdot \nabla(v - h) \geq 0. \tag{2.3.3}$$

This relation is called a *variational inequality*, and it uniquely characterizes the solution to (2.3.1), in particular the coincidence set is completely determined as part of the solution.

In Figure 2.3.1 below, we describe a few instances of solutions to one-dimensional obstacle problems, and in Figure 2.3.2 to higher-dimensional obstacle problems.

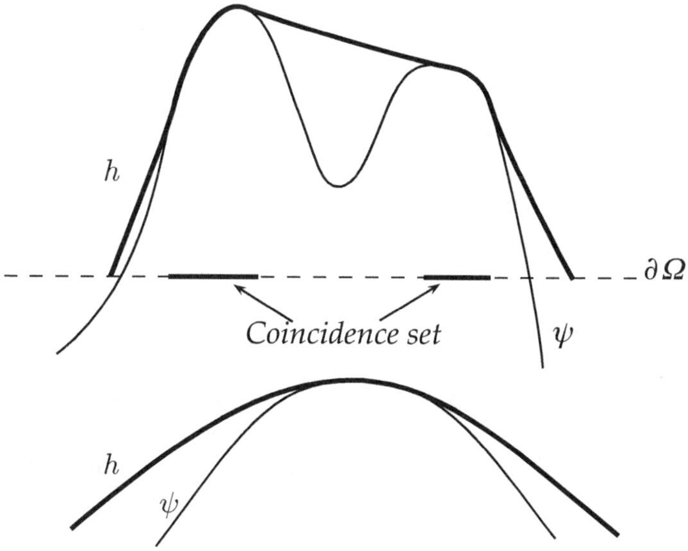

FIGURE 2.3.1. The coincidence set for a one-dimensional obstacle problem

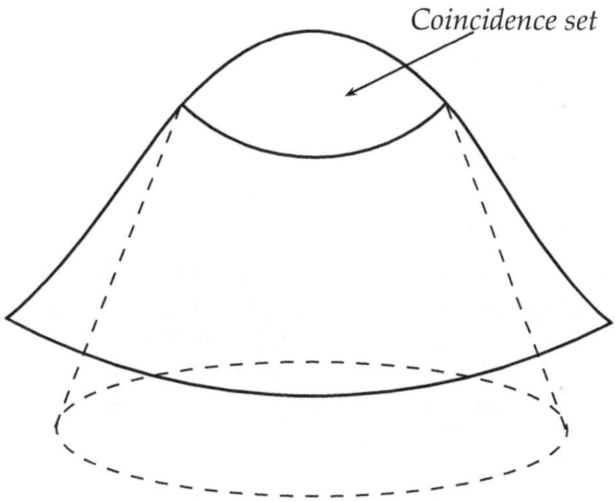

FIGURE 2.3.2. A higher-dimensional obstacle problem

The regularity theory of the solutions to obstacle problems and of their coincidence sets has been developed for many years, culminating with the work of Caffarelli (for a review see [**Caf98**]). This sophisticated PDE theory shows, for example, that the solution h is as regular as ψ up to $C^{1,1}$ [**Fre72**]. The boundary $\partial\Sigma$ of the coincidence set is locally a $C^{1,\alpha}$ graph except for cusps [**Caf98**]. These are points of $\partial\Sigma$ at which, locally, the coincidence set can fit in the region between two parallel planes separated by an arbitrarily small distance (the smallness of the neighborhood depends of course on this desired distance). Figure 2.3.3 gives examples of coincidence sets, a regular one, and one with cusps. Cusps are however nongeneric with respect to the obstacle, as shown in [**FROS20**].

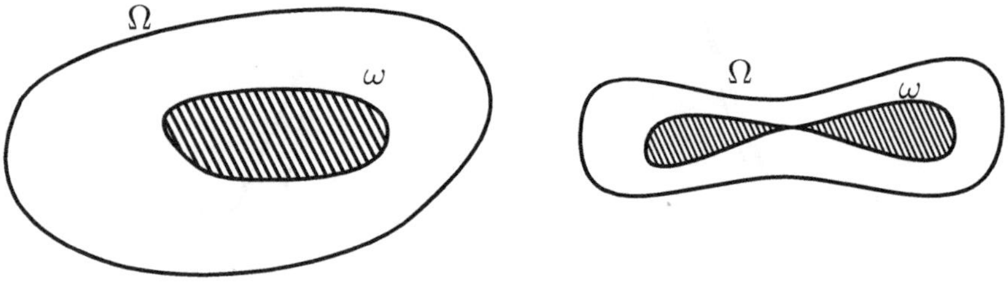

FIGURE 2.3.3. Examples of coincidence sets

Moreover, if ψ is $C^{1,1}$, since ∇h is continuous, the graph of h must lift off from the coincidence set tangentially. This formally leads to the following system of equations, where ω denotes the coincidence set:

$$\begin{cases} -\Delta h = 0 & \text{in } \Omega\backslash\omega \\ h = \psi & \text{in } \omega \\ \frac{\partial h}{\partial \nu} = \frac{\partial \psi}{\partial \nu} & \text{on } \partial\omega \\ h = 0 & \text{on } \Omega. \end{cases}$$

This relation cannot be made rigorous in all cases, because ω is not an open domain, however it gives the right intuition and is correct when ω is regular enough. Note that on the boundary of $\Omega\backslash\omega$ we must have a Dirichlet boundary condition $h = \psi$, together with a Neumann boundary condition $\frac{\partial h}{\partial \nu} = \frac{\partial \psi}{\partial \nu}$. These two boundary conditions make what is called an *overdetermined problem* and this overdetermination explains why there is only one possible coincidence set.

2.3.2. Connection between the two problems. The problem we examined, that of the minimization of $\mathcal{E}$, is phrased in the whole space, and not in a bounded domain. While the minimization problem (2.3.1) may not have a meaning over all $\mathbb{R}^\mathsf{d}$ (because the integral might not converge), the corresponding variational inequality (2.3.3) can still be given a meaning over $\mathbb{R}^\mathsf{d}$ as follows: given $\psi \in H^1_{\mathrm{loc}}(\mathbb{R}^\mathsf{d})$, find $h \in H^1_{\mathrm{loc}}(\mathbb{R}^\mathsf{d})$ such that

$$\forall v \in \mathcal{K}, \quad \int_{\mathbb{R}^\mathsf{d}} \nabla h \cdot \nabla(v - h) \geq 0, \tag{2.3.4}$$

where

$$\mathcal{K} = \left\{ v \in H^1_{\mathrm{loc}}(\mathbb{R}^\mathsf{d}) \text{ such that } v - h \text{ has bounded support and } v \geq \psi \text{ q.e.} \right\}.$$

Solving this is in fact equivalent to the statement that for every $R > 0$, h is the unique solution to

$$\min \left\{ \int_{B_R} |\nabla v|^2, v \in H^1(B_R), v - h \in H^1_0(B_R), v \geq \psi \text{ in } B_R \right\},$$

which replaces (2.3.1). The PDE

$$\min(-\Delta h, h - \psi) = 0 \tag{2.3.5}$$

also still makes sense. We refer to [**SS18**] for more detail on the full space setting.

Provided a "boundary condition at infinity" is added, the problem (2.3.4) is easily seen to have a unique solution: if there are two solutions h_1 and h_2 it suffices to apply (2.3.4) for h_1, with h_2 as a test function, and then reverse the roles of the two and add the two relations to obtain $h_1 = h_2$.

Let us now compare the two problems side by side.

Equilibrium measure: μ_V is characterized by the relations

$$\begin{cases} h^{\mu_V} + V \geq c & \text{quasi everywhere} \\ h^{\mu_V} + V = c & \text{q.e. in the support of } \mu_V. \end{cases} \tag{2.3.6}$$

Obstacle problem:

$$\begin{cases} h \geq \psi & \text{q.e.} \\ h = \psi & \text{q.e. in the coincidence set.} \end{cases} \tag{2.3.7}$$

It is then not surprising to expect a correspondence between the two settings, once one chooses the obstacle to be $\psi = c - V$.

Proposition 2.17 (Equivalence between the minimization of $\mathcal{E}$ and the obstacle problem). *Assume* $\mathsf{s} = \mathsf{d} - 2$*, V is continuous and satisfies* (A2)*. If μ_V is the equilibrium measure associated to the potential V as in Theorem* 2.8*, then its potential h^{μ_V}, as defined in* (2.1.10)*, is the unique solution to the obstacle problem with obstacle $\psi = c - V$ in the sense of* (2.3.4)*. If in addition $V \in C^{1,1}$ then $\mu_V = (\frac{1}{\mathsf{c_d}}\Delta V)\mathbf{1}_\omega$ where ω is the coincidence set $\{h^{\mu_V} = c - V, q.e.\} = \{\zeta = 0, q.e.\}$.*

Note that the converse might not be true, because a solution of the obstacle problem can fail to provide (via taking $-\frac{1}{\mathsf{c_d}}\Delta h$) a *probability* measure, however it does in general when shifting c appropriately.

When one works on a bounded domain, this result can be obtained by observing that the problem of minimizing $\mathcal{E}$ and that of minimizing (2.3.1) are essentially convex duals of each other (see [**Bre72, BS02**]). When working in an infinite domain, the correspondence is probably folklore and could also be worked out by convex duality, but we were not able to find it completely written in the literature, except for [**HM13**], which follows a slightly different formulation. Here, for the proof, we follow the approach of [**ASZ14**] where the result is established in dimension 2 for the particular case of V quadratic (but with more general constraints). The adaptation to any dimension and to general V's is not difficult. Note that further details on the formulation in the whole space are given in [**SS18**], where in this setting the stability of the coincidence set and its precise dependence with respect to the obstacle function is studied.

PROOF OF PROPOSITION 2.17.

Step 1. Let us show that ∇h^{μ_V} is in $L^2_{\text{loc}}(\mathbb{R}^\mathsf{d})$. This is a consequence of the fact that $\mathcal{E}(\mu_V) < \infty$ hence, in view of the assumptions on V, $\iint \mathsf{g}(x - y)\, d\mu_V(x)\, d\mu_V(y) < +\infty$. In the case $\mathsf{s} > 0$, it is a direct consequence of Proposition 2.16, in fact $\nabla h^{\mu_V} \in L^2(\mathbb{R}^\mathsf{d})$. In the case $\mathsf{s} \leq 0$, we consider a reference probability measure $\bar{\mu}$ for which $h^{\bar{\mu}}$ is $C^1_{\text{loc}}(\mathbb{R}^2)$. It suffices to consider for example $\bar{\mu} = \frac{1}{\pi}\mathbf{1}_{B_1}$, the circle law, for which $h^{\bar\mu}$ is radial and can be computed explicitly. Then, let us consider $\rho = \mu_V - \bar\mu$. Since $\int d\rho = 0$, and $\iint \mathsf{g}(x-y)\,d\mu(x)\,d\mu(y) < \infty$ holds for both $\mu = \mu_V$ and $\mu = \bar\mu$, we then obtain by Proposition 2.16 that $\nabla h^{\mu_V} - \nabla h^{\bar\mu} \in L^2(\mathbb{R}^\mathsf{d})$ (we are in the Coulomb case where $\gamma = 0$), from which we deduce ∇h^{μ_V} is also in $L^2_{\mathrm{loc}}(\mathbb{R}^2, \mathbb{R}^2)$, as desired.

Step 2. Let v be admissible in (2.3.4), i.e., belong to $\mathcal{K}$, and set $\varphi = v - h^{\mu_V}$. If φ is smooth and compactly supported, then, integrating by parts, we have

$$\int_{\mathbb{R}^\mathsf{d}} \nabla h^{\mu_V} \cdot \nabla (v - h^{\mu_V}) = \mathsf{c_d} \int_{\mathbb{R}^\mathsf{d}} \varphi\, d\mu_V \geq 0. \tag{2.3.8}$$

Indeed, by (2.1.9), we know that $h^{\mu_V} = \psi$ q.e. in the support of μ_V and by assumption $v \geq \psi$ q.e. in $\mathbb{R}^\mathsf{d}$. Hence φ is q.e. nonnegative on the support of μ_V and the inequality (2.3.8) follows, since μ_V does not charge sets of zero capacity. To obtain (2.3.8) for any $v \in \mathcal{K}$, it suffices to show that the subset of $\mathcal{K}$ consisting of v's for which $v - h^{\mu_V}$ is smooth and compactly supported is dense in $\mathcal{K}$ for the topology of H^1. Fix some v in the admissible set and $R > 1$ such that $v - h^{\mu_V}$ is supported in $B_{R/2}$. Let η_ε be a standard mollifier and χ_R a smooth function supported in B_{2R} with $0 \leq \chi_R \leq 1$ and $\chi_R \equiv 1$ in B_R. One may check that

$$v_{\varepsilon,\delta} = h^{\mu_V} + (v - h^{\mu_V}) * \eta_\varepsilon + \delta \chi_R$$

satisfies that $v_{\varepsilon,\delta} - h^{\mu_V}$ is smooth and approximates v arbitrarily well in H^1 when δ is small enough, and is $\geq \psi$ when ε is chosen small enough relative to δ. This concludes the proof of (2.3.8).

Step 3. We prove the statements about μ_V. First, since the coincidence set ω is closed, its complement is open, and the function h^{μ_V} is harmonic on that set. One can note also that in view of (2.1.9) and the definition of the coincidence set ω, the support of μ_V is included in ω up to a set of capacity 0.

If we assume that $V \in C^{1,1}_{\mathrm{loc}}$, then by Frehse's regularity theorem mentioned above [**Fre72**], h^{μ_V} is also $C^{1,1}_{\mathrm{loc}}$. In particular h^{μ_V} is continuous, and so is V, so the relations (2.1.9) hold pointwise and not only q.e. This means that we have

$$h^{\mu_V} + V = c \quad \text{on } \omega \tag{2.3.9}$$

and $\operatorname{supp} \mu_V \subset \omega$. Also $C^{1,1}_{\mathrm{loc}} = W^{2,\infty}_{\mathrm{loc}}$ hence Δh^{μ_V} and ΔV both make sense as L^∞_{loc} functions, and it suffices to determine μ_V up to sets of measure 0. We already know that $\mu_V = 0$ in the complement of ω since h^{μ_V} is harmonic there, and it suffices to determine it in $\mathring{\omega}$. But taking the Laplacian on both sides of (2.3.9), since $\mu_V = -\frac{1}{\mathsf{c_d}} \Delta h^{\mu_V}$, one finds

$$\mu_V = \frac{1}{\mathsf{c_d}} \Delta V \text{ in } \mathring{\omega},$$

and the result follows. □

Since μ_V is a compactly supported probability measure, we have that $h^{\mu_V} = \int \mathsf{g}(x-y)\,d\mu_V(y)$ asymptotically behaves like $\mathsf{g}(x)$ as $|x| \to \infty$. Since $h^{\mu_V} + V = c$ q.e. in ω and since (A2) holds, it follows that ω must be a bounded, hence compact, set.

We have seen that ω contains, but is not always equal to, the support of μ_V, called the droplet. In [**HM13**], it is discussed how these two sets differ in the two-dimensional case: under a logarithmic growth condition for V, they are equal except at so-called "shallow points."

2.4. The fractional obstacle problem and link with the equilibrium measure in the Riesz case

We recall that the fractional Laplacian can be defined via Fourier transform or by (1.1.15).

2.4.1. The fractional obstacle problem. The study of the fractional obstacle problem is more recent than that of the classical obstacle problem, the first main references are [**CSS08, Sil07**]. The definition of the problem in our setting is

$$\begin{cases} h \geq \psi & \text{in } \mathbb{R}^\mathsf{d} \\ (-\Delta)^\alpha h \geq 0 & \text{in } \mathbb{R}^\mathsf{d} \\ (-\Delta)^\alpha h(x) = 0 & \text{for those } x \text{ such that } h(x) > \psi(x) \\ \lim_{|x|\to\infty} \frac{h}{\mathsf{g}} = 1, \end{cases} \tag{2.4.1}$$

which can also be rewritten as

$$\min\left((-\Delta)^\alpha h, h - \psi\right) = 0, \tag{2.4.2}$$

playing the analog role of (2.3.2). The problem can be expressed variationally as

$$\min_{h \geq \psi} \int |\nabla^\alpha h|^2 := \min_{h\geq \psi} J(h),$$

where

$$J(h) = \iint_{\mathbb{R}^\mathsf{d}\times\mathbb{R}^\mathsf{d}} \frac{|h(x)-h(y)|^2}{|x-y|^{\mathsf{d}+2\alpha}} dxdy.$$

Again the set $\{h = \psi\}$ is called the coincidence set.

This problem is directly related to another previously studied problem, which has an easier visual interpretation: the thin obstacle problem or Signorini problem.

2.4.2. The Signorini or thin obstacle problem. It is an obstacle problem in dimension $\mathsf{d}+1$ with obstacle in $\mathbb{R}^\mathsf{d}$. Formally, given a function ψ over $\mathbb{R}^\mathsf{d}$, the minimization problem is

$$\min\left\{\int_{\mathbb{R}^{\mathsf{d}+1}} |\nabla h|^2,\ h \in H^1(\mathbb{R}^{\mathsf{d}+1}),\ h(x,0) \geq \psi(x) \text{ for all } x \in \mathbb{R}^\mathsf{d}\right\}.$$

Solutions are harmonic functions away from $\mathbb{R}^\mathsf{d} \times \{0\}$. The jump of the normal derivative of h across $\mathbb{R}^\mathsf{d}\times\{0\}$ is equal to (twice) the half-Laplacian of h. This allows to give an interpretation of the fractional obstacle problem when $\alpha = 1/2$. For $\alpha \neq 1/2$, one replaces instead $\int |\nabla h|^2$ by $\int |y|^\gamma |\nabla h|^2$ where y is the last coordinate in $\mathbb{R}^{\mathsf{d}+1}$ and γ is as in (2.2.4).

Now the characterization of the equilibrium measure (2.1.9) can be rewritten as

$$\begin{cases} h^{\mu_V} \geq \psi & \text{in } \mathbb{R}^\mathsf{d} \\ h^{\mu_V} = \psi & \text{in the support of } \mu_V \end{cases}$$

after setting $\psi = c - V$. On the other hand, since the g of (2.0.1) is the kernel of the fractional Laplacian $(-\Delta)^{\frac{\mathsf{d}-\mathsf{s}}{2}}$ up to a constant, we have

$$(-\Delta)^{\frac{\mathsf{d}-\mathsf{s}}{2}} h^{\mu_V} = \mathsf{c}_{\mathsf{d},\mathsf{s}}\mu_V.$$

Thus, setting $\alpha = \frac{\mathsf{d}-\mathsf{s}}{2}$, the potential h^{μ_V} satisfies that either $h^{\mu_V} = \psi$ or $h^{\mu_V}(x) > \psi(x)$ and x is not in the support of μ_V, hence

$$\min\left(h^{\mu_V} - \psi, (-\Delta)^\alpha h^{\mu_V}\right) = 0, \quad \text{in } \mathbb{R}^\mathsf{d}$$

and the condition at infinity is also satisfied, so we have found exactly (2.4.2), or equivalently (2.4.1).

The thin obstacle problem is easier to visualize than the fractional obstacle problem: one considers an obstacle that lives only on $\mathbb{R}^\mathsf{d}$, and extends an elastic sheet above it. The sheet is free away from $\mathbb{R}^\mathsf{d} \times \{0\}$, thus leading to a harmonic (or $\operatorname{div}(|y|^\gamma \nabla)$-free) function away from it. But there is a jump of normal derivative across $\mathbb{R}^\mathsf{d} \times \{0\}$ where the obstacle is touched. The jump of $|y|^\gamma \nabla u$ across $\mathbb{R}^\mathsf{d} \times \{0\}$ is equal to $-\operatorname{div}(|y|^\gamma \nabla u)$ in the sense of distributions, hence to the fractional equilibrium measure, which is supported on the coincidence set. One could also write the analogue of Proposition 2.17.

We have seen how the correspondence between the minimization of $\mathcal{E}$ and the obstacle problem thus allows, via the regularity theory of the obstacle problem, to identify the equilibrium measure in terms of V when the former is regular enough. The known techniques on the obstacle problem [**Caf98**] and on the fractional obstacle problem [**CSS08, CROS17**] also allow for example to analyze the rate at which the solution lifts off from the obstacle: at regular points of the boundary of the coincidence set ω, we have

$$h - \psi = c \operatorname{dist}(x, \omega)^{1+\alpha} + o(\operatorname{dist}(x, \omega)^{1+\alpha}),$$

for some $c > 0$, which gives us information on the growth rate of the function ζ defined in (2.1.12) of the form

$$\zeta(x) \geq c \operatorname{dist}(x, \omega)^{1+\alpha}. \tag{2.4.3}$$

Also, at regular points of $\partial\omega$, the vanishing of the equilibrium measure is at the rate

$$(-\Delta)^\alpha h \sim c' \operatorname{dist}(x, \omega^c)^{1-\alpha}, \tag{2.4.4}$$

where dist denotes the signed distance and c' is equal to c times some constant depending only on α. These regular points are expected to be generic, as seen before this is proved in [**FROS20**] in the case $\alpha = 1$. Moreover, [**CF25**] proved that the behavior (2.4.4) is generic with respect to V. For more details and proofs of these estimates in our particular setting, we also refer to [**RO**] in the appendix of [**PS**].

2.5. The thermal equilibrium measure

We now turn to the study of the minimization of (2.0.4), and justify the existence and uniqueness of its minimizer, which we call the *thermal equilibrium measure*. For this we will need a thermal version of assumption (A2):

$$\int_{\mathbb{R}^\mathsf{d}} \exp\left(-\theta(V(x) + \mathsf{g}_-(x))\right) dx < \infty, \tag{A4}$$

where g_- denotes $\min(\mathsf{g}, 0)$. Note that $\mathsf{g}_- = 0$ for $\mathsf{s} > 0$. We will also denote $\mathsf{g}_+ = \max(\mathsf{g}, 0)$, so $\mathsf{g} = \mathsf{g}_+ + \mathsf{g}_-$.

Proposition 2.18 (Existence and uniqueness of the thermal equilibrium measure). *For every $\theta > 0$, if (A1)–(A4) hold, then $\mathcal{E}_\theta$ is lower semicontinuous, has compact sublevel sets, and has a unique minimizer.*

PROOF. Let us start with the case $\mathsf{s} = 0$. We use (2.1.1) to obtain that for any probability density μ, we have

$$\begin{aligned}\mathcal{E}_\theta(\mu) &\geq \int_{\mathbb{R}^\mathsf{d}} V d\mu - \iint_{|x|\geq|y|} \log\max(|x|,1) d\mu(x)d\mu(y) - C + \frac{1}{\theta}\int_{\mathbb{R}^\mathsf{d}} \mu\log\mu \\ &\geq \int_{\mathbb{R}^\mathsf{d}} V d\mu - \int_{\mathbb{R}^\mathsf{d}} (\log|x|)_+ d\mu(x) - C + \frac{1}{\theta}\int_{\mathbb{R}^\mathsf{d}} \mu\log\mu.\end{aligned}$$

We obtain in all cases (using the fact that $\mathsf{g} \geq 0$ in the cases $\mathsf{s} > 0$, and (2.1.1) in the case $\mathsf{s} < 0$)

$$\mathcal{E}_\theta(\mu) \geq -C + \int_{\mathbb{R}^\mathsf{d}} (V + \mathsf{g}_-) d\mu + \frac{1}{\theta}\int_{\mathbb{R}^\mathsf{d}} \mu\log\mu. \tag{2.5.1}$$

We then observe that the function $\phi(\mu) = \alpha\mu + \frac{1}{\theta}\mu\log\mu$ achieves its minimum at $u = \exp(-\theta\alpha - 1)$ and that the minimum equals $-\frac{1}{\theta}\exp(-\alpha\theta - 1)$. This leads us to introducing

$$u(x) = \exp(-\theta(V + \mathsf{g}_-)(x)), \quad \bar{u} = \int_{\mathbb{R}^\mathsf{d}} u, \quad \rho = \frac{u}{\bar{u}} \tag{2.5.2}$$

which are well-defined thanks to (A4). We may then rewrite

$$\begin{aligned}\mathcal{E}_\theta(\mu) &= \frac{1}{2}\iint_{\mathbb{R}^\mathsf{d}\times\mathbb{R}^\mathsf{d}} (\mathsf{g}(x-y) - (\mathsf{g}_-(x) + \mathsf{g}_-(y)))\, d\mu(x)d\mu(y) \\ &\quad + \int_{\mathbb{R}^\mathsf{d}} (V + \mathsf{g}_- + \frac{1}{\theta}\log\rho) d\mu + \frac{1}{\theta}\int_{\mathbb{R}^\mathsf{d}} \mu\log\frac{\mu}{\rho} \\ &= \frac{1}{2}\iint_{\mathbb{R}^\mathsf{d}\times\mathbb{R}^\mathsf{d}} (\mathsf{g}(x-y) - (\mathsf{g}_-(x) + \mathsf{g}_-(y)))\, d\mu(x)d\mu(y) \\ &\quad - \frac{1}{\theta}\log\bar{u} + \frac{1}{\theta}\int_{\mathbb{R}^\mathsf{d}} \mu\log\frac{\mu}{\rho}.\end{aligned} \tag{2.5.3}$$

The first integral in the right-hand side is bounded below by properties of g (already seen in the proof of (2.1.1)). The last one is recognized as the relative entropy of μ with respect to the probability density ρ, which is always nonnegative. We deduce that $\inf\mathcal{E}_\theta > -\infty$. On the other hand we get $\inf\mathcal{E}_\theta < +\infty$, for instance by taking μ_V as a test probability.

Let now $\{\mu_n\}_n$ be such that $\mathcal{E}_\theta(\mu_n) \leq C$ independent of n. It is well-known that the relative entropy is lower semicontinuous and its sublevel sets are compact, so in view of (2.5.3) we may extract a subsequence (not relabelled) such that $\mu_n \to \mu$ in the weak sense of probabilities. Hence $\mathcal{E}_\theta$ has compact sublevel sets. Moreover, if $\mu_n \to \mu$ we have

$$\liminf_{n\to\infty} \int \mu_n \log\frac{\mu_n}{\rho} \geq \int \mu\log\frac{\mu}{\rho}. \tag{2.5.4}$$

Using the fact that $\mathsf{g}(x-y) - (\mathsf{g}_-(x) + \mathsf{g}_-(y))$ is bounded below, and the same truncation procedure as in the proof of Lemma 2.4, we also have

$$\liminf_{n\to\infty} \iint_{\mathbb{R}^d\times\mathbb{R}^d} (\mathsf{g}(x-y) - (\mathsf{g}_-(x) + \mathsf{g}_-(y)))\, d\mu_n(x)d\mu_n(y)$$
$$\geq \iint_{\mathbb{R}^d\times\mathbb{R}^d} (\mathsf{g}(x-y) - (\mathsf{g}_-(x) + \mathsf{g}_-(y)))\, d\mu(x)d\mu(y),$$

which together with (2.5.4) and (2.5.3) implies that $\liminf_{n\to\infty} \mathcal{E}_\theta(\mu_n) \geq \mathcal{E}_\theta(\mu)$. Thus, taking $\{\mu_n\}_n$ to be a minimizing sequence, we have obtained that $\mathcal{E}_\theta$ is l.s.c. and admits a minimizer.

The uniqueness of a minimizer is by strict convexity: $\mathcal{E}$ itself is strictly convex and $\mu \mapsto \int \mu \log \mu$ as well. □

Remark 2.19. The situation is a bit subtle because even though $\mathcal{E}_\theta(\mu_\theta) < +\infty$ the entropy $\int_{\mathbb{R}^d} \mu \log \mu$ may not be finite, but rather it is the relative entropy $\int_{\mathbb{R}^d} \mu \log \frac{\mu}{\rho}$ that is finite. There is a possible cancellation between V and $\frac{1}{\theta}\log \mu_\theta$ before they get integrated against μ_θ.

We are now led to introducing a new assumption

$$\text{(A5)} \qquad \exists \epsilon > 0 \text{ such that } \int_{\mathbb{R}^d} \exp\left(\epsilon|\mathsf{g}_-| - \theta(V + \mathsf{g}_-)\right) < \infty.$$

Lemma 2.20. *If* (A1)–(A5) *holds, we have*

$$\iint |\mathsf{g}(x-y)|d\mu_\theta(x)d\mu_\theta(y) < \infty \tag{2.5.5}$$

and

$$\int |\mathsf{g}_-|(x)d\mu_\theta(x) < +\infty. \tag{2.5.6}$$

PROOF. **Step 1: Proof of** (2.5.6)**.** We have seen that under (A4) (which is implied by (A5)), $\inf \mathcal{E}_\theta < +\infty$ and μ_θ exists. Returning to (2.5.3) we have

$$\mathcal{E}_\theta(\mu_\theta) = \frac{1}{2}\iint_{\mathbb{R}^d\times\mathbb{R}^d} (\mathsf{g}(x-y) - (\mathsf{g}_-(x) + \mathsf{g}_-(y)))d\mu_\theta(x)d\mu_\theta(y)$$
$$- \frac{1}{\theta}\log \bar{u} + \frac{1}{\theta}\int_{\mathbb{R}^d} \mu_\theta \log \frac{\mu_\theta}{\rho},$$

where ρ and $\bar{u}$ are as in (2.5.2). As seen in the proof of Proposition 2.18 both nonconstant terms are bounded below and we deduce that

$$\iint_{\mathbb{R}^d\times\mathbb{R}^d} (\mathsf{g}(x-y) - (\mathsf{g}_-(x) + \mathsf{g}_-(y)))\, d\mu_\theta(x)d\mu_\theta(y) < +\infty \tag{2.5.7}$$

and

$$\int_{\mathbb{R}^d} \mu_\theta \log \frac{\mu_\theta}{\rho} < +\infty. \tag{2.5.8}$$

By the Donsker-Varadhan lemma (which exploits the fact that the relative entropy and log of exponential moment are dual functions to each other), we can write that for any $\epsilon > 0$

$$\epsilon \int_{\mathbb{R}^d} |\mathsf{g}_-|d\mu_\theta \leq \log \int_{\mathbb{R}^d} e^{\epsilon|\mathsf{g}_-|}d\rho + \int_{\mathbb{R}^d} \mu_\theta \log \frac{\mu_\theta}{\rho}. \tag{2.5.9}$$

Since by (2.5.8) the relative entropy is bounded, by (A5) and definition of ρ in (2.5.2), we obtain (2.5.6).

Step 2: Proof of (2.5.5). In the case $\mathsf{s} > 0$, $\mathsf{g}_- = 0$ so the desired result holds from (2.5.7). In the case $\mathsf{s} \le 0$, it follows from (2.5.7) that

$$\iint \mathsf{g}(x-y)d\mu_\theta(x)d\mu_\theta(y) < +\infty. \tag{2.5.10}$$

Let us show the corresponding lower bound. Using (2.1.1), we may write

$$\begin{aligned}(2.5.11)\quad &\iint \mathsf{g}_-(x-y)d\mu_\theta(x)d\mu_\theta(y) \\ &\ge \iint \Big((\mathsf{g}_-(x)+\mathsf{g}_-(y))\mathbf{1}_{\mathsf{s}<0} + \mathsf{g}_-(x)\wedge \mathsf{g}_-(y)\mathbf{1}_{\mathsf{s}=0}\Big)d\mu_\theta(x)d\mu_\theta(y) \\ &\ge 2\int \mathsf{g}_-(x)d\mu_\theta(x) > -\infty\end{aligned}$$

in view of (2.5.6). In the case $\mathsf{s} < 0$ since g is negative we are done. In the case $\mathsf{s} = 0$, since

$$\iint |\mathsf{g}(x-y)|d\mu_\theta(x)d\mu_\theta(y) = \iint \mathsf{g}(x-y)d\mu_\theta(x)d\mu_\theta(y) - 2\iint \mathsf{g}_-(x-y)d\mu_\theta(x)d\mu_\theta(y),$$

the desired result follows from combining (2.5.10) and (2.5.11). □

As can be guessed formally, when $\theta \to \infty$, μ_θ converges to the (regular) equilibrium measure μ_V. This convergence is studied in detail and made quantitative in the Coulomb case in [**AS22**] and can serve to approximate the solution to the obstacle problem.

By contrast with μ_V, μ_θ is regular and not compactly supported, but always positive in $\mathbb{R}^\mathsf{d}$ with (typically) exponentially decaying tails. A boundary layer lengthscale equal to $\frac{1}{\sqrt{\theta}}$ appears in the convergence, as can be seen in Theorem 2.23 below.

A decay rate in $\exp(-\theta(V+\mathsf{g}_-))$ can be proven by using the maximum principle–based proof provided in [**AS22**] in the Coulomb case (which can be extended without too much difficulty to the Riesz case). The starting point of that argument is to exploit the Euler-Lagrange equation solved by μ_θ minimizing $\mathcal{E}_\theta$, which we now derive.

Proposition 2.21 (Characterization of the thermal equilibrium measure). *Assume V is locally bounded and* (A1)–(A5) *hold. Let $\theta > 0$. If μ_θ is a probability measure that minimizes $\mathcal{E}_\theta$, then it is a measure with a density satisfying*

$$\mu_\theta > 0 \ \text{in } \mathbb{R}^\mathsf{d} \quad a.e.$$

and, for $h^{\mu_\theta} = \mathsf{g} * \mu_\theta$, we have that $h^{\mu_\theta} \in L^1_{\mathrm{loc}}(\mathbb{R}^\mathsf{d})$ and

$$h^{\mu_\theta} + V + \frac{1}{\theta}\log \mu_\theta = c_\theta \quad a.e. \tag{2.5.12}$$

for some constant c_θ. Moreover, the density μ_θ is bounded above and locally bounded below by positive constants. If in addition $V \in C^2$, then μ_θ is continuous and we can get estimates for $\|\mu_\theta\|_{L^\infty}$ in terms of ΔV.

Remark 2.22. As we will see in the proof, the Euler-Lagrange equation (2.5.12) takes an interesting PDE form when differentiated: taking the Laplacian of (2.5.12) we obtain

$$-\frac{1}{\theta}\Delta \log \mu_\theta + \mathsf{c}_{\mathsf{d},\mathsf{s}}(-\Delta)^{1+\frac{\mathsf{s}-\mathsf{d}}{2}}\mu_\theta - \Delta V = 0. \tag{2.5.13}$$

This is called the Kirkwood-Monroe equation in physics, it is an elliptic PDE for $\log \mu_\theta$, with a right-hand side of lower order of differentiation. It is however not a uniformly elliptic PDE, but once written in divergence form as

$$-\frac{1}{\theta}\mathrm{div}\,\frac{\nabla \mu_\theta}{\mu_\theta} = -\mathsf{c}_{\mathsf{d},\mathsf{s}}(-\Delta)^{1+\frac{\mathsf{s}-\mathsf{d}}{2}}\mu_\theta + \Delta V, \tag{2.5.14}$$

one can apply regularity theory for it in the set where μ_θ has good bounds from below (which will naturally be Σ, the support of the equilibrium measure μ_V): this is the strategy of [**AS22**]. One should also point out that in the Coulomb cases, rewriting it as an equation for $u_\theta = \log \mu_\theta$, leads to the elliptic PDE

$$-\Delta u = -\mathsf{c}_{\mathsf{d}}\theta e^u + \theta \Delta V, \tag{2.5.15}$$

which is a generalization of the Kazhdan-Warner equation for prescribed curvature, and coincides with it if ΔV is constant.

Proof Proposition 2.21.

Step 1: Definition of the potential. We first check that $h^{\mu_\theta} \in L^1_{\mathrm{loc}}(\mathbb{R}^{\mathsf{d}})$. Let U be a bounded set in $\mathbb{R}^{\mathsf{d}}$, we have

$$\int_U h^{\mu_\theta} = \int_{\mathbb{R}^{\mathsf{d}}}\int_U \mathsf{g}(x-y)dx d\mu_\theta(y). \tag{2.5.16}$$

If $\mathsf{s} > 0$ it is then straightforward, by integrability and boundedness of g, that the right-hand side is finite. If $\mathsf{s} \le 0$, this follows from (2.1.1) and (2.5.6).

Step 2: Positivity. It is standard that the entropy functional is finite only if μ is a measure which is absolutely continuous with respect to the Lebesgue measure. To show that $\mu_\theta > 0$, we follow [**RSY14**] and assume by contradiction that there exists a set S of nonzero Lebesgue measure on which $\mu_\theta = 0$ and set

$$\nu = \frac{\mu_\theta + \varepsilon \mathbf{1}_S}{1+\varepsilon|S|}.$$

Let us expand out

$$\begin{aligned}\mathcal{E}_\theta\left(\frac{\mu_\theta + \varepsilon \mathbf{1}_S}{1+\varepsilon|S|}\right) &= \mathcal{E}_\theta(\mu_\theta)\\ &\quad - \varepsilon|S|\left(\iint \mathsf{g}(x-y)d\mu_\theta(x)d\mu_\theta(y) + \int V d\mu_\theta + \frac{1}{\theta}\mu_\theta \log \mu_\theta\right)\\ &\quad + \varepsilon \int_S (h^{\mu_\theta} + V) + \frac{|S|}{\theta}\varepsilon \log \varepsilon + O(\varepsilon^2).\end{aligned}$$

By Step 1, since S can be taken as bounded and V takes finite values, we have that $\int_S h^{\mu_\theta} + V < \infty$. We deduce that

$$\mathcal{E}_\theta\left(\frac{\mu_\theta + \varepsilon \mathbf{1}_S}{1+\varepsilon|S|}\right) \le \mathcal{E}_\theta(\mu_\theta) + C\varepsilon + \frac{|S|}{\theta}\varepsilon \log \varepsilon,$$

a contradiction with the minimality of μ_θ if $|S| > 0$ when ε is chosen small enough.

Step 3: Euler-Lagrange equation. For every smooth compactly supported function f such that $\int f d\mu_\theta = 0$ and $t \in \mathbb{R}$ with $|t|$ sufficiently small, $(1+tf)\mu_\theta$ is a probability measure and we may expand

$$\mathcal{E}_\theta(\mu_\theta) \le \mathcal{E}_\theta((1+tf)\mu_\theta)$$

to find

$$t\int_{\mathbb{R}^\mathsf{d}} (h^{\mu_\theta} + V + \frac{1}{\theta}\log\mu_\theta) f d\mu_\theta + O(t^2) \ge 0,$$

where h^{μ_θ} may take infinite values. Since this is true for all small enough $|t|$ and any smooth f with $\int f d\mu_\theta = 0$, and since $\mu_\theta > 0$ almost everywhere, it follows that (2.5.12) holds almost everywhere, for some constant c_θ.

Step 4: Boundedness from above and below. First, we claim that $h^{\mu_\theta} + V$ is bounded below in $\mathbb{R}^\mathsf{d}$. Indeed, we may write

$$h^{\mu_\theta}(x) + V(x) = \int_{\mathbb{R}^\mathsf{d}} (\mathsf{g}(x-y) - \mathsf{g}_-(x)) d\mu_\theta(y) + \mathsf{g}_-(x) + V(x).$$

On the one hand $\mathsf{g}_- + V$ is bounded below by (A2) and definition of g. On the other hand, using (2.1.1) we have

$$\int_{\mathbb{R}^\mathsf{d}} (\mathsf{g}(x-y) - \mathsf{g}_-(x)) d\mu_\theta(y)$$
$$\ge \int_{\mathbb{R}^\mathsf{d}} (\mathsf{g}(x) \wedge \mathsf{g}(y) \mathbf{1}_{\mathsf{s}=0} + (\mathsf{g}(x) + \mathsf{g}(y)) \mathbf{1}_{\mathsf{s}<0} - C - \mathsf{g}_-(x)) d\mu_\theta(y).$$

In the case $\mathsf{s} = 0$ we deduce that

$$\int_{\mathbb{R}^\mathsf{d}} (\mathsf{g}(x-y) - \mathsf{g}_-(x)) d\mu_\theta(y) \ge \int_{|y|\ge|x|} (\mathsf{g}(y) - \mathsf{g}_-(x) - C) d\mu_\theta(y)$$
$$\ge \int_{\mathbb{R}^\mathsf{d}} \mathsf{g}_-(y) d\mu_\theta(y) - C,$$

and the right-hand side is bounded below in view of (2.5.6). In the case $\mathsf{s} < 0$, we find instead

$$\int_{\mathbb{R}^\mathsf{d}} (\mathsf{g}(x-y) - \mathsf{g}_-(x)) d\mu_\theta(y) \ge \int_{\mathbb{R}^\mathsf{d}} (\mathsf{g}(y) - C) d\mu_\theta(y) \ge \int_{\mathbb{R}^\mathsf{d}} \mathsf{g}_-(y) d\mu_\theta(y) - C,$$

and we conclude again by (2.5.6).

From the fact that $h^{\mu_\theta} + V$ is bounded below and (2.5.12) it follows that $\log\mu_\theta$, hence μ_θ is bounded above.

Next, we argue that h^{μ_θ} is bounded above. Using that μ_θ is bounded above, we may write

$$h^{\mu_\theta}(x) = \int_{y\in B(x,1)} \mathsf{g}(x-y) d\mu_\theta(y) + \int_{y\notin B(x,1)} \mathsf{g}(x-y) d\mu_\theta(y)$$
$$\le \|\mu_\theta\|_{L^\infty(\mathbb{R}^\mathsf{d})} \int_{y\in B(0,1)} \mathsf{g}(y) dy + \int_{y\notin B(x,1)} \mathsf{g}_+(x-y) d\mu_\theta(y)$$
$$\le C\|\mu_\theta\|_{L^\infty(\mathbb{R}^\mathsf{d})} + \frac{1}{\mathsf{s}} \mathbf{1}_{\mathsf{s}>0} \int d\mu_\theta,$$

where we used that $\mathsf{g}_+(x) \le 1/\mathsf{s}\mathbf{1}_{\mathsf{s}>0}$ for $|x| \ge 1$ and g is integrable near the origin. We thus deduce that $h^{\mu_\theta} + V$ is locally bounded above, hence in view of (2.5.12), μ_θ is locally bounded below.

Step 5: Continuity and upper bound. Here we give a sketch of how to deduce continuity of μ_θ and estimates for its maximum. If V has bounded second derivatives, we can take the Laplacian of (2.5.12) to find (2.5.13), and once we have shown that μ_θ is locally bounded above and below by positive constants, we may rewrite it as (2.5.14), a uniformly elliptic equation in divergence form for μ_θ. By standard elliptic regularity theory (for instance [**GT01**]) in the Coulomb case $\mathsf{s} = \mathsf{d} - 2$, or by fractional elliptic regularity theory otherwise, using that $\mathsf{s} < \mathsf{d}$, we deduce that μ_θ is as regular as V, in particular μ_θ is continuous.

Once μ_θ is continuous, let $x_0 \in B_R$ be a point where it achieves its maximum m_θ, evaluating (2.5.13) at x_0, by Constantin-Vicol's "nonlinear maximum principle" [**CV12**], we obtain

$$0 \geq \frac{1}{\theta}\Delta \log \mu_\theta(x_0) = \mathsf{c}_{\mathsf{d},\mathsf{s}}(-\Delta)^{1-\frac{\mathsf{d}-\mathsf{s}}{2}}\mu_\theta(x_0) - \Delta V(x_0) \geq C_{\mathsf{d},\mathsf{s}} m_\theta^{\frac{\mathsf{d}+2-(\mathsf{d}-\mathsf{s})}{\mathsf{d}}} - \max_{B_R} \Delta V,$$

hence we deduce an upper bound for m_θ in terms of ΔV. □

We now review the results obtained in [**AS22**] on the Coulomb $\mathsf{d} \geq 2$ case (which extend without too much trouble to $\mathsf{d} = 1$). They were proven under additional assumptions.

We assume that (A1)–(A3) hold so that μ_V exists. We assume in addition $\partial\Sigma \in C^{1,1}$, where we recall Σ is the support of μ_V, and

$$V \in C^2, \tag{2.5.17}$$

(2.5.18)
$$\begin{cases} \int_{|x|\geq 1} \exp\left(-\frac{\theta}{2}V(x)\right) dx < \infty & \text{if } \mathsf{d} \geq 3, \\ \int_{|x|\geq 1} \left(\exp\left(-\frac{\theta}{2}(V(x) - \log|x|)\right) + \exp\left(-\theta(V(x) - \log|x|)\right)|x|\log^2|x|\right) dx < \infty & \text{if } \mathsf{d} = 2, \end{cases}$$

$$\Delta V \geq \alpha > 0 \quad \text{in a neighborhood of } \Sigma. \tag{2.5.19}$$

This last assumption (2.5.19) is placed to use standard results on the obstacle problem [**Caf98**] that ensure that

$$\zeta(x) \geq \alpha \operatorname{dist}(x, \Sigma)^2 \quad \text{in a neighborhood of } \Sigma, \tag{2.5.20}$$

with ζ the function of (2.1.12), and a corresponding upper bound also holds. In particular it implies that $\omega = \Sigma$, i.e., the contact set and the droplet coincide. We now assume in addition that

$$\zeta(x) \geq \alpha \min(\operatorname{dist}(x, \Sigma)^2, 1), \tag{2.5.21}$$

which amounts, up to changing the constant $\alpha > 0$ if necessary, to assume that the solution to the obstacle problem never gets very close to the obstacle, outside of Σ.

Under assumption (2.5.21), the convergence of μ_θ to μ_V is shown to have *a boundary layer of lengthscale* $\theta^{-\frac{1}{2}}$ *near* $\partial\Sigma$ where μ_V is generally discontinuous. This is what we consider as the lengthscale for macroscopic rigidity.

Theorem 2.23 (Convergence of μ_θ to μ_V, see [**AS22**]). *Let* $\mathsf{d} \geq 2$ *and* $\mathsf{s} = \mathsf{d} - 2$. *Assume* (A1)–(A4) *and* (2.5.17)–(2.5.21). *Then* (2.0.4) *has a unique minimizer* μ_θ. *Moreover, there exists* $C(V, \mathsf{d}) > 0$ *such that, for every* $x \in \mathbb{R}^\mathsf{d}$ *and* $\theta \in (2, \infty)$, *we have*

$$0 < \mu_\theta(x) \leq \begin{cases} \min(C, C\exp(-\theta(V(x) - C)) & \text{if } \mathsf{d} \geq 3 \\ \min(C, C\exp(-\theta(V(x) - \log|x| - C))) & \text{if } \mathsf{d} = 2 \end{cases} \tag{2.5.22}$$

$$\mu_\theta(x) > \frac{1}{C} > 0 \quad \textit{for } x \in \Sigma, \tag{2.5.23}$$

$$\exp\left(-\frac{\theta}{C}\operatorname{dist}(x,\Sigma)^2 - C\right) \le \mu_\theta(x) \le \exp\left(-\frac{\theta}{C}\operatorname{dist}(x,\Sigma)^2 + C\right) \tag{2.5.24}$$

in a neighborhood of Σ,

$$\|h^{\mu_\theta} - c_\theta - h^{\mu_V} + c_\infty\|_{L^\infty(\mathbb{R}^{\mathsf{d}})} \le \frac{C}{\theta}, \tag{2.5.25}$$

where c_∞ *is the constant in* (2.1.9), *and*

$$\|\nabla(h^{\mu_V} - h^{\mu_\theta})\|_{L^\infty(\mathbb{R}^{\mathsf{d}})} \le \frac{C}{\sqrt{\theta}}, \tag{2.5.26}$$

$$\mu_\theta(\Sigma^c) \le \frac{C}{\sqrt{\theta}}, \tag{2.5.27}$$

and

$$\left|\int_{\Sigma^c} \mu_\theta \log \mu_\theta\right| \le \frac{C}{\sqrt{\theta}}. \tag{2.5.28}$$

Let m *be an integer* ≥ 2 *such that* $V \in C^{2m,\gamma}$ *for some* $\gamma \in (0,1]$ *and letting* f_k *be defined iteratively by*

$$f_0 = \frac{1}{\mathsf{c_d}}\Delta V, \qquad f_{k+1} = \frac{1}{\mathsf{c_d}}\Delta V + \frac{1}{\theta \mathsf{c_d}}\Delta \log f_k, \tag{2.5.29}$$

we have $f_k \in C^{2(m-k-1),\gamma}(\Sigma)$ *and for every even integer* $n \le 2m-4$ *and* $0 \le \gamma' \le \gamma$, *if* θ *is large enough depending on* m, *we have*

$$\|\mu_\theta - f_{m-2-n/2}\|_{C^{n,\gamma'}(\Sigma)} \le C\theta^{\frac{n+\gamma'}{2}} \exp\left(-C\log^2(\theta \operatorname{dist}(x,\partial\Sigma)^2)\right) + C\theta^{1+n-m+\frac{\gamma'}{2}}. \tag{2.5.30}$$

The functions f_k provide a sequence of improving approximations to μ_θ defined iteratively. Spelling out the iteration we easily find the expansion in powers of $1/\theta$

$$\mu_\theta \simeq \frac{1}{\mathsf{c_d}}\Delta V + \frac{1}{\mathsf{c_d}\theta}\Delta \log \frac{\Delta V}{\mathsf{c_d}} + \frac{1}{\mathsf{c_d}\theta^2}\Delta\left(\frac{\Delta \log \frac{\Delta V}{\mathsf{c_d}}}{\Delta V}\right) + \dots \quad \text{inside } \Sigma, \tag{2.5.31}$$

up to an order dictated by the regularity of V and the size of θ.

Remark 2.24. In the case $\beta = 2$ in the two-dimensional Coulomb case $\mathsf{s} = 0$, $\mathsf{d} = 2$, the large-N expansion of the one-point density $\rho_N^{(1)}$ (as in (1.3.4)) in Σ is known (see [**BF23**] and references therein) to be

$$\rho_N^{(1)}(x) = \frac{1}{\mathsf{c_d}}\Delta V(x) + \frac{1}{\mathsf{c_d}2N}\Delta \log \Delta V(x) + N^{-1}B_2(x) + \dots.$$

Thus comparing with (2.5.31) shows that the thermal equilibrium measure with $\theta = \beta N^{1-\frac{\mathsf{s}}{\mathsf{d}}} = \beta N$ provides, at least in that case, the correct next to leading order correction. In particular, the vanishing lengthscale $\frac{1}{\sqrt{\theta}}$, which is the lengthscale of variation of μ_θ and of decay of its tails away from Σ (the support of μ_V), can be seen as corresponding to a lengthscale of macroscopic rigidity and localization near Σ for the Coulomb/Riesz gas (compare with Theorem 5.33 for the two-dimensional log case, in Section 5.3.3).

In view of (2.5.30), letting $\hat{\Sigma} = \{x \in \Sigma, \mathrm{dist}(x, \partial\Sigma) \geq \theta^{\varepsilon - 1/2}\}$ for some $\varepsilon > 0$, we have uniform bounds

$$\forall \sigma \leq 2m + \gamma - 4, \quad \|\mu_\theta\|_{C^{0,\sigma}(\hat{\Sigma})} \leq C. \tag{2.5.32}$$

The proof of the theorem relies on maximum principle arguments, and on observing that $\frac{\mu_\theta}{f_k} - 1$ solves a divergence form equation which is uniformly elliptic inside Σ then applying regularity theory for elliptic PDE.

CHAPTER 3

The leading order behavior

In this chapter, building on the results of the previous chapter we study the leading order or "mean-field" behavior of the Coulomb or Riesz gas energy (1.1.1) with g given by (2.0.1), and apply it to the Gibbs distribution of (1.1.5), which we recall is

$$d\mathbb{P}_{N,\beta}(X_N) := \frac{1}{Z_{N,\beta}} \exp\left(-\beta N^{-\frac{\mathsf{s}}{\mathsf{d}}} \mathcal{H}_N(X_N)\right) dX_N. \tag{3.0.1}$$

The results are quite standard and in large part adapted from the literature. However, we try to give here a self-contained and general treatment, since results are a bit scattered between the potential theory literature, the probability and statistical mechanics literature and the analysis literature, and not all situations seem to be systematically covered.

The beginning of this chapter is devoted to the analysis of $\mathcal{H}_N$ only, expressed in the framework of Γ-convergence, leading to the mean-field description of its minimizers and their convergence to the equilibrium measure. Then, in Section 3.2 we apply these results to the statistical mechanics model, i.e., to characterizing the states with nonzero temperature under the law (1.1.5). This is done again in terms of the equilibrium or thermal equilibrium measure, and via a large deviations principle (LDP). We have tried to present the most general LDP results, handling all temperature regimes and all interactions $\mathsf{d}-2 \le \mathsf{s} < \mathsf{d}$ at once, under a minimal set of assumptions.

3.1. The case of zero temperature

The energy $\mathcal{H}_N$ is the sum of an interaction term $\sum_{i\neq j} \mathsf{g}(x_i-x_j)$ and a potential term $N\sum_{i=1}^N V(x_i)$. The first term pushes the points to repel and potentially escape to infinity, while the second one confines them. The sum of pairwise interactions is expected to scale like the number of pairs of points, i.e., N^2, while the sum of the potential terms is expected to scale like N times the number of points, i.e., N^2 again. The factor N in front of V in (1.1.1) is thus precisely so that the opposing effects of the repulsion and of the confinement balance each other. This is called the "mean-field scaling", the scaling in which the force acting on each particle is given in terms of the average field generated by the other particles. For general reference on mean-field theory, see statistical mechanics textbooks such as [**Hua63**].

3.1.1. Γ-convergence: General definition. The result we want to show about the leading order behavior of $\mathcal{H}_N$ can be formalized in terms of the notion of Γ-convergence, in the sense of De Giorgi (see [**Bra02**] for an introduction, or [**DM93**] for an advanced reference). It is a notion of convergence for functions (or functionals) which ensures that minimizers tend to minimizers. This notion is popular in the calculus of variations literature and very used in the analysis of sharp-interface and fracture models, dimension reduction for variational problems, homogenization, evolution problems, etc [**Bra02, Bra14**]. Using this formalism here is convenient but not essential.

Let us first give the basic definitions.

Definition 3.1 (Γ-convergence). We say that a sequence $\{F_N\}_N$ of functions on a metric space X Γ-converges to a function $F : X \to (-\infty, +\infty]$ if the following two inequalities hold:

(1) (Γ-lim inf) If $x_N \to x$ in X, then $\liminf_{N\to+\infty} F_N(x_N) \geq F(x)$.
(2) (Γ-lim sup) For all x in X, there is a sequence $\{x_N\}_N$ in X such that $x_N \to x$ and $\limsup_{N\to+\infty} F_N(x_N) \leq F(x)$. Such a sequence is called a *recovery sequence.*

The second inequality is saying that the first one is sharp, since it implies that there is a particular sequence $x_N \to x$ for which the equality $\lim_{N\to+\infty} F_N(x_N) = F(x)$ holds.

Remark 3.2.

(1) In practice a compactness assumption is generally needed and sometimes added in the definition, requiring that if $\{F_N(x_N)\}_N$ is bounded, then $\{x_N\}_N$ has a convergent subsequence. A similar compactness requirement also appears in the definition of a good rate function in large deviations theory (see Definition 3.9 below).
(2) The first inequality is usually proven by functional analysis methods, without making any "ansatz" on the precise form of x_N, whereas the second one is usually obtained by an explicit construction, during which one constructs "by hand" the recovery sequence such that $F_N(x_N)$ has asymptotically less energy than $F(x)$. Note also that by a diagonal argument, one may often reduce to constructing a recovery sequence for a dense subset of x's.
(3) A Γ-limit is always lower semicontinuous. (In particular, a function which is not l.s.c. is a bad candidate for being a Γ-limit.) Thus, a functional is not always its own Γ-limit: in general Γ-lim $F = \bar{F}$ where $\bar{F}$ is the l.s.c. envelope of F.
(4) The notion of Γ-convergence can be generalized to the situation where F_N and F are not defined on the same space. One may instead refer to a sense of convergence of x_N to x which is defined via the convergence of any specific function of x_N to x, which may be a nonlinear function of x_N, cf. [**SS04, JS09**] for instances of this.

We now state the most important property of Γ-convergence: Γ-convergence sends minimizers to minimizers.

Proposition 3.3 (Minimizers converge to minimizers under Γ-convergence). *Assume F_N Γ-converges to F in the sense of Definition* 3.1. *If for every N, x_N minimizes F_N, and if the sequence $\{x_N\}_N$ converges to some x in X as $N \to \infty$, then x minimizes F, and moreover,* $\lim_{N\to+\infty} \min_X F_N = \min_X F$.

Proof. Let $y \in X$. By the Γ-lim sup inequality, there is a recovery sequence $\{y_N\}_N$ converging to y such that $F(y) \geq \limsup_{N\to+\infty} F_N(y_N)$. By minimality of x_N, we have $F_N(y_N) \geq F_N(x_N)$ for each N and by the Γ-lim inf inequality, we have $\liminf_{N\to+\infty} F_N(x_N) \geq F(x)$, hence it follows that $F(y) \geq F(x)$. Since this is true for every y in X, it proves that x is a minimizer of F. The relation $\lim_{N\to+\infty} \min F_N = \min F$ follows from the previous chain of inequalities applied with $y = x$. □

Remark 3.4. An additional compactness assumption as in Remark 3.2 ensures that if $\{\min F_N\}_N$ is bounded then a sequence $\{x_N\}_N$ of minimizers has a limit, up to extraction. That limit must then be a minimizer of F. If moreover it happens that F has a unique minimizer, then the whole sequence $\{x_N\}_N$ must converge to it.

3.1.2. Γ-convergence of the Coulomb gas energy. The example of Γ-convergence of interest to us here concerns the sequence of functions $\{\frac{1}{N^2}\mathcal{H}_N\}_N$ defined as in (1.1.1). The space $\mathcal{P}(\mathbb{R}^{\mathsf{d}})$ of Borel probability measures on $\mathbb{R}^{\mathsf{d}}$ endowed with the topology of weak convergence (i.e., that of the dual of bounded continuous functions in $\mathbb{R}^{\mathsf{d}}$), which is metrizable, will play the role of the metric space X above. We may view $\mathcal{H}_N$ as being defined on $\mathcal{P}(\mathbb{R}^{\mathsf{d}})$ through the map

$$i_N : \begin{cases} (\mathbb{R}^{\mathsf{d}})^N \longrightarrow & \mathcal{P}(\mathbb{R}^{\mathsf{d}}) \\ X_N \mapsto & \widehat{\mu_N}[X_N] := \frac{1}{N}\sum_{i=1}^N \delta_{x_i} \end{cases} \tag{3.1.1}$$

which associates to any configuration of N points the probability measure $\widehat{\mu_N}$ called the *empirical measure*, or *spectral measure* in the context of random matrices. More precisely, we can extend the function $\mathcal{H}_N$ into a function defined for any μ in $\mathcal{P}(\mathbb{R}^{\mathsf{d}})$ by

$$\mathcal{H}_N(\mu) = \begin{cases} \mathcal{H}_N(X_N) & \text{if } \mu \text{ is of the form } \frac{1}{N}\sum_{i=1}^N \delta_{x_i} \\ +\infty & \text{otherwise.} \end{cases} \tag{3.1.2}$$

The first main result that we prove here is that the sequence $\{\frac{1}{N^2}\mathcal{H}_N\}_N$ has the functional $\mathcal{E}$ as its Γ-limit.

Proposition 3.5 (Γ-convergence of $\frac{1}{N^2}\mathcal{H}_N$). *If V satisfies* (A1) *and* (A2), *the sequence $\{\frac{1}{N^2}\mathcal{H}_N\}_N$ of functions (defined on $\mathcal{P}(\mathbb{R}^{\mathsf{d}})$ as above) Γ-converges as $N \to +\infty$ to the function $\mathcal{E}$ defined in* (2.0.3).

A statement and a proof via Γ-convergence for V quadratic in dimension 2 appeared in [**SS07**, Proposition 11.1]. It is not difficult to adapt them to higher dimensions and more general potentials. Similar arguments are also found in the large deviations proofs of [**BAG97, BAZ98, CGZ14**].

The proof of the lower bound uses the same ingredients as the proof of existence of a minimizer of $\mathcal{E}$ in the previous section. The proof of the upper bound can be obtained by constructing a recovery sequence for each measure μ in $\mathcal{P}(\mathbb{R}^{\mathsf{d}})$. By density one may reduce to measures which are in $L^\infty(\mathbb{R}^{\mathsf{d}})$, supported in a cube K and such that the density $\mu(x)$ is bounded below by some $\alpha > 0$ in K, one then cuts cubes into subcubes of size ℓ, $N^{-1/\mathsf{d}} \ll \ell \ll 1$ and places the appropriate number of well distributed points in each subcube K_i in such a way that the difference between $N\mu(K_i)$ and the number of points placed does not exceed 1, so that $\widehat{\mu_N}$ approximates μ. The difficult part is then to show that the contribution of nearby points (or near diagonal terms in $\mathcal{E}$) is not large. This proof is given in full in [**Ser15**, Proposition 2.1]. Here we will use instead a probabilistic argument: the result follows from the existence of a whole set of approximating configurations deriving from the proof of the LDP Theorem 3.17 below in the case with temperature.

In the whole text, when considering sequences of configurations $(x_1, \ldots, x_N)$ we will make the slight abuse of notation that consists in neglecting the dependency of the points $(x_1, \ldots, x_N)$ on N, while one should formally write $(x_{1,N}, \ldots, x_{N,N})$.

PROOF OF PROPOSITION 3.5.

Step 1. Lower bound. From now on, we denote the diagonal of $\mathbb{R}^{\mathsf{d}} \times \mathbb{R}^{\mathsf{d}}$ by $\triangle$ and its complement by $\triangle^c$.

We need to prove that if $\frac{1}{N}\sum_{i=1}^N \delta_{x_i} \to \mu \in \mathcal{P}(\mathbb{R}^{\mathsf{d}})$, then

$$\liminf_{N\to+\infty} \frac{1}{N^2}\mathcal{H}_N(X_N) \geq \mathcal{E}(\mu). \tag{3.1.3}$$

Letting $\widehat{\mu_N}$ denote the empirical measure $\frac{1}{N}\sum_{i=1}^N \delta_{x_i}$, we may write

$$\frac{1}{N^2}\mathcal{H}_N(\widehat{\mu_N}) = \frac{1}{2}\iint_{\triangle^c} \mathsf{g}(x-y)d\widehat{\mu_N}(x)d\widehat{\mu_N}(y) + \int V(x)d\widehat{\mu_N}(x). \tag{3.1.4}$$

In order to handle the singularity of g, as in the proof of Lemma 2.4 let us truncate the singularity of g by writing

$$\iint_{\triangle^c} \mathsf{g}(x-y)d\widehat{\mu_N}(x)d\widehat{\mu_N}(y) \geq \iint (\mathsf{g}(x-y)\wedge M)d\widehat{\mu_N}(x)d\widehat{\mu_N}(y) - \frac{M}{N}, \tag{3.1.5}$$

where $M > 0$ and $\wedge$ still denotes the minimum of two numbers. Indeed one has $\widehat{\mu_N}\otimes\widehat{\mu_N}(\triangle) = \frac{1}{N}$ as soon as the points of the configuration $(x_1, \ldots, x_N)$ are simple (i.e., $x_i \neq x_j$ for $i \neq j$) and if they are not, the assertion is obvious. We may then write that

$$\frac{1}{N^2}\mathcal{H}_N(X_N) \geq \frac{1}{2}\iint (\mathsf{g}(x-y)\wedge M + V(x) + V(y))d\widehat{\mu_N}(x)d\widehat{\mu_N}(y) - \frac{M}{N}. \tag{3.1.6}$$

The function $(x,y) \mapsto \mathsf{g}(x-y)\wedge M$ is continuous hence the function $(x,y) \mapsto \mathsf{g}(x-y)\wedge M + V(x) + V(y)$ is l.s.c. Moreover, it is bounded below arguing as in Remark 2.3, and by taking the limit of (3.1.6) as $N \to +\infty$ one gets, by weak convergence of $\widehat{\mu_N}$ to μ (hence of $\widehat{\mu_N}\otimes\widehat{\mu_N}$ to $\mu\otimes\mu$) that for every $M > 0$

$$\liminf_{N\to+\infty} \frac{1}{2}\iint_{\triangle^c} (\mathsf{g}(x-y) + V(x) + V(y))d\widehat{\mu_N}(x)d\widehat{\mu_N}(y)$$
$$\geq \frac{1}{2}\iint (\mathsf{g}(x-y)\wedge M + V(x) + V(y))d\mu(x)d\mu(y).$$

By the monotone convergence theorem, the (possibly infinite) limit of the right-hand side as $M \to +\infty$ exists and equals $\iint (\mathsf{g}(x-y) + V(x) + V(y))d\mu(x)d\mu(y)$. Combining with (3.1.6), this concludes the proof of the Γ-lim inf convergence (3.1.3).

Step 2. Upper bound. Let μ be a probability density such that $\mathcal{E}(\mu) < +\infty$. As mentioned in Remark 3.2 it suffices to prove the upper bound inequality for a dense subset of probability measures. We may thus assume, without loss of generality, that μ has compact support. Using that μ is a probability density, let us then evaluate

$$\begin{aligned}
&\frac{1}{N^2}\int_{(\mathbb{R}^d)^N}\mathcal{H}_N(X_N)d\mu(x_1)\dots d\mu(x_N)\\
&=\frac{1}{N^2}\left(\frac{1}{2}\sum_{i\neq j}\iint_{(\mathbb{R}^d)^N}\mathsf{g}(x_i-x_j)d\mu^{\otimes N}(X_N)\right.\\
&\qquad\qquad\left.+N\sum_{i=1}^N\int_{(\mathbb{R}^d)^N}V(x_i)d\mu^{\otimes N}(X_N)\right)\\
&=\frac{1}{N^2}\left(\frac{N(N-1)}{2}\iint_{\mathbb{R}^d\times\mathbb{R}^d}\mathsf{g}(x-y)d\mu(x)d\mu(y)+N^2\int_{\mathbb{R}^d}Vd\mu\right)\\
&=\mathcal{E}(\mu)-\frac{1}{2N}\iint_{\mathbb{R}^d\times\mathbb{R}^d}\mathsf{g}(x-y)d\mu(x)d\mu(y)\\
&\le\mathcal{E}(\mu)-\frac{1}{2N}\iint_{\mathbb{R}^d\times\mathbb{R}^d}\mathsf{g}_-(x-y)d\mu(x)d\mu(y).
\end{aligned}$$

Using the definition of g, (2.1.1) and the fact that μ is compactly supported, we find that $\iint_{\mathbb{R}^d\times\mathbb{R}^d}\mathsf{g}_-(x-y)d\mu(x)d\mu(y)>-\infty$. It follows that, $\widehat{\mu_N}$ denoting the empirical measure of X_N, for any $\varepsilon>0$,

$$\begin{aligned}
&\frac{1}{N^2}\int_{\widehat{\mu_N}\in B(\mu,\varepsilon)}\mathcal{H}_N(X_N)d\mu^{\otimes N}(X_N)\\
&\qquad\le\mathcal{E}(\mu)+\frac{C}{N}-\frac{1}{N^2}\int_{\widehat{\mu_N}\notin B(\mu,\varepsilon)}\mathcal{H}_N(X_N)d\mu^{\otimes N}(X_N),
\end{aligned}$$

where $C>0$ depends on μ. Since $\mathcal{H}_N\ge -CN^2$ as a consequence of the Γ-liminf and the fact that $\inf\mathcal{E}>-\infty$ as seen in Lemma 2.4, we may bound

$$-\frac{1}{N^2}\int_{\widehat{\mu_N}\notin B(\mu,\varepsilon)}\mathcal{H}_N(X_N)d\mu^{\otimes N}(X_N)\le C\int_{\widehat{\mu_N}\notin B(\mu,\varepsilon)}d\mu^{\otimes N}(X_N)$$

while, by the law of large numbers,

$$\int_{\widehat{\mu_N}\notin B(\mu,\varepsilon)}d\mu(x_1)\dots d\mu(x_N)=o_N(1).$$

Assembling all these relations, we have obtained that for any $\varepsilon>0$

$$\frac{1}{N^2}\int_{\widehat{\mu_N}\in B(\mu,\varepsilon)}\mathcal{H}_N(X_N)d\mu(x_1)\dots d\mu(x_N)\le\mathcal{E}(\mu)+o_N(1).$$

Thus, by a diagonal argument, we may build a sequence of points such that $\widehat{\mu_N}\to\mu$ as $N\to\infty$ and

$$\limsup_{N\to\infty}\frac{1}{N^2}\mathcal{H}_N(X_N)\le\mathcal{E}(\mu)$$

as desired. □

Remark 3.6. Again we do not really need that g is a Coulombic (or Riesz) kernel, rather we only used (A1)–(A2) and the facts that g is locally bounded below, locally integrable, and l.s.c. away from the origin.

We next derive the consequence of the Γ-convergence Proposition 3.5 given by Proposition 3.3. In order to do so, we must prove the compactness of sequences with suitably bounded energy, as in Remark 3.4.

Lemma 3.7. *Assume that V satisfies* (A1)–(A2). *Let $\{X_N\}_N$ be a sequence of configurations in $(\mathbb{R}^\mathsf{d})^N$, and let $\{\widehat{\mu_N}\}_N$ be the associated empirical measures. Assume $\{\frac{1}{N^2}\mathcal{H}_N(\widehat{\mu_N})\}_N$ is a bounded sequence. Then the sequence $\{\widehat{\mu_N}\}_N$ is tight, and as $N \to \infty$, it converges weakly in $\mathcal{P}(\mathbb{R}^\mathsf{d})$ (up to extraction of a subsequence) to some probability measure μ.*

PROOF. The proof is completely analogous to that of Lemma 2.4. First, by assumption, there exists a constant C independent of N such that $\mathcal{H}_N(X_N) \le CN^2$, and in view of (3.1.4)–(3.1.6) we may write, for every $M > 0$,

$$\begin{aligned} C &\ge \frac{1}{2}\iint (\mathsf{g}(x-y) \wedge M)\, d\widehat{\mu_N}(x)\, d\widehat{\mu_N}(y) - \frac{M}{N} + \int V\, d\widehat{\mu_N} \\ &= \frac{1}{2}\iint (\mathsf{g}(x-y) \wedge M + V(x) + V(y))\, d\widehat{\mu_N}(x)\, d\widehat{\mu_N}(y) - \frac{M}{N}. \end{aligned} \tag{3.1.7}$$

Using Remark 2.3, and arguing as in Lemma 2.4, we get the tightness. □

To conclude, we will make the assumptions on V that ensure both the Γ-convergence result Proposition 3.5 and the existence result Theorem 2.8.

Theorem 3.8 (Convergence of minimizers and minima of $\mathcal{H}_N$). *Assume that V satisfies* (A1)–(A3). *Assume that for each N, X_N is a minimizer of $\mathcal{H}_N$. Then, as $N \to \infty$ we have*

$$\frac{1}{N}\sum_{i=1}^N \delta_{x_i} \to \mu_V \text{ in the weak sense of probability measures} \tag{3.1.8}$$

where μ_V is the unique minimizer of $\mathcal{E}$ as in Theorem 2.8, and

$$\lim_{N\to+\infty} \frac{\mathcal{H}_N(X_N)}{N^2} = \mathcal{E}(\mu_V). \tag{3.1.9}$$

PROOF. Applying the Γ-limsup part of the definition of Γ-convergence, for example to μ_V, ensures that $\limsup_{N\to+\infty} \frac{1}{N^2} \min \mathcal{H}_N$ is bounded above (by $\mathcal{E}(\mu_V)$), hence in particular sequences of minimizers of $\mathcal{H}_N$ satisfy the assumptions of Lemma 3.7. It follows that, up to a subsequence, we have $\widehat{\mu_N} \to \mu$ in the weak sense of probability measures for some $\mu \in \mathcal{P}(\mathbb{R}^\mathsf{d})$. By Propositions 3.5 and 3.3, μ must minimize $\mathcal{E}$, hence, in view of Theorem 2.8, it must be equal to μ_V. This implies that the convergence must hold along the whole sequence. We also get (3.1.9) from Proposition 3.3. □

In the language of statistical mechanics or mean field theory, this result gives the mean-field behavior or average behavior of ground states, and the functional $\mathcal{E}$ is called the mean-field energy functional. It tells us that points are macroscopically distributed according to the probability law μ_V as their number tends to ∞, and we have the leading order asymptotic expansion of the ground state energy

$$\min \mathcal{H}_N \sim N^2 \min \mathcal{E} \quad \text{as } N \to \infty.$$

This is not very precise as it does not give information on the behavior at smaller scales and the patterns followed by the points. Understanding this and going beyond the leading order description is the object of the following chapters.

3.2. The case with temperature: Large deviations principle

At this point, we have understood the mean-field behavior of ground states of the Coulomb/Riesz gas. In this section, we turn for the first time to states with temperature and derive rather easy consequences of the previous sections for the Gibbs measure (1.1.5), via the framework of large deviations which allows to characterize the probability of observing a rare event or a "nontypical" configuration.

3.2.1. Definitions. Let us first recall the basic definitions in the theory of large deviations; for more reference see the textbooks [**dH00, DS89, DZ10, RAS09**].

Definition 3.9 (Rate function). Let X be a complete separable metric space. A rate function is an l.s.c. function $I : X \to [0, +\infty]$, it is called a *good rate function* if its sublevel sets $\{x, I(x) \le \alpha\}$ are compact; see Remark 3.2.

Definition 3.10 (Large deviations). Let $\{P_N\}_N$ be a sequence of Borel probability measures on X and $\{a_N\}_N$ a sequence of positive real numbers diverging to $+\infty$. Let also I be a (good) rate function on X. The sequence $\{P_N\}_N$ is said to satisfy a large deviation principle (LDP) at speed a_N with (good) rate function I if for every Borel set $E \subset X$ the following inequalities hold

$$-\inf_{\overset{\circ}{E}} I \le \liminf_{N\to+\infty} \frac{1}{a_N} \log P_N(E) \le \limsup_{N\to+\infty} \frac{1}{a_N} \log P_N(E) \le -\inf_{\bar{E}} I, \tag{3.2.1}$$

where $\overset{\circ}{E}$ (resp. $\bar{E}$) denotes the interior (resp. the closure) of E for the topology of X.

Formally, it means that $P_N(E)$ behaves roughly like $e^{-a_N \inf_E I}$. The rate function I is the rate of exponential decay of the probability of rare events, and the events with larger probability are the ones on which I is smaller. For this to make sense, $\inf_X I$ must be zero, and the LDP then asserts that all states with $I > \inf_X I$ have exponentially small probability, hence they are rare events and this quantifies their probability.

Definition 3.11 (Exponential tightness). We say that the sequence $\{P_N\}_N$ is exponentially tight (at speed a_N) if for every $\varepsilon > 0$ there exists a compact set K_ε such that

$$\limsup_{N\to\infty} \frac{1}{a_N} \log P_N(K_\varepsilon^c) \le -\varepsilon.$$

Lemma 3.12. *If $\{P_N\}_N$ is exponentially tight at speed a_N, then to prove it satisfies a LDP at speed a_N with rate function I it suffices to prove the upper bound in* (3.2.1) *for every compact set.*

The proof is immediate. As a corollary it suffices to prove (3.2.1) for balls if the space X is metrized.

Corollary 3.13. *If $\{P_N\}_N$ is exponentially tight at speed a_N and if X is metrizable, then P_N satisfies a LDP at speed a_N with (l.s.c.) rate function I if and only if for every $x \in X$*

$$\begin{aligned}-I(x) &\le \liminf_{\varepsilon\to0} \liminf_{N\to\infty} \frac{1}{a_N} \log P_N(B(x,\varepsilon)) \\ &\le \limsup_{\varepsilon\to0} \limsup_{N\to\infty} \frac{1}{a_N} \log P_N(B(x,\varepsilon)) \le -I(x).\end{aligned}$$

This is a consequence of the fact that a compact set can be covered by a finite number of balls and also that a sum of exponentials is dominated by the one with the largest exponent (as in the Laplace or stationary phase method). This is left as an exercise (see also Definitions 2.18, 2.19 in [**RAS09**] and comments around them.)

Remark 3.14. At first sight, Definition 3.10 looks very close to the Γ-convergence

$$\frac{\log p_N}{a_N} \xrightarrow{\Gamma} -I,$$

where p_N is the density of the measure P_N. However, in general there is no equivalence between the two concepts. For example, in order to estimate the quantity

$$\log P_N(E) = \log \int_E p_N(x)dx \tag{3.2.2}$$

it is not sufficient to know the asymptotics of p_N, one also needs to know the size of the volume element $\int_E dx$, which plays a large role in large deviations and usually arises as an entropy term in the rate function. There are however some rigorous connections between Γ-convergence and LDP (see [**Mar18**]).

3.2.2. Heuristics for the LDP of the Riesz gas. We now want to specialize to the Coulomb/Riesz gas ensemble (1.1.5). First, pushing $\mathbb{P}_{N,\beta}$ forward by the map i_N of (3.1.1), we may view it as a probability measure on $\mathcal{P}(\mathbb{R}^\mathsf{d})$.

To explain what LDP we can expect for it, let us start by recalling the Gibbs variational principle which states that a Gibbs measure minimizes the sum of the average energy and the entropy times the temperature, and whose proof is immediate.

Lemma 3.15 (Gibbs variational principle). *Let $dP_\beta(x) = \frac{1}{Z}\exp(-\beta F(x))d\lambda(x)$ be a probability measure on a space X, where λ is a reference measure on X. Then P_β achieves*

$$\min_{P\in\mathcal{P}(X)} \int F(x)dP(x) + \frac{1}{\beta}\int dP(x)\log\frac{dP(x)}{d\lambda(x)} \tag{3.2.3}$$

and the minimum is equal to $-\frac{1}{\beta}\log Z$.

This principle can be applied directly to $\mathbb{P}_{N,\beta}$ with λ the Lebesgue measure on $(\mathbb{R}^\mathsf{d})^N$ and indicates that as $N\to\infty$ the Gibbs measure concentrates on states that minimize the sum of a limit energy and an entropy.

Making this more rigorous is done via large deviations theory, using Sanov's theorem (see for instance [**DZ10, DE97**]), which we now recall in the version appropriate to our context.

Theorem 3.16 (Sanov). *Assume $X_1,\dots,X_N$ are i.i.d. random variables with values in $\mathbb{R}^\mathsf{d}$ defined on a Polish probability space $(X,\mathbb{P})$, with law ρ, i.e., $\mathbb{P}(X_i\in A)=\rho(A)$. Then, for every $A\subset\mathcal{P}(X)$, setting*

$$P_N(A) = \mathbb{P}\left(\frac{1}{N}\sum_{i=1}^N \delta_{X_i}\in A\right),$$

$\{P_N\}_N$ satisfies a LDP at speed N with good rate function

$$\text{ent}[\mu|\rho] = \begin{cases} \int_{\mathbb{R}^d} \mu \log \frac{\mu}{\rho} & \text{if } \mu \ll \rho \\ +\infty & \text{otherwise.} \end{cases} \tag{3.2.4}$$

Here, $\text{ent}[\mu|\rho]$ is called the *relative entropy* of μ with respect to ρ. Informally, Sanov's theorem states that

$$\mathbb{P}\left(\frac{1}{N}\sum_{i=1}^{N} \delta_{X_i} \simeq \mu\right) \simeq \exp\left(-N\int \mu \log \frac{\mu}{\rho}\right)$$

if $\mathbb{P}$ has law ρ.

Let us now think of $\mathbb{P}$ as corresponding to the normalized Lebesgue measure in the set Σ, support of the equilibrium measure μ_V and the X_i's as the points x_i. Then the right-hand side is just a multiple of the regular entropy. This would be the governing function in the case of $\beta \to 0$ where the interaction between the particles disappear and they just become Poissonian (or Bernoulli). In the case with β nonzero, then the result of Proposition 3.5 says that $\mathcal{H}_N \simeq N^2\mathcal{E}$. If $\mathcal{H}_N$ were continuous one could conclude directly by "tilting" or "Varadhan's lemma" that

$$\mathbb{P}_{N,\beta}(\widehat{\mu_N} \simeq \mu) \sim \exp\left(-\beta N^{-\frac{s}{d}} N^2 \mathcal{E}(\mu) - N\int_{\mathbb{R}^d} \mu \log \mu\right).$$

Thus we see that the energy is in competition with the entropy, the two will be comparable in the regime where $\beta N^{1-\frac{s}{d}}$ is proportional to 1, the energy will dominate as soon as $\beta N^{1-\frac{s}{d}} \gg 1$ and the entropy will dominate when $\beta N^{1-\frac{s}{d}} \ll 1$. This leads us to defining an effective temperature that we will use throughout

$$\boxed{\theta := \beta N^{1-\frac{s}{d}}.} \tag{3.2.5}$$

With this new notation, the quantity to minimize is thus

$$-N\left(\theta\mathcal{E}(\mu) + \int_{\mathbb{R}^d} \mu \log \mu\right) = -N\theta\, \mathcal{E}_\theta(\mu),$$

with $\mathcal{E}_\theta$ as in (2.0.4). This shows that its minimizer μ_θ (the thermal equilibrium measure) is an even better mean-field approximation of the empirical measure: even when $\theta \to +\infty$ as $N \to \infty$, it provides a corrected description to μ_V.

3.2.3. Large deviations principle for the Riesz gas. We may now state the LDP for the Gibbs measure associated to the Coulomb or Riesz gas Hamiltonian. This result was proven in the regime of β independent of N, in [**HP00**] (in dimension 1), [**BAG97**] (in dimension 1) and [**BAZ98**] (in dimension 2) for the particular case of a quadratic potential (and $\beta = 2$), see also [**Ber14**] for results in a more general (still determinantal) setting of multidimensional complex manifolds. The papers [**CGZ14, LW20**] recently treated more general singular g's and V's, [**LW20**] containing a result for general k-point interactions. An LDP result for local empirical observables, in the Coulomb gas case, can be found in [**PG24**].

The LDP implies as a corollary the earlier-known law of large numbers that the empirical measure converges to the equilibrium measure, as was for instance first proven in the determinantal case of random matrix ensembles in [**BdMPS95**].

We present here the proof for the Riesz gas in any dimension, general potential, and general $\theta > 0$, which is not more difficult. We will work in the metric space

$\mathcal{P}(\mathbb{R}^{\mathsf{d}})$ equipped with any distance that metrizes weak convergence. The LDP is stated for the push-forward of the Gibbs measure by the map i_N of (3.1.1), which is a probability on $\mathcal{P}(\mathbb{R}^{\mathsf{d}})$. We recall that the functionals $\mathcal{E}$ and $\mathcal{E}_\theta$ were introduced in (2.0.3) and (2.0.4).

Theorem 3.17 (Large deviations principle at leading order for the Riesz gas). *Let $\mathbb{P}_{N,\beta}$ be as in* (1.1.5) *and θ as in* (3.2.5). *Assume that V is finite-valued and satisfies* (A1), (A2) *and* (A5) *for N large enough.*

- *Assume that $\theta \to +\infty$ as $N \to \infty$. Then the sequence $\{i_N \# \mathbb{P}_{N,\beta}\}_N$ of probability measures on $\mathcal{P}(\mathbb{R}^{\mathsf{d}})$ satisfies a large deviations principle at speed $N\theta$ with good rate function $\hat{\mathcal{E}}$ where $\hat{\mathcal{E}} = \mathcal{E} - \min_{\mathcal{P}(\mathbb{R}^{\mathsf{d}})} \mathcal{E}$. Moreover,*

$$\lim_{N \to +\infty} \frac{1}{N\theta} \log Z_{N,\beta} = -\mathcal{E}(\mu_V) = -\min_{\mathcal{P}(\mathbb{R}^{\mathsf{d}})} \mathcal{E}. \tag{3.2.6}$$

- *Assume that θ is independent of N. Then the sequence $\{i_N \# \mathbb{P}_{N,\beta}\}_N$ of probability measures on $\mathcal{P}(\mathbb{R}^{\mathsf{d}})$ satisfies a large deviations principle at speed $N\theta$ with good rate function $\hat{\mathcal{E}}_\theta$ where $\hat{\mathcal{E}}_\theta = \mathcal{E}_\theta - \min_{\mathcal{P}(\mathbb{R}^{\mathsf{d}})} \mathcal{E}_\theta$. Moreover,*

$$\lim_{N \to +\infty} \frac{1}{N\theta} \log Z_{N,\beta} = -\mathcal{E}_\theta(\mu_\theta) = -\min_{\mathcal{P}(\mathbb{R}^{\mathsf{d}})} \mathcal{E}_\theta. \tag{3.2.7}$$

An analogous result for the case $\theta \to 0$ with the entropy as a rate function also naturally holds, except that in order to state it one would have to reduce to a bounded subset.

The heuristic reading of the LDP is that

$$\mathbb{P}_{N,\beta}(E) \approx e^{-N\theta(\min_E \mathcal{E} - \min \mathcal{E})}, \tag{3.2.8}$$

respectively

$$\mathbb{P}_{N,\beta}(E) \approx e^{-N\theta(\min_E \mathcal{E}_\theta - \min \mathcal{E}_\theta)}.$$

As a consequence, if $\theta \gg 1$, or equivalently $\beta \gg N^{\frac{\mathsf{s}}{\mathsf{d}}-1}$, the only likely configurations of points (under $\mathbb{P}_{N,\beta}$) are those for which the empirical measures $\widehat{\mu_N} = \frac{1}{N}\sum_{i=1}^N \delta_{x_i}$ converge to $\mu = \mu_V$, for otherwise $\mathcal{E}(\mu) > \mathcal{E}(\mu_V)$ by uniqueness of the minimizer of $\mathcal{E}$, and the probability decreases exponentially fast according to (3.2.8). Thus, μ_V is not only the limiting distribution of minimizers of $\mathcal{H}_N$, but also the limiting distribution for all "typical" (or likely) configurations. Moreover, we can estimate the probability under $\mathbb{P}_{N,\beta}$ of the nontypical configurations and see that it has exponential decay at speed θN. The temperature plays no role in the rate function as long as $\beta \gg N^{\frac{\mathsf{s}}{\mathsf{d}}-1}$.

The effect of the temperature is felt at the macroscopic scale only if θ does not tend to $+\infty$, i.e., $\beta \le CN^{\frac{\mathsf{s}}{\mathsf{d}}-1}$, which we can consider here a high-temperature regime. Recall that the cases of the classic random matrix ensembles GOE, GUE and Ginibre all correspond to $\mathsf{s} = 0$ and $\beta = 1, 2$, hence they all correspond to a low temperature regime for the criterion $\beta \gg N^{\frac{\mathsf{s}}{\mathsf{d}}-1} = N^{-1}$.

Recalling the corresponding equilibrium measures were given in Example 2.13 in Chapter 2, as a consequence of Theorem 3.17, we have a proof of the law of large numbers, i.e., that the distribution of eigenvalues (more precisely the spectral or empirical measure) has to follow Wigner's semicircle law $\mu_V = \frac{1}{2\pi}\sqrt{4-x^2}\mathbf{1}_{|x|<2}$ for the GUE and GOE, and the circle law $\mu_V = \frac{1}{\pi}\mathbf{1}_{B_1}$ for the Ginibre ensemble, as a consequence of the stronger LDP result. These are the cases originally treated in [**HP00, BAZ98, BAG97**].

In addition, this theorem provides the leading order of the free energy.

The large deviations lower bound can be proven by showing that given a probability measure $\mu \in \mathcal{P}(\mathbb{R}^\mathsf{d})$, there is a sufficiently large volume of empirical measures which approach μ and whose energy does not asymptotically exceed $\mathcal{E}(\mu)$. This is the proof in the original papers, and can also be found in [**Ser15**]. Here, we present a more elegant proof based on Jensen's inequality, inspired from the work of [**GZ19**] and Dupuis.

PROOF. Let us introduce the same notation as in the proof of Proposition 2.18 in (2.5.2), i.e.,

$$u(x) = \exp(-\theta(V+\mathsf{g}_-)), \quad \bar{u} = \int_{\mathbb{R}^\mathsf{d}} u, \quad \rho = \frac{u}{\bar{u}}.$$

This is well-defined by (A4), which is itself a consequence of (A5). Also since (A4) holds for N large enough, in all cases there exists a $\theta_0 > 0$ independent of N such that $\int_{\mathbb{R}^\mathsf{d}} \exp(-\theta_0(V+\mathsf{g}_-)) < +\infty$ and thus we may also define

$$u_0(x) = \exp(-\theta_0(V+\mathsf{g}_-)), \quad \bar{u}_0 = \int_{\mathbb{R}^\mathsf{d}} u_0, \quad \rho_0 = \frac{u_0}{\bar{u}_0}. \tag{3.2.9}$$

Step 1: LDP lower bound. Let us consider μ a probability measure with a density which is positive and continuous over $\mathbb{R}^\mathsf{d}$, such that for N large enough, μ/ρ is bounded below independently of N, and such that $\int |\mathsf{g}_-| d\mu < \infty$.

Let us apply the Jensen-based argument. Starting from (1.1.5) and using (3.2.5), since $\mu > 0$ we may write

(3.2.10)

$$\begin{aligned}
&\mathbb{P}_{N,\beta}(\widehat{\mu_N} \in B(\mu,\varepsilon)) \\
&\quad = \frac{1}{Z_{N,\beta}} \int_{\widehat{\mu_N}\in B(\mu,\varepsilon)} \\
&\quad\times \exp\left(-\theta N^{-1}\mathcal{H}_N(X_N) - \sum_{i=1}^N (\log\frac{\mu}{\rho})(x_i) - \sum_{i=1}^N \log\rho(x_i)\right) d\mu(x_1)\ldots d\mu(x_N).
\end{aligned}$$

Next we apply Jensen's inequality to the integral to obtain

(3.2.11)

$$\begin{aligned}
&\log \mathbb{P}_{N,\beta}(\widehat{\mu_N} \in B(\mu,\varepsilon)) \\
&\quad \geq -\log Z_{N,\beta} + \log \mu^{\otimes N}(\{\widehat{\mu_N} \in B(\mu,\varepsilon)\}) + \frac{1}{\mu^{\otimes N}(\{\widehat{\mu_N}\in B(\mu,\varepsilon)\})} \\
&\quad \times \int_{\widehat{\mu_N}\in B(\mu,\varepsilon)} \left(-\theta N^{-1}\mathcal{H}_N(X_N) - \sum_{i=1}^N (\log\frac{\mu}{\rho})(x_i) - \sum_{i=1}^N \log\rho(x_i)\right) d\mu^{\otimes N}(X_N).
\end{aligned}$$

We then observe that by definition of $\mathcal{H}_N$,

$$\begin{aligned}
&\int_{\widehat{\mu_N}\in B(\mu,\varepsilon)} \left(-\theta N^{-1}\mathcal{H}_N(X_N) - \sum_{i=1}^N (\log\frac{\mu}{\rho})(x_i) - \sum_{i=1}^N \log\rho(x_i)\right) d\mu^{\otimes N}(X_N)\\
&= -\theta \int_{\widehat{\mu_N}\in B(\mu,\varepsilon)} \left(\frac{1}{2N}\sum_{i\neq j}\mathsf{g}(x_i-x_j) + \sum_{i=1}^N V(x_i)\right.\\
&\qquad\qquad \left. + \frac{1}{\theta}\sum_{i=1}^N (\log\frac{\mu}{\rho})(x_i) + \frac{1}{\theta}\sum_{i=1}^N \log\rho(x_i)\right) d\mu^{\otimes N}(X_N)\\
&= N\log\bar{u} - \theta\int_{\widehat{\mu_N}\in B(\mu,\varepsilon)} \left(\frac{1}{2N}\sum_{i\neq j}\mathsf{g}(x_i-x_j) - \sum_{i=1}^N \mathsf{g}_-(x_i)\right.\\
&\qquad\qquad \left. + \frac{1}{\theta}\sum_{i=1}^N (\log\frac{\mu}{\rho})(x_i)\right) d\mu^{\otimes N}(X_N).
\end{aligned}$$

We next rewrite the integral in the right-hand side as the difference between the integral over $(\mathbb{R}^\mathsf{d})^N$ and the integral over $\{\widehat{\mu_N}\notin B(\mu,\varepsilon)\}$. For the integral over the whole space, expanding all the terms in the sums, we find that

(3.2.12)
$$\begin{aligned}
&-\theta\int_{(\mathbb{R}^\mathsf{d})^N} \left(\frac{1}{2N}\sum_{i\neq j}\mathsf{g}(x_i-x_j) - \sum_{i=1}^N \mathsf{g}_-(x_i) + \frac{1}{\theta}\sum_{i=1}^N (\log\frac{\mu}{\rho})(x_i)\right) d\mu^{\otimes N}(X_N)\\
&= -\theta\left(\frac{N(N-1)}{2N}\iint_{\mathbb{R}^\mathsf{d}\times\mathbb{R}^\mathsf{d}} \mathsf{g}(x-y)d\mu(x)d\mu(y) - N\int_{\mathbb{R}^\mathsf{d}} \mathsf{g}_- d\mu + \frac{N}{\theta}\int_{\mathbb{R}^\mathsf{d}} \mu\log\frac{\mu}{\rho}\right)\\
&= -\theta N\mathcal{E}_\theta(\mu) + \frac{\theta}{2}\iint \mathsf{g}(x-y)d\mu(x)d\mu(y) + \theta N\int_{\mathbb{R}^\mathsf{d}} (V+\mathsf{g}_-)d\mu + N\int \mu\log\rho.
\end{aligned}$$

If $\mathsf{s}>0$ we bound the term $\iint \mathsf{g}(x-y)d\mu(x)d\mu(y)$ below by 0, while if $\mathsf{s}\le 0$, we use (2.1.1) and the assumption $\int|\mathsf{g}_-|d\mu<\infty$ to bound it below by $-C$.

Using (2.1.1) we also find that

$$\begin{aligned}
&\theta\int_{\widehat{\mu_N}\notin B(\mu,\varepsilon)} \left(\frac{1}{2N}\sum_{i\neq j}\mathsf{g}(x_i-x_j) - \sum_{i=1}^N \mathsf{g}_-(x_i) + \frac{1}{\theta}\sum_{i=1}^N (\log\frac{\mu}{\rho})(x_i)\right) d\mu^{\otimes N}(X_N)\\
&\ge \theta\int_{\widehat{\mu_N}\notin B(\mu,\varepsilon)} \sum_{i=1}^N \left(-C + \frac{1}{\theta}\log\frac{\mu}{\rho}\right)(x_i) d\mu^{\otimes N}(X_N)\\
&\ge -C\theta N\int_{\widehat{\mu_N}\notin B(\mu,\varepsilon)} d\mu(x_1)\dots d\mu(x_N),
\end{aligned}$$

where we used that by assumption μ/ρ is bounded below independently of N. By the law of large numbers, we have

$$\lim_{N\to\infty}\int_{\widehat{\mu_N}\notin B(\mu,\varepsilon)} d\mu(x_1)\dots d\mu(x_N) = 0,$$

hence the right-hand side above is $o(N\theta)$ and also $\mu^{\otimes N}(\{\widehat{\mu_N}\in B(\mu,\varepsilon)\}) = 1+o_N(1)$.

Combining all the above elements, we have obtained that

(3.2.13)

$$\begin{aligned}\log \mathbb{P}_{N,\beta}(\widehat{\mu_N} \in B(\mu,\varepsilon)) \\ &\geq -\log Z_{N,\beta} - \theta N \mathcal{E}_\theta(\mu) \\ &\quad + (1+o_N(1))\left(N\log \bar{u} + \theta N \int_{\mathbb{R}^{\mathsf{d}}} (V+\mathsf{g}_-)d\mu + N\int \mu \log \rho + o(\theta N)\right) \\ &= -\log Z_{N,\beta} - \theta N \mathcal{E}_\theta(\mu) + o(\theta N).\end{aligned}$$

We then retrace the same steps and use that $\mu_\theta > 0$ as proven in Proposition 2.21, as well as the result of Lemma 2.20, to rewrite

$$\begin{aligned}Z_{N,\beta} = \int_{(\mathbb{R}^{\mathsf{d}})^N} \exp\Bigg(&-\theta N^{-1}\mathcal{H}_N(X_N) \\ &- \sum_{i=1}^N (\log \frac{\mu_\theta}{\rho})(x_i) - \sum_{i=1}^N \log \rho(x_i)\Bigg) d\mu_\theta^{\otimes N}(X_N).\end{aligned}$$

Using Jensen's inequality we obtain in lieu of (3.2.11)

$$\begin{aligned}\log Z_{N,\beta} \geq \int_{(\mathbb{R}^{\mathsf{d}})^N} \Bigg(&-\theta N^{-1}\mathcal{H}_N(X_N) \\ &- \sum_{i=1}^N (\log \frac{\mu_\theta}{\rho})(x_i) - \sum_{i=1}^N \log \rho(x_i)\Bigg) d\mu_\theta^{\otimes N}(X_N).\end{aligned}$$

Proceeding as above and using (3.2.12), we deduce that

$$\log Z_{N,\beta} \geq -\theta N \mathcal{E}_\theta(\mu_\theta) + o(\theta N). \tag{3.2.14}$$

By minimality of μ_θ for $\mathcal{E}_\theta$, we may also write that

$$\mathcal{E}_\theta(\mu_\theta) \leq \mathcal{E}_\theta(\mu_V) = \mathcal{E}(\mu_V) + O(\theta^{-1})$$

hence we have obtained that if $\theta \to +\infty$ as $N \to \infty$,

$$\log Z_{N,\beta} \geq -\theta N \mathcal{E}(\mu_V) + o(\theta N). \tag{3.2.15}$$

Step 2: Large deviations upper bound. Let us start with the case θ fixed. Let $\mu \in \mathcal{P}(\mathbb{R}^{\mathsf{d}})$ such that μ/ρ is continuous and bounded below. By definition of $\mathcal{H}_N$ and of ρ, we may write

(3.2.16)

$$\begin{aligned}&\mathbb{P}_{N,\beta}(\widehat{\mu_N} \in B(\mu,\varepsilon)) \\ &\leq \frac{1}{Z_{N,\beta}} \int_{\widehat{\mu_N}\in B(\mu,\varepsilon)} \exp\Bigg(-\theta\Bigg(\frac{1}{2N}\sum_{i\neq j} \mathsf{g}(x_i-x_j) \\ &\qquad + \sum_{i=1}^N V(x_i) + \frac{1}{\theta}\sum_{i=1}^N \log \mu(x_i)\Bigg)\Bigg) d\mu^{\otimes N}(X_N) \\ &= \frac{1}{Z_{N,\beta}} \int_{\widehat{\mu_N}\in B(\mu,\varepsilon)} \exp\Bigg(-\theta\Bigg(\frac{1}{2N}\sum_{i\neq j} \mathsf{g}(x_i-x_j) \\ &\qquad - \sum_{i=1}^N \mathsf{g}_-(x_i) + \frac{1}{\theta}\sum_{i=1}^N \log \frac{\mu}{\bar{u}\rho}(x_i)\Bigg)\Bigg) d\mu^{\otimes N}(X_N).\end{aligned}$$

Applying Proposition 3.5 to the external potential $-\mathsf{g}_- + \frac{1}{\theta}\log\frac{\mu}{\bar{u}\rho}$ which is l.s.c. bounded below, and the lower semicontinuity of a Γ-liminf, we deduce that if $\widehat{\mu_N} \in B(\mu,\varepsilon)$ we have

$$\liminf_{N\to\infty}\frac{1}{N^2}\left(\frac{1}{2}\sum_{i\neq j}\mathsf{g}(x_i-x_j)+N\sum_{i=1}^N(-\mathsf{g}_-+\frac{1}{\theta}\log\frac{\mu}{\bar{u}\rho})(x_i)\right)$$
$$\geq \frac{1}{2}\iint_{\mathbb{R}^\mathsf{d}\times\mathbb{R}^\mathsf{d}}\mathsf{g}(x-y)d\mu(x)d\mu(y)+\int_{\mathbb{R}^\mathsf{d}}(-\mathsf{g}_-+\frac{1}{\theta}\log\frac{\mu}{\bar{u}\rho})d\mu+o_\varepsilon(1)=\mathcal{E}_\theta(\mu)+o_\varepsilon(1).$$

Therefore, since $-\mathsf{g}_- + \frac{1}{\theta}\log\frac{u}{\bar{u}\rho} = V + \frac{1}{\theta}\log\mu$, we have

$$\log \mathbb{P}_{N,\beta}(\widehat{\mu_N}\in B(\mu,\varepsilon))$$
$$\leq \log\left(\frac{1}{Z_{N,\beta}}\int_{\widehat{\mu_N}\in B(\mu,\varepsilon)}\exp\left(-\theta N(\mathcal{E}_\theta(\mu)+o_{\varepsilon,N}(1))\right)d\mu(x_1)\dots d\mu(x_N)\right)$$
$$+\theta N o_\varepsilon(1).$$

Applying Sanov's theorem (Theorem 3.16) for i.i.d. random variables of law μ we have on the other hand that

$$\limsup_{N\to\infty}\frac{1}{N}\log\left(\int_{\widehat{\mu_N}\in B(\mu,\varepsilon)}\mu(x_1)\dots\mu(x_N)\right)\leq -\inf_{\nu\in\bar{B}(\mu,\varepsilon)}\int\nu\log\frac{\nu}{\mu}\leq 0,$$

thus we conclude in that regime (θ fixed) that

$$\log\mathbb{P}_{N,\beta}(\widehat{\mu_N}\in B(\mu,\varepsilon))\leq -\log Z_{N,\beta}-\theta N\mathcal{E}_\theta(\mu)+o(N)+\theta N o_\varepsilon(1). \tag{3.2.17}$$

Let us now turn to the regime $\theta\to+\infty$. Using (3.2.9) and recalling the definitions in (3.2.9), for any event E, we may write

(3.2.18)
$$\mathbb{P}_{N,\beta}(\widehat{\mu_N}\in E)$$
$$\leq\frac{1}{Z_{N,\beta}}\int_{\widehat{\mu_N}\in E}\exp\Bigg(-\theta\Bigg(\frac{1}{2N}\sum_{i\neq j}\mathsf{g}(x_i-x_j)$$
$$+\sum_{i=1}^N V(x_i)+\frac{1}{\theta}\sum_{i=1}^N\log\rho_0(x_i)\Bigg)\Bigg)d\rho_0^{\otimes N}(X_N)$$
$$=\frac{(\bar{u}_0)^N}{Z_{N,\beta}}\int_{\widehat{\mu_N}\in E}\exp\Bigg(-\theta\Bigg(\frac{1}{2N}\sum_{i\neq j}\mathsf{g}(x_i-x_j)$$
$$+\sum_{i=1}^N(1-\frac{\theta_0}{\theta})V(x_i)-\frac{\theta_0}{\theta}\mathsf{g}_-(x_i)\Bigg)\Bigg)d\rho_0^{\otimes N}(X_N).$$

Applying Proposition 3.5 to the potential $(1-\frac{\theta_0}{\theta})V$ (it depends on N but this is harmless in the proof) and using $\theta\to+\infty$, we deduce that if $\widehat{\mu_N}\in E$ we have

$$\liminf_{N\to\infty}\frac{1}{N^2}\left(\frac{1}{2}\sum_{i\neq j}\mathsf{g}(x_i-x_j)+(1-\frac{\theta_0}{\theta})\sum_{i=1}^N V(x_i)\right)$$
$$\geq\inf_{\mu\in\bar{E}}\left(\frac{1}{2}\iint_{\mathbb{R}^\mathsf{d}\times\mathbb{R}^\mathsf{d}}\mathsf{g}(x-y)d\mu(x)d\mu(y)+\int_{\mathbb{R}^\mathsf{d}}Vd\mu\right)=\inf_{\mu\in\bar{E}}\mathcal{E}(\mu).$$

Therefore,

$$\mathbb{P}_{N,\beta}(\widehat{\mu_N} \in E) \le \frac{(\bar{u}_0)^N}{Z_{N,\beta}} \int_{\widehat{\mu_N}\in E} \exp\left(-\theta N(\inf_{\bar{E}} \mathcal{E} + o_N(1))\right) d\rho_0(x_1)\dots d\rho_0(x_N).$$

Applying Sanov's theorem for i.i.d. random variables of law ρ_0 we obtain on the other hand that

$$\log\left(\int_{\widehat{\mu_N}\in E} d\rho_0(x_1)\dots d\rho_0(x_N)\right) \le -N \inf_{\mu\in\bar{E}} \int_{\mathbb{R}^\mathsf{d}} \mu \log\frac{\mu}{\rho_0} + o(N) \le O(N) = o(\theta N)$$

since $\theta \to +\infty$, and we conclude that

$$\log \mathbb{P}_{N,\beta}(\widehat{\mu_N} \in E) \le -\log Z_{N,\beta} - \theta N \inf_{\bar{E}} \mathcal{E} + o(\theta N). \tag{3.2.19}$$

Step 3: Exponential tightness. Let us first consider the case θ fixed. For any event $E \subset \mathcal{P}(\mathbb{R}^\mathsf{d})$, rewriting as above, we have

$$\begin{aligned}
&\mathbb{P}_{N,\beta}(\widehat{\mu_N} \in E)\\
&= \frac{1}{Z_{N,\beta}} \int_{\widehat{\mu_N}\in E} \exp\Bigg(-\theta N^{-1}\Big(\frac{1}{2}\sum_{i\ne j} \mathsf{g}(x_i - x_j) + N\sum_{i=1}^N V(x_i)\Big)\\
&\qquad\qquad - \sum_{i=1}^N \log\rho(x_i)\Bigg) d\rho^{\otimes N}(X_N)\\
&= \frac{(\bar{u})^N}{Z_{N,\beta}} \int_{\widehat{\mu_N}\in E} \exp\left(-\theta\left(\frac{1}{2N}\sum_{i\ne j} \mathsf{g}(x_i - x_j) - \sum_{i=1}^N \mathsf{g}_-(x_i)\right)\right) d\rho^{\otimes N}(X_N).
\end{aligned}$$

In view of (2.1.1) we may find that

$$\frac{1}{2N}\sum_{i\ne j} \mathsf{g}(x_i - x_j) - \sum_{i=1}^N \mathsf{g}_-(x_i) \ge \frac{1}{2N}\sum_{i=1}^N\sum_{j=1}^N \mathsf{g}(x_i - x_j)_- - \sum_{i=1}^N \mathsf{g}_-(x_i) \ge -CN,$$

so using (3.2.14) to bound $Z_{N,\beta}$, we have obtained that

$$\mathbb{P}_{N,\beta}(\widehat{\mu_N} \in E) \le \exp(CN\theta) \int_{\widehat{\mu_N}\in E} d\rho(x_1)\dots d\rho(x_N). \tag{3.2.20}$$

By Sanov's theorem, Theorem 3.16, applied to i.i.d. random variables of law ρ, we have

$$\limsup_{N\to\infty} \frac{1}{N}\log \int_{\widehat{\mu_N}\in E} d\rho(x_1)\dots d\rho(x_N) \le -\inf_{\nu\in\bar{E}} \int_{\mathbb{R}^\mathsf{d}} \nu\log\frac{\nu}{\rho}.$$

We know that the relative entropy is a good rate function in Sanov's theorem, hence its sublevel sets $K_M = \{\nu \in \mathcal{P}(\mathbb{R}^\mathsf{d}), \int_{\mathbb{R}^\mathsf{d}} \nu\log\frac{\nu}{\rho} \le M\}$ are compact. Applying to $E = (K_M)^c$ we thus have that

$$\limsup_{N\to\infty} \frac{1}{N}\log \mathbb{P}_{N,\beta}(\widehat{\mu_N} \in (K_M)^c) \le -M + C,$$

which proves exponential tightness at speed N (or $N\theta$) in this regime.

In the case $\theta \to +\infty$, exponential tightness is in the same way a direct consequence of (3.2.19) since $\mathcal{E}$ has compact sublevel sets, as seen in Lemma 2.4.

In view of Corollary 3.13, to get an LDP it thus suffices to prove the LDP lower and upper bounds for balls. It also suffices to prove it for balls centered on a dense set, which we take to be the μ's such that $\mu > 0$ is continuous, $\frac{\mu}{\rho}$ is bounded below independently of N and $\int |g_-| d\mu < \infty$. The LDP upper and lower bound for balls thus follows from Steps 1 and 2.

Step 4. Conclusion. Let us first consider the case where θ is fixed. In view of the exponential tightness, we can extend the results of Steps 1 and 2 to arbitrary sets, and not only balls. In particular, applying (3.2.17) to the whole space $\mathcal{P}(\mathbb{R}^d)$ we obtain

$$\log Z_{N,\beta} \leq -\theta N \inf \mathcal{E}_\theta + o(N) + o(\theta N). \tag{3.2.21}$$

Comparing with (3.2.14) we have

$$\log Z_{N,\beta} = -\theta N \mathcal{E}_\theta(\mu_\theta) + o(N) + o(\theta N), \tag{3.2.22}$$

which proves (3.2.7). Inserting into (3.2.13) and (3.2.17), we conclude the proof in this regime.

Let us now consider the case where $\theta \to +\infty$ as $N \to \infty$. Then we note that by definition of $\mathcal{E}_\theta$, $\mathcal{E}_\theta(\mu) = \mathcal{E}(\mu) + o_N(1)$. We can then apply (3.2.19) with E equal to the whole space $\mathcal{P}(\mathbb{R}^d)$ and obtain

$$\log Z_{N,\beta} \leq -\theta N \min \mathcal{E} + o(\theta N).$$

Moreover, (3.2.15) yields the converse inequality, hence (3.2.6) holds. Inserting again into (3.2.13) and (3.2.17) we conclude the proof in that regime. □

Part 2

Modulated electric energy

CHAPTER 4

The modulated electric energy

In this chapter, we introduce and study the next order or "modulated energy" $\mathsf{F}_N(X_N, \mu)$, which naturally appears in the study of Coulomb and Riesz gases, but also in dynamics questions, where it acquired its name "modulated." It will play a crucial role in the remainder of the text. F_N corresponds to the total Coulomb/Riesz interaction of the system of points at X_N neutralized by the background charge $N\mu$. It appears as a next-order energy after an exact "splitting" of $\mathcal{H}_N$ that we present at the beginning of Chapter 5, for μ equal to the equilibrium or thermal equilibrium measure. In this chapter, we focus on F_N and study it for its own sake. Thanks to an electric rewriting as an integral of the form $\int_{\mathbb{R}^\mathsf{d}} |\nabla h|^2$ (for some electric potential h) similar to that of Section 2.2.2, we show the main results that F_N is bounded below, behaves as an effective Riesz distance between the empirical measure and the reference measure μ, and controls discrepancies and fluctuations of linear statistics with respect to μ. Moreover, thanks to the electric rewriting, F_N can be seen as an extensive quantity which *can be localized*, as opposed to a sum of pair interactions. The main tools, that we present in this chapter, are a *renormalization by truncation* of the electric formulation $\int_{\mathbb{R}^\mathsf{d}} |\nabla h|^2$, and a *monotonicity property* with respect to the truncation parameters. The first applications to statistical mechanics will follow in the next chapter and applications to dynamics in Chapter 6.

For $0 < \alpha \le 1$ we let $|\varphi|_{C^\alpha}$ denote the Hölder seminorm

$$|\varphi|_{C^\alpha} = \sup_{x,y} \frac{|\varphi(x) - \varphi(y)|}{|x-y|^\alpha}, \tag{4.0.1}$$

and for $0 < \alpha \le 1$ and k integer, we let $|\varphi|_{C^{k+\alpha}}$ denote the C^α seminorm of $\nabla^{\otimes k}\varphi$, which we will use throughout the text. Note that $\varphi \in C^1$ does not mean that φ is continuously differentiable, but rather that it is Lipschitz.

4.1. Definition and electric representation

4.1.1. Definition. In the whole chapter, we will consider a reference measure (most often a probability) μ and assume that

$$\iint_{\mathbb{R}^\mathsf{d} \times \mathbb{R}^\mathsf{d}} |\mathsf{g}(x-y)| d|\mu|(x) d|\mu|(y) < +\infty. \tag{4.1.1}$$

Definition 4.1 (The modulated energy). Let g be as in (2.0.1). We define for any integrable μ with $\int_{\mathbb{R}^\mathsf{d}} \mu = 1$ satisfying (4.1.1) the modulated energy of the configuration $X_N = (x_1, \ldots, x_N) \in (\mathbb{R}^{\mathsf{d}})^N$ with respect to μ as

(4.1.2)

$$\mathsf{F}_N(X_N, \mu) := \frac{1}{2} \iint_{\mathbb{R}^{\mathsf{d}} \times \mathbb{R}^{\mathsf{d}} \setminus \triangle} \mathsf{g}(x-y)\, d\left(\sum_{i=1}^N \delta_{x_i} - N\mu\right)(x) d\left(\sum_{i=1}^N \delta_{x_i} - N\mu\right)(y),$$

where we recall $\triangle$ denotes the diagonal of $\mathbb{R}^{\mathsf{d}} \times \mathbb{R}^{\mathsf{d}}$.

We defined $\mathsf{F}_N(X_N, \mu)$ for a generic density μ of integral 1. We consider this level of generality at this point because for certain questions pertinent to dynamics and outside the scope of this text, it is useful to define F_N for signed measures μ, however in the situations that we will consider here, μ will in fact be a probability density: in the following chapters on Coulomb and Riesz gases, it will either be the equilibrium measure μ_V or the thermal equilibrium measure μ_θ, according to the choice of approach. We will also soon add the assumption that the density μ is bounded, which is satisfied for the thermal equilibrium measure by Proposition 2.21, and which is true for the equilibrium measure under assumptions of regularity for V; see for instance Example 2.13.

The measure μ should be thought of as a reference measure or neutralizing background charge. This initial formulation of F_N shows it as a sum of pair interactions, or total interaction of the system of charges at x_i's and negative charge $-N\mu$.

We note that another natural normalization of F_N is to consider $\frac{1}{N^2}\mathsf{F}_N$. This way F_N can naturally be thought of as a Coulomb or Riesz (squared) distance between the empirical measure $\widehat{\mu_N} = \frac{1}{N}\sum_{i=1}^N \delta_{x_i}$ and the measure μ. We will see that even though F_N is not necessarily positive, this metric interpretation is approximately correct. In fact, if the diagonal $\triangle$ was not removed, then after Fourier transform $\mathcal{F}$, F_N could be seen as

$$\begin{aligned}
\frac{\mathsf{F}_N(X_N, \mu)}{N^2} &= \int_{\mathbb{R}^{\mathsf{d}}} \mathcal{F}\mathsf{g}(\xi) |\mathcal{F}(\widehat{\mu_N} - \mu)(\xi)|^2 \\
&= C_{\mathsf{d},\mathsf{s}} \int_{\mathbb{R}^{\mathsf{d}}} \frac{1}{|\xi|^{\mathsf{d}-\mathsf{s}}} |\mathcal{F}(\widehat{\mu_N} - \mu)(\xi)|^2 \\
&= C_{\mathsf{d},\mathsf{s}} \|\widehat{\mu_N} - \mu\|^2_{\dot{H}^{\frac{\mathsf{s}-\mathsf{d}}{2}}(\mathbb{R}^{\mathsf{d}})}
\end{aligned} \tag{4.1.3}$$

for some constant $C_{\mathsf{d},\mathsf{s}}$, as in (2.2.8) and (2.2.9). Of course this is not really correct, since Dirac masses generally do not belong to the space $\dot{H}^{\frac{\mathsf{s}-\mathsf{d}}{2}}$. However, this can be given a meaning in a renormalized sense, either by removing the diagonal as done in the definition of F_N, or via truncations as we will see just below. One could also use the term *desingularization*, as in the desingularization of vortices performed in fluid mechanics.

4.1.2. Electric formulation in the Coulomb case. To go further, we use an electrostatic interpretation of the energy F_N, as first used in [**SS15b**], and the rewriting of the energy via truncation, as in [**RS15, PS17**] but using the nearest-neighbor distance truncation as in [**LSZ17, LS18**]. In this section, we first present it in the easier Coulomb case.

Such a computation allows to replace the sum of pairwise interactions of all the charges and background μ by an integral (extensive) quantity, which is easier to handle and can be localized.

Let us first consider the potential generated by the configuration X_N and the background μ, defined by

$$h_N^\mu[X_N](x) := \int_{\mathbb{R}^\mathsf{d}} \mathsf{g}(x-y)d\left(\sum_{i=1}^N \delta_{x_i} - N\mu\right)(y). \tag{4.1.4}$$

We will most often omit the dependence in X_N and μ and simply write h_N when there is no ambiguity. In the Coulomb case, g is (up to the constant c_d), the fundamental solution to Laplace's equation in dimension d, that is $-\Delta \mathsf{g} = \mathsf{c}_\mathsf{d}\delta_0$, so we have

$$-\Delta h_N = \mathsf{c}_\mathsf{d}\left(\sum_{i=1}^N \delta_{x_i} - N\mu\right). \tag{4.1.5}$$

We note that as in Proposition 2.16 and its proof, even if $\mathsf{d} = 1, 2$ where g does not tend to 0 at infinity, $h_N^\mu[X_N]$ always decays at infinity because $\int \mu = 1$ and the system formed by the positive charges at x_i and the background charge $N\mu$ is neutral.

We would like to rewrite $\mathsf{F}_N(X_N, \mu)$ defined in (4.1.2) via Green's formula as

$$\mathsf{F}_N(X_N, \mu) = \frac{1}{2\mathsf{c}_\mathsf{d}}\int_{\mathbb{R}^\mathsf{d}} h_N(-\Delta h_N) = \frac{1}{2\mathsf{c}_\mathsf{d}}\int_{\mathbb{R}^\mathsf{d}} |\nabla h_N|^2, \tag{4.1.6}$$

exactly as done in Proposition 2.16, with the boundary terms at infinity vanishing thanks to the above noted decay. This rewriting of the energy of electrostatic charges as the Dirichlet energy of the potential is the well known operation of "carré du champ" [**Cha**, **Hir78**].

However this formal computation is not correct due to the singularities of h_N at the points x_i, which make the integral diverge if $\mathsf{s} \geq 0$, or equivalently due to the fact that this computation neglects the issue of the diagonal.

To remedy this, we use a truncation procedure which allows to give a renormalized meaning to this integral. Giving an electric formulation in a renormalized fashion was first done in the context of Coulomb gases in [**SS15b**], inspired by [**BBH94**] in the context of Ginzburg-Landau vortices (see also [**Ser15**] for a discussion). Using a mollification to make sense of this divergent integral is a natural idea. In the Coulomb case, what we describe below is another avatar of Onsager's lemma (see [**LS10**, **RS15**]). However, here we make use of two specific points: first we do not mollify the interaction but only the singular charge distribution (and not the background charge μ). Second we use a regularization scale η_i which may depend on the point x_i hence on the whole configuration X_N. This will provide more flexibility, in particular when two points get very close. More specifically, we will crucially use the nearest-neighbor distance as a truncation parameter, as first used in [**LS18**] and inspired by [**GP77**]. We denote by $\mathbb{R}_+$ the set of nonnegative real numbers.

Definition 4.2 (Truncated potentials, Coulomb case)**.** For any number $\eta > 0$, let us denote

$$\mathsf{f}_\eta(x) := (\mathsf{g}(x) - \mathsf{g}(\eta))_+, \qquad \mathsf{g}_\eta := \mathsf{g} - \mathsf{f}_\eta, \tag{4.1.7}$$

where $(\cdot)_+$ denotes the positive part of a number, and we naturally also view g as a function of $\mathbb{R}$, i.e., $\mathsf{g}(\eta)$ means $\frac{1}{\mathsf{s}}\eta^{-\mathsf{s}}$ or $-\log\eta$.

For any $\vec{\eta} = (\eta_1, \ldots, \eta_N) \in \mathbb{R}_+^N$, and any function h satisfying a relation of the form

$$-\Delta h = \mathsf{c_d} \left(\sum_{i=1}^N \delta_{x_i} - N\mu \right), \tag{4.1.8}$$

we define the truncated potential

$$h_{\vec{\eta}} := h - \sum_{i=1}^N \mathsf{f}_{\eta_i}(\cdot - x_i). \tag{4.1.9}$$

Let us point out that f_η is supported in $B(0,\eta)$, and that $\mathsf{g}_\eta = \min(\mathsf{g}, \mathsf{g}(\eta))$ is a truncation of the Coulomb kernel. We note here that we could choose more regular truncated potentials, for instance any g_η which is smooth and radial and coincides with g outside $B(0,\eta)$. The one we choose has the advantage of being constant in $B(0,\eta)$.

Definition 4.3 (Smeared charges, Coulomb case)**.** We denote

$$\delta_{x_0}^{(\eta)} := -\frac{1}{\mathsf{c_d}} \Delta \mathsf{g}_\eta(\cdot - x_0). \tag{4.1.10}$$

It is the uniform measure of mass 1 supported on $\partial B(x_0, \eta)$.

Since g_η is constant in $B(0,\eta)$, and equal to g, which is harmonic, outside $B(0,\eta)$, the distribution $\Delta \mathsf{g}_\eta$ can only be supported on $\partial B(0,\eta)$. By radial symmetry it must have a uniform density on $\partial B(0,\eta)$. To check that $\int \delta_0^{(\eta)} = 1$, one can either compute it explicitly or use Green's formula to say that for $R > \eta$,

$$\int_{\partial B(0,R)} \frac{\partial \mathsf{g}}{\partial \nu} = \int_{\partial B(0,R)} \frac{\partial \mathsf{g}_\eta}{\partial \nu} = \int_{B(0,R)} \Delta \mathsf{g} = \int_{B(0,R)} \Delta \mathsf{g}_\eta,$$

thus in view of (1.1.11) and (4.1.10), we conclude that $\int_{\mathbb{R}^\mathsf{d}} d\delta_0^{(\eta)} = 1$ and $\int_{\mathbb{R}^\mathsf{d}} d\delta_{x_0}^{(\eta)} = 1$.

We may next notice that in view of (4.1.10) and the fact that g_η and g coincide outside $B(0,\eta)$, we have

$$\mathsf{g}_\eta = \mathsf{g} * \delta_0^{(\eta)}, \tag{4.1.11}$$

and thus by (4.1.7),

$$\mathsf{f}_\eta = \mathsf{g} * \left(\delta_0 - \delta_0^{(\eta)} \right). \tag{4.1.12}$$

This is a rephrasing of what physicists call Newton's theorem: the Coulomb potential generated by a point charge and the Coulomb potential generated by a radial smearing of that point charge coincide outside the smearing region (here, g and g_η coincide outside $B(0,\eta)$). Moreover, the Coulomb potential of the smeared charge is smaller in the smearing region than the original one (here $\mathsf{g}_\eta \le \mathsf{g}$ everywhere). Mathematically, this is a consequence of the mean-value (in)equality for (sub)harmonic functions.

Taking the Laplacian of (4.1.12), it follows that

$$-\Delta \mathsf{f}_\eta = \mathsf{c_d} \left(\delta_0 - \delta_0^{(\eta)} \right). \tag{4.1.13}$$

We note that in view of (4.1.13) and (4.1.12), the function $h_{N,\vec{\eta}}$ defined via (4.1.4) and (4.1.9), i.e.,

$$h_{N,\vec{\eta}} = h_N - \sum_{i=1}^N \mathsf{f}_\eta(\cdot - x_i)$$

then satisfies

$$h_{N,\vec{\eta}} = \mathsf{g} * \left(\sum_{i=1}^N \delta_{x_i}^{(\eta_i)} - N\mu\right)$$

and

$$-\Delta h_{N,\vec{\eta}} = \mathsf{c_d}\left(\sum_{i=1}^N \delta_{x_i}^{(\eta_i)} - N\mu\right). \tag{4.1.14}$$

In the case $\mathsf{s} \le 0$, which we only consider in dimension $\mathsf{d} = 1$, the renormalization is not needed at all—however, we can still do it (for the sake of uniformity of notation).

We can then show the following exact representation for F_N, which may be viewed as a "renormalized" version of Proposition 2.16.

Lemma 4.4 (Electric representation of the modulated energy—Coulomb case). *Assume* $\mathsf{s} = \mathsf{d} - 2$. *Let* μ *be integrable such that* $\int_{\mathbb{R}^\mathsf{d}} \mu = 1$ *and* (4.1.1) *holds. For any* X_N *in* $(\mathbb{R}^\mathsf{d})^N$ *pairwise distinct configuration, we have*

$$\mathsf{F}_N(X_N, \mu) = \frac{1}{2\mathsf{c_d}} \lim_{\eta_i \to 0} \left(\int_{\mathbb{R}^\mathsf{d}} |\nabla h_{N,\vec{\eta}}|^2 - \mathsf{c_d} \sum_{i=1}^N \mathsf{g}(\eta_i) \right). \tag{4.1.15}$$

PROOF. We apply the result of Proposition 2.16 to $\mu_+ = \frac{1}{N}\sum_{i=1}^N \delta_{x_i}^{(\eta_i)}$ and $\mu_- = \mu$. Thanks to assumption (4.1.1), the assumptions of that proposition are satisfied and we deduce that

(4.1.16)

$$\iint_{\mathbb{R}^\mathsf{d}\times\mathbb{R}^\mathsf{d}} \mathsf{g}(x-y)d\left(\sum_{i=1}^N \delta_{x_i}^{(\eta_i)} - N\mu\right)(x)d\left(\sum_{i=1}^N \delta_{x_i}^{(\eta_i)} - N\mu\right)(y) = \frac{1}{\mathsf{c_d}} \int_{\mathbb{R}^\mathsf{d}} |\nabla h_{N,\vec{\eta}}|^2.$$

Next, we observe that by translation-invariance

$$\iint \mathsf{g}(x-y)d\delta_{x_i}^{(\eta_i)}(x)d\delta_{x_i}^{(\eta_i)}(y) = \iint \mathsf{g}(x-y)d\delta_0^{(\eta_i)}(x)d\delta_0^{(\eta_i)}(y) = \int \mathsf{g}_{\eta_i} d\delta_0^{(\eta_i)} = \mathsf{g}(\eta_i).$$

Here we have used (4.1.11) and the fact that $\mathsf{g}_\eta = \mathsf{g}$ on $\partial B(0,\eta)$. We may thus write that

$$\lim_{\vec{\eta}\to 0}\left[\iint_{\mathbb{R}^\mathsf{d}\times\mathbb{R}^\mathsf{d}} \mathsf{g}(x-y)d\left(\sum_{i=1}^N \delta_{x_i}^{(\eta_i)} - N\mu\right)(x)d\left(\sum_{i=1}^N \delta_{x_i}^{(\eta_i)} - N\mu\right)(y) - \sum_{i=1}^N \mathsf{g}(\eta_i)\right]$$
$$= \iint_{\triangle^c} \mathsf{g}(x-y)d\left(\sum_{i=1}^N \delta_{x_i} - N\mu\right)(x)d\left(\sum_{i=1}^N \delta_{x_i} - N\mu\right)(y)$$

and we deduce in view of (4.1.16) that (4.1.15) holds. □

4.1.3. Electric formulation in the Riesz case. In the Riesz or one-dimensional logarithmic case, the above representation is not true because g is not the fundamental solution to the Laplacian. However, we can use the extension representation as described in Section 2.2.1 to view g as the fundamental solution of a divergence-form operator in an extended space. This is the approach that was proposed in the one-dimensional log case in [**SS15a**], and in [**PS17**] for the more general Riesz cases (2.0.1). Let us go through the details, using the notation of Section 2.2.1. We recall that we use an extension to $\mathbb{R}^{\mathsf{d}+\mathsf{k}}$ where $\mathsf{k}=0$ in the Coulomb case $\mathsf{s}=\mathsf{d}-2$ and $\mathsf{k}=1$ in all other Riesz cases. The most important fact is that for any distribution f on $\mathbb{R}^{\mathsf{d}}$, the potential

$$\mathsf{g}*f(x)=\int_{\mathbb{R}^{\mathsf{d}}}\mathsf{g}(x-x')f(x')dx'$$

can naturally be extended into a potential on $\mathbb{R}^{\mathsf{d}+1}$ (if $\mathsf{k}=1$)

$$h^f(x,y):=\int_{\mathbb{R}^{\mathsf{d}}}\mathsf{g}((x,y)-(x',0))f(x')dx'$$

which satisfies

$$-\mathrm{div}\,(|y|^\gamma\nabla h^f)=\mathsf{c}_{\mathsf{d},\mathsf{s}}f\delta_{\mathbb{R}^{\mathsf{d}}}\quad\text{in }\mathbb{R}^{\mathsf{d}+\mathsf{k}},\tag{4.1.17}$$

where $\mathsf{c}_{\mathsf{d},\mathsf{s}}$ is defined in (1.1.13) and γ is given by (2.2.5).

Definition 4.5 (Truncated potentials)**.** For any number $\eta>0$, we denote

$$\mathsf{f}_\eta:=(\mathsf{g}(x)-\mathsf{g}(\eta))_+,\qquad \mathsf{g}_\eta:=\mathsf{g}-\mathsf{f}_\eta.\tag{4.1.18}$$

We also naturally extend $\mathsf{g},\mathsf{g}_\eta$ and f_η into radial functions of $\mathbb{R}^{\mathsf{d}+\mathsf{k}}$. For any $\vec{\eta}=(\eta_1,\dots,\eta_N)\in\mathbb{R}_+^N$, and any function h satisfying a relation of the form

$$-\mathrm{div}\,(|y|^\gamma\nabla h)=\mathsf{c}_{\mathsf{d},\mathsf{s}}\left(\sum_{i=1}^N\delta_{x_i}-N\mu\delta_{\mathbb{R}^{\mathsf{d}}}\right)\quad\text{in }\mathbb{R}^{\mathsf{d}+\mathsf{k}},\tag{4.1.19}$$

we define the truncated potential

$$h_{\vec{\eta}}:=h-\sum_{i=1}^N\mathsf{f}_{\eta_i}(\cdot-x_i)\quad\text{in }\mathbb{R}^{\mathsf{d}+\mathsf{k}}.\tag{4.1.20}$$

Again, f_η is supported in the ball $B(0,\eta)$ of $\mathbb{R}^{\mathsf{d}+\mathsf{k}}$, and $\mathsf{g}_\eta=\min(\mathsf{g},\mathsf{g}(\eta))$.

Definition 4.6 (Smeared charges)**.** We denote

$$\delta_{x_0}^{(\eta)}:=-\frac{1}{\mathsf{c}_{\mathsf{d},\mathsf{s}}}\mathrm{div}\,(|y|^\gamma\nabla\mathsf{g}_\eta(x-x_0))\,.\tag{4.1.21}$$

It is a measure of mass 1 supported on $\partial B(x_0,\eta)$, sphere of center x_0 and radius η in $\mathbb{R}^{\mathsf{d}+\mathsf{k}}$.

The fact that $\delta_{x_0}^{(\eta)}$ is a probability measure supported on $\partial B(0,\eta)$ can be argued as in the Coulomb case, or by direct computation: one sees that its density on the $(\mathsf{d}+\mathsf{k})$-dimensional sphere is equal to $\frac{1}{\mathsf{c}_{\mathsf{d},\mathsf{s}}}\frac{|y|^\gamma}{|z|^{\mathsf{s}+1}}=\frac{1}{\mathsf{c}_{\mathsf{d},\mathsf{s}}}\frac{|y|^\gamma}{\eta^{\mathsf{s}+1}}$ where $z=(x,y)\in\mathbb{R}^{\mathsf{d}+\mathsf{k}}$, hence it is generally *not* uniform.

Contrary to the Coulomb case, it is not true that $\mathsf{g}_\eta=\mathsf{g}*\delta_0^{(\eta)}$ (the two functions coincide however on $\mathbb{R}^{\mathsf{d}}\times\{0\}$ as proven just below) because $\delta_0^{(\eta)}$ is not supported in $\mathbb{R}^{\mathsf{d}}\times\{0\}$, while g is the fundamental solution to the $-\mathrm{div}\,(|y|^\gamma\nabla\cdot)$ operator only for functions supported in $\mathbb{R}^\mathsf{d} \times \{0\}$. This corresponds to the fact that Newton's theorem or the mean-value theorem does not hold for Riesz interactions. We will, however, not need this fact. Instead, we will use that in view of (2.2.6), (4.1.18) and (4.1.21) we have

$$-\mathrm{div}\left(|y|^\gamma \nabla \mathsf{f}_\eta\right) = \mathsf{c}_{\mathsf{d},\mathsf{s}} \left(\delta_0 - \delta_0^{(\eta)}\right) \quad \text{in } \mathbb{R}^{\mathsf{d}+\mathsf{k}}. \tag{4.1.22}$$

We now check that as claimed g_η and $\mathsf{g} * \delta_0^{(\eta)}$ coincide on $\mathbb{R}^\mathsf{d} \times \{0\}$.

Lemma 4.7. *If $w, z \in \mathbb{R}^\mathsf{d} \times \{0\}$ then*

$$\int_{\mathbb{R}^{\mathsf{d}+\mathsf{k}}} \mathsf{g}(x-w) d\delta_z^{(\eta)}(x) = \mathsf{g}_\eta(w-z). \tag{4.1.23}$$

Proof. By translation invariance, it suffices to consider $z = 0$. By definition (4.1.21) and by (2.2.6), using integration by parts we have

$$\begin{aligned}
\int_{\mathbb{R}^{\mathsf{d}+\mathsf{k}}} \mathsf{g}(x-w) d(\delta_0^{(\eta)} - \delta_0)(x) &= -\frac{1}{\mathsf{c}_{\mathsf{d},\mathsf{s}}} \int_{\mathbb{R}^{\mathsf{d}+\mathsf{k}}} \mathsf{g}(x-w) \mathrm{div}\left(|y|^\gamma \nabla (\mathsf{g}_\eta - \mathsf{g})\right)(x) dx \\
&= \frac{1}{\mathsf{c}_{\mathsf{d},\mathsf{s}}} \int_{\mathbb{R}^{\mathsf{d}+\mathsf{k}}} \nabla \mathsf{g}(x-w) \cdot \nabla (\mathsf{g}_\eta - \mathsf{g})(x) |y|^\gamma dx \\
&= -\frac{1}{\mathsf{c}_{\mathsf{d},\mathsf{s}}} \int_{\mathbb{R}^{\mathsf{d}+\mathsf{k}}} \mathrm{div}\left(|y|^\gamma \nabla \mathsf{g}(x-w)\right) (\mathsf{g}_\eta - \mathsf{g})(x) dx \\
&= \int_{\mathbb{R}^{\mathsf{d}+\mathsf{k}}} (\mathsf{g}_\eta - \mathsf{g})(x) d\delta_w(x) = (\mathsf{g}_\eta - \mathsf{g})(w).
\end{aligned}$$

In the integration by parts, we can check that the boundary terms vanished at infinity using the same reasoning as in the proof of Proposition 2.16. The result follows. □

As explained above, given $x_1, \ldots, x_N \in \mathbb{R}^\mathsf{d}$, if $\mathsf{k} = 1$ we identify them with the points $(x_1, 0), \ldots, (x_N, 0)$ in $\mathbb{R}^{\mathsf{d}+\mathsf{k}}$, and we may then define the potentials h_N and truncated potentials $h_{N,\vec{\eta}}$ in $\mathbb{R}^{\mathsf{d}+\mathsf{k}}$ by

$$h_N^\mu[X_N] = \mathsf{g} * \left(\sum_{i=1}^N \delta_{x_i} - N\mu\delta_{\mathbb{R}^\mathsf{d}}\right) \qquad h_{N,\vec{\eta}}[X_N] := h_N - \sum_{i=1}^N \mathsf{f}_{\eta_i}(\cdot - x_i), \tag{4.1.24}$$

later on simply denoted h_N and $h_{N,\vec{\eta}}$. Since g is naturally extended to a function in $\mathbb{R}^{\mathsf{d}+\mathsf{k}}$, these potentials make sense as functions in $\mathbb{R}^{\mathsf{d}+\mathsf{k}}$. In view of (4.1.17), (2.2.6) and (4.1.21), h_N solves

$$-\mathrm{div}\left(|y|^\gamma \nabla h_N\right) = \mathsf{c}_{\mathsf{d},\mathsf{s}} \left(\sum_{i=1}^N \delta_{x_i} - N\mu\delta_{\mathbb{R}^\mathsf{d}}\right) \quad \text{in } \mathbb{R}^{\mathsf{d}+\mathsf{k}}, \tag{4.1.25}$$

while $h_{N,\vec{\eta}}$ solves

$$-\mathrm{div}\left(|y|^\gamma \nabla h_{N,\vec{\eta}}\right) = \mathsf{c}_{\mathsf{d},\mathsf{s}} \left(\sum_{i=1}^N \delta_{x_i}^{(\eta_i)} - N\mu\delta_{\mathbb{R}^\mathsf{d}}\right) \quad \text{in } \mathbb{R}^{\mathsf{d}+\mathsf{k}}. \tag{4.1.26}$$

Remark 4.8. In the case $\mathsf{d} = 1$, $\mathsf{s} = 0$ we have $\gamma = 0$, and h_N is nothing else than the harmonic extension to $\mathbb{R}^2$, away from the real axis, of the potential defined in dimension 1 by the analogue of (4.1.4). This is closely related to the *Stieltjes transform*, a commonly used object in random matrix theory; see [**AGZ10**]. In the cases (2.0.1), the situation is the same, except for the presence of the $|y|^\gamma$ weight.

The electric representation presented in the previous section for the Coulomb case goes through without change, if one replaces

$$\int_{\mathbb{R}^{\mathsf{d}}} |\nabla h_{N,\vec{\eta}}|^2$$

with

$$\int_{\mathbb{R}^{\mathsf{d}+\mathsf{k}}} |y|^\gamma |\nabla h_{N,\vec{\eta}}|^2.$$

As in Section 2.2.1, everything can be written in a unified way encompassing Coulomb and Riesz cases (2.0.1) by working in $\mathbb{R}^{\mathsf{d}+\mathsf{k}}$ and using (2.2.5).

We will frequently use the following bounds.

Lemma 4.9. *For any $\eta > 0$ and any integer $m \geq 0$ with $m < \mathsf{d} - \mathsf{s}$, we have*

$$\int_{\mathbb{R}^{\mathsf{d}}} |\nabla^{\otimes m} \mathsf{f}_\eta| \leq C \eta^{\mathsf{d}-\mathsf{s}-m} \tag{4.1.27}$$

with $C > 0$ depending only on d, s *and* m.

These can be proven using for instance the explicit form

$$\mathsf{f}_\eta(x) = \begin{cases} \left(-\log\left(\frac{|x|}{\eta}\right)\right)_+ & \mathsf{s} = 0, \\ \frac{\eta^{-\mathsf{s}}}{\mathsf{s}}\left(\frac{\eta^{\mathsf{s}} - |x|^{\mathsf{s}}}{|x|^{\mathsf{s}}}\right)_+ & \mathsf{s} > 0, \\ -\frac{\eta^{-\mathsf{s}}}{\mathsf{s}}\left(\frac{|x|^{\mathsf{s}} - \eta^{\mathsf{s}}}{|x|^{\mathsf{s}}}\right)_+ & \mathsf{s} < 0. \end{cases} \tag{4.1.28}$$

We may now state the generalization of Lemma 4.4.

Lemma 4.10 (Electric representation—general Riesz case)**.** *Assume* $\mathsf{s} \in [\mathsf{d}-2, \mathsf{d})$. *Let* $\mathsf{k} = 0$ *if* $\mathsf{s} = \mathsf{d} - 2$ *and* $\mathsf{k} = 1$ *otherwise. Let γ be given by* (2.2.5). *Assume μ is a probability measure satisfying* (4.1.1). *For any pairwise distinct X_N in* $(\mathbb{R}^{\mathsf{d}})^N$, *we have*

$$\mathsf{F}_N(X_N, \mu) = \frac{1}{2\mathsf{c}_{\mathsf{d},\mathsf{s}}} \lim_{\eta_i \to 0} \left(\int_{\mathbb{R}^{\mathsf{d}+\mathsf{k}}} |y|^\gamma |\nabla h_{N,\vec{\eta}}|^2 - \mathsf{c}_{\mathsf{d},\mathsf{s}} \sum_{i=1}^N \mathsf{g}(\eta_i) \right). \tag{4.1.29}$$

Proof. Using the assumption (4.1.1), we apply the result of Proposition 2.16 to $\mu_+ = \frac{1}{N}\sum_{i=1}^N \delta_{x_i}^{(\eta_i)}$ and $\mu_- = \mu$ and obtain

$$\begin{aligned} \iint_{\mathbb{R}^{\mathsf{d}+\mathsf{k}} \times \mathbb{R}^{\mathsf{d}+\mathsf{k}}} \mathsf{g}(x-y) d\left(\sum_{i=1}^N \delta_{x_i}^{(\eta_i)} - N\mu\delta_{\mathbb{R}^{\mathsf{d}}}\right)(x) d\left(\sum_{i=1}^N \delta_{x_i}^{(\eta_i)} - N\mu\delta_{\mathbb{R}^{\mathsf{d}}}\right)(y) \\ = \frac{1}{\mathsf{c}_{\mathsf{d},\mathsf{s}}} \int_{\mathbb{R}^{\mathsf{d}+\mathsf{k}}} |y|^\gamma |\nabla h_{N,\vec{\eta}}|^2. \end{aligned} \tag{4.1.30}$$

By definition (4.1.24) we next rewrite this, via an integration by parts, as

$$\int_{\mathbb{R}^{\mathsf{d}+\mathsf{k}}} |y|^\gamma |\nabla h_{N,\vec{\eta}}|^2 = \mathsf{c}_{\mathsf{d},\mathsf{s}} \int_{\mathbb{R}^{\mathsf{d}+\mathsf{k}}} \left(h_N - \sum_{i=1}^N \mathsf{f}_{\eta_i}(\cdot - x_i)\right) d\left(\sum_{i=1}^N \delta_{x_i}^{(\eta_i)} - N\mu\delta_{\mathbb{R}^{\mathsf{d}}}\right).$$

Using that $\mathsf{f}_\eta(\cdot - x_i)$ vanishes on $\partial B(x_i, \eta_i) = \mathrm{supp}(\delta_{x_i}^{(\eta_i)})$, we obtain that if η_i are small enough that the balls $B(x_i, \eta_i)$ are disjoint, we have

$$\int_{\mathbb{R}^{\mathsf{d}+\mathsf{k}}} |y|^\gamma |\nabla h_{N,\vec{\eta}}|^2 = \mathsf{c}_{\mathsf{d},\mathsf{s}} \int_{\mathbb{R}^{\mathsf{d}+\mathsf{k}}} h_N \, d\left(\sum_{i=1}^N \delta_{x_i}^{(\eta_i)} - N\mu\right) + N\mathsf{c}_{\mathsf{d},\mathsf{s}} \sum_{i=1}^N \int_{\mathbb{R}^{\mathsf{d}+\mathsf{k}}} \mathsf{f}_{\eta_i}(\cdot - x_i)\, d(\mu\delta_{\mathbb{R}^\mathsf{d}}).$$

Rewriting h_N again via (4.1.24), and using that $\int \mathsf{g}\delta_0^{(\eta)} = \mathsf{g}(\eta)$, we deduce that if the η_i's are small enough, then

(4.1.31)

$$\begin{aligned}
&\int_{\mathbb{R}^{\mathsf{d}+\mathsf{k}}} |y|^\gamma |\nabla h_{N,\vec{\eta}}|^2 \\
&= \mathsf{c}_{\mathsf{d},\mathsf{s}} \iint_{\mathbb{R}^{\mathsf{d}+\mathsf{k}}\times\mathbb{R}^{\mathsf{d}+\mathsf{k}}} \mathsf{g}(x-y) d\left(\sum_{i=1}^N \delta_{x_i} - N\mu\delta_{\mathbb{R}^\mathsf{d}}\right)(x) d\left(\sum_{i=1}^N \delta_{x_i}^{(\eta_i)} - N\mu\delta_{\mathbb{R}^\mathsf{d}}\right)(y) \\
&\quad + N\mathsf{c}_{\mathsf{d},\mathsf{s}} \sum_{i=1}^N \int_{\mathbb{R}^{\mathsf{d}+\mathsf{k}}} \mathsf{f}_{\eta_i}(\cdot - x_i)\, d(\mu\delta_{\mathbb{R}^\mathsf{d}}) \\
&= \mathsf{c}_{\mathsf{d},\mathsf{s}} \Bigg[\iint_{\mathbb{R}^{\mathsf{d}+\mathsf{k}}\times\mathbb{R}^{\mathsf{d}+\mathsf{k}}} \mathsf{g}(x-y) d\left(\sum_{i=1}^N \delta_{x_i}^{(\eta_i)} - N\mu\delta_{\mathbb{R}^\mathsf{d}}\right)(y) \\
&\qquad \times d\left(\sum_{j\neq i} \delta_{x_j} - N\mu\delta_{\mathbb{R}^\mathsf{d}}\right)(x) + \sum_{i=1}^N \mathsf{g}(\eta_i)\Bigg] \\
&\quad + N\mathsf{c}_{\mathsf{d},\mathsf{s}} \sum_{i=1}^N \int_{\mathbb{R}^{\mathsf{d}+\mathsf{k}}} \mathsf{f}_{\eta_i}(\cdot - x_i)\, d(\mu\delta_{\mathbb{R}^\mathsf{d}}).
\end{aligned}$$

Then, we use that $\mathsf{f}_{\eta_i}(x - x_i)\mu(x)$ converges monotonically to 0 as $\eta_i \to 0$ to write that $\int_{\mathbb{R}^{\mathsf{d}+\mathsf{k}}} \mathsf{f}_{\eta_i}(\cdot - x_i) d(\mu\delta_{\mathbb{R}^\mathsf{d}}) \to 0$. Taking the limit of (4.1.31) as $\eta_i \to 0$, we deduce that (4.1.29) holds. □

The limit in (4.1.29) is not quantitative. We will show next that equality in (4.1.29) holds for η_i small enough *without having to take a limit*, thanks to a monotonicity property that will bring other consequences. In particular, all fine energy controls will be obtained below by leveraging this monotonicity.

4.2. Monotonicity with respect to truncation and consequences

The following proposition, first proven in [**LS18**, Prop. 2.3] and [**Ser20**, Prop. 3.3], gives an exact representation for F_N and shows a monotonicity property that increasing the truncation parameters η_i can only decrease the value of the quantity whose limit is taken in the right-hand side of (4.1.29). The idea is natural when one recalls Newton's theorem in the Coulomb case: when smearing charges radially, the potential they generate can only decrease, hence the interaction energy between smeared charges is smaller than between the discrete charges. Moreover, there is equality in case the smearing balls are disjoint, since the potential generated by a radially smeared charge in a ball coincides with that of the discrete charge outside of the ball. A more precise way to write this is that if g is Coulomb, then using the mean-value or Newton's theorem, we have

$$
\begin{aligned}
(4.2.1)\quad \iint \mathsf{g}(x-y)d\delta_{x_1}^{(\eta_1)}(x)d\delta_{x_2}^{(\eta_2)}(y) &\le \iint \mathsf{g}(x-y)d\delta_{x_1}(x)d\delta_{x_2}^{(\eta_2)}(y)\\
&\le \iint \mathsf{g}(x-y)d\delta_{x_1}(x)d\delta_{x_2}(y) = \mathsf{g}(x_1-x_2)
\end{aligned}
$$

with equality if and only if $B(x_1,\eta_1)$ and $B(x_2,\eta_2)$ are disjoint. The choice of truncation procedure made in Definition 4.5 allows to have the same property also in the Riesz case, despite the absence of the mean-value theorem. The following proposition shows how to apply this idea to the context of F_N, moreover it retains a positive term corresponding to the difference between the right and left-hand side in (4.2.1), which encodes the small-scale interactions that are erased by smearing.

Proposition 4.11 (Electric formulation and monotonicity property). *Assume* (2.0.1). *Let* μ *be a bounded probability density on* $\mathbb{R}^{\mathsf{d}}$ *satisfying* (4.1.1), *and* X_N *be in* $(\mathbb{R}^{\mathsf{d}})^N$. *For any* $\vec{\eta}\in\mathbb{R}_+^N$ *we have*

$$
\begin{aligned}
(4.2.2)\quad \frac{1}{2}\sum_{i\neq j}(\mathsf{g}(x_i-x_j)-\mathsf{g}(\eta_i))_+ &\le \mathsf{F}_N(X_N,\mu) - \Bigg(\frac{1}{2\mathsf{c}_{\mathsf{d},\mathsf{s}}}\int_{\mathbb{R}^{\mathsf{d}+\mathsf{k}}}|y|^\gamma|\nabla h_{N,\vec{\eta}}|^2\\
&\quad -\frac{1}{2}\sum_{i=1}^N \mathsf{g}(\eta_i) - N\sum_{i=1}^N\int_{\mathbb{R}^{\mathsf{d}}}\mathsf{f}_{\eta_i}(x-x_i)d\mu(x)\Bigg),
\end{aligned}
$$

where f_η *is defined in* (4.1.18). *Moreover, there* is equality *if the* $B(x_i,\eta_i)$*'s are all disjoint.*

In view of (4.1.27), if η_i is small, we may consider $\sum_{i=1}^N\int_{\mathbb{R}^{\mathsf{d}}}\mathsf{f}_{\eta_i}(x-x_i)d\mu(x)$ as a small error. More precisely, we can use that

$$
(4.2.3)\qquad \left|\sum_{i=1}^N\int_{\mathbb{R}^{\mathsf{d}}}\mathsf{f}_{\eta_i}(x-x_i)d\mu(x)\right| \le C\|\mu\|_{L^\infty}\sum_{i=1}^N \eta_i^{\mathsf{d}-\mathsf{s}},
$$

where $C>0$ depends only on d,s, which follows from (4.1.28).

Remark 4.12 (The case $\mathsf{s}<0$). In the case $\mathsf{s}<0$ no normalization is needed, we may take $\eta_i=0$ for all i, and we then obtain

$$
(4.2.4)\qquad \mathsf{F}_N(X_N,\mu) = \frac{1}{2\mathsf{c}_{\mathsf{d},\mathsf{s}}}\int_{\mathbb{R}^{\mathsf{d}+\mathsf{k}}}|y|^\gamma|\nabla h_N|^2.
$$

We will prove the following more refined statement, from which Proposition 4.11 follows.

Lemma 4.13 (Monotonicity with respect to the truncation parameter). *Let* U *be a domain in* $\mathbb{R}^{\mathsf{d}}$ *and* u *solve*

$$
(4.2.5)\qquad -\mathrm{div}\,(|y|^\gamma\nabla u) = \mathsf{c}_{\mathsf{d},\mathsf{s}}\left(\sum_{i=1}^N\delta_{x_i} - N\mu\delta_{\mathbb{R}^{\mathsf{d}}}\right) \quad \text{in } U\times\mathbb{R}^{\mathsf{k}},
$$

and let $u_{\vec\alpha}, u_{\vec\eta}$ be truncated fields as in (4.1.20). Assume $\alpha_i \le \eta_i$ for each i. Letting I_N denote $\{i, \alpha_i \neq \eta_i\}$, assume that for each $i \in I_N$ we have $B(x_i, \eta_i) \subset U$. Then

(4.2.6)
$$\int_{U\times\mathbb{R}^k} |y|^\gamma |\nabla u_{\vec\eta}|^2 - \mathsf{c_{d,s}} \sum_{i\in I_N} \mathsf{g}(\eta_i) - 2N\mathsf{c_{d,s}} \sum_{i\in I_N} \int_U \mathsf{f}_{\eta_i}(x-x_i)d\mu(x)$$
$$- \left(\int_{U\times\mathbb{R}^k} |y|^\gamma |\nabla u_{\vec\alpha}|^2 - \mathsf{c_{d,s}} \sum_{i\in I_N} \mathsf{g}(\alpha_i) - 2N\mathsf{c_{d,s}} \sum_{i\in I_N} \int_U \mathsf{f}_{\alpha_i}(x-x_i)d\mu(x) \right) \le 0.$$

Moreover, there is equality if the $B(x_i, \eta_i)$'s are disjoint from all the other $B(x_j, \eta_j)$'s for each $i \in I_N$.

Proof. For any $\alpha \le \eta$, let us denote $\mathsf{f}_{\alpha,\eta} := \mathsf{f}_\alpha - \mathsf{f}_\eta$, where f is as in (4.1.18). Observe that $\mathsf{f}_{\alpha,\eta}$ vanishes outside $B(0,\eta)$ and

$$\mathsf{g}(\eta) - \mathsf{g}(\alpha) \le \mathsf{f}_{\alpha,\eta} \le 0,$$

while, in view of (4.1.22),

$$-\mathrm{div}\left(|y|^\gamma \nabla \mathsf{f}_{\alpha,\eta}\right) = \mathsf{c_{d,s}} \left(\delta_0^{(\eta)} - \delta_0^{(\alpha)} \right). \tag{4.2.7}$$

Using the fact that by definition and (4.1.20) we have

$$\nabla u_{\vec\eta}(z) - \nabla u_{\vec\alpha}(z) = \sum_{i\in I_N} \nabla \mathsf{f}_{\alpha_i,\eta_i}(z - x_i),$$

we compute

$$\begin{aligned} T &:= \int_{U\times\mathbb{R}^k} |y|^\gamma |\nabla u_{\vec\eta}|^2 - \int_{U\times\mathbb{R}^k} |y|^\gamma |\nabla u_{\vec\alpha}|^2 \\ &= 2\int_{U\times\mathbb{R}^k} |y|^\gamma (\nabla u_{\vec\eta} - \nabla u_{\vec\alpha}) \cdot \nabla u_{\vec\alpha} + \int_{U\times\mathbb{R}^k} |y|^\gamma |\nabla u_{\vec\eta} - \nabla u_{\vec\alpha}|^2 \\ &= 2\sum_{i\in I_N} \int_{U\times\mathbb{R}^k} |y|^\gamma \nabla \mathsf{f}_{\alpha_i,\eta_i}(\cdot - x_i) \cdot \nabla u_{\vec\alpha} \\ &\quad + \sum_{i,j\in I_N} \int_{U\times\mathbb{R}^k} |y|^\gamma \nabla \mathsf{f}_{\alpha_i,\eta_i}(\cdot - x_i) \cdot \nabla \mathsf{f}_{\alpha_j,\eta_j}(\cdot - x_j). \end{aligned} \tag{4.2.8}$$

If $B(x_i, \eta_i) \subset U$ the function $\mathsf{f}_{\alpha_i,\eta_i}(\cdot - x_i)$ vanishes on $\partial(U \times \mathbb{R}^k)$, and we can integrate by parts without getting any boundary contribution. With the help of (4.2.5) and (4.2.7) we thus obtain

$$
\begin{aligned}
T = {} & 2\mathsf{c}_{\mathsf{d},\mathsf{s}} \sum_{i\in I_N} \int_{U\times\mathbb{R}^{\mathsf{k}}} \mathsf{f}_{\alpha_i,\eta_i}(\cdot - x_i) d\Big(\sum_{j=1}^N \delta_{x_j}^{(\alpha_j)} - N d\mu\, \delta_{\mathbb{R}^{\mathsf{d}}}\Big) \\
& + \mathsf{c}_{\mathsf{d},\mathsf{s}} \sum_{i,j\in I_N} \int_{U\times\mathbb{R}^{\mathsf{k}}} \mathsf{f}_{\alpha_i,\eta_i}(\cdot - x_i) d\left(\delta_{x_j}^{(\eta_j)} - \delta_{x_j}^{(\alpha_j)}\right) \\
= {} & \mathsf{c}_{\mathsf{d},\mathsf{s}} \sum_{i\in I_N} \int_{U\times\mathbb{R}^{\mathsf{k}}} \mathsf{f}_{\alpha_i,\eta_i}(\cdot - x_i) d\left(\sum_{j=1}^N \delta_{x_j}^{(\alpha_j)} + \delta_{x_j}^{(\eta_j)}\right) \\
& - 2N\mathsf{c}_{\mathsf{d},\mathsf{s}} \sum_{i\in I_N} \int_U \mathsf{f}_{\alpha_i,\eta_i}(\cdot - x_i) d\mu \\
= {} & \sum_{j=1}^N \sum_{i\in I_N, i\neq j} \mathsf{c}_{\mathsf{d},\mathsf{s}} \int_{\mathbb{R}^{\mathsf{d}}} \mathsf{f}_{\alpha_i,\eta_i}(\cdot - x_i) d(\delta_{x_j}^{(\alpha_j)} + \delta_{x_j}^{(\eta_j)}) \\
& + \mathsf{c}_{\mathsf{d},\mathsf{s}} \sum_{i\in I_N} \int_{\mathbb{R}^{\mathsf{d}+\mathsf{k}}} \mathsf{f}_{\alpha_i,\eta_i}(\cdot - x_i) d(\delta_{x_i}^{(\alpha_i)} + \delta_{x_i}^{(\eta_i)}) \\
& - 2N\mathsf{c}_{\mathsf{d},\mathsf{s}} \sum_{i\in I_N} \int_U \mathsf{f}_{\alpha_i,\eta_i}(\cdot - x_i) d\mu.
\end{aligned} \tag{4.2.9}
$$

Since $\mathsf{f}_{\alpha_i,\eta_i} \le 0$, the first term in the right-hand side is nonpositive, and is zero if the $B(x_i,\eta_i)$'s with $i \in I_N$ are disjoint from the other balls. For the diagonal terms, we note that

$$\int_{\mathbb{R}^{\mathsf{d}+\mathsf{k}}} \mathsf{f}_{\alpha_i,\eta_i}(\cdot - x_i) \left(\delta_{x_i}^{(\alpha_i)} + \delta_{x_i}^{(\eta_i)}\right) = -(\mathsf{g}(\alpha_i) - \mathsf{g}(\eta_i))$$

by definition of $\mathsf{f}_{\alpha,\eta}$ and the fact that $\delta_0^{(\alpha)}$ is a measure of mass 1 on $\partial B(0,\alpha)$. Since $\mathsf{f}_{\alpha_i,\eta_i} = \mathsf{f}_{\alpha_i} - \mathsf{f}_{\eta_i}$, this finishes the proof of (4.2.6) after rearranging terms. □

To prove Proposition 4.11, we take advantage of the nonpositive term in the right-hand side of (4.2.9).

Proof of Proposition 4.11. Let us apply Lemma 4.13 to $u = h_N$ on $U = \mathbb{R}^{\mathsf{d}}$. Let us then return to the nonpositive first term in the right-hand side of (4.2.9) and bound it above and below via

$$
\begin{aligned}
& \sum_{i\neq j} (\mathsf{g}_{\eta_i}(|x_i - x_j| + \alpha_j) - \mathsf{g}(|x_i - x_j| - \alpha_j))_- \\
& \quad \le \sum_{i\neq j} \int_{\mathbb{R}^{\mathsf{d}+\mathsf{k}}} \mathsf{f}_{\alpha_i,\eta_i}(x - x_i) d(\delta_{x_j}^{(\alpha_j)} + \delta_{x_j}^{(\eta_j)}) \\
& \quad \le \sum_{i\neq j} \int_{\mathbb{R}^{\mathsf{d}+\mathsf{k}}} (\mathsf{g}_{\eta_i}(x - x_i) - \mathsf{g}_{\alpha_i}(x - x_i))\, d\delta_{x_j}^{(\alpha_j)} \\
& \quad \le \sum_{i\neq j} (\mathsf{g}(\eta_i) - \mathsf{g}_{\alpha_i}(|x_i - x_j| + \alpha_j))_-\,,
\end{aligned} \tag{4.2.10}
$$

where we used the definition of $\mathsf{f}_{\alpha_i,\eta_i}$ and the properties of g_α, in particular the fact that it is radially decreasing. Combining the previous relations and (4.2.9), we find

$$\begin{aligned}
&\mathsf{c}_{\mathsf{d},\mathsf{s}} \sum_{i\neq j} \left(\mathsf{g}_{\alpha_i}(|x_i - x_j| + \alpha_j) - \mathsf{g}(\eta_i)\right)_+ \\
&\le \left(\int_{\mathbb{R}^{\mathsf{d}+\mathsf{k}}} |y|^\gamma |\nabla h_{N,\vec{\alpha}}|^2 - \mathsf{c}_{\mathsf{d},\mathsf{s}} \sum_{i=1}^N \mathsf{g}(\alpha_i) - 2N\mathsf{c}_{\mathsf{d},\mathsf{s}} \sum_{i=1}^N \int_{\mathbb{R}^\mathsf{d}} \mathsf{f}_{\alpha_i}(x - x_i) d\mu \right) \\
&\quad - \left(\int_{\mathbb{R}^{\mathsf{d}+\mathsf{k}}} |y|^\gamma |\nabla h_{N,\vec{\eta}}|^2 - \mathsf{c}_{\mathsf{d},\mathsf{s}} \sum_{i=1}^N \mathsf{g}(\eta_i) - 2N\mathsf{c}_{\mathsf{d},\mathsf{s}} \sum_{i=1}^N \int_{\mathbb{R}^\mathsf{d}} \mathsf{f}_{\eta_i}(x - x_i) d\mu \right).
\end{aligned}$$

Letting all $\alpha_i \to 0$, the limit of the first term in the right-hand side is $\mathsf{F}_N(X_N, \mu)$ in view of (4.1.29), while that of the left-hand side is $\mathsf{c}_{\mathsf{d},\mathsf{s}} \sum (\mathsf{g}(x_i - x_j) - \mathsf{g}(\eta_i))_+$. This finishes the proof of (4.2.2). □

As a corollary, we obtain that even though F_N has no sign in general, it is bounded below by a constant much smaller than N^2. Choosing η_i large in (4.2.2) will make the lower bound in $\mathsf{g}(\eta_i)$ larger, however it will increase the error terms $\int \mathsf{f}_{\eta_i}(x - x_i) d\mu(x)$ which are bounded by (4.2.3). Optimizing leads to finding η_i such that $\mathsf{g}(\eta_i) = N\eta_i^{\mathsf{d}-\mathsf{s}}$ which leads to the choice $\eta_i = N^{-\frac{1}{\mathsf{d}}}$, which is the natural microscale. However, if we assume that $\mu \in L^\infty$, a sharper dependence in $\|\mu\|_{L^\infty}$ is obtained by choosing $\eta_i = \lambda$ with

$$\lambda := (N\|\mu\|_{L^\infty})^{-\frac{1}{\mathsf{d}}}, \tag{4.2.11}$$

which is a natural distance lengthscale.

Corollary 4.14. *Let μ be a probability density satisfying* (4.1.1). *There exists $C > 0$ depending only on d and s such that for any $X_N \in (\mathbb{R}^\mathsf{d})^N$, we have*

$$\mathsf{F}_N(X_N, \mu) + \left(\frac{N}{2\mathsf{d}} \log(N\|\mu\|_{L^\infty}) \right) \mathbf{1}_{\mathsf{s}=0} \ge -C\|\mu\|_{L^\infty}^{\frac{\mathsf{s}}{\mathsf{d}}} N^{1+\frac{\mathsf{s}}{\mathsf{d}}} \mathbf{1}_{\mathsf{s}\ge 0}. \tag{4.2.12}$$

PROOF. It suffices to apply (4.2.2). Discarding nonnegative terms and using (4.2.3), we obtain

$$\begin{aligned}
\mathsf{F}_N(X_N, \mu) &\ge -\frac{1}{2} \sum_{i=1}^N \mathsf{g}(\eta_i) - N \sum_{i=1}^N \int_{\mathbb{R}^\mathsf{d}} \mathsf{f}_{\eta_i}(x - x_i) d\mu(x) \\
&\ge -\frac{1}{2} \sum_{i=1}^N \mathsf{g}(\eta_i) - CN\|\mu\|_{L^\infty} \sum_{i=1}^N \eta_i^{\mathsf{d}-\mathsf{s}}.
\end{aligned}$$

If $\mathsf{s} \ge 0$, choosing $\eta_i = \lambda$ yields the result. If $\mathsf{s} < 0$, the result is known from Remark 4.12 (or obtained again by taking instead $\eta_i = 0$). □

Remark 4.15. We will see below in Corollary 5.26 that this estimate is sharp.

Remark 4.16. We can work with μ less regular than L^∞ as long as we can control the $\int \mathsf{f}_\alpha d\mu$ terms. For instance suitable L^p integrability of μ suffices.

Remark 4.17. We see that the choice of η_i can be refined a bit: instead of (4.2.2), if we assume $\mu \in C^\alpha$, $\alpha > 0$, $\mu \ge 0$ and write

$$\mu(x) = \mu(x_i) + O(|\mu|_{C^\alpha} |x - x_i|^\alpha)$$

we have

$$\left|\int \mathsf{f}_{\eta_i}(x-x_i)d\mu(x)\right| \le \mu(x_i)\eta_i^{\mathsf{d}-\mathsf{s}} + |\mu|_{C^\alpha}\eta_i^{\mathsf{d}-\mathsf{s}+\alpha}.$$

Optimizing leads to choosing $\eta_i = (N\mu(x_i))^{-\frac{1}{\mathsf{d}}}$ which yields

$$\begin{aligned}(4.2.13)\qquad \mathsf{F}_N(X_N,\mu) + \left(\sum_{i=1}^N \frac{1}{2\mathsf{d}}\log(N\mu(x_i))\right)\mathbf{1}_{\mathsf{s}=0}\\ \ge -CN^{\frac{\mathsf{s}}{\mathsf{d}}}\sum_{i=1}^N \mu(x_i)^{\frac{\mathsf{s}}{\mathsf{d}}} - N^{\frac{\mathsf{s}-\alpha}{\mathsf{d}}}|\mu|_{C^\alpha}\sum_{i=1}^N \mu(x_i)^{-\frac{\mathsf{d}-\mathsf{s}+\alpha}{\mathsf{d}}}.\end{aligned}$$

We then define an important particular choice of truncation parameters, which we think of as the *nearest-neighbor distance* for x_i.

Definition 4.18 (Nearest-neighbor distance). If $X_N = (x_1,\dots,x_N)$ is a N-tuple of points in $\mathbb{R}^\mathsf{d}$ we denote for all $i=1,\dots,N$,

$$(4.2.14)\qquad \mathsf{r}_i := \frac{1}{4}\min\left(\min_{j\neq i}|x_i-x_j|,\lambda\right),$$

where λ is as in (4.2.11). We will denote $h_{N,\mathsf{r}}$ for $h_{N,\vec{\eta}}$ with the choice $\eta_i = \mathsf{r}_i$.

The next result shows that even though there is a cancellation between the very large terms $\int |y|^\gamma|\nabla h_{N,\vec{\eta}}|^2$ and $2\mathsf{c}_{\mathsf{d},\mathsf{s}}\sum \mathsf{g}(\eta_i)$ in the singular case $\mathsf{s}\ge 0$, a very interesting choice of η_i is r_i because the r_i's are small enough that the balls are disjoint and there is still equality in (4.2.2), and large enough that *each* of the terms $\int |y|^\gamma|\nabla h_{N,\vec{\eta}}|^2$ and $2\mathsf{c}_{\mathsf{d},\mathsf{s}}\sum \mathsf{g}(\mathsf{r}_i)$ is *separately* controlled by the energy. A localized version of this result will be later given in Proposition 4.28. A different proof is provided in [**HCRS25**], however it cannot be localized.

Proposition 4.19 (Minimal distance and truncated energy controls). *Let μ be a bounded probability density satisfying* (4.1.1). *Assume $\mathsf{s}\in[(\mathsf{d}-2)_+,\mathsf{d})$. Given any pairwise distinct configuration $X_N\in(\mathbb{R}^\mathsf{d})^N$, it holds that*

(4.2.15)

$$\begin{cases}\displaystyle\sum_{i=1}^N \mathsf{g}(\mathsf{r}_i) \le C\left(\mathsf{F}_N(X_N,\mu) + \|\mu\|_{L^\infty}^{\frac{\mathsf{s}}{\mathsf{d}}}N^{1+\frac{\mathsf{s}}{\mathsf{d}}}\right) & \text{if } \mathsf{s}>0\\ \displaystyle\sum_{i=1}^N \mathsf{g}\left(\frac{\mathsf{r}_i}{\lambda}\right) \le 2\left(\mathsf{F}_N(X_N,\mu) + N\frac{\log(N\|\mu\|_{L^\infty})}{2\mathsf{d}}\mathbf{1}_{\mathsf{s}=0}\right) + C\|\mu\|_{L^\infty}^{\frac{\mathsf{s}}{\mathsf{d}}}N^{1+\frac{\mathsf{s}}{\mathsf{d}}} & \text{if } \mathsf{s}=0,\end{cases}$$

and

(4.2.16)

$$\int_{\mathbb{R}^{\mathsf{d}+\mathsf{k}}}|y|^\gamma|\nabla h_{N,\mathsf{r}}|^2 \le C\left(\mathsf{F}_N(X_N,\mu) + N\frac{\log(N\|\mu\|_{L^\infty})}{2\mathsf{d}}\mathbf{1}_{\mathsf{s}=0}\right) + C\|\mu\|_{L^\infty}^{\frac{\mathsf{s}}{\mathsf{d}}}N^{1+\frac{\mathsf{s}}{\mathsf{d}}}$$

for some $C>0$ depending only on d *and* s.

Proof. Let us apply (4.2.2) with $\eta_i = \lambda$ from (4.2.11) and observe that, for each i, by definition (4.2.14) there exists $j\neq i$ such that $(\mathsf{g}(x_i-x_j)-\mathsf{g}(\lambda))_+ = (\mathsf{g}(4\mathsf{r}_i)-\mathsf{g}(\lambda))_+$. Using (4.2.3), we may thus write that

(4.2.17)

$$\frac{1}{2}\sum_{i=1}^N(\mathsf{g}(4\mathsf{r}_i)-\mathsf{g}(\lambda))_+ \le \mathsf{F}_N(X_N,\mu) - \frac{1}{2\mathsf{c}_{\mathsf{d},\mathsf{s}}}\int_{\mathbb{R}^{\mathsf{d}+\mathsf{k}}}|y|^\gamma|\nabla h_{\vec{\eta}}|^2 + \frac{1}{2}N\mathsf{g}(\lambda) + C\|\mu\|_{L^\infty}^{\frac{\mathsf{s}}{\mathsf{d}}}N^{1+\frac{\mathsf{s}}{\mathsf{d}}}.$$

Then (4.2.15) follows after rearranging terms and noting that in the case $\mathsf{s} = 0$ we have $\|\mu\|_{L^\infty}^{\mathsf{s}/\mathsf{d}} N^{1+\mathsf{s}/\mathsf{d}} = N$, which can absorb the $-\sum_{i=1}^N \mathsf{g}(4)$ term. Let us next choose $\eta_i = \mathsf{r}_i$ in (4.2.2). Using that $\mathsf{r}_i \le \lambda$, this yields

$$0 \le \mathsf{F}_N(X_N,\mu) - \frac{1}{2\mathsf{c}_{\mathsf{d},\mathsf{s}}} \int_{\mathbb{R}^{\mathsf{d}+\mathsf{k}}} |y|^\gamma |\nabla h_{N,\mathsf{r}}|^2 + \frac{1}{2}\sum_{i=1}^N \mathsf{g}(\mathsf{r}_i) + C\|\mu\|_{L^\infty}^{\frac{\mathsf{s}}{\mathsf{d}}} N^{1+\frac{\mathsf{s}}{\mathsf{d}}}.$$

Combining with (4.2.15), and in the case $\mathsf{s} = 0$, writing $\mathsf{g}(\mathsf{r}_i) = \mathsf{g}(\mathsf{r}_i/\lambda) + \mathsf{g}(\lambda)$, (4.2.16) follows. □

4.3. Coercivity of the electric energy

Here, we prove that the modulated energy does metrize the convergence of μ_N to μ and acts as an effective Coulomb or Riesz distance. In the case $\mathsf{s} < 0$, by Remark 4.12 it is exactly a Coulomb/Riesz distance, equal to the (square of the) $\dot H^{\frac{\mathsf{s}-\mathsf{d}-\mathsf{k}}{2}}(\mathbb{R}^{\mathsf{d}+\mathsf{k}})$ norm of $\sum_i \delta_{x_i} - N\mu$. As in (2.2.8) and (4.1.3), in view of (4.1.30) and Plancherel's theorem, we have

$$\begin{aligned}\int_{\mathbb{R}^{\mathsf{d}+\mathsf{k}}} |y|^\gamma |\nabla h_{N,\mathsf{r}}|^2 &= \mathsf{c}_{\mathsf{d},\mathsf{s}} \iint_{\mathbb{R}^{\mathsf{d}+\mathsf{k}}\times\mathbb{R}^{\mathsf{d}+\mathsf{k}}} \mathsf{g}(x-y) d\left(\sum_{i=1}^N \delta_{x_i}^{(\mathsf{r}_i)} - N\mu\delta_{\mathbb{R}^\mathsf{d}}\right)(x) \\ &\qquad\qquad \times d\left(\sum_{i=1}^N \delta_{x_i}^{(\mathsf{r}_i)} - N\mu\delta_{\mathbb{R}^\mathsf{d}}\right)(y) \\ &= C_{\mathsf{d},\mathsf{s}} \left\|\sum_{i=1}^N \delta_{x_i}^{(\mathsf{r}_i)} - N\mu\delta_{\mathbb{R}^\mathsf{d}}\right\|^2_{\dot H^{\frac{\mathsf{s}-\mathsf{d}-\mathsf{k}}{2}}(\mathbb{R}^{\mathsf{d}+\mathsf{k}})},\end{aligned}$$

where the homogeneous Sobolev seminorm $\dot H^m(\mathbb{R}^\mathsf{d})$ is defined by (2.2.9).

Since by (4.2.16) F_N controls $\int_{\mathbb{R}^{\mathsf{d}+\mathsf{k}}} |y|^\gamma |\nabla h_{N,\mathsf{r}}|^2$, it thus controls this fractional Sobolev seminorm of $\sum_{i=1}^N \delta_{x_i}^{(\mathsf{r}_i)} - N\mu\delta_{\mathbb{R}^\mathsf{d}}$. In particular, in the Coulomb case where $\mathsf{k}=0$ and $\mathsf{s}=\mathsf{d}-2$, we thus control $\sum_{i=1}^N \delta_{x_i}^{(\mathsf{r}_i)} - N\mu$ in $\dot H^{-1}(\mathbb{R}^\mathsf{d})$. This is the way in which F_N can be seen as the square of a Coulomb/Riesz distance.

To control $\sum_{i=1}^N \delta_{x_i} - N\mu$, it then suffices to estimate $\sum_i \delta_{x_i}^{(\mathsf{r}_i)} - \delta_{x_i}$ which is easily controlled by the fact that the r_i's are small, more precisely smaller than $\lambda = (N\|\mu\|_{L^\infty})^{-1/\mathsf{d}}$. Since in general $\sum_i \delta_{x_i}$ does not belong to $\dot H^{\frac{\mathsf{s}-\mathsf{d}-\mathsf{k}}{2}}(\mathbb{R}^{\mathsf{d}+\mathsf{k}})$, we can only get a control of $\sum_{i=1}^N \delta_{x_i} - N\mu$ in a weaker space, which $\dot H^{\frac{\mathsf{s}-\mathsf{d}-\mathsf{k}}{2}}(\mathbb{R}^{\mathsf{d}+\mathsf{k}})$ embeds into. There are several possible choices.

As a first possibility, we give the following control against test functions, i.e., bounds on Lipschitz linear statistics. It is a control in terms of $h_{N,\mathsf{r}}$ but by (4.2.16) this amounts to a control by F_N.

Lemma 4.20 (The modulated energy controls the fluctuations). *Let φ be a function with bounded support and assume that $\Omega \subset \mathbb{R}^\mathsf{d}$ contains a λ-neighborhood of this support. For any configuration $X_N \in (\mathbb{R}^\mathsf{d})^N$, letting h_N be defined as in* (4.1.24), *r_i as in* (4.2.14), *and letting I_Ω denote $\{i, x_i \in \Omega\}$ and $\#I_\Omega$ its cardinality, for any $0 < \alpha \le 1$, we have*

- *in the Coulomb case,*

(4.3.1)
$$\left|\int_{\mathbb{R}^{\mathsf{d}}} \varphi \left(\sum_{i=1}^N \delta_{x_i} - N d\mu\right)\right| \le C\|\nabla\varphi\|_{L^2(\Omega)}\|\nabla h_{N,\mathsf{r}}\|_{L^2(\Omega)} + \# I_\Omega |\varphi|_{C^\alpha} N^{-\frac{\alpha}{\mathsf{d}}}\|\mu\|_{L^\infty}^{-\frac{\alpha}{\mathsf{d}}},$$

- *in the Riesz case,*

$$\left|\int_{\mathbb{R}^{\mathsf{d}}} \varphi \left(\sum_{i=1}^N \delta_{x_i} - N d\mu\right)\right| \le C\|\varphi\|_{\dot H^{\frac{\mathsf{d}-\mathsf{s}}{2}}} \left(\int_{\mathbb{R}^{\mathsf{d}+\mathsf{k}}} |y|^\gamma |\nabla h_{N,\mathsf{r}}|^2\right)^{\frac12} + C\# I_\Omega \lambda^{\mathsf{d}-\mathsf{s}} \|(-\Delta)^{\frac{\mathsf{d}-\mathsf{s}}{2}}\varphi\|_{L^\infty}, \tag{4.3.2}$$

- *in a localized way, for any $\varepsilon \ge \lambda$ as in* (4.2.11),

(4.3.3)
$$\left|\int_{\mathbb{R}^{\mathsf{d}}} \varphi \left(\sum_{i=1}^N \delta_{x_i} - N d\mu\right)\right| \le C\left(\varepsilon^{\gamma-1}\|\varphi\|_{L^2(\Omega)}^2 + \varepsilon^{\gamma+1}\|\nabla\varphi\|_{L^2(\Omega)}^2\right)^{\frac12}\left(\int_{\Omega\times[-2\varepsilon,2\varepsilon]^{\mathsf{k}}} |y|^\gamma |\nabla h_{N,\mathsf{r}}|^2\right)^{\frac12} + \# I_\Omega |\varphi|_{C^\alpha} N^{-\frac{\alpha}{\mathsf{d}}}\|\mu\|_{L^\infty}^{-\frac{\alpha}{\mathsf{d}}},$$

where $C > 0$ depends only on d *and* s.

PROOF. The Coulomb case is very straightforward. Integrating (4.1.14) against φ and using Green's formula and the Cauchy-Schwarz inequality, we have

(4.3.4)
$$\left|\int_\Omega \varphi\, d\Big(\sum_{i=1}^N \delta_{x_i}^{(\mathsf{r}_i)} - N\mu\Big)\right| = \frac{1}{\mathsf{c}_{\mathsf{d}}}\left|\int_\Omega \nabla h_{N,\mathsf{r}}\cdot\nabla\varphi\right| \le \frac{1}{\mathsf{c}_{\mathsf{d}}}\|\nabla\varphi\|_{L^2(\Omega)}\|\nabla h_{N,\mathsf{r}}\|_{L^2(\Omega)}.$$

On the other hand, since by definition $\mathsf{r}_i \le \frac{\lambda}{4}$ for each i, we have

$$\left|\int_\Omega \varphi\, d\Big(\sum_{i=1}^N (\delta_{x_i} - \delta_{x_i}^{(\mathsf{r}_i)})\Big)\right| \le \# I_\Omega |\varphi|_{C^\alpha}\lambda^\alpha, \tag{4.3.5}$$

hence by definition of λ we get the result.

Let us now turn to the Riesz case. Let $\tilde\varphi$ be the $\frac12(\mathsf{d}-\mathsf{s})$-harmonic extension of φ to $\mathbb{R}^{\mathsf{d}+\mathsf{k}}$. Following [**CS07**, Sec. 2.4], it can be defined by

$$\tilde\varphi(x,y) = \mathsf{c}_{\mathsf{d},\mathsf{s}}\int_{\mathbb{R}^{\mathsf{d}}} \frac{|y|^{\mathsf{d}-\mathsf{s}}}{(|x-x'|^2+|y|^2)^{\frac{2\mathsf{d}-\mathsf{s}}{2}}}\varphi(x')dx' \quad \text{for } (x,y)\in\mathbb{R}^{\mathsf{d}}\times\mathbb{R}.$$

It is such [**CS07**, (3.7)] that

$$\int_{\mathbb{R}^{\mathsf{d}+\mathsf{k}}} |y|^\gamma|\nabla\tilde\varphi|^2 \le C\|\varphi\|^2_{\dot H^{\frac{\mathsf{d}-\mathsf{s}}{2}}(\mathbb{R}^{\mathsf{d}})} \tag{4.3.6}$$

and

$$-\mathrm{div}\,(|y|^\gamma\nabla\tilde\varphi) = 2\left((-\Delta)^{\frac{\mathsf{d}-\mathsf{s}}{2}}\varphi\right)\delta_{\mathbb{R}^{\mathsf{d}}} \quad \text{in } \mathbb{R}^{\mathsf{d}+1}. \tag{4.3.7}$$

Integrating (4.1.26) against φ and using Green's formula and Cauchy-Schwarz, we obtain

$$\begin{aligned}\left|\int_{\mathbb{R}^{d+k}} \tilde{\varphi}\, d\Big(\sum_{i=1}^N \delta_{x_i}^{(r_i)} - N\mu\delta_{\mathbb{R}^d}\Big)\right| &= \frac{1}{c_{d,s}}\left|\int_{\mathbb{R}^{d+k}} |y|^\gamma \nabla h_{N,r}\cdot\nabla\tilde{\varphi}\right| \\ &\le \frac{1}{c_{d,s}}\left(\int_{\mathbb{R}^{d+k}} |y|^\gamma|\nabla\tilde{\varphi}|^2\right)^{\frac12}\left(\int_{\mathbb{R}^{d+k}} |y|^\gamma|\nabla h_{N,r}|^2\right)^{\frac12} \\ &\le C\|\varphi\|_{\dot{H}^{\frac{d-s}{2}}}\left(\int_{\mathbb{R}^{d+k}} |y|^\gamma|\nabla h_{N,r}|^2\right)^{\frac12}. \end{aligned} \tag{4.3.8}$$

On the other hand, using (4.1.22), we may write

$$\int_{\mathbb{R}^{d+k}} \tilde{\varphi}\, d\left(\sum_{i=1}^N \delta_{x_i} - \delta_{x_i}^{(r_i)}\right) = -\frac{1}{c_{d,s}}\sum_{i=1}^N \int_{\mathbb{R}^{d+k}} \tilde{\varphi}\,\mathrm{div}\left(|y|^\gamma \nabla f_{r_i}(x-x_i)\right).$$

Using integration by parts and (4.3.7), it follows that

$$\begin{aligned}\left|\int_{\mathbb{R}^{d+k}} \tilde{\varphi}\, d\left(\sum_{i=1}^N \delta_{x_i} - \delta_{x_i}^{(r_i)}\right)\right| &= \frac{2}{c_{d,s}}\left|\sum_{i=1}^N \int_{\mathbb{R}^d} f_{r_i}(x-x_i)(-\Delta)^{\frac{d-s}{2}}\varphi\right| \\ &\le C\sum_{i\in I_\Omega} r_i^{d-s}\|(-\Delta)^{\frac{d-s}{2}}\varphi\|_{L^\infty}, \end{aligned} \tag{4.3.9}$$

where we used (4.1.27). Since $r_i \le \lambda$, assembling the relations, we deduce that (4.3.2) holds.

We next turn to the localized version. For that we define a different extension

$$\tilde{\varphi}(x,y) = \varphi(x)\chi(y),$$

where χ is a cutoff function, equal to 1 for $|y| < \varepsilon$ and vanishing for $|y| \ge 2\varepsilon$ with $|\nabla\chi| \le \frac{1}{\varepsilon}$, where ε is any number $\ge \lambda$. In lieu of (4.3.8), we get

$$\begin{aligned} &\left|\int_{\mathbb{R}^{d+k}} \tilde{\varphi}\, d\Big(\sum_{i=1}^N \delta_{x_i}^{(r_i)} - N\mu\delta_{\mathbb{R}^d}\Big)\right| \\ &\le \frac{1}{c_{d,s}}\left(\int_{\Omega\times[-2\varepsilon,2\varepsilon]^k} |y|^\gamma|\nabla\tilde{\varphi}|^2\right)^{\frac12}\left(\int_{\Omega\times[-2\varepsilon,2\varepsilon]^k} |y|^\gamma|\nabla h_{N,r}|^2\right)^{\frac12} \\ &\le C\left(\int_\Omega |\varphi|^2\int_\varepsilon^{2\varepsilon}\frac{|y|^\gamma}{\varepsilon^2}dy + \int_\Omega|\nabla\varphi|^2\int_0^{2\varepsilon}|y|^\gamma dy\right)^{\frac12} \\ &\quad\times\left(\int_{\Omega\times[-2\varepsilon,2\varepsilon]^k} |y|^\gamma|\nabla h_{N,r}|^2\right)^{\frac12} \\ &\le C\left(\varepsilon^{\gamma-1}\|\varphi\|^2_{L^2(\Omega)} + \varepsilon^{\gamma+1}\|\nabla\varphi\|^2_{L^2(\Omega)}\right)^{\frac12}\left(\int_{\Omega\times[-2\varepsilon,2\varepsilon]^k} |y|^\gamma|\nabla h_{N,r}|^2\right)^{\frac12}. \end{aligned} \tag{4.3.10}$$

In lieu of (4.3.9), since $\varepsilon \ge \lambda$ we simply write

$$\left|\int_{\mathbb{R}^{d+k}} \tilde{\varphi}\, d\left(\sum_{i=1}^N \delta_{x_i} - \delta_{x_i}^{(r_i)}\right)\right| \le \#I_\Omega|\varphi|_{C^\alpha}\lambda^\alpha.$$

We conclude that (4.3.3) holds. □

Remark 4.21. Other bounds than (4.3.2), requiring less regularity of φ can be obtained by putting more derivatives on $\mathsf{f}_{\mathsf{r}_i}$ in (4.3.9). They lead to a worse power of λ.

By duality, the bound (4.3.2) allows to deduce a bound on $\sum_{i=1}^N \delta_{x_i} - N\mu$ in a negative Sobolev norm, which shows that $\frac{1}{N^2}\mathsf{F}_N$ does metrize the convergence of μ_N to μ. In **[HCRS25]** a control in a stronger Sobolev norm is proven by a different method.

Corollary 4.22. *For any $\sigma > \frac{\mathsf{d}}{2} + \mathsf{d} - \mathsf{s}$,*

$$\left\| \sum_{i=1}^N \delta_{x_i} - N\mu \right\|_{H^{-\sigma}(\mathbb{R}^{\mathsf{d}})} \le C_\sigma \left(\int_{\mathbb{R}^{\mathsf{d}+\mathsf{k}}} |y|^\gamma |\nabla h_{N,\mathsf{r}}|^2 \right)^{\frac12} + CN^{\frac{\mathsf{s}}{\mathsf{d}}} \|\mu\|_{L^\infty}^{-1+\frac{\mathsf{s}}{\mathsf{d}}}, \tag{4.3.11}$$

where $H^{-\sigma}$ is the dual of H^σ (the standard Sobolev space). In particular, if $\frac{1}{N^2}\mathsf{F}_N(X_N,\mu) \to 0$ as $N \to \infty$, we have that

$$\frac{1}{N}\sum_{i=1}^N \delta_{x_i} \to \mu \quad \textit{in } H^{-\sigma}(\mathbb{R}^{\mathsf{d}}). \tag{4.3.12}$$

PROOF. Indeed, the Sobolev embedding implies that $H^\sigma(\mathbb{R}^{\mathsf{d}}) \subset C^{\mathsf{d}-\mathsf{s}}(\mathbb{R}^{\mathsf{d}})$ and also $H^\sigma(\mathbb{R}^{\mathsf{d}}) \subset \dot{H}^{\frac{\mathsf{d}-\mathsf{s}}{2}}(\mathbb{R}^{\mathsf{d}})$ from which, starting from (4.3.2) or (4.3.1) we deduce (4.3.11). For the second assertion, we combine (4.3.11) and (4.2.16) to obtain

$$\begin{aligned} \left\| \sum_{i=1}^N \delta_{x_i} - N\mu \right\|_{H^{-\sigma}}^2 &\le C \left(\mathsf{F}_N(X_N,\mu) + N \frac{\log(N\|\mu\|_{L^\infty})}{2\mathsf{d}} \mathbf{1}_{\mathsf{s}=0} \right) \\ &\quad + C\|\mu\|_{L^\infty}^{\frac{\mathsf{s}}{\mathsf{d}}} N^{1+\frac{\mathsf{s}}{\mathsf{d}}} + CN^{\frac{2\mathsf{s}}{\mathsf{d}}} \|\mu\|_{L^\infty}^{-2+\frac{2\mathsf{s}}{\mathsf{d}}}, \end{aligned}$$

hence the result after dividing by N^2, using that $\mathsf{s} < \mathsf{d}$. □

We may also obtain L^p bounds on the gradient of the potential ∇h_N. In that respect we have the following estimate, which in turn provides a control of $\sum_{i=1}^N \delta_{x_i} - N\mu$ in view of (4.1.25). If $\mathsf{s} < 0$ this is not needed since we have (4.2.4), hence an L^2 control.

Proposition 4.23 (Control of the electric potential by the modulated energy). *Assume $\mathsf{s} \ge 0$. Let $X_N \in (\mathbb{R}^{\mathsf{d}})^N$. With the same notation as above, letting B_R be the ball of radius R centered at 0 in $\mathbb{R}^{\mathsf{d}+\mathsf{k}}$, for every p such that $1 \le p < \frac{\mathsf{d}+\mathsf{k}}{\mathsf{s}+1}$ and $p \le 2$, we have,*

$$\|\nabla h_N\|_{L^p(B_R)} \le C_{p,R} \left(\int_{B_R} |y|^\gamma |\nabla h_{N,\mathsf{r}}|^2 \right)^{\frac12} + C_p (\# I_{B_R})^{\frac1p} \lambda^{\frac{\mathsf{d}+\mathsf{k}}{p} - \mathsf{s} - 1}. \tag{4.3.13}$$

PROOF. Let us start from (4.1.24), which gives

$$\nabla h_N = \nabla h_{N,\mathsf{r}} + \sum_{i=1}^N \nabla \mathsf{f}_{\mathsf{r}_i}(\cdot - x_i). \tag{4.3.14}$$

The function $\mathsf{f}_{\mathsf{r}_i}$, which is supported in $B(0,\mathsf{r}_i)$, has the same singularity at 0 as $\mathsf{g}(x)$ (since $\mathsf{s} \ge 0$) and $\nabla \mathsf{f}_{\mathsf{r}_i}$ a singularity in $\frac{1}{|x|^{\mathsf{s}+1}}$. Thus, $\nabla \mathsf{f}_{\mathsf{r}_i}$ is in $L^p(\mathbb{R}^{\mathsf{d}+\mathsf{k}})$ if and only if $p(\mathsf{s}+1) < \mathsf{d}+\mathsf{k}$, and if so

$$\int_{\mathbb{R}^{\mathsf{d}+\mathsf{k}}} |\nabla \mathsf{f}_{\mathsf{r}_i}|^p \le \int_{B(0,\mathsf{r}_i)} \frac{dx}{|x|^{p(\mathsf{s}+1)}} \le C_p \mathsf{r}_i^{\mathsf{d}+\mathsf{k}-p\mathsf{s}-p}. \tag{4.3.15}$$

Since the balls $B(x_i, \mathsf{r}_i)$ are disjoint and $\mathsf{r}_i \le \lambda$, we deduce from (4.3.14) that

$$\|\nabla h_N\|_{L^p(B_R)} \le \|\nabla h_{N,\mathsf{r}}\|_{L^p(B_R)} + C_p \left(\# I_{B_R} \lambda^{\mathsf{d}+\mathsf{k}-p\mathsf{s}-p}\right)^{\frac{1}{p}} \tag{4.3.16}$$

In addition, by Hölder's inequality, if $p < \min(2, \frac{\mathsf{d}+\mathsf{k}}{\mathsf{s}+1})$, we have

$$\int_{B_R} |\nabla h_{N,\mathsf{r}}|^p \le \left(\int_{B_R} \left(\frac{1}{|y|^\gamma}\right)^{\frac{p}{2-p}}\right)^{1-\frac{p}{2}} \left(\int_{B_R} |y|^\gamma |\nabla h_{N,\mathsf{r}}|^2\right)^{\frac{p}{2}}.$$

Since $\gamma = \mathsf{s}+2-\mathsf{k}-\mathsf{d}$ in (2.2.5), we note that the condition $p < \frac{\mathsf{d}+\mathsf{k}}{\mathsf{s}+1}$ always implies (since $\mathsf{s} < \mathsf{d}$) that $\frac{\gamma p}{2-p} < 1$ hence the first integral in the right-hand side converges. We deduce that under this condition

$$\int_{B_R} |\nabla h_{N,\mathsf{r}}|^p \le C_{p,R} \left(\int_{B_R} |y|^\gamma |\nabla h_{N,\mathsf{r}}|^2\right)^{\frac{p}{2}}$$

and (4.3.13) follows. □

This result can be localized as we will see in Section 4.5 below.

4.4. Discrepancy bounds

We now prove that the electric energy $\int |y|^\gamma |\nabla h_{N,\mathsf{r}}|^2$, hence via (4.2.16) the modulated energy F_N, controls the discrepancy (as discussed in Section 1.3), defined as follows.

Definition 4.24. Given a measurable set Ω in $\mathbb{R}^\mathsf{d}$, we define the discrepancy in Ω of the configuration X_N as

$$D(\Omega) := \int_\Omega d\left(\sum_{i=1}^N \delta_{x_i} - N\mu\right). \tag{4.4.1}$$

If Ω is a set of finite perimeter (see, e.g., [**EG15, Mag12**]), we let $d(x)$ denote the signed distance function to Ω, which is positive in the complement of Ω and negative inside Ω. For any $\delta \in \mathbb{R}$ (positive or negative), we let

$$\Omega_\delta = \{x \in \mathbb{R}^\mathsf{d}, d(x) < \delta\}. \tag{4.4.2}$$

We denote $|\Omega|$ for the volume of Ω and $|\partial\Omega|$ for its perimeter. The proofs are adapted from [**RS15**, Lemma 4.6], [**PS17**, Lemma 2.2], [**LS18**].

Lemma 4.25 (Control of charge discrepancy, general domain). *Let X_N be a configuration in $(\mathbb{R}^\mathsf{d})^N$, let h_N be associated via (4.1.4), and let Ω be a set of finite perimeter. We have, for a constant $C > 0$ depending only on d and s,*

- *in the Coulomb case, if $D(\Omega) \ge 0$, for any $\lambda < \delta \le 1$,*

$$\left(D(\Omega) - N\|\mu\|_{L^\infty}|\Omega_\delta \backslash \Omega|\right)_+^2 \le C\frac{|\partial\Omega_\delta|}{\delta} \int_{\Omega_\delta\backslash\Omega} |\nabla h_{N,\mathsf{r}}|^2 \tag{4.4.3}$$

- *in the Coulomb case, if $D(\Omega) \le 0$, for any $-1 \le \delta < -\lambda$,*

$$\left(D(\Omega) + N\|\mu\|_{L^\infty}|\Omega \backslash \Omega_\delta|\right)_-^2 \le C\frac{|\partial\Omega|}{|\delta|} \int_{\Omega\backslash\Omega_\delta} |\nabla h_{N,\mathsf{r}}|^2 \tag{4.4.4}$$

- *in the Riesz case, if* $D(\Omega) \geq 0$*, for any* $\lambda < \delta \leq 1$*,*

$$(4.4.5)\qquad \left(D(\Omega) - N\|\mu\|_{L^\infty}|\Omega_\delta \backslash \Omega|\right)_+^2 \leq C\left(\frac{|\Omega_\delta|}{|\partial\Omega_\delta|}\right)^\gamma \frac{|\Omega_\delta|}{\delta}\int_{(\Omega_\delta\backslash\Omega)\times[-\frac{|\Omega_\delta|}{|\partial\Omega_\delta|}-\delta,\frac{|\Omega_\delta|}{|\partial\Omega_\delta|}+\delta]^{\mathsf{k}}}|y|^\gamma|\nabla h_{N,\mathsf{r}}|^2$$

- *in the Riesz case, if* $D(\Omega) \leq 0$*, for any* $-1 \leq \delta < -\lambda$*,*

$$(4.4.6)\qquad \left(D(\Omega) + N\|\mu\|_{L^\infty}|\Omega \backslash \Omega_\delta|\right)_-^2 \leq C\left(\frac{|\Omega|}{|\partial\Omega|}\right)^\gamma \frac{|\Omega|}{|\delta|}\int_{(\Omega\backslash\Omega_\delta)\times[-\frac{|\Omega|}{|\partial\Omega|}-|\delta|,\frac{|\Omega|}{|\partial\Omega|}+|\delta|]^{\mathsf{k}}}|y|^\gamma|\nabla h_{N,\mathsf{r}}|^2.$$

Lemma 4.26 (Control of charge discrepancy, case of a ball)**.** *Let* B_R *be a ball of radius* $R > 2\lambda$*. If* $D(B_R) \geq 0$ *then either* $D(B_R) \leq CN^{1-\frac{1}{\mathsf{d}}}R^{\mathsf{d}-1}\|\mu\|_{L^\infty}^{1-\frac{1}{\mathsf{d}}}$ *or*

$$(4.4.7)\qquad \frac{D(B_R)^2}{R^{\mathsf{s}}}\min\left(1, \frac{D(B_R)}{R^{\mathsf{d}}\|\mu\|_{L^\infty}}\right) \leq C\int_{(B_{R+\delta}\backslash B_R)\times[-C(R+\delta),C(R+\delta)]^{\mathsf{k}}}|y|^\gamma|\nabla h_{N,\mathsf{r}}|^2,$$

with

$$\delta = \min\left(\frac{R}{2}, \left(R^{\mathsf{d}} + \frac{D(B_R)}{2C_0N\|\mu\|_{L^\infty}}\right)^{\frac{1}{\mathsf{d}}} - R\right),$$

while if $D(B_R) \leq 0$ *then either* $|D(B_R)| \leq CN^{1-\frac{1}{\mathsf{d}}}R^{\mathsf{d}-1}\|\mu\|_{L^\infty}^{1-\frac{1}{\mathsf{d}}}$ *or*

$$(4.4.8)\qquad \frac{D(B_R)^2}{R^{\mathsf{s}}}\left|\min\left(1, \frac{|D(B_R)|}{R^{\mathsf{d}}\|\mu\|_{L^\infty}}\right)\right| \leq C\int_{(B_R\backslash B_{R+\delta})\times[-C(R+\delta),C(R+\delta)]^{\mathsf{k}}}|y|^\gamma|\nabla h_{N,\mathsf{r}}|^2,$$

with

$$\delta = \max\left(-\frac{R}{2}, \left(R^{\mathsf{d}} + \frac{D(B_R)}{2C_0N\|\mu\|_{L^\infty}}\right)^{\frac{1}{\mathsf{d}}} - R\right),$$

where $C, C_0 > 0$ *depend only on* d *and* s.

Note that $|D(B_R)|$ should be compared to the rescaled volume NR^{d} or rescaled perimeter $N^{1-\frac{1}{\mathsf{d}}}R^{\mathsf{d}-1}$.

PROOF. We will prove both lemmas at once.

Step 1. The Coulomb positive case. Let us first assume $D(\Omega) \geq 0$. We are going to take advantage of the charge excess by examining the energy outside Ω.

By definition of the r_i (4.2.14), for $t > \frac{1}{2}\lambda$ we have

$$\int_{d(x)<t} d\left(\sum_{i=1}^N \delta_{x_i}^{(\mathsf{r}_i)} - N\mu\right) \geq D(\Omega) - N\int_{0<d(x)<t} d\mu.$$

Thus, with Green's formula, if $\frac{1}{2}\lambda < t < \delta$,

$$(4.4.9)\qquad -\int_{d(x)=t}\frac{\partial h_{N,\mathsf{r}}}{\partial\nu} = -\int_{d(x)<t}\Delta h_{N,\mathsf{r}} = \mathsf{c_d}\int_{d(x)<t} d\left(\sum_{i=1}^N \delta_{x_i}^{(\mathsf{r}_i)} - N\mu\right) \geq \mathsf{c_d}\left(D(\Omega) - N\|\mu\|_{L^\infty}|\Omega_\delta\backslash\Omega|\right).$$

On the other hand, the co-area formula (see [**EG15,Mag12**]) gives (since $|\nabla d| = 1$) that

(4.4.10)

$$\int_{\frac{1}{2}\lambda<d(x)<\delta} |\nabla h_{N,\mathsf{r}}|^2 = \int_{\frac{1}{2}\lambda}^{\delta} \left(\int_{d(x)=t} |\nabla h_{N,\mathsf{r}}|^2 \right) dt \geq \int_{\frac{1}{2}\lambda}^{\delta} \int_{d(x)=t} \left| \frac{\partial h_{N,\mathsf{r}}}{\partial \nu} \right|^2 dt.$$

But the Cauchy-Schwarz inequality gives

$$\int_{d(x)=t} \left| \frac{\partial h_{N,\mathsf{r}}}{\partial \nu} \right|^2 \geq \left(\int_{d(x)=t} \frac{\partial h_{N,\mathsf{r}}}{\partial \nu} \right)^2 \frac{1}{|\{d(x)=t\}|}. \tag{4.4.11}$$

Combining (4.4.9)–(4.4.11), we thus obtain

$$\int_{\frac{1}{2}\lambda<d(x)<\delta} |\nabla h_{N,\mathsf{r}}|^2 \geq \mathsf{c}_{\mathsf{d}}^2 \left(D(\Omega) - N\|\mu\|_{L^\infty} |\Omega_\delta \backslash \Omega| \right)_+^2 \int_{\frac{1}{2}\lambda}^{\delta} \frac{dt}{|\{d(x)=t\}|}. \tag{4.4.12}$$

Using that $|\{d=t\}| \leq C|\partial\Omega_\delta|$ for $t \leq \delta \leq 1$, we deduce that

$$C \int_{\Omega_\delta \backslash \Omega} |\nabla h_{N,\mathsf{r}}|^2 \geq \left(D(\Omega) - N\|\mu\|_{L^\infty} |\Omega_\delta \backslash \Omega| \right)_+^2 \frac{\delta}{|\partial\Omega_\delta|} \tag{4.4.13}$$

if $\delta > \lambda$, which proves (4.4.3).

Specializing (4.4.12) to Ω a ball of radius $R > 2\lambda$, we find

$$C \int_{B_{R+\delta} \backslash B_R} |\nabla h_{N,\mathsf{r}}|^2 \geq \left(D(B_R) - C_0 N \|\mu\|_{L^\infty} ((R+\delta)^{\mathsf{d}} - R^{\mathsf{d}}) \right)_+^2 \int_{\lambda/2}^{\delta} \frac{dt}{(R+t)^{\mathsf{d}-1}},$$

where $C, C_0 > 0$ depend only on d. We now choose

(4.4.14)

$$\delta := \min \left(\frac{R}{2}, \left(R^{\mathsf{d}} + \frac{D(B_R)}{2C_0 N \|\mu\|_{L^\infty}} \right)^{\frac{1}{\mathsf{d}}} - R \right) \geq R \min \left(\frac{1}{2}, C \frac{D(B_R)}{N R^{\mathsf{d}} \|\mu\|_{L^\infty}} \right)$$

so that the term $C_0 N \|\mu\|_{L^\infty} ((R+\delta)^{\mathsf{d}} - R^{\mathsf{d}})$ appearing above is $< \frac{1}{2} D(B_R)$. Either $\delta \leq \lambda$ in which case $D(B_R) \leq C\lambda N R^{\mathsf{d}-1} \|\mu\|_{L^\infty}$, or $\delta \geq \lambda$ and we then obtain

$$C \int_{B_{R+\delta} \backslash B_R} |\nabla h_{N,\mathsf{r}}|^2 \geq D(B_R)^2 \delta R^{1-\mathsf{d}} \geq D(B_R)^2 R^{2-\mathsf{d}} \min \left(\frac{1}{2}, C \frac{D(B_R)}{N R^{\mathsf{d}} \|\mu\|_{L^\infty}} \right).$$

This proves (4.4.7) in this case.

Step 2. The Coulomb negative case. Let us next turn to the case $D(\Omega) \leq 0$, still in the Coulomb case. We now find the excess energy inside Ω. For $t < -\lambda/2$ we have

$$\int_{d(x)<t} d \left(\sum_{i=1}^N \delta_{x_i}^{(\mathsf{r}_i)} - N\mu \right) \leq D(\Omega) + N \int_{t<d(x)<0} d\mu.$$

Thus

$$\begin{aligned} - \int_{d(x)=t} \frac{\partial h_{N,\mathsf{r}}}{\partial \nu} = - \int_{d(x)<t} \Delta h_{N,\mathsf{r}} &= \mathsf{c}_{\mathsf{d}} \int_{d(x)<t} d \left(\sum_{i=1}^N \delta_{x_i}^{(\mathsf{r}_i)} - N\mu \right) \\ &\leq D(\Omega) + N\|\mu\|_{L^\infty} |\{t < d(x) < 0\}|. \end{aligned} \tag{4.4.15}$$

Arguing as in the case $D(\Omega) \geq 0$ we deduce that for $\delta < -\lambda$

$$\int_{\delta<d(x)<-\frac{1}{2}\lambda} |\nabla h_{N,\mathsf{r}}|^2 \geq \left(D(\Omega) + N\|\mu\|_{L^\infty} |\Omega \backslash \Omega_\delta| \right)_-^2 \int_{\delta}^{-\frac{1}{2}\lambda} \frac{dt}{|\{d(x)=t\}|}. \tag{4.4.16}$$

We conclude as above that (4.4.3) holds.

Specializing now to a ball B_R of radius $R > 2\lambda$ we find

$$\int_{B_R \setminus B_{R+\delta}} |\nabla h_{N,\mathsf{r}}|^2 \geq \left(D(\Omega) + C_0 N\|\mu\|_{L^\infty}(R^{\mathsf{d}} - (R+\delta)^{\mathsf{d}})\right)_-^2 \int_\delta^{-\frac{1}{2}\lambda} \frac{dt}{(R+t)^{\mathsf{d}-1}}.$$

Choosing

$$\delta := \max\left(\left(R^{\mathsf{d}} + \frac{D(B_R)}{2C_0 N\|\mu\|_{L^\infty}}\right)^{\frac{1}{\mathsf{d}}} - R, -\frac{1}{2}R\right) \leq -R\min\left(\frac{1}{2}, C\frac{|D(B_R)|}{NR^{\mathsf{d}}\|\mu\|_{L^\infty}}\right),$$

we then obtain that either $\delta \geq -\lambda$ in which case $|D(B_R)| \leq C\lambda N R^{\mathsf{d}-1}\|\mu\|_{L^\infty}$ or $\delta < -\lambda$ and

$$C\int_{B_R \setminus B_{R+\delta}} |\nabla h_{N,\mathsf{r}}|^2 \geq D(B_R)^2 R^{2-\mathsf{d}} \min\left(1, C\frac{|D(B_R)|}{N\|\mu\|_{L^\infty} R^{\mathsf{d}}}\right),$$

concluding the proof of (4.4.8) in the Coulomb case.

Step 3. The Riesz positive case. The relation (4.4.9) can be replaced by

(4.4.17)
$$\begin{aligned} -\int_{\partial(\{d(x)<t\}\times[-m-t,m+t])} |y|^\gamma \frac{\partial h_{N,\mathsf{r}}}{\partial \nu} &= -\int_{\{d(x)<t\}\times[-m-t,m+t]} \operatorname{div}(|y|^\gamma \nabla h_{N,\mathsf{r}})\delta_{\mathbb{R}^{\mathsf{d}}} \\ &\geq D(\Omega) - N\|\mu\|_{L^\infty}|\Omega_\delta \setminus \Omega| \end{aligned}$$

for $t > \frac{1}{2}\lambda$, where $m \geq 0$ is to be determined later, and (4.4.10)–(4.4.11) are replaced by

$$\begin{aligned} &\int_{\{\frac{1}{2}\lambda<d(x)<\delta\}\times[-m-\delta,m+\delta]} |y|^\gamma |\nabla h_{N,\mathsf{r}}|^2 \\ &\geq \int_{\frac{1}{2}\lambda}^\delta \left(\int_{\partial(\{d(x)<t\}\times[-m-t,m+t])} |y|^\gamma \frac{\partial h_{N,\mathsf{r}}}{\partial \nu}\right)^2 \left(\int_{\partial(\{d(x)<t\}\times[-m-t,m+t])} |y|^\gamma\right)^{-1} dt \\ &\geq (D(\Omega) - N\|\mu\|_{L^\infty}|\Omega_\delta\setminus\Omega|)_+^2 \int_{\frac{1}{2}\lambda}^\delta \left(\int_{\partial(\{d(x)<t\}\times[-m-t,m+t])} |y|^\gamma\right)^{-1} dt. \end{aligned}$$

We next evaluate that for $t \in [\frac{1}{2}\lambda, \delta]$

$$\begin{aligned} &\int_{\partial(\{d(x)<t\}\times[-m-t,m+t])} |y|^\gamma \\ &\quad = \frac{2}{\gamma+1}(m+t)^{\gamma+1}|\{d=t\}| + 2|\Omega_t|(m+t)^\gamma \\ &\quad \leq C(m+\delta)^{\gamma+1}|\partial\Omega_\delta| + 2|\Omega_\delta|(m+t)^\gamma. \end{aligned}$$

Let us now take $m = \frac{|\Omega_\delta|}{|\partial\Omega_\delta|}$ to balance the two terms, and note that m is bounded below independently of $\delta \leq 1$. We then obtain

$$\int_{\partial(\{d(x)<t\}\times[-m-t,m+t])} |y|^\gamma \leq C\left(m^{\gamma+1}|\partial\Omega_\delta| + m^\gamma|\Omega_\delta|\right) = C\left(\frac{|\Omega_\delta|}{|\partial\Omega_\delta|}\right)^\gamma |\Omega_\delta|.$$

Inserting into (4.4.17), it follows that

$$(4.4.18)\qquad C\int_{(\Omega_\delta\setminus\Omega)\times[-\frac{|\Omega_\delta|}{|\partial\Omega_\delta|}-\delta,\frac{|\Omega_\delta|}{|\partial\Omega_\delta|}+\delta]^{\mathsf{k}}}|y|^\gamma|\nabla h_{N,\mathsf{r}}|^2 \geq (D(\Omega)-N\|\mu\|_{L^\infty}|\Omega_\delta\setminus\Omega|)_+^2\frac{\delta}{|\Omega_\delta|}\left(\frac{|\Omega_\delta|}{|\partial\Omega_\delta|}\right)^{-\gamma},$$

where we recall that $\gamma=\mathsf{s}-\mathsf{d}+1$ in the non-Coulomb case. This proves (4.4.5).

Let us now specialize to the case of a ball B_R, $R>2\lambda$, assuming $D(B_R)\geq 0$. We obtain

$$C\int_{(B_{R+\delta}\setminus B_R)\times[-\frac{|B_{R+\delta}|}{|\partial B_{R+\delta}|}-\delta,\frac{|B_{R+\delta}|}{|\partial B_{R+\delta}|}+\delta]^{\mathsf{k}}}|y|^\gamma|\nabla h_{N,\mathsf{r}}|^2 \geq \left(D(B_R)-C_0N\|\mu\|_{L^\infty}((R+\delta)^{\mathsf{d}}-R^{\mathsf{d}})\right)_+^2\delta(R+\delta)^{-\gamma-\mathsf{d}}.$$

Choosing

$$\delta:=\min\left(\left(R^{\mathsf{d}}+\frac{D(B_R)}{2C_0N\|\mu\|_{L^\infty}}\right)^{\frac{1}{\mathsf{d}}}-R,\frac{1}{2}R\right)\geq R\min\left(\frac{1}{2},C\frac{D}{NR^{\mathsf{d}}\|\mu\|_{L^\infty}}\right)$$

we then obtain that either $\delta\leq\lambda$ in which case $D(B_R)\leq C\lambda NR^{\mathsf{d}-1}\|\mu\|_{L^\infty}$ or

$$C\int_{(B_{R+\delta}\setminus B_R)\times\mathbb{R}^{\mathsf{k}}}|y|^\gamma|\nabla h_{N,\mathsf{r}}|^2\geq D(B_R)^2\delta R^{-\gamma-\mathsf{d}}$$

hence the result (4.4.7) follows in this case after noticing that $-\gamma-\mathsf{d}=-\mathsf{s}-1$.

The Riesz negative case is similar. □

4.5. Localized version of the energy

One of the features of F_N is that once in electric formulation, it can naturally be localized, which will be important for obtaining results at meso- and microscale later. Let us now define this localized version of the energy.

First, for any subset Ω of $\mathbb{R}^{\mathsf{d}}$, we define a variant of the nearest neighbor distance, relative to the set Ω (we may say relative to $\partial\Omega$):

$$(4.5.1)\qquad \tilde{\mathsf{r}}_i:=\frac{1}{4}\begin{cases}\min\left(\min_{j\neq i}|x_i-x_j|,\lambda\right) & \text{if } \mathrm{dist}(x_i,\partial\Omega)\geq 2\lambda,\\ \lambda & \text{if } \mathrm{dist}(x_i,\partial\Omega)\leq\lambda\\ t\min\left(\min_{j\neq i}|x_i-x_j|,\lambda\right)+(1-t)\lambda & \text{if } \mathrm{dist}(x_i,\partial\Omega)=(1+t)\lambda,\ t\in[0,1].\end{cases}$$

Definition 4.27 (Localized version of the energy). Given $\Omega\subset\mathbb{R}^{\mathsf{d}}$, and $X_N\in(\mathbb{R}^{\mathsf{d}})^N$, we let $I_\Omega=\{i,x_i\in\Omega\}$ and define the localized modulated energy by

$$(4.5.2)\qquad \mathsf{F}_N^\Omega(X_N,\mu):=\frac{1}{2\mathsf{c}_{\mathsf{d},\mathsf{s}}}\left(\int_{\Omega\times\mathbb{R}^{\mathsf{k}}}|y|^\gamma|\nabla h_{N,\tilde{\mathsf{r}}}|^2-\mathsf{c}_{\mathsf{d},\mathsf{s}}\sum_{i\in I_\Omega}\mathsf{g}(\tilde{\mathsf{r}}_i)\right) - N\sum_{i\in I_\Omega}\int_{\mathbb{R}^{\mathsf{d}}}\mathsf{f}_{\tilde{\mathsf{r}}_i}(x-x_i)d\mu(x),$$

where we denote $h_{N,\tilde{\mathsf{r}}}$ for $h_{N,\vec{\eta}}$ with $\eta_i=\tilde{\mathsf{r}}_i$ as in (4.5.1).

Note that points that lie outside Ω but near $\partial\Omega$ affect the truncation in $h_{N,\tilde{\mathsf{r}}}$.

In view of Lemma 4.13, changing for all points the radii r_i into $\tilde{\mathsf{r}}_i$ relative to Ω (which is the same as relative to Ω^c) can only decrease the computed value of F_N, hence we find the important superadditivity property

$$\mathsf{F}_N(X_N,\mu) \geq \mathsf{F}_N^\Omega(X_N,\mu) + \mathsf{F}_N^{\Omega^c}(X_N,\mu). \tag{4.5.3}$$

We next give localized versions of some results of the previous sections.

First, we have the following localized version of Proposition 4.19, from [**RS24b**].

Proposition 4.28 (Localized minimal distance and truncated electric energy controls). *Assume* $\mathsf{s} \in [(\mathsf{d}-2)_+,\mathsf{d})$. *Let* $\Omega \subset \mathbb{R}^\mathsf{d}$ *and let now*

$$\lambda = (N\|\mu\|_{L^\infty(\Omega)})^{-\frac{1}{\mathsf{d}}}.$$

Denote $\hat{\Omega} = \{x \in \mathbb{R}^\mathsf{d}, \operatorname{dist}(x,\Omega) \leq \frac{1}{4}\lambda\}$. *For any* $\vec{\eta}$ *satisfying* $\frac{1}{2}\tilde{\mathsf{r}}_i \leq \eta_i \leq \tilde{\mathsf{r}}_i$ *for every* $1 \leq i \leq N$ *and* $\eta_i = \tilde{\mathsf{r}}_i$ *if* $\operatorname{dist}(x_i,\partial\Omega) \leq \eta_i$, *it holds that*

$$2\left(\mathsf{F}_N^\Omega(X_N,\mu) - \left(\frac{\#I_\Omega \log\lambda}{2}\right)\mathbf{1}_{\mathsf{s}=0}\right) + C\#I_\Omega N^{\frac{\mathsf{s}}{\mathsf{d}}}\|\mu\|_{L^\infty(\hat{\Omega})}\|\mu\|_{L^\infty(\Omega)}^{-1+\frac{\mathsf{s}}{\mathsf{d}}} \geq \begin{cases} \dfrac{1}{C}\displaystyle\sum_{i\in I_\Omega} \mathsf{g}(\eta_i) & \text{if } \mathsf{s}\neq 0 \\ \displaystyle\sum_{i\in I_\Omega} \mathsf{g}(\eta_i/\lambda) & \text{if } \mathsf{s}=0 \end{cases} \tag{4.5.4}$$

and

$$\int_{\Omega\times\mathbb{R}^\mathsf{k}} |y|^\gamma|\nabla h_{N,\vec{\eta}}|^2 \leq C\left(\mathsf{F}_N^\Omega(X_N,\mu) - \left(\frac{\#I_\Omega \log\lambda}{2}\right)\mathbf{1}_{\mathsf{s}=0}\right) + C\#I_\Omega N^{\frac{\mathsf{s}}{\mathsf{d}}}\|\mu\|_{L^\infty(\hat{\Omega})}\|\mu\|_{L^\infty(\Omega)}^{-1+\frac{\mathsf{s}}{\mathsf{d}}}, \tag{4.5.5}$$

where $C>0$ *depends only on* d *and* s.

For any $\vec{\eta}$ *satisfying* $\frac{1}{2}\mathsf{r}_i \leq \eta_i \leq \mathsf{r}_i$ *for every* $1 \leq i \leq N$, *it holds that*

$$\frac{1}{\mathsf{c}_{\mathsf{d},\mathsf{s}}}\int_{\Omega\times[-\lambda,\lambda]^\mathsf{k}} |y|^\gamma|\nabla h_{N,\vec{\eta}}|^2 + C\#I_\Omega N^{\frac{\mathsf{s}}{\mathsf{d}}}\|\mu\|_{L^\infty(\hat{\Omega})}\|\mu\|_{L^\infty(\Omega)}^{-1+\frac{\mathsf{s}}{\mathsf{d}}} \geq \begin{cases} \dfrac{1}{C}\displaystyle\sum_{i\in I_\Omega,\operatorname{dist}(x_i,\partial\Omega)\geq\frac{1}{4}\lambda} \mathsf{g}(\eta_i) & \text{if } \mathsf{s}\neq 0 \\ \displaystyle\sum_{i\in I_\Omega,\operatorname{dist}(x_i,\partial\Omega)\geq\frac{1}{4}\lambda} \mathsf{g}(\eta_i/\lambda) & \text{if } \mathsf{s}=0. \end{cases} \tag{4.5.6}$$

Now that we have (4.5.5) as a localization of (4.2.16), we can use it in conjunction with the local controls such as Lemma 4.20 and 4.25.

In order to prove this proposition let us take another look at the monotonicity property.

Lemma 4.29. *For* $\Omega \subset \mathbb{R}^\mathsf{d}$, *denoting temporarily*

$$\mathcal{F}^{\vec{\alpha}} := \frac{1}{2\mathsf{c}_{\mathsf{d},\mathsf{s}}}\left(\int_{\Omega\times\mathbb{R}^\mathsf{k}} |y|^\gamma|\nabla h_{N,\vec{\alpha}}|^2 - \mathsf{c}_{\mathsf{d},\mathsf{s}}\sum_{i\in I_\Omega}\mathsf{g}(\alpha_i) - 2N\mathsf{c}_{\mathsf{d},\mathsf{s}}\sum_{i\in I_\Omega}\int_{\mathbb{R}^\mathsf{d}} \mathsf{f}_{\alpha_i}(x-x_i)d\mu(x)\right), \tag{4.5.7}$$

Let $\vec{\eta}$, $\vec{\alpha}$, $\vec{\zeta}$ be such that $\zeta_i \le \eta_i \le \alpha_i$ for all i, with $\alpha_i = \eta_i$ if $\mathrm{dist}(x_i, \partial\Omega) \le \alpha_i$ and $\zeta_i = \eta_i$ if $\eta_i > \mathsf{r}_i$ or if $\mathrm{dist}(x_i, \partial\Omega) \le \eta_i$, then we have

$$\mathcal{F}^{\vec{\eta}} - \mathcal{F}^{\vec{\alpha}} \ge \frac{1}{2} \sum_{\substack{i,j \in I_\Omega, i \ne j \\ \mathrm{dist}(x_i, \partial\Omega) \ge \alpha_i}} \left(\mathsf{g}_{\zeta_i}(|x_i - x_j| + \zeta_j) - \mathsf{g}(\alpha_i)\right)_+ . \tag{4.5.8}$$

In particular for $\eta_i = \tilde{\mathsf{r}}_i$ and $\alpha_i \ge \eta_i$ such that $\alpha_i = \tilde{\mathsf{r}}_i$ if $\mathrm{dist}(x_i, \partial\Omega) \le \alpha_i$, we have

$$\mathsf{F}_N^\Omega(X_N, \mu) - \mathcal{F}^{\vec{\alpha}} \ge \frac{1}{2} \sum_{\substack{i,j \in I_\Omega, i \ne j \\ \mathrm{dist}(x_i, \partial\Omega) \ge \alpha_i, \tilde{\mathsf{r}}_i \le \mathsf{r}_i}} \left(\mathsf{g}(x_i - x_j) - \mathsf{g}(\alpha_i)\right)_+ . \tag{4.5.9}$$

PROOF. Let us consider $\vec{\eta}$ and $\vec{\zeta}$ such that $\zeta_i \le \eta_i$, and $\zeta_i = \eta_i$ if $B(x_i, \eta_i) \not\subset \Omega$. Then (4.2.9) can be rewritten as

$$\mathcal{F}^{\vec{\eta}} - \mathcal{F}^{\vec{\zeta}} = \frac{1}{2} \sum_{i \ne j} \int_{\mathbb{R}^{\mathsf{d}+\mathsf{k}}} \mathsf{f}_{\zeta_i, \eta_i}(z - x_i) d\left(\delta_{x_j}^{(\eta_i)} + \delta_{x_j}^{(\zeta_j)}\right)(z), \tag{4.5.10}$$

where we recall $\mathsf{f}_{\zeta,\eta} = \mathsf{f}_\zeta - \mathsf{f}_\eta$. The couples $i \ne j$ that contribute to sum on the right-hand side are those for which $\zeta_i \ne \eta_i$ and $B(x_i, \eta_i)$ intersects $B(x_j, \eta_j)$. Moreover, there is no contribution for points that do not satisfy $B(x_i, \eta_i) \subset \Omega$.

In particular, if $\eta_i \le \mathsf{r}_i$ for i such that $\eta_i \ne \zeta_i$, then the growing balls are disjoint, and we find that the right-hand side of (4.5.10) vanishes hence

$$\mathcal{F}^{\vec{\eta}} = \mathcal{F}^{\vec{\zeta}}. \tag{4.5.11}$$

Applying now (4.5.10) to $\vec{\alpha}$ and $\vec{\zeta}$ such that $\zeta_i \le \alpha_i$ with $\zeta_i = \alpha_i$ if $B(x_i, \alpha_i) \not\subset \Omega$, using the monotonicity of g and the definition of $\mathsf{f}_\alpha, \mathsf{f}_\zeta$, we find from (4.5.10), as in (4.2.10), that

$$\mathcal{F}^{\vec{\zeta}} - \mathcal{F}^{\vec{\alpha}} \ge \frac{1}{2} \sum_{\substack{i,j \in I_\Omega, i \ne j \\ B(x_i, \alpha_i) \subset \Omega}} \left(\mathsf{g}_{\zeta_i}(|x_i - x_j| + \zeta_j) - \mathsf{g}(\alpha_i)\right)_+ . \tag{4.5.12}$$

Combining (4.5.11) and (4.5.12), the result (4.5.8) follows. The relation (4.5.9) follows from (4.5.2) after choosing

$$\zeta_i = \begin{cases} 0 & \text{if } \eta_i \le \mathsf{r}_i \\ \eta_i & \text{if } \eta_i > \mathsf{r}_i \text{ or } \mathrm{dist}(x_i, \partial\Omega) \le \eta_i. \end{cases}$$

□

PROOF OF PROPOSITION 4.28. We apply the previous lemma, choosing $\alpha_i = \alpha := \frac{1}{4}\lambda$ for every i, $\zeta_i = \eta_i = \tilde{\mathsf{r}}_i$. These choices satisfy the assumptions of the lemma by definition of $\tilde{\mathsf{r}}_i$ in (4.5.1). We also notice that in $\mathcal{F}^{\vec{\eta}} - \mathcal{F}^{\vec{\alpha}}$, the integrals can be reduced to $\Omega \times [-\lambda, \lambda]^{\mathsf{k}}$ because $\nabla h_{N,\vec{\eta}}$ and $\nabla h_{N,\vec{\alpha}}$ coincide outside $\mathbb{R}^{\mathsf{d}} \times [-\lambda, \lambda]^{\mathsf{k}}$.

Applying inequality (4.5.8) and discarding nonnegative terms, by definition of $\tilde{\mathsf{r}}_i$ (4.5.1) and (4.5.2), we obtain

$$\mathsf{F}_N^\Omega(X_N,\mu) \geq -\frac{1}{2}\sum_{i\in I_\Omega} \mathsf{g}(\alpha) - N\|\mu\|_{L^\infty(\hat{\Omega})}\#I_\Omega\|\mathsf{f}_\alpha\|_{L^1} + \frac{1}{2}\sum_{\substack{i,j\in I_\Omega, i\neq j\\ \mathrm{dist}(x_i,\partial\Omega)\geq\alpha}} \left(\mathsf{g}_{\tilde{\mathsf{r}}_i}(|x_i-x_j|+\tilde{\mathsf{r}}_j)-\mathsf{g}(\alpha)\right)_+. \tag{4.5.13}$$

Let us first consider the case $\min_{j\neq i}|x_i-x_j|>\lambda$ in that case by definition of $\tilde{\mathsf{r}}_i$, we have $4\tilde{\mathsf{r}}_i=\lambda>\alpha$, and

$$\forall j\neq i, \qquad \left(\mathsf{g}_{\tilde{\mathsf{r}}_i}(|x_i-x_j|+\tilde{\mathsf{r}}_j)-\mathsf{g}(\alpha)\right)_+ = 0 > \mathsf{g}(4\tilde{\mathsf{r}}_i)-\mathsf{g}(\alpha) \geq \mathsf{g}(6\tilde{\mathsf{r}}_i)-\mathsf{g}(\alpha). \tag{4.5.14}$$

Next consider the case $\min_{j\neq i}|x_i-x_j|\leq\lambda$ and let j achieve the minimum, so that $|x_i-x_j|\leq 4\tilde{\mathsf{r}}_i$. We claim that $\tilde{\mathsf{r}}_j\leq 2\tilde{\mathsf{r}}_i$. If $\tilde{\mathsf{r}}_j\leq\tilde{\mathsf{r}}_i$ there is nothing to prove, so let us assume the converse. By triangle inequality, we have $\mathrm{dist}(x_j,\partial\Omega)\geq \mathrm{dist}(x_i,\partial\Omega)-4\tilde{\mathsf{r}}_i$, so $t_j\geq t_i-\frac{4\tilde{\mathsf{r}}_i}{\lambda}$ where t_i and t_j are the corresponding t's in the definition (4.5.1). It follows that $\tilde{\mathsf{r}}_j\leq\tilde{\mathsf{r}}_i-\frac{\tilde{\mathsf{r}}_i}{\lambda}(|x_i-x_j|-\lambda)\leq 2\tilde{\mathsf{r}}_i$, proving the claim.

In all cases, there exists $j\neq i$ such that, by definition and monotonicity of $\mathsf{g}_{\tilde{\mathsf{r}}_i}$,

$$\left(\mathsf{g}_{\tilde{\mathsf{r}}_i}(|x_i-x_j|+\tilde{\mathsf{r}}_j)-\mathsf{g}(\alpha)\right)_+ \geq \mathsf{g}(6\tilde{\mathsf{r}}_i)-\mathsf{g}(\alpha). \tag{4.5.15}$$

Keeping only that j in the sum, it follows from (4.5.13), in view of (4.1.27) applied to the factor $\|\mathsf{f}_\alpha\|_{L^1}$ from (4.5.13), that if $\mathsf{s}\neq 0$,

$$\frac{1}{2}\sum_{\substack{i\in I_\Omega\\ \mathrm{dist}(x_i,\partial\Omega)\geq\alpha}} \mathsf{g}(6\tilde{\mathsf{r}}_i) \leq \mathsf{F}_N^\Omega(X_N,\mu)+\#I_\Omega\mathsf{g}(\alpha)+CN\|\mu\|_{L^\infty(\hat{\Omega})}\#I_\Omega\alpha^{\mathsf{d}-\mathsf{s}}. \tag{4.5.16}$$

Now in view of our choice of α and the definition of $\tilde{\mathsf{r}}_i$, if $x_i\in\Omega$ with $\mathrm{dist}(x_i,\partial\Omega)<\alpha$, then $\tilde{\mathsf{r}}_i=\alpha$ by definition. Hence, reinserting such points, we have

$$\frac{1}{2}\sum_{i\in I_\Omega}\mathsf{g}(6\tilde{\mathsf{r}}_i)\leq \mathsf{F}_N^\Omega(X_N,\mu)+\#I_\Omega\mathsf{g}(\alpha)+C\|\mu\|_{L^\infty(\hat{\Omega})}\#I_\Omega N\alpha^{\mathsf{d}-\mathsf{s}}+\frac{1}{2}\#I_\Omega\mathsf{g}(6\alpha). \tag{4.5.17}$$

Inserting the definition of α into this inequality and in view of our requirement that $\frac{1}{2}\tilde{\mathsf{r}}_i\leq\eta_i\leq\tilde{\mathsf{r}}_i$, we conclude that (4.5.4) holds if $\mathsf{s}\neq 0$. If $\mathsf{s}=0$, starting from (4.5.13) and using the same reasoning and the properties of the log, we arrive at the inequality

$$\sum_{i\in I_\Omega}\mathsf{g}(6\tilde{\mathsf{r}}_i/\alpha)\leq 2\left(\mathsf{F}_N^\Omega(X_N,\mu)+\frac{\#I_\Omega}{2}\mathsf{g}(\alpha)+C\|\mu\|_{L^\infty(\hat{\Omega})}\#I_\Omega N\alpha^{\mathsf{d}-\mathsf{s}}\right)+\#I_\Omega\mathsf{g}(6),$$

and the conclusion follows as well, absorbing $\mathsf{g}(C)$ terms into $\|\mu\|_{L^\infty}^{\mathsf{s}/\mathsf{d}}=1$.

For the proof of (4.5.6), for $\eta_i\in[\frac{1}{2}\tilde{\mathsf{r}}_i,\tilde{\mathsf{r}}_i]$ we let $\alpha_i=\eta_i$ if $\mathrm{dist}(x_i,\partial\Omega)\leq\frac{1}{4}\lambda$, $\alpha_i=\frac{1}{4}\lambda$ otherwise, and $\zeta_i=\eta_i$, and obtain from (4.5.8), discarding a nonnegative term, that

$$\frac{1}{2\mathsf{c_d}}\int_{\Omega\times[-\lambda,\lambda]^{\mathsf{k}}}|y|^\gamma|\nabla h_{N,\vec{\eta}}|^2-\frac{1}{2}\sum_{i\in I_\Omega,\alpha_i\neq\eta_i}(\mathsf{g}(\eta_i)-\mathsf{g}(\alpha_i))$$
$$\geq\frac{1}{2}\sum_{\substack{i,j\in I_\Omega,i\neq j\\ \mathrm{dist}(x_i,\partial\Omega)\geq\alpha_i}}(\mathsf{g}_{\eta_i}(|x_i-x_j|+\eta_j)-\mathsf{g}(\alpha_i))_+-CN\|\mu\|_{L^\infty(\hat{\Omega})}\sum_{i\in I_\Omega}\|\mathsf{f}_{\alpha_i}\|_{L^1} \tag{4.5.18}$$

The rest of the proof is entirely analogous.

We next turn to showing (4.5.5). Let us choose $\alpha_i=\eta_i$ (with $\eta_i\in[\frac{1}{2}\tilde{\mathsf{r}}_i,\tilde{\mathsf{r}}_i]$) in (4.5.9), where we replace the right-hand side by 0. Using that $\tilde{\mathsf{r}}_i\leq\frac{1}{4}\lambda$, we deduce, using again (4.1.27), that

$$\mathsf{F}_N^\Omega(X_N,\mu)$$
$$\geq\frac{1}{2\mathsf{c_{d,s}}}\left(\int_{\Omega\times[-\lambda,\lambda]^{\mathsf{k}}}|y|^\gamma|\nabla h_{N,\vec{\eta}}|^2-\mathsf{c_{d,s}}\sum_{i\in I_\Omega}\mathsf{g}(\eta_i)\right)-C\#I_\Omega\|\mu\|_{L^\infty(\hat{\Omega})}\|\mu\|_{L^\infty(\Omega)}^{-1+\frac{\mathsf{s}}{\mathsf{d}}}N^{\frac{\mathsf{s}}{\mathsf{d}}}, \tag{4.5.19}$$

and in view of (4.5.4), (4.5.5) follows. In the case where $\mathsf{s}=0$, we write $\mathsf{g}(\eta_i)=\mathsf{g}(\eta_i/\lambda)+\mathsf{g}(\lambda)$, and then apply (4.5.4). □

CHAPTER 5

Splittings, concentration, and separation estimates

In Chapter 3, we examined the leading order behavior of the Hamiltonian $\mathcal{H}_N$ of (1.1.1) with g as in (2.0.1), which can be summarized as follows.

- The minimal energy $\min \mathcal{H}_N$ behaves like $N^2 \min \mathcal{E}$, where $\mathcal{E}$ is the mean-field limit energy, defined on the set of probability measures of $\mathbb{R}^{\mathsf{d}}$.
- If each $(x_1, \dots, x_N)$ minimizes $\mathcal{H}_N$, the empirical measure
$$\widehat{\mu_N} = \frac{1}{N} \sum_{i=1}^{N} \delta_{x_i}$$
converges weakly to the unique minimizer μ_V of $\mathcal{E}$, also known as Frostman's equilibrium measure, which can be characterized via an obstacle problem.
- This behavior also holds for typical states under the Gibbs measure law (1.1.5) when $\theta \to \infty$, where $\theta = \beta N^{1-\frac{\mathsf{s}}{\mathsf{d}}}$, except with a very small probability determined by a large deviation principle.

The following questions thus arise naturally:

(1) What lies beyond the term $N\theta\mathcal{E}(\mu_V)$, respectively $N\theta\mathcal{E}_\theta(\mu_\theta)$ in the expansion of $\mathcal{H}_N$ or of the free energy $-\frac{1}{\beta}\log Z_{N,\beta}$?

(2) What is the optimal *microscopic* distribution of the points?

To study these questions, we wish to zoom or blow-up the configurations by the factor $N^{1/\mathsf{d}}$ (the inverse of the typical distance between two points or microscale), so that the points are well-separated, and find a way of expanding the Hamiltonian to next order. This very simple "splitting", or quadratic expansion, was first introduced in the context of Coulomb gases in [**SS15b**].

This will directly lead us to the next order (or modulated electric) energy F_N of Chapter 4, whose coercivity properties we can then exploit.

Henceforth, we will make the assumption that V satisfies (A1)–(A5), thus the Frostman equilibrium measure μ_V and the thermal equilibrium measure μ_θ uniquely exist, and we assume that μ_V is absolutely continuous with respect to the Lebesgue measure and has a bounded density.

We present two splitting formulas: one with respect to the equilibrium measure, which is the appropriate one when studying energy minimizers, and one with respect to the thermal equilibrium measure, more appropriate when studying the Gibbs measure. They lead, after subtracting off constant leading order terms, to F_N becoming the main energy of the system. Combining with the results of Chapter 4 on F_N, one easily deduces energy and free energy bounds and first concentration estimates.

Let us point out that the splitting and in particular the electric formulation of the Gibbs measure have been instrumental in a number of studies that go beyond this text: they have served for instance to obtain DLR equations for the one-dimensional log gas in [**DHLM21, Leb24**], obtain CLT for the linear statistics of the sine-beta process [**Leb21**], and to study the maximum of the log-gas potential in [**LLZ24, Pei25**].

At the end of the chapter we will describe separation results for energy minimizers, and mention other approaches to such results in the literature which apply to the Gibbs measure as well, in particular the isotropic averaging method of [**Tho24**].

5.1. Splitting the Hamiltonian

5.1.1. Splitting with respect to the equilibrium measure. The splitting consists in an exact formula that separates the leading (N^2) order term in $\mathcal{H}_N$ from next order terms. It suffices to expand around μ_V by writing $\widehat{\mu_N} = \mu_V + (\widehat{\mu_N} - \mu_V)$. We note that since μ_V is compactly supported, as seen in the proof of Theorem 2.8, it satisfies (4.1.1) and thus $\mathsf{F}_N(\cdot, \mu_V)$ is well defined.

Lemma 5.1 (Splitting formula). *Assume μ_V is absolutely continuous with respect to the Lebesgue measure. For any N and any $X_N \in (\mathbb{R}^\mathsf{d})^N$ we have*

$$\mathcal{H}_N(X_N) = N^2\mathcal{E}(\mu_V) + N\sum_{i=1}^N \zeta(x_i) + \mathsf{F}_N(X_N, \mu_V), \tag{5.1.1}$$

where $\triangle$ denotes the diagonal of $\mathbb{R}^\mathsf{d} \times \mathbb{R}^\mathsf{d}$, $\mathcal{E}$ is as in (2.0.3)*, ζ as in* (2.1.12) *and F_N as in* (4.1.2).

Proof. We may write

$$\begin{aligned}
\mathcal{H}_N(X_N) &= \frac{1}{2}\sum_{i\neq j} \mathsf{g}(x_i - x_j) + N\sum_{i=1}^N V(x_i) \\
&= \frac{N^2}{2}\iint_{\triangle^c} \mathsf{g}(x-y)d\widehat{\mu_N}(x)d\widehat{\mu_N}(y) + N^2\int_{\mathbb{R}^\mathsf{d}} V d\widehat{\mu_N}(x) \\
&= \frac{N^2}{2}\iint_{\triangle^c} \mathsf{g}(x-y)d\mu_V(x)d\mu_V(y) + N^2\int_{\mathbb{R}^\mathsf{d}} V d\mu_V \\
&\quad + N\iint_{\triangle^c} \mathsf{g}(x-y)d\mu_V(x)d\left(\sum_{i=1}^N \delta_{x_i} - N\mu_V\right)(y) \\
&\quad + N\int_{\mathbb{R}^\mathsf{d}} V d\left(\sum_{i=1}^N \delta_{x_i} - N\mu_V\right) \\
&\quad + \frac{1}{2}\iint_{\triangle^c} \mathsf{g}(x-y)d\left(\sum_{i=1}^N \delta_{x_i} - N\mu_V\right)(x)d\left(\sum_{i=1}^N \delta_{x_i} - N\mu_V\right)(y).
\end{aligned} \tag{5.1.2}$$

We now recall that ζ was defined in (2.1.12) by

$$\zeta = h^{\mu_V} + V - c = \int_{\mathbb{R}^\mathsf{d}} \mathsf{g}(\cdot - y)\, d\mu_V(y) + V - c. \tag{5.1.3}$$

We may then rewrite the medium line in the right-hand side of (5.1.2) as

$$\begin{aligned}
N \iint_{\triangle^c} \mathsf{g}(x-y) d\mu_V(x) d\left(\sum_{i=1}^N \delta_{x_i} - N\mu_V\right)(y) + N \int_{\mathbb{R}^\mathsf{d}} V d\left(\sum_{i=1}^N \delta_{x_i} - N\mu_V\right) \\
= N \int_{\mathbb{R}^\mathsf{d}} (h^{\mu_V} + V) d\left(\sum_{i=1}^N \delta_{x_i} - N\mu_V\right) = N \int_{\mathbb{R}^\mathsf{d}} (\zeta + c) d\left(\sum_{i=1}^N \delta_{x_i} - N\mu_V\right) \\
= N^2 \int_{\mathbb{R}^\mathsf{d}} \zeta d\widehat{\mu_N} - N^2 \int_{\mathbb{R}^\mathsf{d}} \zeta d\mu_V + Nc \int_{\mathbb{R}^\mathsf{d}} d\left(\sum_{i=1}^N \delta_{x_i} - N\mu_V\right) = N^2 \int_{\mathbb{R}^\mathsf{d}} \zeta d\widehat{\mu_N}.
\end{aligned}$$

The last equality is due to the facts that $\zeta = 0$ μ_V-a.e. as seen in (2.1.9) and that $\widehat{\mu_N}$ and μ_V are both probability measures. We also have to notice that since μ_V is absolutely continuous with respect to the Lebesgue measure, we may include the diagonal back into the domain of integration. By that same argument, one may recognize in the first line of the right-hand side of (5.1.2), the quantity $N^2\mathcal{E}(\mu_V)$. One then recognizes the last line of (5.1.2) as $\mathsf{F}_N(X_N, \mu_V)$. This concludes the proof. □

The function ζ then appears as an effective confinement potential, which only acts outside $\Sigma = \operatorname{supp} \mu_V$ and localizes the particles to Σ. We discuss this localization further in Section 5.3.1 below. With this splitting, we may also rewrite the Gibbs measure (1.1.5) as

(5.1.4)
$$d\mathbb{P}_{N,\beta}(X_N) = \frac{e^{-\theta N \mathcal{E}(\mu_V)}}{Z_{N,\beta}} \exp\left(-\beta N^{-\frac{\mathsf{s}}{\mathsf{d}}}\left(\mathsf{F}_N(X_N, \mu_V) + N\sum_{i=1}^N \zeta(x_i)\right)\right) dX_N.$$

This leads us to introducing reduced measures and partition functions as follows:

$$(5.1.5) \quad \tilde{\mathsf{K}}_{N,\beta}(\mu, \zeta) := \int_{(\mathbb{R}^\mathsf{d})^N} \exp\left(-\beta N^{-\frac{\mathsf{s}}{\mathsf{d}}}\left(\mathsf{F}_N(X_N, \mu) + N\sum_{i=1}^N \zeta(x_i)\right)\right) dX_N$$

and

(5.1.6)
$$\tilde{\mathbb{Q}}_{N,\beta}(\mu, \zeta) := \frac{1}{\tilde{\mathsf{K}}_{N,\beta}(\mu, \zeta)} \exp\left(-\beta N^{-\frac{\mathsf{s}}{\mathsf{d}}}\left(\mathsf{F}_N(X_N, \mu) + N\sum_{i=1}^N \zeta(x_i)\right)\right) dX_N,$$

which make sense for any probability density μ for which $\mathsf{F}_N(\cdot, \mu)$ is defined, and ζ growing fast enough at infinity. With this notation, we may rewrite

$$(5.1.7) \qquad d\mathbb{P}_{N,\beta}(X_N) = d\tilde{\mathbb{Q}}_{N,\beta}(\mu_V, \zeta), \qquad Z_{N,\beta} = e^{-\theta N \mathcal{E}(\mu_V)} \tilde{\mathsf{K}}_{N,\beta}(\mu_V, \zeta),$$

with the ζ of (2.1.12).

5.1.2. Splitting with respect to the thermal equilibrium measure. It is advantageous to split with respect to μ_θ, the minimizer of (2.0.4), where $\theta = \beta N^{1-\frac{\mathsf{s}}{\mathsf{d}}}$, which may depend on N. As explained in Chapter 2, when $\theta \gg 1$, μ_θ can be considered as an N-dependent deterministic correction to μ_V.

We recall that under (A1)–(A5), in view of Propositions 2.18 and 2.21, the thermal equilibrium measure μ_θ minimizing (2.0.4) exists and satisfies

$$h^{\mu_\theta} + V + \frac{1}{\theta} \log \mu_\theta = c_\theta \quad \text{in } \mathbb{R}^\mathsf{d}, \tag{5.1.8}$$

where c_θ is a constant. Moreover, thanks to (2.5.5), we have that (4.1.1) holds and $\mathsf{F}_N(\cdot, \mu_\theta)$ is well-defined.

Lemma 5.2 (Splitting formula with the thermal equilibrium measure). *For any configuration $X_N \in (\mathbb{R}^\mathsf{d})^N$, we have*

$$\mathcal{H}_N(X_N) = N^2 \mathcal{E}_\theta(\mu_\theta) - \frac{N}{\theta} \sum_{i=1}^N \log \mu_\theta(x_i) + \mathsf{F}_N(X_N, \mu_\theta), \tag{5.1.9}$$

where $\mathcal{E}_\theta$ is as in (2.0.4) and F_N as in (4.1.2).

PROOF. Similarly as in (5.1.2), it suffices to rewrite $\mathcal{H}_N(X_N)$ as

$$\mathcal{H}_N(X_N) = \frac{1}{2} \iint_{\triangle^c} \mathsf{g}(x-y) d\left(\sum_{i=1}^N \delta_{x_i}\right)(x) d\left(\sum_{i=1}^N \delta_{x_i}\right)(y) \\ + N \int_{\mathbb{R}^\mathsf{d}} V(x) d\left(\sum_{i=1}^N \delta_{x_i}\right)(x),$$

and expand the integral after writing $\sum_{i=1}^N \delta_{x_i} = N\mu_\theta + \left(\sum_{i=1}^N \delta_{x_i} - N\mu_\theta\right)$, to find

$$\begin{aligned} \mathcal{H}_N(X_N) &= N^2 \mathcal{E}(\mu_\theta) + N \iint_{\triangle^c} \mathsf{g}(x-y) d\left(\sum_{i=1}^N \delta_{x_i} - N\mu_\theta\right)(x) d\mu_\theta(y) \\ &\quad + N \int_{\mathbb{R}^\mathsf{d}} V(x) d\left(\sum_{i=1}^N \delta_{x_i} - N\mu_\theta\right) + \mathsf{F}_N(X_N, \mu_\theta) \\ &= N^2 \mathcal{E}(\mu_\theta) + N \int_{\mathbb{R}^\mathsf{d}} (h^{\mu_\theta} + V) d\left(\sum_{i=1}^N \delta_{x_i} - N\mu_\theta\right) + \mathsf{F}_N(X_N, \mu_\theta). \end{aligned}$$

Inserting (5.1.8), we obtain

$$\mathcal{H}_N(X_N) = N^2 \mathcal{E}(\mu_\theta) - \frac{N}{\theta}\left(\sum_{i=1}^N \log \mu_\theta(x_i) - N \int_{\mathbb{R}^\mathsf{d}} \mu_\theta \log \mu_\theta\right) + \mathsf{F}_N(X_N, \mu_\theta),$$

hence the result by definition of $\mathcal{E}_\theta$ (2.0.4). □

This splitting works out nicely when inserted into the Gibbs measure definition: it then yields as an alternate to (5.1.4)

$$d\mathbb{P}_{N,\beta}(X_N) = \frac{e^{-\theta N \mathcal{E}_\theta(\mu_\theta)}}{Z_{N,\beta}} \exp\left(-\beta N^{-\frac{\mathsf{s}}{\mathsf{d}}} \mathsf{F}_N(X_N, \mu_\theta)\right) d\mu_\theta(x_1) \ldots d\mu_\theta(x_N). \tag{5.1.10}$$

The measure is thus made absolutely continuous with respect to the probability measure $\mu_\theta^{\otimes N}$ instead of the (nonintegrable) Lebesgue measure dX_N in (5.1.4), and a confinement potential is no longer needed. The only disadvantage of this more precise description is that μ_θ still depends on N, but this is remedied by the fact that we can have precise estimates for μ_θ in Σ, as seen in Theorem 2.23. We can thus rewrite $\mathbb{P}_{N,\beta}$ as

$$d\mathbb{P}_{N,\beta}(X_N) = \frac{1}{\mathsf{K}_{N,\beta}(\mu_\theta)} \exp\left(-\beta N^{-\frac{\mathsf{s}}{\mathsf{d}}} \mathsf{F}_N(X_N, \mu_\theta)\right) d\mu_\theta(x_1)\dots d\mu_\theta(x_N), \tag{5.1.11}$$

where we let, for any probability density μ,

$$\mathsf{K}_{N,\beta}(\mu) := \int_{(\mathbb{R}^\mathsf{d})^N} \exp\left(-\beta N^{-\frac{\mathsf{s}}{\mathsf{d}}} \mathsf{F}_N(X_N, \mu)\right) d\mu(x_1)\dots d\mu(x_N). \tag{5.1.12}$$

Definition 5.3. The *modulated Gibbs measure*[1] with respect to a reference probability density μ is defined by

$$d\mathbb{Q}_{N,\beta}(\mu) := \frac{1}{\mathsf{K}_{N,\beta}(\mu)} \exp\left(-\beta N^{-\frac{\mathsf{s}}{\mathsf{d}}} \mathsf{F}_N(X_N, \mu)\right) d\mu^{\otimes N}(X_N). \tag{5.1.13}$$

Using the definition, we have thus obtained

$$d\mathbb{P}_{N,\beta} = d\mathbb{Q}_{N,\beta}(\mu_\theta), \qquad Z_{N,\beta} = \exp\left(-\theta N \mathcal{E}_\theta(\mu_\theta)\right) \mathsf{K}_{N,\beta}(\mu_\theta). \tag{5.1.14}$$

In both cases, we have thus reduced to studying F_N, whose main properties were explored in the previous chapter.

When one wishes to go back and forth between the two representations (5.1.14) and (5.1.11), one may use that for $\mu > 0$,

(5.1.15)

$$\begin{aligned}
&\frac{\tilde{\mathsf{K}}_{N,\beta}(\mu,\zeta)}{\mathsf{K}_{N,\beta}(\mu)} \\
&= \frac{1}{\mathsf{K}_{N,\beta}(\mu)} \int_{(\mathbb{R}^\mathsf{d})^N} \exp\left(-\beta N^{-\frac{\mathsf{s}}{\mathsf{d}}}\left(\mathsf{F}_N(X_N,\mu) + N\sum_{i=1}^N \zeta(x_i)\right) - \sum_{i=1}^N \log\mu(x_i)\right) d\mu^{\otimes N}(X_N) \\
&= \mathbb{E}_{\mathbb{Q}_{N,\beta}(\mu)}\left(\exp\left(-\beta N^{1-\frac{\mathsf{s}}{\mathsf{d}}} \sum_{i=1}^N \zeta(x_i) - \sum_{i=1}^N \log\mu(x_i)\right)\right) \\
&= \exp\left(-\theta N \int_{\mathbb{R}^\mathsf{d}} \zeta d\mu - N\int_{\mathbb{R}^\mathsf{d}} \mu\log\mu\right) \mathbb{E}_{\mathbb{Q}_{N,\beta}(\mu)}\left(e^{\mathrm{Fluct}_\mu(-\theta\zeta - \log\mu)}\right),
\end{aligned}$$

where

$$\mathrm{Fluct}_\mu(\xi) := \sum_{i=1}^N \xi(x_i) - N\int_{\mathbb{R}^\mathsf{d}} \xi d\mu.$$

A relative entropy control can be deduced from the rewriting of the Gibbs measure by splitting combined with the lower bound for F_N. It ensures that the Gibbs measure is close to $\mu_\theta^{\otimes N}$ when $\beta \ll 1$. This observation, recorded below, is due to Zhenfu Wang [**Wan24**].

Remark 5.4 (Concentration in relative entropy around the thermal equilibrium measure)**.** Assume $\mathsf{s} > 0$. Let the normalized relative entropy H_N be defined[2] by

$$H_N(f_N|\mu^{\otimes N}) = N\int_{(\mathbb{R}^\mathsf{d})^N} f_N \log\frac{f_N}{\mu^{\otimes N}} dX_N = N\mathrm{ent}(f_N|\mu^{\otimes N}). \tag{5.1.16}$$

---

[1]This is by analogy with the modulated energy, but it could also be called a relative Gibbs measure. Its connection with the modulated energy and modulated free energy is explored in [**RS25b**].

[2]It is the quantity $\frac{1}{N^2}H_N$ which quantifies the convergence of f_N to $\mu^{\otimes N}$; see the discussion at the beginning of Section 6.2.5.

Then we have

$$\frac{1}{N^2} H_N(\mathbb{Q}_{N,\beta}(\mu)|\mu^{\otimes N}) \le C\beta, \tag{5.1.17}$$

for C depending only on d, s and $\|\mu\|_{L^\infty}$. In particular,

$$\frac{1}{N^2} H_N(\mathbb{P}_{N,\beta}|\mu_\theta^{\otimes N}) \le C\beta, \tag{5.1.18}$$

where μ_θ is the thermal equilibrium measure.

PROOF. By definition of the relative entropy and (5.1.13), we have

$$\frac{\mathbb{Q}_{N,\beta}(\mu)}{\mu^{\otimes N}} = \frac{1}{\mathsf{K}_{N,\beta}(\mu)} \exp\left(-\beta N^{-\frac{\mathsf{s}}{\mathsf{d}}} \mathsf{F}_N(X_N, \mu)\right),$$

hence by definition (5.1.16) we find that

$$H_N(\mathbb{Q}_{N,\beta}(\mu)|\mu^{\otimes N}) + H_N(\mu^{\otimes N}|\mathbb{Q}_{N,\beta}(\mu))$$
$$= -\beta N^{1-\frac{\mathsf{s}}{\mathsf{d}}} \left(\int_{(\mathbb{R}^\mathsf{d})^N} \mathsf{F}_N(X_N,\mu) d\mathbb{Q}_{N,\beta}(\mu) - \int_{(\mathbb{R}^\mathsf{d})^N} \mathsf{F}_N(X_N,\mu) d\mu^{\otimes N} \right).$$

For the second integral in the right-hand side, we compute as in (5.2.25) below that

$$\int_{(\mathbb{R}^\mathsf{d})^N} \mathsf{F}_N(X_N,\mu) d\mu^{\otimes N} = -\frac{N}{2} \iint \mathsf{g}(x-y) d\mu(x) d\mu(y) \le 0$$

under the assumption $\mathsf{s} > 0$, while for the first, we insert the lower bound (4.2.12), written as

$$\mathsf{F}_N(X_N,\mu) \ge -CN^{1+\frac{\mathsf{s}}{\mathsf{d}}}.$$

By nonnegativity of the relative entropy, we thus obtain as claimed

$$H_N(\mathbb{Q}_{N,\beta}(\mu)|\mu^{\otimes N}) \le C\beta N^2.$$

The second relation follows by (5.1.14). □

5.2. Free energy bounds and concentration

5.2.1. Lower bound for the (free) energy. The properties of F_N shown in Chapter 4 allow us to immediately deduce a few statements. The first is a bound on the free energy obtained from using that F_N is (almost) bounded below. Combining (5.1.1), Corollary 4.14 and the fact that $\zeta \ge 0$, we easily obtain

Corollary 5.5 (First energy lower bound). *If $\mu_V \in L^\infty(\mathbb{R}^\mathsf{d})$, then for any configuration X_N in $(\mathbb{R}^\mathsf{d})^N$, we have*

$$\mathcal{H}_N(X_N) \ge N^2 \mathcal{E}(\mu_V) + N \sum_{i=1}^N \zeta(x_i) - \left(\frac{N}{2\mathsf{d}} \log(N\|\mu_V\|_{L^\infty})\right) \mathbf{1}_{\mathsf{s}=0} - C\|\mu_V\|_{L^\infty}^{\frac{\mathsf{s}}{\mathsf{d}}} N^{1+\frac{\mathsf{s}}{\mathsf{d}}} \mathbf{1}_{\mathsf{s}\ge 0}, \tag{5.2.1}$$

where C depends only on d and s. In particular

$$\min \mathcal{H}_N \ge N^2 \mathcal{E}(\mu_V) - \left(\frac{N}{2\mathsf{d}} \log(N\|\mu_V\|_{L^\infty})\right) \mathbf{1}_{\mathsf{s}=0} - C\|\mu_V\|_{L^\infty}^{\frac{\mathsf{s}}{\mathsf{d}}} N^{1+\frac{\mathsf{s}}{\mathsf{d}}} \mathbf{1}_{\mathsf{s}\ge 0}. \tag{5.2.2}$$

This lower bound easily translates into a lower bound for the free energy $-\frac{1}{\beta}\log Z_{N,\beta}$ or the generic modulated free energy $-\frac{1}{\beta}\log \mathsf{K}_{N,\beta}(\mu)$ of (5.1.12).

Corollary 5.6 (Upper bound for the partition function—thermal equilibrium measure version). *Assume that μ and μ_θ are L^∞ probability densities. Then for all $\beta > 0$, and for N large enough, we have*

$$\log \mathsf{K}_{N,\beta}(\mu) \le \beta\Big(\frac{N}{2\mathsf{d}}\log(N\|\mu\|_{L^\infty})\Big)\mathbf{1}_{\mathsf{s}=0} + C\beta N\|\mu\|_{L^\infty}^{\frac{\mathsf{s}}{\mathsf{d}}}\mathbf{1}_{\mathsf{s}\ge 0} \tag{5.2.3}$$

and

$$-\frac{1}{\beta}\log Z_{N,\beta} \ge N^{2-\frac{\mathsf{s}}{\mathsf{d}}}\mathcal{E}_\theta(\mu_\theta) - \Big(\frac{N}{2\mathsf{d}}\log(N\|\mu_\theta\|_{L^\infty})\Big)\mathbf{1}_{\mathsf{s}=0} - CN\|\mu_\theta\|_{L^\infty}^{\frac{\mathsf{s}}{\mathsf{d}}}\mathbf{1}_{\mathsf{s}\ge 0}, \tag{5.2.4}$$

where C depends only on s and d.

PROOF. It suffices to insert the result of Corollary 4.14 into (5.1.12) or (5.1.10). □

We next turn to the analogous result expressed with the regular equilibrium measure.

Corollary 5.7 (Upper bound for the partition function—regular equilibrium measure version). *Assume* (A1)–(A4) *so that μ_V exists and is compactly supported. Assume also that μ_V has a bounded density. Then for all $\beta > 0$ and for N large enough, we have*

$$-\frac{1}{\beta}\log Z_{N,\beta} \ge N^{2-\frac{\mathsf{s}}{\mathsf{d}}}\mathcal{E}(\mu_V) - \Big(\frac{N}{2\mathsf{d}}\log(N\|\mu_V\|_{L^\infty})\Big)\mathbf{1}_{\mathsf{s}=0} - CN\|\mu_V\|_{L^\infty}^{\frac{\mathsf{s}}{\mathsf{d}}}\mathbf{1}_{\mathsf{s}\ge 0} - \frac{N}{\beta}C_\zeta, \tag{5.2.5}$$

where C depends only on s and d, and C_ζ depends on the integral in (A4) *and on ζ.*

PROOF. We insert (5.2.1) into (1.1.6) and obtain

$$\log Z_{N,\beta} \le -\beta N^{2-\frac{\mathsf{s}}{\mathsf{d}}}\mathcal{E}(\mu_V) + \beta\frac{N}{2\mathsf{d}}\log(N\|\mu_V\|_{L^\infty})\mathbf{1}_{\mathsf{s}=0} + \beta CN\|\mu_V\|_{L^\infty}^{\frac{\mathsf{s}}{\mathsf{d}}}\mathbf{1}_{\mathsf{s}\ge 0}$$
$$+ \log\int_{(\mathbb{R}^{\mathsf{d}})^N}\exp\left(-\theta\sum_{i=1}^N\zeta(x_i)\right)dX_N.$$

By separation of variables,

$$\log\int_{(\mathbb{R}^{\mathsf{d}})^N}\exp\left(-\theta\sum_{i=1}^N\zeta(x_i)\right)dX_N = N\log\int_{\mathbb{R}^{\mathsf{d}}}\exp(-\theta\zeta(x))dx.$$

In view of (2.1.14) we have $\zeta \ge \mathsf{g}+V-C$, thus in view of (A4) the last integral above converges. Either θ is fixed and this is a fixed constant, or $\theta \to +\infty$ as $N \to \infty$ and since $\zeta \ge 0$ the integral in the right-hand side converges monotonically to $|\{\zeta = 0\}| = |\omega|$. Thus in all cases, we obtain the result. □

Remark 5.8. As discussed in the proof, C_ζ is bounded below by a positive constant independent of N, thus the estimate is less good than (5.2.4), particularly when β is small, illustrating again that the thermal equilibrium measure provides a more precise description than the regular equilibrium measure.

To obtain a converse bound on the free energy, it will be more convenient to work in the "blown-up scale" which we now introduce.

5.2.2. Rescaling the energy by blow up. We now define electric potentials in blown-up coordinates $x' = N^{1/\mathsf{d}}x$, which turn out to be more natural in several instances. For a measure μ we let $\mu'(x) = \mu(xN^{-1/\mathsf{d}})$ and observe that if μ is a probability density then $\int_{\mathbb{R}^\mathsf{d}} \mu' = N$. We let

$$x'_i = N^{1/\mathsf{d}}x_i, \quad X'_N = (x'_1, \dots, x'_N) \tag{5.2.6}$$

be the blown-up configuration. Analogously to (4.1.2), we may define

$$\mathsf{F}(X'_N, \mu') = \frac{1}{2}\iint_{\triangle^c} \mathsf{g}(x-y)d\left(\sum_{i=1}^N \delta_{x'_i} - \mu'\right)(x)d\left(\sum_{i=1}^N \delta_{x'_i} - \mu'\right)(y), \tag{5.2.7}$$

where this time the subscript N is absent. Changing variables in (5.2.7) we find the rescaling formula for the modulated energy

$$N^{-\frac{\mathsf{s}}{\mathsf{d}}}\left(\mathsf{F}_N(X_N, \mu) + \left(\frac{N}{2\mathsf{d}}\log N\right)\mathbf{1}_{\mathsf{s}=0}\right) = \mathsf{F}(X'_N, \mu'). \tag{5.2.8}$$

This explains the recurrent appearance of the additive term $\left(\frac{N}{2\mathsf{d}}\log N\right)\mathbf{1}_{\mathsf{s}=0}$ in the logarithmic case $\mathsf{s}=0$, and in that case one should think of the full sum $\mathsf{F}_N(X_N, \mu) + \left(\frac{N}{2\mathsf{d}}\log N\right)\mathbf{1}_{\mathsf{s}=0}$ as being the energy, instead of just $\mathsf{F}_N(X_N, \mu)$.

We have seen in Remark 4.15 that we expect $\mathsf{F}_N(X_N, \mu) + \left(\frac{N}{2\mathsf{d}}\log N\right)\mathbf{1}_{\mathsf{s}=0}$ to be of order $N^{1+\frac{\mathsf{s}}{\mathsf{d}}}$, this way in view of (5.2.8) the blown-up energy $\mathsf{F}(X'_N, \mu')$ will be of order N. In other words, $\mathsf{F}(X'_N, \mu')$ then becomes proportional to the number of points, or to the volume $N^{1/\mathsf{d}}|\Sigma|$ effectively occupied by the zoomed gas.

We may now drop the primes and consider the following general definition.

Definition 5.9 (Blown-up scale modulated electric energy)**.** For any configuration $X_N \in (\mathbb{R}^\mathsf{d})^N$ and any density μ with $\int_{\mathbb{R}^\mathsf{d}} \mu = N$, the blown-up scale modulated energy is defined by

$$\mathsf{F}(X_N, \mu) := \frac{1}{2}\iint_{\triangle^c} \mathsf{g}(x-y)d\left(\sum_{i=1}^N \delta_{x_i} - \mu\right)(x)d\left(\sum_{i=1}^N \delta_{x_i} - \mu\right)(y). \tag{5.2.9}$$

We may now define a general blown-up scale partition function and modulated Gibbs measure.

Definition 5.10 (Blown-up modulated Gibbs measure and partition function)**.** Let μ be a bounded density with $\int_{\mathbb{R}^\mathsf{d}} \mu = N$ an integer, we let

$$\mathsf{K}_\beta(\mu) := N^{-N}\int_{(\mathbb{R}^\mathsf{d})^N} \exp\left(-\beta\mathsf{F}(X_N, \mu)\right) d\mu^{\otimes N}(X_N) \tag{5.2.10}$$

and

$$d\mathbb{Q}_\beta(\mu) := \frac{1}{N^N\mathsf{K}_\beta(\mu)} \exp\left(-\beta\mathsf{F}(X_N, \mu)\right) d\mu^{\otimes N}(X_N), \tag{5.2.11}$$

where we omit N from the notation since we can recover it from $\int_{\mathbb{R}^\mathsf{d}} \mu$.

In view of (5.2.8) we have

$$\log \mathsf{K}_{N,\beta}(\mu) = \log \mathsf{K}_\beta(\mu') + \beta\left(\frac{N}{2\mathsf{d}}\log N\right)\mathbf{1}_{\mathsf{s}=0}, \tag{5.2.12}$$

where μ' is the blown-up of μ as above.

The upper bound of (5.2.3) becomes in blown-up scale

$$\log \mathsf{K}_\beta(\mu) \le C\beta N \|\mu\|_{L^\infty}^{\frac{\mathsf{s}}{\mathsf{d}}} \mathbf{1}_{\mathsf{s}\ge 0}. \tag{5.2.13}$$

The choice of temperature/energy scaling with $\beta N^{-\frac{\mathsf{s}}{\mathsf{d}}}$ made in (1.1.5) becomes more transparent after blow-up: when (5.2.8) is inserted into (1.1.5), we may write using the above definition that

$$d\mathbb{P}_{N,\beta}(X_N) = \frac{1}{N^N \mathsf{K}_\beta(\mu_\theta')} \exp\left(-\beta \mathsf{F}(X_N, \mu_\theta')\right) d(\mu_\theta')^{\otimes N}(X_N) = d\mathbb{Q}_\beta(\mu_\theta'). \tag{5.2.14}$$

We can give an electric interpretation of F that is similar to the one given in Chapter 4.

Analogously to (4.1.4), define

$$h_N^\mu[X_N] := \int_{\mathbb{R}^\mathsf{d}} \mathsf{g}(\cdot - y) d\left(\sum_{i=1}^N \delta_{x_i} - \mu\right)(y) \tag{5.2.15}$$

$$h_{N,\vec{\eta}}^\mu[X_N] := \int_{\mathbb{R}^\mathsf{d}} \mathsf{g}(\cdot - y) d\left(\sum_{i=1}^N \delta_{x_i}^{(\eta_i)} - \mu\right)(y).$$

Dropping the $[X_N]$ and μ, they solve (in extended space)

$$-\mathrm{div}\left(|y|^\gamma \nabla h_N\right) = \mathsf{c_d}\left(\sum_{i=1}^N \delta_{x_i} - \mu\delta_{\mathbb{R}^\mathsf{d}}\right) \quad \text{in } \mathbb{R}^{\mathsf{d}+\mathsf{k}}, \tag{5.2.16}$$

$$-\mathrm{div}\left(|y|^\gamma \nabla h_{N,\vec{\eta}}\right) = \mathsf{c_d}\left(\sum_{i=1}^N \delta_{x_i}^{(\eta_i)} - \mu\delta_{\mathbb{R}^\mathsf{d}}\right) \quad \text{in } \mathbb{R}^{\mathsf{d}+\mathsf{k}}. \tag{5.2.17}$$

Similarly as (4.2.14) we define instead

$$\mathsf{r}_i := \frac{1}{4}\min\left(\min_{i\neq j} |x_i - x_j|, \|\mu\|_{L^\infty}^{-1/\mathsf{d}}\right). \tag{5.2.18}$$

The lower bound for F deduced from (4.2.12) is in this setting

$$\mathsf{F}(X_N, \mu) \ge -C\|\mu\|_{L^\infty}^{\frac{\mathsf{s}}{\mathsf{d}}} N \mathbf{1}_{\mathsf{s}\ge 0}, \tag{5.2.19}$$

with C depending only on d and s. It is of order N, as announced.

We next record here the equivalent of the discrepancy bounds of Lemma 4.25 at the blown-up scale.

Lemma 5.11 (Control of charge discrepancy, blown-up scale). *Let X_N be a configuration in $(\mathbb{R}^\mathsf{d})^N$, let h_N be associated via (5.2.16), and let Ω be a set of finite perimeter and Ω_δ as in (4.4.2). Let $D(\Omega) = \int_\Omega d(\sum_{i=1}^N \delta_{x_i} - \mu)$. We have, for a constant $C > 0$ depending only on d and s,*

- *in the Coulomb case, if $D(\Omega) \ge 0$, for any $\|\mu\|_{L^\infty}^{-1/\mathsf{d}} < \delta \le N^{1/\mathsf{d}}$,*

$$\left(D(\Omega) - \|\mu\|_{L^\infty}|\Omega_\delta \backslash \Omega|\right)_+^2 \le C\frac{|\partial\Omega_\delta|}{\delta}\int_{\Omega_\delta\backslash\Omega} |\nabla h_{N,\mathsf{r}}|^2 \tag{5.2.20}$$

- *in the Coulomb case, if $D(\Omega) \le 0$, for any $-N^{1/\mathsf{d}} \le \delta < -\|\mu\|_{L^\infty}^{-1/\mathsf{d}}$,*

$$\left(D(\Omega) + \|\mu\|_{L^\infty}|\Omega \backslash \Omega_\delta|\right)_-^2 \le C\frac{|\partial\Omega|}{\delta}\int_{\Omega\backslash\Omega_\delta} |\nabla h_{N,\mathsf{r}}|^2 \tag{5.2.21}$$

- *in the Riesz case, if* $D(\Omega) \geq 0$*, for any* $\|\mu\|_{L^\infty}^{-1/\mathsf{d}} < \delta \leq N^{1/\mathsf{d}}$,

$$\left(D(\Omega) - \|\mu\|_{L^\infty}|\Omega_\delta \backslash \Omega|\right)_+^2 \leq C\left(\frac{|\Omega_\delta|}{|\partial\Omega_\delta|}\right)^\gamma \frac{|\Omega_\delta|}{\delta} \int_{(\Omega_\delta\backslash\Omega)\times[-\frac{|\Omega_\delta|}{|\partial\Omega_\delta|}-\delta, \frac{|\Omega_\delta|}{|\partial\Omega_\delta|}+\delta]^\mathsf{k}} |y|^\gamma |\nabla h_{N,\mathsf{r}}|^2 \tag{5.2.22}$$

- *in the Riesz case, if* $D(\Omega) \leq 0$*, for any* $-N^{1/\mathsf{d}} \leq \delta < -\|\mu\|_{L^\infty}^{-1/\mathsf{d}}$,

$$\left(D(\Omega) + \|\mu\|_{L^\infty}|\Omega \backslash \Omega_\delta|\right)_-^2 \leq C\left(\frac{|\Omega|}{|\partial\Omega|}\right)^\gamma \frac{|\Omega|}{\delta} \int_{(\Omega\backslash\Omega_\delta)\times[-\frac{|\Omega|}{|\partial\Omega|}-|\delta|, \frac{|\Omega|}{|\partial\Omega|}+|\delta|]^\mathsf{k}} |y|^\gamma |\nabla h_{N,\mathsf{r}}|^2. \tag{5.2.23}$$

5.2.3. Upper bound for the free energy, thermal equilibrium version. To complement this lower bound, we present an upper bound based on Garcia-Zelada's Jensen argument [**GZ19**] (as used in the proof of Theorem 3.17). This is done in the blown-up scale. As an alternate, the proof of Proposition 5.18 below will be done at the original scale.

Lemma 5.12. *If* $0 < \mathsf{s} < \mathsf{d}$*, for any bounded nonnegative density* μ *such that* $\int_{\mathbb{R}^\mathsf{d}} \mu = N$*, we have*

$$\log \mathsf{K}_\beta(\mu) \geq 0. \tag{5.2.24}$$

PROOF. Starting from (5.2.10) and using Jensen's inequality, we may then write

$$\log \mathsf{K}_\beta(\mu) \geq -\frac{\beta}{(\int_{\mathbb{R}^\mathsf{d}} \mu)^N} \int_{(\mathbb{R}^\mathsf{d})^N} \mathsf{F}(X_N, \mu) d\mu^{\otimes N}(X_N).$$

We next insert (5.2.9) to obtain, since $\int_{\mathbb{R}^\mathsf{d}} \mu = N$,

$$\begin{aligned} \int_{(\mathbb{R}^\mathsf{d})^N} \mathsf{F}(X_N, \mu) d\mu^{\otimes N}(X_N) &= \frac{1}{2} \int_{(\mathbb{R}^\mathsf{d})^N} \Bigg(\sum_{i \neq j} \mathsf{g}(x_i - x_j) - 2\sum_{i=1}^N \int_{\mathbb{R}^\mathsf{d}} \mathsf{g}(x_i - y) d\mu(y) \\ &\qquad + \iint_{\mathbb{R}^\mathsf{d}\times\mathbb{R}^\mathsf{d}} \mathsf{g}(x-y) d\mu(x) d\mu(y) \Bigg) d\mu^{\otimes N}(X_N) \\ &= \frac{1}{2}\left(N(N-1)N^{N-2} - 2NN^{N-1} + N^N\right) \iint_{(\mathbb{R}^\mathsf{d})^2} \mathsf{g}(x-y) d\mu(x) d\mu(y) \\ &= -\frac{N^{N-1}}{2} \iint_{(\mathbb{R}^\mathsf{d})^2} \mathsf{g}(x-y) d\mu(x) d\mu(y). \end{aligned} \tag{5.2.25}$$

It follows that

$$\log \mathsf{K}_\beta(\mu) \geq \frac{\beta}{2N} \iint_{(\mathbb{R}^\mathsf{d})^2} \mathsf{g}(x-y) d\mu(x) d\mu(y). \tag{5.2.26}$$

Since s is assumed to be positive, $\mathsf{g} \geq 0$ and this yields the desired result. □

The case $\mathsf{s} \leq 0$ is more delicate. The above proof can be seen to give (after a change of scales, assuming μ compactly supported for instance) that $\log \mathsf{K}_\beta(\mu) \geq -C\beta N \mathsf{g}(N^{\frac{1}{\mathsf{d}}})$, which is not the optimal bound. A better bound, proportional to N, can be obtained by a form of superadditivity property of $\log \mathsf{K}_\beta$ (see also Chapter 8) after partitioning the support of μ into regions of size that should be taken to depend on β if one seeks a sharp estimate as $\beta \to 0$. We present here a much simpler proof than the one originally given in [**AS21**] and that gives a better result if $\mathsf{d} \geq 3$.

In order to do so, we need extra assumptions which ensure that μ's tails are thin. Let us define

$$\chi(\beta) := \begin{cases} 1 & \text{if } \mathsf{s} > 0 \text{ or } \beta \geq 1 \\ \beta^{\frac{\mathsf{s}}{\mathsf{d}-\mathsf{s}}} |\log \beta| + 1 & \text{if } \mathsf{s} \leq 0 \text{ and } \beta \leq 1, \end{cases} \tag{5.2.27}$$

a quantity which will appear repeatedly.

The assumptions are:

(5.2.28) There exists a set Λ with piecewise C^1 boundary, s.t. $\mu \geq m > 0$ in Λ with (if $\mu(\Lambda^c) \neq 0$ and $\mathsf{s} \leq 0$)

$$\frac{1}{\mu(\Lambda^c)} \iint_{(\Lambda^c)^2} \mathsf{g}(x-y) d\mu(x) d\mu(y) \geq -C\chi(\beta) N. \tag{5.2.29}$$

In Theorem 2.23 we have seen that, in the Coulomb cases, under suitable assumptions on V, Σ being the compact support of μ_V, from (2.5.23) and (2.5.27) we have $\mu_\theta(\Sigma^c) \leq \frac{C}{\sqrt{\theta}} = C\beta^{-1/2} N^{\frac{\mathsf{s}-\mathsf{d}}{2\mathsf{d}}}$ and $\mu_\theta \geq m > 0$ (independent of N) in Σ. Moreover by scaling, if μ is the blown-up of μ_θ and Σ' the blown-up of Σ, remembering that we are treating the case $\mathsf{s} \leq 0$ and using (2.1.1), we have

(5.2.30)

$$\begin{aligned} &\iint_{((\Sigma')^c)^2} \mathsf{g}(x-y) d\mu(x) d\mu(y) \\ &= N^{2-\frac{\mathsf{s}}{\mathsf{d}}} \iint_{(\Sigma^c)^2} \mathsf{g}(x-y) d\mu_\theta(x) d\mu_\theta(y) - N^2 \log N^{1/\mathsf{d}} \mu_\theta(\Sigma^c)^2 \mathbf{1}_{\mathsf{s}=0} \\ &\geq 2N^{2-\frac{\mathsf{s}}{\mathsf{d}}} \iint_{(\Sigma^c)^2} (\mathsf{g}(x)_- \wedge \mathsf{g}_-(y) - C) \, d\mu_\theta(x) d\mu_\theta(y) - N^2 \log N^{1/\mathsf{d}} \mu_\theta(\Sigma^c)^2 \mathbf{1}_{\mathsf{s}=0} \\ &\geq -CN^{2-\frac{\mathsf{s}}{\mathsf{d}}} \mu_\theta(\Sigma^c) - N^2 \log N^{1/\mathsf{d}} \mu_\theta(\Sigma^c)^2 \mathbf{1}_{\mathsf{s}=0}, \end{aligned}$$

where we used (2.5.6), which is true if (A5) holds. Therefore,

$$\begin{aligned} &\frac{1}{\mu((\Sigma')^c)} \iint_{((\Sigma')^c)^2} \mathsf{g}(x-y) d\mu(x) d\mu(y) \\ &= \frac{1}{N\mu_\theta(\Sigma^c)} \iint_{((\Sigma')^c)^2} \mathsf{g}(x-y) d\mu(x) d\mu(y) \geq -CN^{1-\frac{\mathsf{s}}{\mathsf{d}}} - C\beta^{-\frac{1}{2}} N^{\frac{\mathsf{s}-\mathsf{d}}{2\mathsf{d}}+1} \log N \mathbf{1}_{\mathsf{s}=0}. \end{aligned} \tag{5.2.31}$$

We may check that in view of (5.2.27), the right-hand side is $\geq -C\chi(\beta)N$ as long as $\beta \geq \theta_0 N^{\frac{\mathsf{s}}{\mathsf{d}}-1}$ for some $\theta_0 > 0$, which is equivalent to $\theta \geq \theta_0 > 0$, hence we can consider always satisfied. Thus, these assumptions are verified for the blown-up thermal equilibrium measure with $\Lambda = \Sigma'$, at least in the Coulomb case.

To partition the support, we may use the following tiling lemma, adapted from [**SS15a**, Lemma 6.5], that we will use several times in later chapters.

Lemma 5.13 (Tiling with quantized cells). *Assume the set Λ is such that $\partial\Lambda$ is piecewise C^1, and the density μ is bounded below in Λ by a constant $m > 0$. There exists a constant $C > 0$ depending on m, such that, given any $R \geq 1$, there* *exists for any $n \in \mathbb{N}^*$ a collection $\mathcal{K}$ of closed hyperrectangles in Λ with disjoint interiors, whose sidelengths are between R and $R+R^{1-\mathsf{d}}m^{-1}$, and which are such that*

$$\{x \in \Lambda, \operatorname{dist}(x, \partial\Lambda) \geq CR\} \subset \bigcup_{K \in \mathcal{K}} K := \Lambda_{\text{int}} \subset \{x \in \Lambda, \operatorname{dist}(x, \partial\Lambda) \geq R\}, \tag{5.2.32}$$

and

$$\forall K \in \mathcal{K}, \quad \int_K \mu \in \mathbb{N}. \tag{5.2.33}$$

Moreover, if Λ is a hyperrectangle with $\int_\Lambda \mu$ an integer, then $\mathcal{K}$ can be constructed so that

$$\bigcup_{K \in \mathcal{K}} K = \Lambda. \tag{5.2.34}$$

Finally if Λ is the $N^{1/\mathsf{d}}$-blown up of a domain with C^1 boundary, for N large enough, the same tiling into (5.2.34) with (5.2.33) can be performed with K's belonging to the larger class of sets of piecewise C^1 boundary and diameter bounded by CR.

PROOF. Let us first look at dimension $\mathsf{d} = 1$. Then Λ is an interval, say (a, b) over which $\mu \geq m$. We let $t_0 = a$, and for each $k \geq 1$, t_k be the minimal value $\geq t_{k-1} + R$ such that

$$\int_{t_{k-1}}^{t_k} \mu(x)dx \in \mathbb{N}.$$

By definition we have

$$m(t_k - (t_{k-1} + R)) \leq \int_{t_{k-1}+R}^{t_k} \mu(x)dx \leq 1,$$

hence

$$R \leq t_k - t_{k-1} \leq R + \frac{1}{m}.$$

We take $\mathcal{K}$ to be the family of intervals (t_{k-1}, t_k) and obtain the result.

Let us now consider $\mathsf{d} \geq 2$. For each $\vec{k}$, we let $t_{\vec{k},0} = -\infty$ and define by induction $t_{\vec{k},l}$ to be the smallest $t \geq t_{\vec{k},l-1} + R$ such that

$$\int_{k_1 R}^{(k_1+1)R} dx_1 \ldots \int_{k_{\mathsf{d}-1}R}^{(k_{\mathsf{d}-1}+1)R} dx_{\mathsf{d}-1} \int_{t_{\vec{k},l-1}}^{t} du\, \mu(x_1, \ldots, x_{\mathsf{d}-1}, u) \in \mathbb{N}.$$

We next consider the hyperrectangles $[k_1 R, (k_1+1)R] \times \cdots \times [k_{\mathsf{d}-1}R, (k_{\mathsf{d}-1}+1)R] \times [t_{\vec{k},l}, t_{\vec{k},l+1}]$. If such a rectangle K is entirely contained in Λ, then we have $\mu(x) \geq m$ for each $x \in K$, and so by definition of $t_{\vec{k},l}$ we must have

$$\begin{aligned} &mR^{\mathsf{d}-1}(t_{\vec{k},l} - (t_{\vec{k},l-1} + R)) \\ &\quad \leq \int_{k_1 R}^{(k_1+1)R} dx_1 \ldots \int_{k_{\mathsf{d}-1}R}^{(k_{\mathsf{d}-1}+1)R} dx_{\mathsf{d}-1} \int_{t_{\vec{k},l-1}+R}^{t_{\vec{k},l}} du\, \mu(x_1, \ldots, x_{\mathsf{d}-1}, u) \leq 1, \end{aligned} \tag{5.2.35}$$

hence

$$R \leq t_{\vec{k},l} - t_{\vec{k},l-1} \leq R + \frac{1}{mR^{\mathsf{d}-1}}. \tag{5.2.36}$$

We then define $\mathcal{K}$ to be the collection of such hyperrectangles which are entirely included in $\{x \in \Lambda, \mathrm{dist}(x, \partial\Lambda) > R\}$. By construction, (5.2.33) is satisfied, the sidelengths of the rectangle are as desired by (5.2.36), and (5.2.32) holds by arguing as above in (5.2.35) and using the regularity of $\partial\Lambda$.

The case of Λ equal to a hyperrectangle is even easier and follows from the same proof.

Let us now turn to the last assertion, for which we sketch a proof provided by Etienne Sandier. First, we return to the blown-down scale where Λ is a domain with C^1 boundary. We then wish to tile with cells of size $O(RN^{-1/\mathsf{d}})$ in such a way that the μ-mass of each cell belongs to $\frac{1}{N}\mathbb{N}$. Since $\partial\Lambda$ is assumed C^1 in this case, it admits an outer normal unit vector $\vec{n}$ at each point. For $x \in \partial\Lambda$, we may define

$$\tilde{\mu}(x) = \int_{t \geq 0, x - t\vec{n} \in \Lambda \setminus \Lambda_{\mathrm{int}}} \mu(x - t\vec{n})dt,$$

which is finite for a.e. $x \in \partial\Lambda$, and defines a positive density on $\partial\Lambda$. It then suffices to find a partitioning of $\partial\Lambda$ into cells $\mathcal{C}_i$ of diameter bounded by CR such that $\int_{\mathcal{C}_i} \tilde{\mu} \in \frac{1}{N}\mathbb{N}$. To do so, we start from a Whitney triangulation of $\partial\Sigma$ with cells of diameter $O(R)$, and denote $\mathcal{K}_i$ the cells of the triangulation. We do not have $\int_{\mathcal{K}_i} \tilde{\mu} \in \frac{1}{N}\mathbb{N}$, however we can achieve it by a small perturbation of the triangulation: consider for a given face F_j of the triangulation, $F_j^t = F_j + t\vec{n_j}\chi_j$, where $\vec{n_j}$ is a normal to the face and χ_j is a cutoff function vanishing on the boundary of F_j, and for any $\vec{t}$ let $\mathcal{K}_i^{\vec{t}}$ denote the cells of the triangulation with faces equal to $F_j^{t_j}$. Consider then the map

$$\Phi : \vec{t} \mapsto \left(\tilde{\mu}(\mathcal{K}_i^{\vec{t}})\right)_i.$$

We claim that its differential at 0 has maximal rank. Indeed, let $\vec{t}$ be such that $t_k = \delta_{kj}$. Then $D\Phi(0)\vec{t}$ is proportional to $(0, \dots, 1, -1, \dots, 0)$, where the coefficients of the ith entry is nonzero only iff F_j is a face adjacent to the cell $\mathcal{K}_i$ (the mass can only change in those cells, and it increases on one side and decreases by the same amount on the other). Composing such variations, one can obtain that the range of $D\Phi(0)$ contains all vectors of the form $(0, \dots, 1, \dots, -1, \dots, 0)$, and by linearity all vectors whose sum of entries equals zero. Thus $D\Phi(0)$ has maximal rank, hence Φ is surjective from a neighborhood of 0 to a neighborhood of $(\tilde{\mu}(\mathcal{K}_i))_i$. For N large enough, that neighborhood contains a vector whose entries all belong to $\frac{1}{N}\mathbb{N}$, and thus we find the existence of a $\vec{t_0}$ such that the partitioning $\mathcal{K}_i^{\vec{t_0}}$ of $\partial\Lambda$ satisfies the requirements. □

We now proceed to obtaining a lower bound for $\log \mathsf{K}_\beta$ in the case $\mathsf{s} \leq 0$.

Proposition 5.14. *Assume $\mathsf{s} \leq 0$. Then, if μ is a bounded nonnegative density such that $\int_{\mathbb{R}^\mathsf{d}} \mu = N$ and (5.2.28)–(5.2.29) hold for Λ either a hyperrectangle or the $N^{1/\mathsf{d}}$-blown up of a fixed set with C^1 boundary, and if*

$$\theta \geq \theta_0 > 0, \tag{5.2.37}$$

where θ is as in (3.2.5), we have

$$\log \mathsf{K}_\beta(\mu) \geq -C\beta\chi(\beta)N, \tag{5.2.38}$$

where χ is as in (5.2.27), and $C > 0$ depends only on $\mathsf{d}, \mathsf{s}, \|\mu\|_{L^\infty}, \theta_0$ and the constants in the assumptions.

Remark 5.15. The assumption that Λ is either a hyperrectangle or the $N^{1/\mathsf{d}}$-blown up of a fixed set with C^1 boundary is placed so that we can apply Lemma 5.13 and partition Λ completely into nondegenerate quantized cells. However, this is an overkill: we could instead for instance require that the set Λ_{int} given by Lemma 5.13 (which is already partitioned), is such that (5.2.29) holds for it.

PROOF. Let us apply Lemma 5.13 to Λ as in the assumptions, and let us denote Q_i, $i = 1, \dots, p$ the cells of diameter $\le CR$ constructed via the last statement in the lemma (so they are not necessarily hyperrectangles, and partition all of Λ) and $n_i = \int_{Q_i} d\mu$ integer. We also let $\mu_i = \mu \mathbf{1}_{Q_i}$. Finally, we let $\mu_0 := \mu - \sum_{i=1}^p \mu_i = \mu \mathbf{1}_{\Lambda^c}$, for which $n_0 := \int \mu_0$ must also be integer.

Returning to the definition (5.2.10), and decomposing μ into $\mu = \sum_{i=0}^p \mu_i$ we may write

$$\mu^{\otimes N} = (\sum_{i=0}^p \mu_i)^{\otimes N} \ge \frac{N!}{n_0! \dots n_p!} \prod_{i=0}^p \mu_i^{\otimes n_i}$$

hence

$$\begin{aligned}
\text{(5.2.39)} \qquad \mathsf{K}_\beta(\mu) &\ge N^{-N} \frac{N!}{n_0! \dots n_p!} \int \exp\left(-\beta \mathsf{F}(X_N, \mu)\right) d\left(\prod_{i=0}^p \mu_i^{\otimes n_i}\right)(X_N) \\
&= \frac{N! \prod_{i=0}^p n_i^{n_i}}{N^N \prod_{i=0}^p n_i!} \int \exp\left(-\beta \mathsf{F}(X_N, \mu)\right) d\left(\prod_{i=0}^p \frac{\mu_i^{\otimes n_i}}{n_i^{n_i}}\right)(X_N).
\end{aligned}$$

Thus, using Jensen's inequality, we find

$$\text{(5.2.40)} \qquad \log \mathsf{K}_\beta(\mu) \ge \log \frac{N! \prod_{i=0}^p n_i^{n_i}}{N^N \prod_{i=0}^p n_i!} - \beta \int \mathsf{F}(X_N, \mu) d\left(\prod_{i=0}^p \frac{\mu_i^{\otimes n_i}}{n_i^{n_i}}\right)(X_N).$$

First, with Stirling's formula and since $\sum_i n_i = N$, we have

$$\text{(5.2.41)} \qquad \log \frac{N! \prod_{i=0}^p n_i^{n_i}}{N^N \prod_{i=0}^p n_i!} = \sum_{i=0}^p O(\log n_i).$$

Secondly, we separate the points into batches by letting

$$I_i = \left\{ \sum_{k<i} n_k + 1, \dots, \sum_{k \le i} n_k \right\}$$

so that $\{1, \dots, N\} = \cup_{i=0}^p I_i$ and $X_N^i = \{x_k\}_{k \in I_i}$ and we may expand F as

(5.2.42)

$$\begin{aligned}
\mathsf{F}(X_N, \mu) &= \sum_{i=0}^p \mathsf{F}(X_N^i, \mu_i) \\
&\quad + \frac{1}{2} \sum_{0 \le i \ne j \le p} \left(\sum_{k \in I_i, l \in I_j} \mathsf{g}(x_k - x_l) - \sum_{k \in I_i} \int \mathsf{g}(x_k - y) d\mu_j(y) \right. \\
&\quad \left. - \sum_{k \in I_j} \int \mathsf{g}(x_k - y) d\mu_i(y) + \iint \mathsf{g}(x - y) d\mu_i(x) d\mu_j(y) \right).
\end{aligned}$$

Integrating against $d\left(\prod_{i=0}^{p} \frac{\mu_i^{\otimes n_i}}{n_i^{n_i}}\right)$ we find

$$\int \mathsf{F}(X_N,\mu) d\left(\prod_{i=0}^{p} \frac{\mu_i^{\otimes n_i}}{n_i^{n_i}}\right)(X_N) = \sum_{i=0}^{p} \frac{1}{n_i^{n_i}} \int \mathsf{F}(X_N^i,\mu_i) d\mu_i^{\otimes n_i}(X_N^i) \tag{5.2.43}$$

after noticing that the integral of the second line in (5.2.42) vanishes. Computing as in (5.2.25), we thus find

$$\int \mathsf{F}(X_N,\mu) d\left(\prod_{i=0}^{p} \frac{\mu_i^{\otimes n_i}}{n_i^{n_i}}\right)(X_N) \geq -\frac{1}{2}\sum_{i=0}^{p} \frac{1}{n_i} \iint \mathsf{g}(x-y) d\mu_i(x) d\mu_i(y). \tag{5.2.44}$$

Combining with (5.2.40) and (5.2.41), we have obtained

$$\log \mathsf{K}_\beta(\mu) \geq \frac{\beta}{2}\sum_{i=0}^{p} \frac{1}{n_i} \iint \mathsf{g}(x-y) d\mu_i(x) d\mu_i(y) + O(\log n_i).$$

This can be bounded below by

$$\begin{aligned}\log \mathsf{K}_\beta(\mu) \geq \frac{\beta}{2}\frac{1}{n_0} &\iint_{(\Lambda^c)^2} \mathsf{g}(x-y) d\mu(x) d\mu(y) - C\log n_0 \\ &+ \sum_{i=1}^{p}\left(-\frac{\beta}{2} n_i |\mathsf{g}(\mathrm{diam}(\mathrm{supp}\ \mu_i))| - C\log n_i\right).\end{aligned}$$

Since the support of μ_i is of size $\in [R, CR]$ for $i \geq 1$, and $n_i = O(R^{\mathsf{d}})$ since μ has bounded density, we have $p \leq C\frac{N}{R^{\mathsf{d}}}$ and using also (5.2.29), we arrive at

$$\begin{aligned}\log \mathsf{K}_\beta(\mu) &\geq -C\frac{N}{R^{\mathsf{d}}}\left(\frac{\beta}{2}R^{\mathsf{d}}|\mathsf{g}(R)| + C\log R\right) - C\beta\chi(\beta)N - C\log N \\ &\geq -CN\left(\frac{\log R}{R^{\mathsf{d}}} - \beta \mathsf{g}(R)\right) - C\beta\chi(\beta)N - C\log N,\end{aligned} \tag{5.2.45}$$

where C depends only on $\mathsf{s} \leq 0, \mathsf{d}$ and the constants in the assumptions. Optimizing over R leads to the choice

$$R = \begin{cases} 1 & \text{if } \beta \geq 1 \\ \beta^{-\frac{1}{\mathsf{d}-\mathsf{s}}} & \text{if } \beta \leq 1 \end{cases} \tag{5.2.46}$$

for which we find by definition (5.2.27)

$$-CN\left(\frac{\log R}{R^{\mathsf{d}}} - \beta\mathsf{g}(R)\right) \geq -CN\beta^{1+\frac{\mathsf{s}}{\mathsf{d}-\mathsf{s}}}|\log \beta| = -CN\beta\chi(\beta).$$

On the other hand $\log N$ can be absorbed into $\beta\chi(\beta)N$ because the assumption (5.2.37) implies that $C\beta\chi(\beta) \geq \frac{\log N}{N}$. The conclusion then follows from inserting into (5.2.45). □

This concludes the proof of the converse to (5.2.13) in all cases.

Corollary 5.16 (Bound on the reduced free energy)**.** *Let μ be a bounded nonnegative density with $\int_{\mathbb{R}^{\mathsf{d}}} \mu = N$. If $\mathsf{s} \leq 0$, assume* (5.2.28)–(5.2.29) *and* (5.2.37), *with*

Λ either a hyperrectangle or the $N^{1/\mathsf{d}}$-blown up of a fixed set with C^1 boundary. Then

$$|\log \mathsf{K}_\beta(\mu)| \le C\beta\chi(\beta)N, \tag{5.2.47}$$

where $C>0$ depends only on s,d, $\|\mu\|_{L^\infty}$ and the constants in the assumptions.

We can translate this into a bound at the original scale, and inserting into (5.1.14), an approximation for the original free energy, which improves (3.2.6)–(3.2.7).

Corollary 5.17 (First approximation of the free energy). *Assume μ_θ satisfies the assumptions of the previous corollary,*[3] *where θ as in* (2.0.4). *Then we have*

$$\left|\log Z_{N,\beta} + \theta N \mathcal{E}_\theta(\mu_\theta) - \left(\frac{\beta}{2\mathsf{d}} N\log N\right)\mathbf{1}_{\mathsf{s}=0}\right| \le C\beta\chi(\beta)N, \tag{5.2.48}$$

where C depends only on d,s, $\|\mu_\theta\|_{L^\infty}$ and the constants in the assumptions.

We emphasize that these bounds are valid for all β and N (within the range (5.2.37) for $\mathsf{s}\le 0$), so that we may let β depend on N.

Note that the assumptions on the regularity of $\partial\Lambda$ could be relaxed according to Remark 5.15.

5.2.4. Upper bound for the free energy, equilibrium measure version.

We next describe how to obtain the analogous results to the previous subsection for the free energy relative to the standard equilibrium measure as in (5.1.7).

Proposition 5.18. *Assume μ is a bounded probability density with support Σ, a set with C^1 boundary, and that $\zeta = 0$ in Σ. Then, letting $\tilde{\mathsf{K}}_{N,\beta}(\mu,\zeta)$ be as in* (5.1.5), *we have*

$$\log \tilde{\mathsf{K}}_{N,\beta}(\mu,\zeta) \ge -N\int \mu\log\mu + \left(\frac{\beta N}{2\mathsf{d}}\log N\right)\mathbf{1}_{\mathsf{s}=0} - C\beta\chi(\beta)N, \tag{5.2.49}$$

and in particular, if μ_V exists and has compact support, then for all β, we have

$$\log Z_{N,\beta} \ge -\theta N \mathcal{E}_\theta(\mu_V) + \left(\frac{\beta N}{2\mathsf{d}}\log N\right)\mathbf{1}_{\mathsf{s}=0} - C\beta\chi(\beta)N, \tag{5.2.50}$$

where $C>0$ depends only on d, s and $\|\mu\|_{L^\infty}$, resp. $\|\mu_V\|_{L^\infty}$.

Proof. Starting from (5.1.5) and using that $\zeta=0$ in Σ, the support of μ, we have

$$\begin{aligned}\tilde{\mathsf{K}}_{N,\beta}(\mu,\zeta) &\ge \int_{(\Sigma)^N} \exp\left(-\beta N^{-\frac{\mathsf{s}}{\mathsf{d}}}\mathsf{F}_N(X_N,\mu)\right) dX_N\\ &= \int_{(\Sigma)^N} \exp\left(-\beta N^{-\frac{\mathsf{s}}{\mathsf{d}}}\mathsf{F}_N(X_N,\mu) - \sum_{i=1}^N \log\mu(x_i)\right) d\mu^{\otimes N}(X_N).\end{aligned}$$

Applying Jensen's inequality, we then find

$$\log \tilde{\mathsf{K}}_{N,\beta}(\mu,\zeta) \ge \int_{\Sigma^N}\left(-\beta N^{-\frac{\mathsf{s}}{\mathsf{d}}}\mathsf{F}_N(X_N,\mu) - \sum_{i=1}^N\log\mu(x_i)\right) d\mu^{\otimes N}(X_N). \tag{5.2.51}$$

[3] We have seen that this is proven at least in the Coulomb case on the basis of Theorem 2.23.

Reasoning as in (5.2.25) (except we are now working at the usual scale) we obtain

$$\text{(5.2.52)} \quad \log \tilde{\mathsf{K}}_{N,\beta}(\mu,\zeta) \geq -\beta N^{-\frac{\mathsf{s}}{\mathsf{d}}}\left(-\frac{N}{2}\iint_{\Sigma^2} \mathsf{g}(x-y)d\mu(x)d\mu(y)\right) - N\int \mu\log\mu.$$

If $\mathsf{s} > 0$ then $\mathsf{g} \geq 0$ and the result follows. If $\mathsf{s} \leq 0$, we argue as in the proof of Proposition 5.14 and split $\mu = \sum_{i=1}^p \mu_i$ where μ_i is supported in Q_i, cell of diameter $\leq RN^{-\frac{1}{\mathsf{d}}}$, with the Q_i's forming a partition of Σ, $N\int_{Q_i}\mu_i = n_i$, an integer. We then obtain, as in (5.2.40),

$$\text{(5.2.53)} \quad \log \tilde{\mathsf{K}}_{N,\beta}(\mu,\zeta) \geq \log \frac{N!\prod_{i=1}^p (n_i/N)^{n_i}}{\prod_{i=1}^p n_i!} + \int \left(-\beta N^{-\frac{\mathsf{s}}{\mathsf{d}}}\mathsf{F}_N(X_N,\mu) - \sum_{i=1}^N \log\mu(x_i)\right) d\left(\prod_{i=1}^p \frac{\mu_i^{\otimes n_i}}{(n_i/N)^{n_i}}\right)(X_N).$$

First, using Stirling's formula and $\sum_i n_i = N$, we have

$$\text{(5.2.54)} \qquad \log \frac{N!\prod_{i=1}^p (n_i/N)^{n_i}}{\prod_{i=1}^p n_i!} = \log \frac{N!\prod_{i=1}^p n_i^{n_i}}{N^N\prod_{i=1}^p n_i!} = \sum_{i=1}^p O(\log n_i).$$

Separating the points into batches, and integrating, we obtain as in (5.2.43) and (5.2.44) that

$$\begin{aligned}\text{(5.2.55)} \quad & \int \mathsf{F}_N(X_N,\mu)d\left(\prod_{i=1}^p \frac{\mu_i^{\otimes n_i}}{(n_i/N)^{n_i}}\right)(X_N) \\ &= \sum_{i=1}^p \frac{N^{n_i}}{n_i^{n_i}}\int \mathsf{F}_N(X_N^i,\mu_i)d\mu_i^{\otimes n_i}(X_N^i) \\ &= \frac{1}{2}\sum_{i=1}^p \frac{N^{n_i}}{n_i^{n_i}}\left(n_i(n_i-1)(\frac{n_i}{N})^{n_i-2} - 2n_iN(\frac{n_i}{N})^{n_i-1} + N^2(\frac{n_i}{N})^{n_i}\right) \\ &\qquad \times \iint \mathsf{g}(x-y)d\mu_i(x)d\mu_i(y) \\ &= -\frac{1}{2}\sum_{i=1}^p \frac{N^2}{n_i}\iint \mathsf{g}(x-y)d\mu_i(x)d\mu_i(y).\end{aligned}$$

Also

$$\int \sum_{i=1}^N \log\mu(x_i)d\left(\prod_{i=1}^p \frac{\mu_i^{\otimes n_i}}{(n_i/N)^{n_i}}\right)(X_N) = \sum_{i=1}^p n_i\frac{N}{n_i}\int \log\mu\, d\mu_i = N\int \mu\log\mu.$$

It follows that
(5.2.56)

$$\log \tilde{\mathsf{K}}_{N,\beta}(\mu,\zeta) \geq \sum_{i=1}^p O(\log n_i) + \frac{\beta}{2}N^{-\frac{\mathsf{s}}{\mathsf{d}}}\sum_{i=1}^p \frac{N^2}{n_i}\iint \mathsf{g}(x-y)d\mu_i(x)d\mu_i(y) - N\int \mu\log\mu.$$

The sizes of the cells being $RN^{-1/\mathsf{d}}$, and thus $n_i = O(R^{\mathsf{d}})$ and $p \le CNR^{-\mathsf{d}}$, we find

$$\log \tilde{\mathsf{K}}_{N,\beta}(\mu,\zeta) \tag{5.2.57}$$
$$\ge -C\frac{N}{R^{\mathsf{d}}}\log R + \frac{\beta}{2}N^{-\frac{\mathsf{s}}{\mathsf{d}}}\sum_{i=1}^{p}\frac{N^2}{n_i}\frac{n_i^2}{N^2}\left(N^{\frac{\mathsf{s}}{\mathsf{d}}}\mathsf{g}(R) + \frac{1}{\mathsf{d}}\log N\mathbf{1}_{\mathsf{s}=0}\right) - N\int \mu\log\mu$$
$$\ge -N\int \mu\log\mu + \left(\frac{\beta N}{2\mathsf{d}}\log N\right)\mathbf{1}_{\mathsf{s}=0} - CN\left(\frac{N}{R^{\mathsf{d}}}\log R - \frac{\beta}{2}\mathsf{g}(R)\right).$$

Optimizing and choosing R as in (5.2.46), we obtain the result and (5.2.50) follows from (5.1.7) after grouping terms in the form (2.0.4). □

Combining with the converse bound in Corollary 5.7, we obtain the following equilibrium measure version of Corollary 5.17. As explained above, the estimate is less precise that that former one as β gets small.

Corollary 5.19 (First approximation of the free energy—equilibrium measure case). *Assume* (A1)–(A4) *so that μ_V exists and is compactly supported with support with C^1 boundary. Assume also that μ_V has a bounded density. Then for all $\beta > 0$, and for N large enough, we have*

$$\left|\log Z_{N,\beta} + \theta N\mathcal{E}_\theta(\mu_V) - \left(\frac{\beta N}{2\mathsf{d}}\log N\right)\mathbf{1}_{\mathsf{s}=0}\right| \le C\beta\chi(\beta)N + NC_\zeta, \tag{5.2.58}$$

where C depends only on s, d and $\|\mu_V\|_{L^\infty}$, and C_ζ depends only on the quantity in (A4) *and on ζ.*

Again, the assumptions on the regularity of $\partial\Lambda$ could be relaxed according to Remark 5.15.

5.2.5. Concentration bounds and minimal energy bound. We can directly deduce from the above a first concentration bound, in the form of a control in exponential moments of the next-order energy.

5.2.5.1. *Concentration around the thermal equilibrium measure.*

Corollary 5.20 (First concentration bound, in blown-up scale). *Let μ be a bounded nonnegative density with $\int_{\mathbb{R}^{\mathsf{d}}}\mu = N$ and $\mathbb{Q}_\beta$ as in* (5.2.11)*, then, under the same assumptions as Corollary* 5.16*, we have*

$$\left|\log \mathbb{E}_{\mathbb{Q}_\beta(\mu)}\left(\exp\frac{\beta}{2}\mathsf{F}(X_N,\mu)\right)\right| \le C\beta\chi(\beta)N, \tag{5.2.59}$$

where C depends only on d,s and $\|\mu\|_{L^\infty}$.

Proof. It suffices to rewrite

$$\mathbb{E}_{\mathbb{Q}_\beta(\mu)}\left(\exp\frac{\beta}{2}\mathsf{F}(X_N,\mu)\right)$$
$$= \frac{1}{\mathsf{K}_\beta(\mu)}\int_{(\mathbb{R}^{\mathsf{d}})^N}\exp\left(-\beta\mathsf{F}(X_N,\mu) + \frac{\beta}{2}\mathsf{F}(X_N,\mu)\right)d\mu^{\otimes N}(X_N)$$
$$= \frac{\mathsf{K}_{\beta/2}(\mu)}{\mathsf{K}_\beta(\mu)},$$

which we bound thanks to (5.2.47). □

This estimate means that $\mathsf{F}(X_N, \mu)$ is bounded in exponential moments by CN (modulo the $\chi(\beta)$ at low β in the case $\mathsf{s} \le 0$) i.e the next order energy is of the order of the number of points / of the blown-up volume occupied by the particles. This is to be compared with the leading order energy which scales like N^2.

We can then translate this into an estimate at the original scale. For this, we need to restate the assumptions at that scale, using (5.2.30).

(5.2.60)
There exists a set Λ with C^1 boundary or a hyperrectangle, s.t. $\mu \ge m > 0$ in Λ, with

$$\frac{N^{-\frac{\mathsf{s}}{\mathsf{d}}}}{\mu(\Lambda^c)} \iint_{(\Lambda^c)^2} \mathsf{g}(x-y)d\mu(x)d\mu(y) - \frac{\log N}{\mathsf{d}}\mu(\Lambda^c)\mathbf{1}_{\mathsf{s}=0} \ge -C\chi(\beta)N. \tag{5.2.61}$$

Corollary 5.21 (First concentration bound—thermal equilibrium measure version). *For all* (2.0.1), *for any N and β, and any probability measure μ with bounded density, and if $\mathsf{s} \le 0$ assuming in addition* (5.2.60), (5.2.61), *and* (5.2.37), *we have*

(5.2.62)
$$\left|\log \mathbb{E}_{\mathbb{Q}_{N,\beta}(\mu)}\left(\exp \frac{\beta N^{-\frac{\mathsf{s}}{\mathsf{d}}}}{2}\left(\mathsf{F}_N(X_N,\mu) + \left(\frac{N}{2\mathsf{d}}\log N\right)\mathbf{1}_{\mathsf{s}=0}\right)\right)\right| \le C\beta\chi(\beta)N.$$

If these assumptions are satisfied for $\mu = \mu_\theta$,[4] *we have*

(5.2.63)
$$\left|\log \mathbb{E}_{\mathbb{P}_{N,\beta}}\left(\exp \frac{\beta N^{-\frac{\mathsf{s}}{\mathsf{d}}}}{2}\left(\mathsf{F}_N(X_N,\mu_\theta) + \left(\frac{N}{2\mathsf{d}}\log N\right)\mathbf{1}_{\mathsf{s}=0}\right)\right)\right| \le C\beta\chi(\beta)N,$$

where $C > 0$ depends only on s, d and $\|\mu\|_{L^\infty}$, resp. $\|\mu_\theta\|_{L^\infty}$, and the constants in the assumptions.

In view of the results of Lemma 4.20, the bound (5.2.62) shows that $\widehat{\mu_N}$ concentrates near μ_θ, with explicit bounds:

Corollary 5.22 (Concentration around the equilibrium measure—Local laws at the macroscopic scale). *Let φ be a regular function, then, under the same assumptions,*

$$\left|\log \mathbb{E}_{\mathbb{Q}_{N,\beta}(\mu)}\left(\exp \frac{\beta}{C\|\varphi\|^2} N^{2-\frac{\mathsf{s}}{\mathsf{d}}}\left(\int_{\mathbb{R}^\mathsf{d}} \varphi\, d(\widehat{\mu_N} - \mu)\right)^2\right)\right| \le C\beta\chi(\beta)N, \tag{5.2.64}$$

where C depends only on s, d and $\|\mu\|_{L^\infty}$, and

$$\|\varphi\| = \begin{cases} \|\nabla\varphi\|_{L^2} + \|\nabla\varphi\|_{L^\infty} & \text{if } \mathsf{s} = \mathsf{d}-2 \\ \|\varphi\|_{\dot{H}^{\frac{\mathsf{d}-\mathsf{s}}{2}}} + \|(-\Delta)^{\frac{\mathsf{d}-\mathsf{s}}{2}}\varphi\|_{L^\infty} & \text{if } \mathsf{s} \ne \mathsf{d}-2. \end{cases}$$

PROOF. It suffices to insert (4.3.1) or (4.3.2) combined with (4.2.16) into (5.2.63). □

More precise concentration results but in W_1 distance, using "Coulomb-transport inequalities", are obtained in [**CHM18**] for a fixed β regimes. Analogous concentration near μ_θ in the small β regime is established in [**PG23**].

[4] We have seen that this is true in all cases $\mathsf{s} > 0$ and in the $\mathsf{d} = 2$ Coulomb case under the assumption (5.2.37) and (A1)–(A5)

We then also reexpress these concentration estimates in terms of the charge discrepancies of Section 4.4. Combining the result of Lemma 4.26 with (4.2.16) and inserting into (5.2.62), we obtain

Corollary 5.23 (First discrepancy bound). *Let B_R be any ball of radius R in $\mathbb{R}^{\mathsf{d}}$, and let $D(B_R)$ be as in* (4.4.1). *Under the same assumptions, we have*

(5.2.65)
$$\left|\log \mathbb{E}_{\mathbb{Q}_{N,\beta}(\mu)}\left(\exp\left(\beta C^{-1}N^{-\frac{\mathsf{s}}{\mathsf{d}}}\frac{D(B_R)^2}{R^{\mathsf{s}}}\left|\min\left(1,\frac{D(B_R)}{\|\mu\|_{L^\infty(B_R)}R^{\mathsf{d}}}\right)\right|\right)\right)\right| \le C\beta\chi(\beta)N,$$

for some $C>0$ depending only on d,s and $\|\mu\|_{L^\infty}$.

Again, this gives the result for the Gibbs measure by using (5.1.14).

We will see in Chapter 8 a localized version of the results of Corollary 5.22 and 5.23, which we call *local laws.*

5.2.5.2. *Concentration around the equilibrium measure.* We now give the analog statements relative to the equilibrium measure, relying on Corollary 5.19.

Corollary 5.24 (First concentration bound—equilibrium measure version). *For g as in* (2.0.1), *assume* (A1)–(A4) *so that μ_V exists and is compactly supported. Assume also that μ_V has a bounded density. Then for all $\beta>0$, we have*

(5.2.66)
$$\left|\log \mathbb{E}_{\mathbb{P}_{N,\beta}}\left(\exp\frac{\beta N^{-\frac{\mathsf{s}}{\mathsf{d}}}}{2}\left(\mathsf{F}_N(X_N,\mu_V)+\left(\frac{N}{2\mathsf{d}}\log N\right)\mathbf{1}_{\mathsf{s}=0}+N\sum_{i=1}^N\zeta(x_i)\right)\right)\right|$$
$$\le C\beta\chi(\beta)N+C_\zeta N,$$

where $C>0$ depends only on s,d and $\|\mu_V\|_{L^\infty}$, and C_ζ depends on (A4) *and ζ.*

PROOF. We start from (5.1.7) and (5.1.5) to write

$$\mathbb{E}_{\mathbb{P}_{N,\beta}}\left(\exp\frac{\beta N^{-\frac{\mathsf{s}}{\mathsf{d}}}}{2}\left(\mathsf{F}_N(X_N,\mu_V)+N\sum_{i=1}^N\zeta(x_i)\right)\right)=\frac{\tilde{\mathsf{K}}_{N,\beta/2}(\mu_V,\zeta)}{\tilde{\mathsf{K}}_{N,\beta}(\mu_V,\zeta)}.$$

We next bound the log of the ratio in the right-hand side using (5.2.5) and (5.2.49) and deduce the result. □

Similarly as Corollary 5.22 we obtain a bound for linear statistics.

Corollary 5.25 (Concentration around the equilibrium measure—local laws at the macroscopic scale). *Let φ be a regular function, then, under the same assumptions,*

(5.2.67)
$$\left|\log \mathbb{E}_{\mathbb{P}_{N,\beta}}\left(\exp\frac{\beta}{C\|\varphi\|^2}N^{2-\frac{\mathsf{s}}{\mathsf{d}}}\left(\int_{\mathbb{R}^{\mathsf{d}}}\varphi\, d(\widehat{\mu_N}-\mu_V)\right)^2\right)\right| \le C\beta\chi(\beta)N+C_\zeta N,$$

where C depends only on s,d and $\|\mu_V\|_{L^\infty}$, and

$$\|\varphi\|=\begin{cases}\|\nabla\varphi\|_{L^2}+\|\nabla\varphi\|_{L^\infty} & \text{if } \mathsf{s}=\mathsf{d}-2\\ \|\varphi\|_{\dot{H}^{\frac{\mathsf{d}-\mathsf{s}}{2}}}+\|(-\Delta)^{\frac{\mathsf{d}-\mathsf{s}}{2}}\varphi\|_{L^\infty} & \text{if } \mathsf{s}\neq\mathsf{d}-2.\end{cases}$$

Of course an analogue to Corollary 5.23 can also be written down.

5.2.5.3. *Bound on the minimal energy.* We finally note that Lemma 5.12 and Proposition 5.14 provide an easy (probabilistic) way to prove an upper bound on the minimal modulated energy. Of course, this can be combined with the splitting formula (5.1.1) or (5.1.9) to provide a bound for the minimum of $\mathcal{H}_N$.

Corollary 5.26.

Original scale. *Let μ be a bounded probability density over $\mathbb{R}^{\mathsf{d}}$. In the case $\mathsf{s} \le 0$, assume also that μ satisfies* (5.2.60)–(5.2.61) *for $\beta = 1$. Then we have*

$$\min \left(\mathsf{F}_N(\cdot, \mu) + \left(\frac{N}{2\mathsf{d}} \log N \right) \mathbf{1}_{\mathsf{s}=0} \right) \le C N^{1+\frac{\mathsf{s}}{\mathsf{d}}} \mathbf{1}_{\mathsf{s}\le 0} \tag{5.2.68}$$

for some $C > 0$ depending only on d, $\|\mu\|_{L^\infty}$ and the constants in the assumptions.

Blown-up scale. *Let μ be a bounded density such that $\int_{\mathbb{R}^{\mathsf{d}}} \mu = N$. In the case $\mathsf{s} \le 0$, assume also that μ satisfies* (5.2.28)–(5.2.29) *for $\beta = 1$. Then*

$$\min \mathsf{F}(\cdot, \mu) \le C N \mathbf{1}_{\mathsf{s}\le 0} \tag{5.2.69}$$

for some $C > 0$ depending only on d, s, $\|\mu\|_{L^\infty}$ and the constants in the assumptions.

PROOF. Applying Lemma 5.12 and Proposition 5.14 with $\beta = 1$ we have

$$\log \left(\int \exp \left(-N^{-\frac{\mathsf{s}}{\mathsf{d}}} \left(\mathsf{F}_N(X_N, \mu) + \left(\frac{N}{2\mathsf{d}} \log N \right) \mathbf{1}_{\mathsf{s}=0} \right) \right) d\mu^{\otimes N}(X_N) \right) \ge -C N \mathbf{1}_{\mathsf{s}\le 0}$$

with $C > 0$ depending only on d, V. A mean-value argument ensures that there exists X_N such that $\mathsf{F}_N(X_N, \mu) \le C N^{1+\frac{\mathsf{s}}{\mathsf{d}}} \mathbf{1}_{\mathsf{s}\le 0}$ for the same C. □

The bound (5.2.69) can also be proven (as first done in [**PS17**], for instance) by constructing a test configuration with the help of the subadditivity of Chapter 7; see Chapter 8 for more detail.

5.3. Localization, separation, and discrepancy

We now turn to other methods, in particular involving maximum principle ideas, that allow to obtain more refined estimates on localization, point separation and discrepancy.

5.3.1. Localization. The splitting formula (5.1.1) indicates that ζ acts as an effective confinement potential, which vanishes in the droplet, i.e., the set ω equal to the zero set of ζ, which contains Σ (with possibly strict inclusion). The concentration result Corollary 5.24 provides an average localization result. Indeed, inserting the lower bound (4.2.12) into (5.2.66) we obtain

Corollary 5.27 (Average localization)**.** *Assume* (A1)–(A4) *so that μ_V exists and is compactly supported. Assume also that μ_V has a bounded density. Then for all $\beta > 0$, we have*

$$\left| \log \mathbb{E}_{\mathbb{P}_{N,\beta}} \left(\exp \frac{\beta N^{1-\frac{\mathsf{s}}{\mathsf{d}}}}{2} \sum_{i=1}^N \zeta(x_i) \right) \right| \le C\beta\chi(\beta) N + C_\zeta N, \tag{5.3.1}$$

where $C > 0$ depends only on s, d and $\|\mu_V\|_{L^\infty}$, and C_ζ depends on (A4) *and ζ.*

By Markov's inequality, this allows to bound from above the probability that $\sum_i \zeta(x_i)$ becomes large, which allows to control the distance to Σ when estimates such as (2.5.21) and (2.4.3), coming from the analysis of the (fractional) obstacle problem, are available.

In the two-dimensional Coulomb gas, a stronger result is proven in [**Ame21**], we will discuss it below, see also [**CHM18**] for the Coulomb case in any $\mathsf{d} \geq 2$.

We now prove perfect localization in the droplet for minimizers, i.e., that all the points belong to Σ, the support of μ_V. The simple argument (found in [**RNS15**] in the two-dimensional case) relies on the maximum principle.

Theorem 5.28 (Perfect localization for minimizers). *Assume* $\mathsf{s} = \mathsf{d} - 2$. *Let* X_N *minimize* $\mathcal{H}_N$. *Then for all* i, $x_i \in \Sigma$, *the support of* μ_V.

PROOF. The main point is that for minimizers, each point is at the minimum of the potential generated by the other points. Indeed, for any x, we may write that

$$\mathcal{H}_N(X_N) \leq \mathcal{H}_N(\widehat{X_N}),$$

where the points of $\widehat{X_N}$ are defined by

$$\hat{x}_j = \begin{cases} x_j & \text{if } j \neq i \\ x & \text{if } j = i. \end{cases}$$

Spelling out the definition of the energy, this implies by direct comparison that

$$\sum_{j\neq i} \mathsf{g}(x_i - x_j) + NV(x_i) \leq \sum_{j\neq i} \mathsf{g}(x - x_j) + NV(x) \tag{5.3.2}$$

or in other words

$$h_i(x_i) \leq h_i(x), \tag{5.3.3}$$

where

$$h_i := \sum_{j\neq i} \mathsf{g}(\cdot - x_j) + NV = \sum_{j\neq i} \mathsf{g}(\cdot - x_j) + N(\zeta - h^{\mu_V} + c)$$

using (2.1.12). We then notice that in Σ^c, $\sum_{j\neq i} \mathsf{g}(\cdot - x_j)$ is superharmonic and h^{μ_V} is harmonic. Thus $\sum_{j\neq i} \mathsf{g}(\cdot - x_j) - Nh^{\mu_V} = \mathsf{g} * (\sum_{j\neq i} \delta_{x_j} - N\mu_V)$, which is asymptotic to $-\mathsf{g}$ at ∞ since μ_V is compactly supported, can only achieve its strict minimum relative to Σ^c on $\partial\Sigma$ (after distinguishing the cases $\mathsf{s} \leq 0$ and $\mathsf{s} > 0$). Meanwhile $\zeta + c$ is also minimized in Σ. So we conclude that $\min_{\mathbb{R}^\mathsf{d}\setminus\Sigma} h_i$ is only achieved on $\partial\Sigma$, and thus if $x_i \in \mathbb{R}^\mathsf{d}\setminus\Sigma$, we obtain a contradiction with (5.3.3). □

In the case of general $\mathsf{s} \in (\mathsf{d} - 2, \mathsf{d})$, the result is still true and the argument is similar, except we use the maximum principle for the operator $\operatorname{div}(|y|^\gamma \nabla \cdot)$ in $\mathbb{R}^{\mathsf{d}+1}\setminus(\Sigma \times \{0\})$, see [**PS17**] for the full proof.

5.3.2. Isotropic averaging and consequences. We now continue with maximum principle based arguments, and present a result showing that points of energy minimizers are well-separated. Separation results for Fekete points but also subsequently for Riesz s-energy minimizers have been established in very general geometries [**Dah78, BDS14, HRSV19**]. The result we present below was shown in the two-dimensional Coulomb case in [**AOC12**], and recently extended to the low temperature regime $\beta \geq C \log N$ in [**AR23**], still in the two-dimensional Coulomb case. A variant of the short proof below was given in [**RNS15**] in the d-dimensional

Coulomb case and in [**PS17**] in the Riesz case $\mathsf{s} \in [\mathsf{d}-2, \mathsf{d})$, using a maximum-principle unpublished argument of Lieb [**Lie09**]. We now present a reformulation of the argument that works in the Coulomb case and relies on *isotropic averaging.* The isotropic averaging method was introduced in the context of the Gibbs measure in [**Leb**], and made into a systematic and powerful method, that we discuss below, by Thoma in [**Tho24**]. An idea of similar flavor was recently used in [**CH25**] to obtain a short proof of the bound on smooth linear statistics fluctuations in the Coulomb gas.

Theorem 5.29 (Point separation for minimizers in the Coulomb case by isotropic averaging). *Assume* $\mathsf{s} = \mathsf{d} - 2$. *Assume* V *is such that* $(\Delta V)_+ \in L^\infty$. *Let* X_N *be a minimizer of* $\mathcal{H}_N$. *There exists an explicit constant* $C > 0$ *depending only on* d *such that*

$$\min_{i \neq j} |x_i - x_j| \geq \frac{C}{(N\|(-\Delta V)_+\|_{L^\infty})^{1/\mathsf{d}}}. \tag{5.3.4}$$

PROOF. The core of the isotropic averaging is very similar to the monotonicity argument in Section 4.2. Let us use the same notation, in particular $\delta_x^{(\eta)}$. If X_N minimizes $\mathcal{H}_N$ then its energy is less than that obtained by isotropically averaging over the position of one point, call it x_i. In other words, averaging (5.3.2) over x, we may write that

$$\sum_{j \neq i} (\mathsf{g}(x_i - x_j) - \int \mathsf{g}(x - x_j) d\delta_{x_i}^{(\eta)}(x) \leq N \int (V(x) - V(x_i)) d\delta_{x_i}^{(\eta)}(x).$$

Using that $\mathsf{g} * \delta_0^{(\eta)} = \mathsf{g}_\eta$ by definition, the radial nature of $\delta_{x_i}^{(\eta)}$, and a Taylor expansion for V around x_i, we deduce that

$$\sum_{j \neq i} \mathsf{g}(x_i - x_j) - \mathsf{g}_\eta(x_i - x_j) \leq N\eta^2 \|(\Delta V)_+\|_{L^\infty}. \tag{5.3.5}$$

We now recall that $\mathsf{g}_\eta \leq \mathsf{g}$ and $\mathsf{g}_\eta(x) = \mathsf{g}(\eta)$ for $|x| \leq \eta$, thus the sum contains only nonnegative terms, and restricting it to the points such that $|x_i - x_j| \leq \frac{1}{2}\eta$ (supposing that such points exist) we obtain

$$\sum_{j \neq i, |x_i - x_j| \leq \frac{1}{2}\eta} \mathsf{g}(\frac{\eta}{2}) - \mathsf{g}(\eta) \leq N\eta^2 \|(\Delta V)_+\|_{L^\infty}.$$

We may now choose

$$\eta < \begin{cases} (2N\|(\Delta V)_+\|_{L^\infty})^{-1} & \text{if } \mathsf{d} = 1 \\ (N\|(\Delta V)_+\|_{L^\infty}/(\log 2))^{-\frac{1}{2}} & \text{if } \mathsf{d} = 2 \\ \left(\frac{2^{\mathsf{d}-1}-1}{N\|(\Delta V)_+\|_{L^\infty}}\right)^{\frac{1}{\mathsf{d}}} & \text{if } \mathsf{d} \geq 3 \end{cases}$$

to obtain a contradiction, thus implying that there is no other point in $B(x_i, \frac{1}{2}\eta)$ than x_i. □

The fact that V can contain a possibly singular arbitrary superharmonic part, without changing the estimate (because it does not change $(-\Delta V)_+$) is useful when considering particles subjected to an external potential itself generated by point charges, and for "incompressibility estimates" in the Laughlin phase as in [**LRY18**], this is taken advantage of in [**Tho24**].

The article [**Tho24**] provides a temperature-dependent version of the argument in the proof above by estimating quantitatively the entropic effect of the isotropic averaging (in other words, the change of volume in phase space due to it). Thanks to these quantitative estimates, Thoma is able to obtain, for the Coulomb case in arbitrary dimension

- "overcrowding estimates" phrased at the blown-up scale that are valid up to the boundary and confirm the predictions of [**JLM93**] (see Section 10.3 for the statement of the prediction).

Theorem 5.30 ([**Tho24**]). *Assume* $\mathsf{s} = \mathsf{d} - 2$, $\mathsf{d} \geq 2$. *For any* β, *any ball* B_R *of radius* R, *if* $Q \geq CR^{\mathsf{d}} + C\beta^{-1}R^{\mathsf{d}-2}$, *then*

$$\mathbb{P}_{N,\beta}\left(\#\left(\{X'_N\} \cap B_R\right) \geq Q\right) \leq \begin{cases} \exp\left(-\frac{\beta}{4}Q^2 \log \frac{Q}{R^2} + C\beta Q^2 + CQ\right) & \mathsf{d} = 2 \\ \exp\left(-C\beta R^{2-\mathsf{d}} Q(Q-1)\right) & \mathsf{d} \geq 3. \end{cases}$$

- estimates for the size of the minimal gap $\eta := \min_{i \neq j} |x_i - x_j|$ between particles, which in the case $\mathsf{d} = 2$ takes the form of the sharp estimate

$$\mathbb{P}_{N,\beta}(\eta N^{1/\mathsf{d}} \leq \rho N^{-\frac{1}{2+\beta}}) \leq C\rho^{2+\beta} \quad \forall \rho > 0, \tag{5.3.6}$$

  where C is independent of N and one can let $\beta \to \infty$ as $N \to \infty$ to retrieve the result of Theorem 5.29, which is analogous to the freezing regime of [**Ame18, AR23**]. Note that in the case of the two-dimensional Coulomb gas at $\beta = 2$, precise asymptotics for large gap probabilities are known [**For92, Cha24**], and [**Cha25**] recently established the law of the smallest gap (which is typically of order $N^{-3/4}$), k-th smallest gap, and smallest gap locations (which are Poissonnian).
- uniform upper bounds on the k-point correlation function $\rho_N^{(k)}$ as defined in (1.3.4) which in the case $k = 1$ provide so-called *Wegner estimates.*
- non-number-rigidity in dimension $\mathsf{d} \geq 3$ and the possibility of number rigidity in dimension $\mathsf{d} = 2$ in [**Tho23**].

Less sharp separation estimates in the bulk (with temperature) are also given in Corollary 8.12.

A sort of opposite question to that of obtaining charge overcrowding probabilities is to estimate the probability of a hole (without any point). This involves potential theory and balayage of measure and is addressed in [**Adh18, Cha23b**]; see also [**GN18**] for a review of such questions.

A corollary of Theorem 5.30 is that [**Tho23**] also gives the existence of a limiting point process to the shifted blown-up configuration $\{X'_N - x\}$, for *any* $x \in \mathbb{R}^{\mathsf{d}}$ (and not only for x in the bulk) up to subsequences. Besides this result, the existence of limiting point process was only known in the one-dimensional logarithmic case (with the sine-β process [**KN04, VV09**]) and the two-dimensional Ginibre ensemble case for which V is quadratic and $\beta = 2$, converging to the Ginibre point process, and the one-dimensional circular Riesz gas for all β in [**Boua**].

There is a large statistical mechanics literature (see [**GLM80, GLM78, Mar88**] and references therein) from the 1970s on sum rules and various relations for correlation functions of interacting particle systems, in particular Kirkwood-Salzbourg, BBGKY, KMS, DLR equations. These can be shown to be equivalent relations in the case of regular interaction kernels but in the case of singular interactions like the Coulomb one, the existence of solutions to these hierarchies was not known. The existence of a limiting point process, though up to subsequences, is thus important toward putting these ideas on firmer ground.

5.3.3. Discrepancy, separation, and localization estimates by complex analysis. Obtaining bounds on the charge discrepancies (also called Beurling-Landau densities in the context of sampling and interpolation theory) as defined in (4.4.1) has always been one of the important goals of the analysis of Coulomb systems. We described in Section 4.4 how the electric formulation rapidly provides discrepancy estimates, which are however suboptimal. In the series of works [**AOC12, AR23, MR23**], tools from complex analysis and interpolation theory are used, and the role of the electric formulation is played by Lagrange interpolation polynomials and reproducing kernel estimates, which are however restricted to the two-dimensional situation. We now present the Lagrange interpolation polynomial approach originating in [**Ame21**].

For any i, introduce the Lagrange interpolant

$$L_i(x) = \prod_{j\neq i} \frac{(x - x_j)}{(x_i - x_j)} \frac{e^{-NV(x)}}{e^{-NV(x_i)}}, \tag{5.3.7}$$

for which one observes that $L_i(x_i) = 1$ and $L_i(x_j) = 0$ for any $j \neq i$. Also, since we are in the two-dimensional Coulomb case, by (1.1.1) we may rewrite L_i as

$$L_i(x) = \frac{e^{-\mathcal{H}_N(\widehat{X_N})}}{e^{-\mathcal{H}_N(X_N)}}, \tag{5.3.8}$$

where the points of $\widehat{X_N}$ are as above defined by

$$\hat{x}_j = \begin{cases} x_j & \text{if } j \neq i \\ x & \text{if } j = i. \end{cases}$$

Thus, if X_N are Fekete points (minimizing $\mathcal{H}_N$) then $L_i(x) \leq 1$ for every x. Controls of L_i express rigidity of the configuration. For such controls, the crucial fact is that, under the assumption that V is analytic, the function L_i is also analytic away from the points of the configuration. Using complex analysis estimates based on Cauchy's formula, which replace maximum principle arguments, this allows to obtain regularity for L_i, for instance control $\|L_i\|_{L^\infty}$ and $\|\nabla L_i\|_{L^\infty}$ by weaker norms of L_i. For instance once $\|L_i\|_{L^\infty} \leq C$ and $\|\nabla L_i\|_{L^\infty} \leq CN^{1/2}$ is proven, one can easily deduce separation of Fekete points as in Theorem 5.29: indeed, for X_N Fekete points, by the above remarks, for $i \neq j$,

$$1 = |L_i(x_j) - L_i(x_i)| \leq \|\nabla L_i\|_{L^\infty}|x_i - x_j| \leq CN^{1/2}|x_i - x_j|,$$

which yields the separation lower bound $|x_i - x_j| \geq N^{-1/2}/C$.

Let us now give a rough idea of how the Lagrange interpolation idea can be used in the situation with temperature, the goal still being to show controls on norms of L_i, encoding some rigidity of the configurations.

It follows from (5.3.8) and definition (1.1.5) that for any test function φ,

$$\begin{aligned} \mathbb{E}_{\mathbb{P}_{N,\beta}}\left(\varphi(x, x_i)L_i(x)^\beta\right) &= \frac{1}{Z_{N,\beta}} \int_{(\mathbb{R}^2)^N} \varphi(x, x_i) \frac{e^{-\beta\mathcal{H}_N(\widehat{X_N})}}{e^{-\beta\mathcal{H}_N(X_N)}} e^{-\beta\mathcal{H}_N(X_N)} dX_N \\ &= \frac{1}{Z_{N,\beta}} \int_{(\mathbb{R}^2)^N} \varphi(x, x_i) e^{-\beta\mathcal{H}_N(\widehat{X_N})} dX_N \end{aligned} \tag{5.3.9}$$

and integrating in x we obtain

$$\begin{aligned}\int \mathbb{E}_{\mathbb{P}_{N,\beta}}\left(\varphi(x,x_i)L_i(x)^\beta\right)dx &= \frac{1}{Z_{N,\beta}}\int_{(\mathbb{R}^2)^N\times\mathbb{R}^2}\varphi(x,x_i)e^{-\beta\mathcal{H}_N(\widehat{X_N})}dX_N\,dx\\ &= \frac{1}{Z_{N,\beta}}\int_{(\mathbb{R}^2)^N\times\mathbb{R}^2}\varphi(x,x_i)e^{-\beta\mathcal{H}_N(\widehat{X_N})}d\widehat{X_N}\,dx_i\\ &= \int\varphi(x,y)d\rho_N^{(1)}(x)dy,\end{aligned}$$

where $\rho_N^{(1)}$ is the one-point correlation function as defined in (1.3.4).

Applying for instance with $\varphi(x,y)=\mathbf{1}_A(y)$ one obtains

$$\mathbb{E}_{\mathbb{P}_{N,\beta}}\left(\mathbf{1}_A(x_i)\|L_i\|^\beta_{L^\beta(\mathbb{R}^2)}\right)=|A|, \tag{5.3.10}$$

where $|A|$ denotes the measure of the set.

We now give an example of a maximum-principle based lemma from [**Ame21**].

Lemma 5.31. *Let q be a holomorphic polynomial with degree $\le N-1$, then, letting $f=qe^{-NV}$ we have*

$$f(x)\le\|f\|_{L^\infty}e^{-N\zeta(x)} \tag{5.3.11}$$

and, letting $s>\max_{\overline{U}}(\Delta V)$ where U is some bounded open neighborhood of Σ, we have for any $x\in U$,

$$|f(x)|^\beta\le e^{Cs\beta}\fint_{B(x,\frac{1}{\sqrt{N}})}|f|^\beta, \tag{5.3.12}$$

where ζ is the function of (2.1.12), and C depends only on U.

Proof. Let us start with (5.3.11). We let $u=\frac{1}{N}\log(\frac{|q|}{\|f\|_{L^\infty}})$ and remark that, since the logarithm of the modulus of a holomorphic function is harmonic, u is harmonic, hence

$$\Delta(u+h^{\mu_V}-c)=-\mathsf{c_d}\mu_V\le 0,$$

where $h^{\mu_V}=\mathsf{g}*\mu_V$ and c is the constant in (2.1.12). Moreover, by the assumption on the degree of q, $u(x)\ge\log|x|+O(1)$ as $|x|\to\infty$, hence, since h^{μ_V} behaves like $-\log|x|$ at ∞, we have $u+h^{\mu_V}-c\ge O(1)$. But a bounded below superharmonic function in the plane is constant by the maximum principle. So we have $u+h^{\mu_V}-c=cst$. Moreover by definition of f we have $qe^{-NV}\le\|f\|_{L^\infty}$ hence after taking the log of the modulus of both sides, $u\le V$ and thus $u+h^{\mu_V}-c\le\zeta$ by (2.1.12). In particular $u+h^{\mu_V}-c\le 0$ in ω so the constant above must be ≤ 0. It follows that $u+h^{\mu_V}-c\le 0$ hence $u\le V-\zeta$, hence the result (5.3.11) after taking the exponential.

For the second result (5.3.12), let $F(x)=|f(x)|^\beta e^{\frac12 s\beta|x|^2}$. By definition of f and choice of s, we have

$$\Delta\log F(x)\ge-\beta N\Delta V(x)+s\beta N\ge 0.$$

Using that $\Delta\log F=\frac{1}{F}\Delta F-\frac{1}{F^2}|\nabla F|^2$, we find that F is subharmonic in U. By the mean-value inequality for subharmonic functions it follows that for N large enough

$$F(x)\le\fint_{B(x,N^{-1/\mathsf{d}})}F\le e^{Cs\beta}\fint_{B(x,N^{-1/\mathsf{d}})}|f|^\beta,$$

which implies the result. □

Applying this lemma to $q(x) = \frac{\prod_{j\neq i}(x-x_j)}{\prod_{j\neq i}(x_i-x_j)e^{-NV(x_j)}}$ and returning to (5.3.7) one immediately deduces that

Corollary 5.32. *Under the same assumptions, we have*

$$|L_i(x)| \leq \|L_i\|_{L^\infty} e^{-N\zeta(x)} \tag{5.3.13}$$

and for any $x \in U$,

$$|L_i(x)|^\beta \leq e^{Cs\beta} N \int_{B(x,\frac{1}{\sqrt{N}})} |L_i|^\beta. \tag{5.3.14}$$

The relation (5.3.10) allows to control the $L^\beta(\mathbb{R}^2)$ norm of L_i with large probability: it suffices to find a bounded set A that contains all the points except with small probability (this is provided by a first, weak, confinement), then argue that (5.3.10) yields

$$\mathbb{P}_{N,\beta}(\|L_i\|_{L^\beta} > \lambda) \leq \mathbb{P}_{N,\beta}(\|L_i\|_{L^\beta} > \lambda \mathbf{1}_A(x_i)) + \mathbb{P}_{N,\beta}(\mathbf{1}_{A^c}(x_i))$$

with both terms in the right-hand side being small if λ is large. Then, once $\|L_i\|_{L^\beta}$ is bounded, up to a small probability event, (5.3.14) allows to upgrade this into an L^∞ bound $\|L_i\|_{L^\infty} \leq CN^{\frac{1}{\beta}}$. We can then reinsert this into (5.3.13) to obtain

$$|L_i(x)| \leq Ce^{-N\zeta(x)} N^{\frac{1}{\beta}}.$$

But since $L_i(x_i) = 1$, taking the logarithm we deduce that for every i

$$\zeta(x_i) \leq \frac{1}{\beta}\frac{\log N}{N} + \frac{C}{N}$$

up to a small probability event. Together with (2.5.21) it provides the following strong localization result.

Theorem 5.33 (Strong localization, [**Ame21**]). *Assume* $\mathsf{d} = 2$ *and* $\mathsf{s} = 0$. *The points of X_N belong to*

$$\left\{x, \mathrm{dist}(x, \Sigma) \leq C\sqrt{\frac{\log N}{\beta N}}\right\}$$

except on an event with probability $o_N(1)$.

We refer to [**Ame21**] for a more general and precise statement which contains correction terms and is valid as long as $\beta \gg \frac{\log N}{N}$. Such results were known earlier for the $\mathsf{d} = 1$ logarithmic case, at least if V is quadratic.

The same starting point (controls of L^β norms of L_i, except with small probability) is used in [**AR23**] and [**MR23**] to obtain strong discrepancy estimates, still in the two-dimensional Coulomb case. One may say that in these works, Lagrange interpolants are combined with reproducing kernel representations and techniques from sampling and interpolation and spectral analysis of Toeplitz operators, while in [**Tho24**] this is replaced by isotropic averaging combined with the electric formulation to deduce control on discrepancies.

In [**MR23**] these techniques are pushed to obtain a new "freezing" regime for $\beta \geq c \log N$. More precisely, it is shown, that when $\beta \geq c \log N$, the orders of the separation and discrepancy are the same as that of energy minimizers (the $\beta = \infty$ case), while for β of order 1, they are known to fluctuate more [**LR10**] while the point separation is not bounded below independently of N (see for instance (5.3.6)).

In [**AMR22**] the same results are shown in the one-dimensional logarithmic case $\mathsf{s} = 0$ in the case of quadratic V. Using known discrepancy and fluctuation results, the authors can in addition show that the freezing regime is exactly $\beta \geq c \log N$.

CHAPTER 6

The commutator estimate and application to dynamics

We continue here our study of the modulated energy F_N, begun in Chapter 4, and make a detour through the study of dynamics of Coulomb and Riesz systems, in particular to the question of deriving mean-field limits for dynamics of the type (1.1.8), (1.1.9) (with possibly $\beta = \infty$) via the *modulated energy method.*

At the core of the method is a functional inequality on F_N that we call a commutator estimate, and which will also be important in Chapter 9 for studying fluctuations of Coulomb gases.

We start by presenting the functional inequality in Section 6.1, and then turn to the dynamics in Section 6.2. Again, we consider throughout the interaction range (2.0.1).

6.1. The functional inequality

For the questions mentioned above (dynamics, fluctuations), we need to consider how F_N varies when the points are transported by a perturbation of identity $\Phi_t := I + tv(x)$, while μ is pushed-forward by the same map Φ_t. We easily observe that

$$(6.1.1)\quad \frac{d}{dt}\Big|_{t=0} \mathsf{F}_N\left((I+tv)^{\oplus N}(X_N), (I+tv)\#\mu\right) \\ = \frac{1}{2}\int_{(\mathbb{R}^\mathsf{d})^2\setminus\triangle} (v(x)-v(y))\cdot\nabla\mathsf{g}(x-y)\,d\Big(\sum_{i=1}^N \delta_{x_i} - N\mu\Big)^{\otimes 2}(x,y),$$

where $(I+tv)^{\oplus N}(X_N) = (x_1+tv(x_1),\dots,x_N+tv(x_N))$. More generally, for any $n\ge 1$,

$$(6.1.2)\quad \frac{d^n}{dt^n}\Big|_{t=0} \mathsf{F}_N((I+tv)^{\oplus N}(X_N),(I+tv)\#\mu) = \mathsf{A}_n(X_N,\mu,v),$$

where we let

$$(6.1.3)\quad \mathsf{A}_n(X_N,\mu,v) := \frac{1}{2}\int_{(\mathbb{R}^\mathsf{d})^2\setminus\triangle} \nabla^{\otimes n}\mathsf{g}(x-y) : (v(x)-v(y))^{\otimes n}\,d\Big(\sum_{i=1}^N \delta_{x_i} - N\mu\Big)^{\otimes 2}(x,y),$$

where : denotes the inner product between the tensors. We will discuss the control of such next order quantities in Section 6.1.3 below.

The functional inequality we were referring to shows that A_1, the first variation of F_N along the transport by a Lipschitz vector field v is controlled by F_N itself. It first appeared in [**Ser20**] after a partial result in [**Due16**]. It was then recognized in [**Ros20**] that it could be seen as a commutator estimate and thanks to this point of view it was generalized in [**NRS22**] to a broader class of interactions. We now present the sharp version of the estimate as obtained in [**RS24b**]. The notation used is that of Section 4.5, in particular F_N^Ω corresponds to the modulated energy localized in a subset Ω. The estimate is *localized* in the sense that if the vector field v is localized in Ω, say a small region of $\mathbb{R}^\mathsf{d}$, then the variation of energy is bounded in terms of the localized energy F_N^Ω only. This is unlike typical commutator estimates found in the harmonic analysis or PDE literature, and it will be essential for the analysis of fluctuations of Coulomb/Riesz gases at mesoscales.

Theorem 6.1 (Sharp commutator estimate). *There exists a constant $C > 0$ depending only on* d *and* s *such that the following holds. Let $\mu \in L^\infty(\mathbb{R}^\mathsf{d})$ be a probability density satisfying* (4.1.1). *Let $v : \mathbb{R}^\mathsf{d} \to \mathbb{R}^\mathsf{d}$ be a Lipschitz vector field and let Ω be a closed set containing a 3λ-neighborhood of* $\operatorname{supp} \nabla v$, *where $\lambda = (N\|\mu\|_{L^\infty})^{-1/\mathsf{d}} < 1$.*[1] *For any pairwise distinct configuration $X_N \in (\mathbb{R}^\mathsf{d})^N$, it holds that*

$$\left| \int_{(\mathbb{R}^\mathsf{d})^2 \setminus \triangle} (v(x) - v(y)) \cdot \nabla \mathsf{g}(x-y) d\Big(\sum_{i=1}^N \delta_{x_i} - N\mu \Big)^{\otimes 2}(x,y) \right| \tag{6.1.4}$$
$$\le C \|\nabla v\|_{L^\infty} \left(\mathsf{F}_N^\Omega(X_N, \mu) - \# I_\Omega \Big(\frac{\log \lambda}{2} \Big) \mathbf{1}_{\mathsf{s}=0} + C \|\mu\|_{L^\infty}^{\frac{\mathsf{s}}{\mathsf{d}}} \# I_\Omega N^{\frac{\mathsf{s}}{\mathsf{d}}} \right),$$

where F_N^Ω is as in (4.5.2).

In a first reading, the reader may simply take $\Omega = \mathbb{R}^\mathsf{d}$ and $\mathsf{F}_N^\Omega = \mathsf{F}_N$, which will suffice for the dynamics. An examination of the proof also reveals that we can replace $\mathsf{F}_N^\Omega(X_N, \mu) - \# I_\Omega \big(\frac{\log \lambda}{2} \big) \mathbf{1}_{\mathsf{s}=0}$ in the right-hand side by $\int_{\Omega \times [-\ell, \ell]^\mathsf{k}} |y|^\gamma |\nabla h_{N,\mathsf{r}}|^2$, where ℓ is the size of a ball containing Ω, details can also be found in [**RS24b**].

As seen in [**NRS22**], (6.1.4) can be seen as a commutator estimate, because, letting $f = \sum_{i=1}^N \delta_{x_i} - N\mu$ and ignoring the excision of the diagonal, we may rewrite the integral in the left-hand side of (6.1.4) as

$$\int v \cdot \nabla (\mathsf{g} * f) - \mathsf{g} * (\operatorname{div}(vf)) = \left\langle f, \left[v, \frac{\nabla}{(-\Delta)^{\frac{\mathsf{d}-\mathsf{s}}{2}}} \right] f \right\rangle_{L^2},$$

where $[\cdot,\cdot]$ denotes here (and only here) the commutator between two operators. Although commutators of this type have been studied in the harmonic analysis literature (see [**SSS19**] and references therein), the estimates there only apply to divergence-free vector fields, are not localizable, and do not quite provide what we need. In [**NRS22**], this general commutator point of view was exploited and a proof that covers more general interactions g that have singularities of Riesz type (but without being necessarily exactly the Riesz kernel) for any $\mathsf{s} \in [0, \mathsf{d})$, was provided in the nonlocalized case (i.e., for $\Omega = \mathbb{R}^\mathsf{d}$), without giving the sharp additive rate however. Recently, [**HCRS25**] provided a different proof valid for all $\mathsf{s} \in [0, \mathsf{d})$ and for a class of similar interactions, still in the nonlocalized case, but providing the sharp additive rate $N^{1+\frac{\mathsf{s}}{\mathsf{d}}}$.

Here, we give the simplest proof, which works when considering all the exact Riesz kernels (2.0.1).

[1] In [**RS24b**], we present a slightly more precise version where the definition of λ is instead $(N\|\mu\|_{L^\infty(\Omega)})^{-1/\mathsf{d}}$.

6.1.1. Proof outline. The proof relies crucially on the electric formulation in extended space (we use the same notation as in Section 4.1.3) and the notion of *stress-energy tensor*, a standard notion in mechanics and calculus of variations.

Definition 6.2 (Stress-energy tensor). Given two (regular enough) distributions f and w, we define the cross stress-energy tensor as

$$[\nabla h^f, \nabla h^w]_{ij} := |y|^\gamma \left(\partial_i h^f \partial_j h^w + \partial_j h^f \partial_i h^w - \nabla h^f \cdot \nabla h^w \delta_{ij}\right), \tag{6.1.5}$$

where δ_{ij} is the Kronecker symbol and $h^f = \mathsf{g} * f$ is naturally extended to $\mathbb{R}^{\mathsf{d}+\mathsf{k}} = \{(x,y), x \in \mathbb{R}^\mathsf{d}, y \in \mathbb{R}\}$ if $\mathsf{k} = 1$.

Strictly speaking, the stress-energy tensor associated to the potential h^f (naturally extended to a function on $\mathbb{R}^{\mathsf{d}+\mathsf{k}}$) is the quantity $[\nabla h^f, \nabla h^f]$, but it is convenient, as in [**Ser20**], to view it as a bilinear function.

Our interest in the stress-energy tensor stems from the relation

$$\operatorname{div}[\nabla h^f, \nabla h^f] = \nabla h^f \operatorname{div}(|y|^\gamma \nabla h^f) = -\mathsf{c}_{\mathsf{d},\mathsf{s}} \nabla h^f f, \tag{6.1.6}$$

which is true by direct computation if f is sufficiently regular. Here the divergence of the tensor is a vector, whose components are equal to the divergence of the rows/columns of the tensor. This way, nonlinear terms of the form $\nabla h^f f$ can be rewritten in divergence form. Moreover by definition (6.1.5), the term $[\nabla h^f, \nabla h^f]$ is pointwise controlled by the energy density:

$$|[\nabla h^f, \nabla h^f]| \le |y|^\gamma |\nabla h^f|^2. \tag{6.1.7}$$

Arguing formally, if one neglects the issue of the diagonal terms in both sides of (6.1.4), after desymmetrizing the left-hand side of (6.1.4) and using that g is even, one can rewrite it as

$$\iint_{\triangle^c} v(x) \cdot \nabla \mathsf{g}(x-y) df(x) df(y) - \iint_{\triangle^c} v(y) \cdot \nabla \mathsf{g}(x-y) df(x) df(y) = 2 \int_{\mathbb{R}^\mathsf{d}} v \cdot \nabla h^f df,$$

with $f = \sum_{i=1}^N \delta_{x_i} - N\mu$. Using (6.1.6) and an integration by parts, this can be reinterpreted as $\frac{2}{\mathsf{c}_{\mathsf{d},\mathsf{s}}} \int \nabla v : [\nabla h^f, \nabla h^f]$, and thanks to the pointwise control (6.1.7), we have bounded the left-hand side of (6.1.4) as follows:

$$\iint_{\mathbb{R}^\mathsf{d} \times \mathbb{R}^\mathsf{d}} (v(x) - v(y)) \cdot \nabla \mathsf{g}(x-y) df(x) df(y) \le C \|\nabla v\|_{L^\infty} \int_{\mathbb{R}^{\mathsf{d}+\mathsf{k}}} |y|^\gamma |\nabla h^f|^2, \tag{6.1.8}$$

where, in view of (4.1.29), we formally recognize in the right-hand side the electric rewriting of $\mathsf{F}_N(X_N, \mu)$. Of course, the main problem is that this neglects the issue of the diagonal, and that the estimate needs to be properly *renormalized* on both sides. The truncation method presented in Chapter 4, as well as all the properties demonstrated there, provide the right tool to do so. More specifically, the key is to take the *point-dependent* truncation parameters r_i (minimal distances) as defined in (4.2.14) and to apply (6.1.8) to $f = \sum_{i=1}^N \delta_{x_i}^{(\mathsf{r}_i)} - N\mu$ instead. Then, the right-hand side equals $\int_{\mathbb{R}^{\mathsf{d}+\mathsf{k}}} |y|^\gamma |\nabla h_{N,\mathsf{r}}|^2$, which, crucially, we are able to control by F_N itself (without needing to substract off a renormalization term!) thanks to (4.2.16). Then, what remains to be done, which is the most delicate part of the proof, is to evaluate the renormalization error of the left-hand side of (6.1.8) (i.e., the error made when replacing $\sum_{i=1}^N \delta_{x_i}$ by $\sum_{i=1}^N \delta_{x_i}^{(\mathsf{r}_i)}$) in terms of the r_i and $\sum_i \mathsf{g}(\mathsf{r}_i)$, which we also control thanks to (4.2.15).

To make this strategy rigorous, let us start by reformulating a useful identity from [**Ser20**, Lemma 4.3].

Lemma 6.3. *Let $v : \mathbb{R}^{\mathsf{d}} \to \mathbb{R}^{\mathsf{d}}$ be a Lipschitz vector field. Let us extend it trivially into a vector-field on $\mathbb{R}^{\mathsf{d}+\mathsf{k}}$, still denoted v, by letting it depend only on the first d coordinates, and have vanishing last component (we note that the extension has the same Lipschitz norm as the original vector field). For any test functions f and w in the Schwartz class of $\mathbb{R}^{\mathsf{d}}$, it holds that*

$$\int_{(\mathbb{R}^{\mathsf{d}})^2} (v(x) - v(y)) \cdot \nabla \mathsf{g}(x-y) df(x) dw(y) = \frac{1}{\mathsf{c}_{\mathsf{d},\mathsf{s}}} \int_{\mathbb{R}^{\mathsf{d}+\mathsf{k}}} \nabla v : [\nabla h^f, \nabla h^w]. \tag{6.1.9}$$

Applying (6.1.9) to $w = f$ and desymmetrizing the left-hand side, we obtain

$$\int_{\mathbb{R}^{\mathsf{d}+\mathsf{k}}} \nabla v : [\nabla h^f, \nabla h^f] = 2\mathsf{c}_{\mathsf{d},\mathsf{s}} \int_{(\mathbb{R}^{\mathsf{d}})^2} v(x) \cdot \nabla \mathsf{g}(x-y) df(x) df(y), \tag{6.1.10}$$

where the right-hand side also equals $2\mathsf{c}_{\mathsf{d},\mathsf{s}} \int_{\mathbb{R}^{\mathsf{d}}} v \cdot \nabla h^f df$ by definition of h^f. Hence, we have obtained the rigorous version of the claimed relation (6.1.6).

6.1.2. Proof of Theorem 6.1. First, we may assume that the points of X_N are pairwise distinct, otherwise the right-hand side is $+\infty$. Let us consider $\vec{\eta}$ such that for every i, $\eta_i \le \mathsf{r}_i$. In particular, we always have $\eta_i \le \frac{1}{4}\lambda$ by (4.2.14). Let

$$h_N^i(x) := h_N(x) - \mathsf{g}(x - x_i), \tag{6.1.11}$$

be the potential in $\mathbb{R}^{\mathsf{d}}$ generated by the configuration with x_i removed and μ, naturally extended to $\mathbb{R}^{\mathsf{d}+\mathsf{k}}$, and observe in view of (4.1.20) that if the balls $\{B(x_i, \eta_i)\}_{i=1}^N$ are pairwise disjoint, then

$$\nabla h_{N,\vec{\eta}} = \nabla h_N^i + \nabla \mathsf{g}_{\eta_i}(\cdot - x_i) = \begin{cases} \nabla h_N & \text{outside } \cup_{i=1}^N B(x_i, \eta_i) \\ \nabla h_N^i & \text{in } B(x_i, \eta_i). \end{cases} \tag{6.1.12}$$

From this property, it follows that

$$\int_{\mathbb{R}^{\mathsf{d}+\mathsf{k}}} |y|^\gamma |\nabla h_{N,\vec{\eta}}|^2 = \int_{\mathbb{R}^{\mathsf{d}+\mathsf{k}} \setminus \cup_i B(x_i,\eta_i)} |y|^\gamma |\nabla h_N|^2 + \sum_{i=1}^N \int_{B(x_i,\eta_i)} |y|^\gamma |\nabla h_N^i|^2. \tag{6.1.13}$$

We now turn to the proof of the estimate (6.1.4).

Step 1. Rewriting the left-hand side. Desymmetrizing the left-hand side of (6.1.4), we may rewrite it as

$$\begin{aligned} I &:= \int_{(\mathbb{R}^{\mathsf{d}})^2 \setminus \triangle} (v(x) - v(y)) \cdot \nabla \mathsf{g}(x-y) d\Big(\sum_{i=1}^N \delta_{x_i} - N\mu\Big)^{\otimes 2}(x,y) \\ &= 2\sum_{i=1}^N \int_{\mathbb{R}^{\mathsf{d}+\mathsf{k}}} v(x_i) \cdot \nabla \mathsf{g}(x_i - x) d\Big(\sum_{j \ne i} \delta_{x_j} - N\mu\delta_{\mathbb{R}^{\mathsf{d}}}\Big)(x) \\ &\quad - 2N \int_{\mathbb{R}^{\mathsf{d}} \times \mathbb{R}^{\mathsf{d}+\mathsf{k}}} v(x) \cdot \nabla \mathsf{g}(x-y) d\mu(x) d\Big(\sum_{i=1}^N \delta_{x_i} - N\mu\delta_{\mathbb{R}^{\mathsf{d}}}\Big)(y) \end{aligned}$$

$$= 2\sum_{i=1}^N v(x_i) \cdot \nabla h_N^i(x_i) - 2N \int_{\mathbb{R}^{\mathsf{d}}} v \cdot \nabla h_N d\mu. \tag{6.1.14}$$

Using (6.1.12) and $h_N = h_{N,\vec{\eta}} + \sum_{i=1}^N \mathsf{f}_{\eta_i}(\cdot - x_i)$, we next decompose I as $\mathrm{Term}_1 + \mathrm{Term}_2 + \mathrm{Term}_3$, where

(6.1.15) $$\mathrm{Term}_1 := 2\int_{\mathbb{R}^{\mathsf{d}+\mathsf{k}}} v \cdot \nabla h_{N,\vec{\eta}}\, d\Big(\sum_{i=1}^N \delta_{x_i}^{(\eta_i)} - N\mu\delta_{\mathbb{R}^{\mathsf{d}}}\Big),$$

(6.1.16) $$\mathrm{Term}_2 := 2\sum_{i=1}^N \int_{\mathbb{R}^{\mathsf{d}+\mathsf{k}}} (v(x_i) - v)\cdot \nabla h_N^i\, d\delta_{x_i}^{(\eta_i)}$$
$$+2\sum_{i=1}^N \int_{\mathbb{R}^{\mathsf{d}+\mathsf{k}}} (v(x_i)-v)\cdot \nabla \mathsf{g}_{\eta_i}(\cdot - x_i)\, d\delta_{x_i}^{(\eta_i)} + 2N\sum_{i=1}^N \int_{\mathbb{R}^{\mathsf{d}}} (v(x_i)-v)\cdot\nabla \mathsf{f}_{\eta_i}(\cdot - x_i)\, d\mu,$$

and

(6.1.17)
$$\mathrm{Term}_3 := 2\sum_{i=1}^N \int_{\mathbb{R}^{\mathsf{d}+\mathsf{k}}} v(x_i)\cdot\nabla h_N^i\, d\Big(\delta_{x_i} - \delta_{x_i}^{(\eta_i)}\Big) - 2\sum_{i=1}^N \int_{\mathbb{R}^{\mathsf{d}+\mathsf{k}}} v(x_i)\cdot \nabla \mathsf{g}_{\eta_i}(\cdot - x_i)\, d\delta_{x_i}^{(\eta_i)}$$
$$- 2N\sum_{i=1}^N \int_{\mathbb{R}^{\mathsf{d}}} v(x_i)\cdot \nabla \mathsf{f}_{\eta_i}(\cdot - x_i)\, d\mu.$$

Step 2. Showing that $\mathrm{Term}_3 = 0$. By definition (6.1.11) of h_N^i, we may write

(6.1.18) $$\mathrm{Term}_3 = 2\sum_{i=1}^N \sum_{j\neq i} \int_{\mathbb{R}^{\mathsf{d}+\mathsf{k}}} v(x_i)\cdot \nabla \mathsf{g}(\cdot - x_j)\, d\Big(\delta_{x_i} - \delta_{x_i}^{(\eta_i)}\Big)$$
$$- 2N\sum_{i=1}^N \int_{\mathbb{R}^{\mathsf{d}+\mathsf{k}}} \int_{\mathbb{R}^{\mathsf{d}}} v(x_i)\cdot\nabla \mathsf{g}(\cdot - y)\, d\mu(y)\, d\Big(\delta_{x_i} - \delta_{x_i}^{(\eta_i)}\Big)$$
$$- 2\sum_{i=1}^N \int_{\mathbb{R}^{\mathsf{d}+\mathsf{k}}} v(x_i)\cdot\nabla \mathsf{g}_{\eta_i}(\cdot - x_i)\, d\delta_{x_i}^{(\eta_i)} - 2N\sum_{i=1}^N \int_{\mathbb{R}^{\mathsf{d}}} v(x_i)\cdot\nabla \mathsf{f}_{\eta_i}(\cdot - x_i)\, d\mu.$$

Thanks to (4.1.23) and (4.1.18), we have for $i \neq j$,

(6.1.19)
$$\int_{\mathbb{R}^{\mathsf{d}+\mathsf{k}}} \nabla \mathsf{g}(\cdot - x_j)\, d\Big(\delta_{x_i} - \delta_{x_i}^{(\eta_i)}\Big) = \nabla \mathsf{g}(x_i - x_j) - \nabla \mathsf{g}_{\eta_i}(x_i - x_j) = \nabla \mathsf{f}_{\eta_i}(x_i - x_j).$$

The right-hand side vanishes because $|x_i - x_j| > \eta_i$ for $j \neq i$ (by assumption that $\eta_i \leq \mathsf{r}_i$) and f_{η_i} vanishes outside of $B(0, \eta_i)$. Thus, the first line of (6.1.18) vanishes. By the same reasoning, the second line of (6.1.18) equals

(6.1.20) $$-2N\sum_{i=1}^N \int_{\mathbb{R}^{\mathsf{d}}} v(x_i)\cdot\nabla \mathsf{f}_{\eta_i}(x_i - y)\, d\mu(y),$$

thus it cancels with the last term on the third line of (6.1.18). It remains to show that for every i,

(6.1.21) $$\int_{\mathbb{R}^{\mathsf{d}+\mathsf{k}}} v(x_i)\cdot \nabla \mathsf{g}_{\eta_i}(\cdot - x_i)\, d\delta_{x_i}^{(\eta_i)} = 0.$$

This is true because by definition (4.1.21) and (6.1.6),

$$\begin{aligned}\nabla \mathsf{g}_{\eta_i}(\cdot - x_i)\delta_{x_i}^{(\eta_i)} &= -\frac{1}{\mathsf{c}_{\mathsf{d},\mathsf{s}}}\nabla \mathsf{g}_{\eta_i}(\cdot - x_i)\mathrm{div}\,(|y|^\gamma \nabla \mathsf{g}_{\eta_i}(\cdot - x_i)) \\ &= -\frac{1}{2\mathsf{c}_{\mathsf{d},\mathsf{s}}}\mathrm{div}\,[\nabla \mathsf{g}_{\eta_i}(\cdot - x_i), \nabla \mathsf{g}_{\eta_i}(\cdot - x_i)].\end{aligned} \tag{6.1.22}$$

Step 3. Estimating Term_1**.** Using Lemma 6.3, we may rewrite Term_1 as

$$\mathrm{Term}_1 = \int_{\hat{\Omega}\times\mathbb{R}^{\mathsf{k}}} \nabla v : [\nabla h_{N,\vec{\eta}}, \nabla h_{N,\vec{\eta}}], \tag{6.1.23}$$

where we let $\hat{\Omega} := \{x, \mathrm{dist}(x, \mathrm{supp}\nabla v) \le \lambda\}$, and we use again the definition (6.1.5). It now follows from the pointwise bound (6.1.7) that

$$|\mathrm{Term}_1| \le C\|\nabla v\|_{L^\infty} \int_{\hat{\Omega}\times\mathbb{R}^{\mathsf{k}}} |y|^\gamma |\nabla h_{N,\vec{\eta}}|^2. \tag{6.1.24}$$

Step 4. Estimating Term_2**.** Note that if i is such that $\mathrm{dist}(x_i, \mathrm{supp}\,\nabla v) > \frac{1}{4}\lambda$, since $\eta_i \le \frac14\lambda$, the corresponding terms in Term_2 vanish. For the remaining i, using the mean-value theorem on $v - v(x_i)$ and the explicit form of the probability measure $\delta_{x_j}^{(\eta_j)}$ given after Definition 4.6, we see that

$$\begin{aligned}2\sum_{i=1}^N &\left| \int_{\mathbb{R}^{\mathsf{d}+\mathsf{k}}} (v(x_i) - v)\cdot \nabla h_N^i \, d\delta_{x_i}^{(\eta_i)}\right| \\ &\le C\|\nabla v\|_{L^\infty} \sum_{i, \mathrm{dist}(x_i, \mathrm{supp}\,\nabla v)\le \frac14\lambda} \eta_i^{-\mathsf{s}} \int_{\partial B(x_i,\eta_i)} |y|^\gamma |\nabla h_N^i| d\mathcal{H}^{\mathsf{d}+\mathsf{k}-1},\end{aligned} \tag{6.1.25}$$

where $\partial B(x_i, \eta_i)$ is the sphere in $\mathbb{R}^{\mathsf{d}+\mathsf{k}}$ and $\mathcal{H}^{\mathsf{d}+\mathsf{k}-1}$ is the Hausdorff measure. Similarly,

$$2\sum_{i=1}^N \left|\int_{\mathbb{R}^{\mathsf{d}+\mathsf{k}}} (v - v(x_i))\cdot \nabla \mathsf{g}_{\eta_i}(\cdot - x_i) d\delta_{x_i}^{(\eta_i)}\right| \le C\|\nabla v\|_{L^\infty} \sum_{i, \mathrm{dist}(x_i, \mathrm{supp}\,\nabla v)\le\frac14\lambda} \eta_i^{-\mathsf{s}}. \tag{6.1.26}$$

Finally, recalling that f_η is supported in $B(0, \eta)$, we have

$$2\sum_{i=1}^N \left|\int_{\mathbb{R}^{\mathsf{d}+\mathsf{k}}} (v - v(x_i))\cdot \nabla \mathsf{f}_{\eta_i}(\cdot - x_i) d\mu\right| \le C\|\nabla v\|_{L^\infty}\|\mu\|_{L^\infty} \sum_{i, \mathrm{dist}(x_i, \mathrm{supp}\nabla v)\le\frac14\lambda} \eta_i^{\mathsf{d}-\mathsf{s}}. \tag{6.1.27}$$

Combining (6.1.25), (6.1.26), (6.1.27) yields

$$\begin{aligned}|\mathrm{Term}_2| &\le C\|\nabla v\|_{L^\infty} \sum_{i, \mathrm{dist}(x_i, \mathrm{supp}\,\nabla v)\le\frac14\lambda} \\ &\times \left(\eta_i^{-\mathsf{s}} + N\|\mu\|_{L^\infty}\eta_i^{\mathsf{d}-\mathsf{s}} + \eta_i^{-\mathsf{s}} \int_{\partial B(x_i,\eta_i)} |y|^\gamma |\nabla h_N^i| d\mathcal{H}^{\mathsf{d}+\mathsf{k}-1}\right).\end{aligned} \tag{6.1.28}$$

Step 5. Conclusion. Combining the estimates (6.1.24) and (6.1.28), we have found that there exists a constant $C > 0$ depending only on d, s, such that for every choice of $\vec{\eta}$ satisfying $\eta_i \le \mathsf{r}_i$, we have

(6.1.29)
$$|I| \le C\|\nabla v\|_{L^\infty}\Bigg[\int_{\hat{\Omega}\times\mathbb{R}^{\mathsf{k}}}|y|^\gamma|\nabla h_{N,\vec{\eta}}|^2 + \sum_{i,\mathrm{dist}(x_i,\mathrm{supp}\,\nabla v)\le\frac14\lambda}\Bigg(\eta_i^{-\mathsf{s}} + N\|\mu\|_{L^\infty}\eta_i^{\mathsf{d}-\mathsf{s}} + \eta_i^{-\mathsf{s}}\int_{\partial B(x_i,\eta_i)}|y|^\gamma|\nabla h_N^i|d\mathcal{H}^{\mathsf{d}+\mathsf{k}-1}\Bigg)\Bigg].$$

For each $t \in [\frac12, 1]$, we apply this relation with $\eta_i = t\mathsf{r}_i$ and then average both sides of the resulting inequality over $t \in [\frac12, 1]$. The average of the last term on the right-hand side becomes

$$\le C \sum_{i,\mathrm{dist}(x_i,\mathrm{supp}\,\nabla v)\le\frac14\lambda} \mathsf{r}_i^{-\mathsf{s}-1}\int_{B(x_i,\mathsf{r}_i)}|y|^\gamma|\nabla h_N^i|. \tag{6.1.30}$$

Applying the Cauchy-Schwarz inequality, we can also estimate

$$\int_{B(x_i,\mathsf{r}_i)}|y|^\gamma|\nabla h_N^i| \le \left(\int_{B(x_i,\mathsf{r}_i)}|y|^\gamma|\nabla h_N^i|^2\right)^{\frac12}\left(\int_{B(0,\mathsf{r}_i)}|y|^\gamma\right)^{\frac12}. \tag{6.1.31}$$

Since $\gamma > -1$ the last integral is convergent and bounded by $C\mathsf{r}_i^{\mathsf{d}+1+\gamma}$ if $\mathsf{k} = 1$ or $C\mathsf{r}_i^{\mathsf{d}}$ otherwise, which is always $C\mathsf{r}_i^{\mathsf{s}+2}$ by (2.2.5). Using (6.1.13) and Cauchy-Schwarz inequality, the average of the last term on the right-hand side of (6.1.29) is thus bounded by

$$C\sum_{i,\mathrm{dist}(x_i,\mathrm{supp}\,\nabla v)\le\frac14\lambda}\mathsf{r}_i^{-\frac{\mathsf{s}}{2}}\left(\int_{B(x_i,\mathsf{r}_i)}|y|^\gamma|\nabla h_N^i|^2\right)^{\frac12}$$
$$\le C\Bigg(\sum_{i,\mathrm{dist}(x_i,\mathrm{supp}\,\nabla v)\le\frac14\lambda}\mathsf{r}_i^{-\mathsf{s}}\Bigg)^{\frac12}\left(\int_{\{\mathrm{dist}(\cdot,\mathrm{supp}\,\nabla v)\le\lambda\}\times\mathbb{R}^{\mathsf{k}}}|y|^\gamma|\nabla h_{N,\mathsf{r}}|^2\right)^{\frac12}$$
$$\le C\Bigg(\sum_{i,\mathrm{dist}(x_i,\mathrm{supp}\,\nabla v)\le\frac14\lambda}\mathsf{r}_i^{-\mathsf{s}} + \int_{\{\mathrm{dist}(\cdot,\mathrm{supp}\,\nabla v)\le\lambda\}\times\mathbb{R}^{\mathsf{k}}}|y|^\gamma|\nabla h_{N,\mathsf{r}}|^2\Bigg). \tag{6.1.32}$$

Inserting this estimate into (6.1.29), we obtain, after averaging over $t \in [\frac12, 1]$,

$$|I| \le C\|\nabla v\|_{L^\infty}\Bigg(\fint_{t\in[\frac12,1]}\int_{\hat{\Omega}\times\mathbb{R}^{\mathsf{k}}}|y|^\gamma|\nabla h_{N,t\mathsf{r}}|^2 + \int_{\hat{\Omega}\times\mathbb{R}^{\mathsf{k}}}|y|^\gamma|\nabla h_{N,\mathsf{r}}|^2 + \sum_{i,x_i\in\hat{\Omega}}\mathsf{r}_i^{-\mathsf{s}} + N\|\mu\|_{L^\infty}\sum_{i,x_i\in\hat{\Omega}}\mathsf{r}_i^{\mathsf{d}-\mathsf{s}}\Bigg). \tag{6.1.33}$$

Recall the assumption that Ω contains a closed 3λ-neighborhood of $\mathrm{supp}\,\nabla v$, so that the condition $\mathrm{dist}(x, \mathrm{supp}\,\nabla v) \le \lambda$ implies $x \in \Omega$ and $\mathrm{dist}(x, \partial\Omega) \ge 2\lambda$. Thus, for points in $\hat{\Omega}$, we have $\mathsf{r}_i = \tilde{\mathsf{r}}_i$ in the notation (4.5.1). Using the estimates (4.5.4), (4.5.5) on the right-hand side of (6.1.33) and recalling $\mathsf{r}_i \le \lambda = (\|\mu\|_{L^\infty}N)^{-1/\mathsf{d}}$, we conclude that

$$(6.1.34)\quad |I| \le C\|\nabla v\|_{L^\infty}\left(\mathsf{F}_N^\Omega(X_N,\mu) - \Big(\frac{\# I_\Omega \log\lambda}{2}\Big)\mathbf{1}_{\mathsf{s}=0} + C\|\mu\|_{L^\infty}^{\frac{\mathsf{s}}{\mathsf{d}}}\# I_\Omega N^{\frac{\mathsf{s}}{\mathsf{d}}}\right),$$

which completes the proof.

Remark 6.4. In the local Riesz case (where dimension extension is needed) we may assume that the vector field v is supported in some ball of radius ℓ, and we may extend the vector field $v:\mathbb{R}^\mathsf{d}\to\mathbb{R}^\mathsf{d}$ to $\mathbb{R}^{\mathsf{d}+\mathsf{k}}$ by

$$(6.1.35)\qquad \tilde v(x,z) := \chi\Big(\frac{z}{\ell}\Big)(v(x),0), \qquad \forall (x,z)\in\mathbb{R}^{\mathsf{d}+\mathsf{k}},$$

where χ is a smooth bump function satisfying

$$(6.1.36)\qquad \chi(z)=1 \quad \text{if } |z|\le \frac12, \qquad \chi(z)=0 \quad \text{if } |z|\ge 1.$$

In contrast to the trivial extension used above, this choice of $\tilde v$ is also localized in the z coordinate. Moreover, if $\operatorname{supp} v$ is contained in an open ball B of radius ℓ in $\mathbb{R}^\mathsf{d}$, then $\operatorname{supp}\tilde v$ is contained in an open ball $\tilde B$ of radius $C\ell$ in $\mathbb{R}^{\mathsf{d}+\mathsf{k}}$.

Since v vanishes at some point x_0, if $\ell<\infty$, the mean-value theorem implies that

$$(6.1.37)\qquad \|v\|_{L^\infty} \le C\ell\|\nabla v\|_{L^\infty}.$$

We thus have $\|\nabla\tilde v\|_{L^\infty}\le C\|\nabla v\|_{L^\infty}$ and in the same way

$$\|\nabla^{\otimes m}\tilde v\|_{L^\infty}\le C\|\nabla^{\otimes m} v\|_{L^\infty}.$$

Running through the proof with v replaced by $\tilde v$, all integrals become reduced to integrals over $\operatorname{supp}\nabla v\times[-\ell,\ell]^\mathsf{k}$. It will be useful to record another way of writing the result, which is a consequence of (6.1.33)

$$(6.1.38)\quad \left|\int_{(\mathbb{R}^\mathsf{d})^2\setminus\triangle}(v(x)-v(y))\cdot\nabla\mathsf{g}(x-y)d\Big(\sum_{i=1}^N\delta_{x_i}-N\mu\Big)^{\otimes 2}(x,y)\right|$$
$$\le C\|\nabla v\|_{L^\infty}\left(\int_{\tilde\Omega\times[-\ell,\ell]^\mathsf{k}}|y|^\gamma|\nabla h_{N,\mathsf{r}}|^2 + \|\mu\|_{L^\infty}^{\frac{\mathsf{s}}{\mathsf{d}}}\# I_\Omega N^{\frac{\mathsf{s}}{\mathsf{d}}}\right),$$

where $\tilde\Omega := \{x\in\Omega, \operatorname{dist}(x,\partial\Omega)\ge\lambda\}$.

Proof. First, we directly deduce from (6.1.33), with the localization mentioned above, that

$$\left|\int_{(\mathbb{R}^\mathsf{d})^2\setminus\triangle}(v(x)-v(y))\cdot\nabla\mathsf{g}(x-y)d\Big(\sum_{i=1}^N\delta_{x_i}-N\mu\Big)^{\otimes 2}(x,y)\right|$$
$$\le C\|\nabla v\|_{L^\infty}\left(\fint_{t\in[\frac12,1]}\int_{\hat\Omega\times[-\ell,\ell]^\mathsf{k}}|y|^\gamma|\nabla h_{N,t\mathsf{r}}|^2\right.$$
$$+\int_{\hat\Omega\times[-\ell,\ell]^\mathsf{k}}|y|^\gamma|\nabla h_{N,\mathsf{r}}|^2$$
$$\left.+\sum_{i,x_i\in\hat\Omega}\mathsf{r}_i^{-\mathsf{s}} + \|\mu\|_{L^\infty}^{\frac{\mathsf{s}}{\mathsf{d}}}\# I_\Omega N^{\frac{\mathsf{s}}{\mathsf{d}}}\right),$$

where $\hat{\Omega} = \{x \in \Omega, \mathrm{dist}(x, \partial\Omega) \geq 2\lambda\}$. By definition (4.1.24) we have

$$h_{N,t\mathsf{r}} = h_{N,\mathsf{r}} + \sum_{i=1}^{N} \mathsf{f}_{\mathsf{r}_i, t\mathsf{r}_i}(x - x_i)$$

using again the notation $\mathsf{f}_{\alpha,\eta} = \mathsf{f}_\alpha - \mathsf{f}_\eta$. Thus, using the triangle inequality and the fact that the supports of the $\mathsf{f}_{\mathsf{r}_i,t\mathsf{r}_i}(x - x_i)$ are disjoint, and $t \in [\frac{1}{2}, 1]$, we have

(6.1.39)
$$\begin{aligned}
&\int_{\hat{\Omega}\times[-\ell,\ell]^{\mathsf{k}}} |y|^\gamma |\nabla h_{N,t\mathsf{r}}|^2 \\
&\leq 2\int_{\hat{\Omega}\times[-\ell,\ell]^{\mathsf{k}}} |y|^\gamma |\nabla h_{N,\mathsf{r}}|^2 + 2 \sum_{i, \mathrm{dist}(x_i, \hat{\Omega}) \leq \frac{1}{4}\lambda} \int_{\mathbb{R}^{\mathsf{d}}} |y|^\gamma |\nabla \mathsf{f}_{\mathsf{r}_i, t\mathsf{r}_i}(x - x_i)|^2 \\
&\leq 2\int_{\hat{\Omega}\times[-\ell,\ell]^{\mathsf{k}}} |y|^\gamma |\nabla h_{N,\mathsf{r}}|^2 + C \sum_{i, \mathrm{dist}(x_i, \hat{\Omega}) \leq \frac{1}{4}\lambda} \mathsf{r}_i^{-\mathsf{s}}.
\end{aligned}$$

In view of (4.5.6), we also have

$$\sum_{i, \mathrm{dist}(x_i, \partial\hat{\Omega}) \leq \frac{1}{4}\lambda} \mathsf{r}_i^{-\mathsf{s}} \leq C\left(\int_{\tilde{\Omega}\times[-\ell,\ell]^{\mathsf{k}}} |y|^\gamma |\nabla h_{N,\mathsf{r}}|^2 + \#I_\Omega \|\mu\|_{L^\infty}^{\frac{\mathsf{s}}{\mathsf{d}}} N^{\frac{\mathsf{s}}{\mathsf{d}}}\right)$$

The result follows. □

6.1.3. Next-order commutator estimates. In the study of fluctuations in Chapter 10, we will need a second order version of the commutator estimate, controlling (6.1.2) with $n = 2$ instead of (6.1.1), in a localizable way. Such a version was first obtained in [**LS18**] in the two-dimensional Coulomb case, also subsequently in a nonlocalized way in [**Ros20**], and then revisited in [**Ser20**] in the Coulomb case, but the estimate is only sharp in the case $\mathsf{d} = 2, \mathsf{s} = 0$. In [**NRS22**], second order commutator estimates were also proven for all the Riesz cases, with suboptimal error terms. The proof relied on commutator estimates. Finally, estimates for all order n, with sharp error terms and which are localizable have been obtained in [**RS24b**].

We now state the result from [**RS24b**] valid for all (2.0.1), and refer the reader to that paper for the proof, which is considerably more involved than that of Theorem 6.1.

Theorem 6.5 (Sharp next-order commutator estimates [**RS24b**]). *Let μ be a probability measure with a bounded density, satisfying* (4.1.1). *Assume that $v : \mathbb{R}^{\mathsf{d}} \to \mathbb{R}^{\mathsf{d}}$ is smooth, that Ω' is a ball of radius ℓ containing a 2λ-neighborhood of the support of ∇v and Ω contains a 5ℓ-neighborhood of Ω', where $\ell > 2\lambda$. For any*

$n \ge 2$, we have, for N large enough,
(6.1.40)

$$|\mathsf{A}_n(X_N,\mu,v)| \le C\Bigg(\sum_{p=0}^{n}(\ell\|\nabla^2 v\|_{L^\infty})^p \times \sum_{\substack{1\le c_1,\dots,c_{n-p}\\ c_i \text{ integer}\\ n-p\le c_1+\dots+c_{n-p}\le 2n}} \lambda^{-(n-p)+\sum_{k=1}^{p} c_{n-k}}\|\nabla^{\otimes c_1} v\|_{L^\infty}\cdots\|\nabla^{\otimes c_{n-p}} v\|_{L^\infty}\Bigg) \times \left(\int_{\Omega\times[-\ell,\ell]^{\mathsf{k}}}|y|^\gamma|\nabla h_{N,\mathsf{r}}|^2 + C\#I_\Omega\|\mu\|_{L^\infty}^{\frac{\mathsf{s}}{\mathsf{d}}}N^{\frac{\mathsf{s}}{\mathsf{d}}}\right).$$

where $C>0$ depends only on d, s, and n.

The factor in front of the energy has the natural scaling in ℓ. For instance, in a typical situation of application, $\|\nabla^m v\|_{L^\infty} \le M\ell^{-m}$, where ℓ is the lengthscale of variation of v. In that situation, using that $\ell \ge \lambda$, the theorem yields

$$|\mathsf{A}_n(X_N,\mu,v)| \le C\ell^{-n}\Big(\int_{\Omega\times[-\ell,\ell]^{\mathsf{k}}}|y|^\gamma|\nabla h_{N,\mathsf{r}}|^2 + C\#I_\Omega\|\mu\|_{L^\infty}^{\frac{\mathsf{s}}{\mathsf{d}}}N^{\frac{\mathsf{s}}{\mathsf{d}}}\Big),$$

where C depends on M.

We may note that for the one-dimensional Coulomb case, it is immediate from the definition (6.1.3) that $\mathsf{A}_2 \equiv 0$, since $\mathsf{g}''(x-y) = \delta_0(x-y)$.

6.2. Application to dynamics

6.2.1. Mean-field limits. As mentioned above, the commutator estimate has found two important applications, one is to central limit theorems for fluctuations in Coulomb gases, which we will discuss in Chapters 9 and 10, one is to dynamics, via the modulated energy method.

The question of proving mean-field limits for interacting systems is a classical one in statistical physics, and has attracted much attention in mathematics. We refer the reader to the surveys [**CD22, Jab14, JW17, Gol22**] and references therein.

The question is to understand the limit as $N \to \infty$ of the empirical measure

$$\mu_N^t := \frac{1}{N}\sum_{i=1}^{N}\delta_{x_i^t} \tag{6.2.1}$$

associated to a solution $X_N^t := (x_1^t,\dots,x_N^t)$ of a system of ordinary differential equations (ODEs) of the form

$$\begin{cases} \dot{x}_i = -\dfrac{1}{N}\mathbb{M}\nabla_{x_i}\mathcal{H}_N(x_1,\dots,x_N), & i=1,\dots,N,\\ x_i(0) = x_i^0. \end{cases} \tag{6.2.2}$$

Here $\mathbb{M}$ is a fixed matrix that satisfies $\langle \mathbb{M}\xi,\xi\rangle \ge 0$. The case $\mathbb{M} = I$ corresponds to the gradient-flow dynamics. The case where $\mathbb{M}$ is antisymmetric corresponds to conservative dynamics, including for instance the important "point-vortex system" in fluids. The energy $\mathcal{H}_N$ is of the form (1.1.1) but for possibly general interaction potential g (more general evolutions are also considered). Studying the same evolutions with added noise or diffusion

$$dx_i = -\frac{1}{N}\nabla_{x_i}\mathcal{H}_N(x_1,\dots,x_N)dt + \sqrt{\frac{2}{\theta}}dW_i, \tag{6.2.3}$$

or

$$dx_i = -\frac{1}{N}\mathbb{M}\nabla_{x_i}\mathcal{H}_N(x_1,\dots,x_N)dt + \sqrt{\frac{2}{\theta}}dW_i, \tag{6.2.4}$$

with W_i being N independent Brownian motions and $\theta > 0$ an inverse temperature, is also very interesting, as discussed in the introduction.

If the points x_i^0, which themselves depend on N, are such that μ_N^0 converges to some regular measure μ^0, then a formal derivation leads to expecting that for $t > 0$, μ_N^t converges to the solution of the Cauchy problem with initial data μ^0 for the limiting continuity equation

$$\partial_t\mu = \operatorname{div}(((\mathbb{M}\nabla \mathsf{g}) * \mu)\mu), \tag{6.2.5}$$

or, in the case with noise,

$$\partial_t\mu = \operatorname{div}(((\mathbb{M}\nabla \mathsf{g}) * \mu)\mu) + \frac{1}{\theta}\Delta\mu. \tag{6.2.6}$$

Proving the convergence in law of the empirical measure is more or less equivalent (see [**CD22**]) to proving *propagation of chaos* (a notion introduced by Kac in kinetic theory). One says that there is propagation of chaos if, when particles are initially distributed according to a probability density $f_N^0(x_1,\dots,x_N)$ in a tensorized form

$$f_N^0(x_1,\dots,x_N) = \mu^0(x_1)\dots\mu^0(x_N),$$

(in other words, initial particles positions are iid with law μ^0) then at later times, f_N^t is *approximately* in tensorized form $\mu^t(x_1)\dots\mu^t(x_N)$. The same questions also apply to second-order evolutions of the form (1.1.10), which lead to Vlasov-type kinetic equations.

To rigorously establish convergence, various methods have been put forward: classical trajectorial methods [**McK67, Szn91**] for the case with noise, relative entropy method which works well in the presence of noise (see [**JW18, BJS23, BDJ24**] and [**Lac23**], and the related method in [**HCR25**], for the best results to date), and methods that rely on finding a good metric, such as a Wasserstein distance, to measure the distance from μ_N^t to μ^t and its time evolution. Such methods have allowed to treat some singular interactions, but not as singular as the Coulomb, let alone super-Coulomb Riesz, interaction, which remained open until recently.

The crucial point is that the modulated energy $\mathsf{F}_N(X_N,\mu)$ acts as a good distance to measure the distance of the empirical measure to the expected limit, or to quantify a *weak-strong uniqueness* result at the level of the mean-field equation (6.2.5). This modulated energy method was introduced in [**Ser17**] in the closely related context of dynamics of Ginzburg-Landau vortices in (1.2.4)–(1.2.5). It was then first adapted to the context of Coulomb and Riesz gases in dimension $\mathsf{d} \le 2$ in [**Due16**]. The full Coulomb and Riesz cases with (1.1.2) without noise was finally treated in [**Ser20**] thanks to the commutator estimate of Theorem 6.1. It gives quantitative convergence of the empirical measure to the solution of (6.2.5) in modulated energy. Let us state a main result.

Theorem 6.6 (Mean-field convergence for Coulomb and Riesz first-order dynamics). *Assume* $\mathsf{d}-2 \le \mathsf{s} < \mathsf{d}$. *Assume* μ^0 *is a probability density satisfying*

$$\int_{\mathbb{R}^{\mathsf{d}}} |\mathsf{g}_-| d\mu^0 < \infty, \tag{6.2.7}$$

and is such that μ^t*, solution of* (6.2.5) *with initial data* μ^0*, satisfies on some interval* $[0,T]$*,*

$$\sup_{t\in[0,T]} \|\mu^t\|_{L^\infty} < \infty, \qquad \sup_{t\in[0,T]} \|\nabla^{\otimes 2}\mathsf{g} * \mu^t\|_{L^\infty} < \infty, \tag{6.2.8}$$

then, letting X_N^t *be the solution to* (6.2.2)*, if* $\mathsf{F}_N(X_N^0, \mu^0) + N\left(\frac{\log N}{2\mathsf{d}}\right)\mathbf{1}_{\mathsf{s}=0} = o(N^2)$*, we have*

$$\frac{1}{N}\sum_{i=1}^N \delta_{x_i^t} \rightharpoonup \mu^t \quad \text{for every } t \in [0,T],$$

with convergence in modulated energy or by Corollary 4.22 *in* $H^{-\sigma}(\mathbb{R}^{\mathsf{d}})$. *More precisely we have that* (6.2.14) *holds in* $[0,T]$.

The condition (6.2.8) boils down to a question of regularity of the solution to (6.2.5) and, if such a regularity is known, can then be reduced to a condition on the initial distribution μ^0. We refer to [**Ser20**] for a discussion of the regularity results proven in the literature.

Examining the proof, particularly the last application of Gronwall's lemma that yields (6.2.14), thanks to the fact that the commutator estimate (6.1.4) is made explicit in its μ-dependence, we obtain a uniform in time rate of convergence provided $\int_0^\infty \|\nabla^{\otimes 2}\mathsf{g} * \mu^s\|_{L^\infty} ds < \infty$, i.e., provided the solution μ^t has sufficient decay as $t \to \infty$, which is a purely PDE question. We know how to prove this in the setting of the torus [**CdCRS25**], and also in the case of added diffusion in the sub-Coulomb Riesz case in the whole space [**RS23**]. In both instances this strong decay of the solution allows to obtain uniform-in-time convergence.

The modulated energy method for dynamics was later improved, allowing to treat more general interactions [**NRS22, HCRS25**], or relaxing assumptions on the regularity of the solution [**Ros22, Ros20**]. Other applications to dynamics, including quantum dynamics, all relying on the commutator estimate Theorem 6.1, have been given in [**HKI21, Ros23, BP23, GP21, M2́4, RS25a**].

6.2.2. Weak-strong uniqueness proof. Let us now present the short proof of the weak-strong uniqueness principle for (6.2.5) as it will be a model for the main proof. We focus on the dissipative case $\mathbb{M} = I$ (the conservative one is an easy adaptation) and on the Coulomb case for simplicity, the Riesz case is the same, using the extension procedure and adding the $|y|^\gamma$ weight.

Let μ_1 and μ_2 be two solutions to (6.2.5) and $h_i = \mathsf{g} * \mu_i$ the associated potentials, which solve (2.1.16). Let us compute

$$\begin{aligned}
\partial_t \int_{\mathbb{R}^d} |\nabla(h_1 - h_2)|^2 &= 2\mathsf{c_d} \int_{\mathbb{R}^d} (h_1 - h_2)\partial_t(\mu_1 - \mu_2) \\
&= 2\mathsf{c_d} \int_{\mathbb{R}^d} (h_1 - h_2)\mathrm{div}\,(\mu_1 \nabla h_1 - \mu_2 \nabla h_2) \\
&= -2\mathsf{c_d} \int_{\mathbb{R}^d} (\nabla h_1 - \nabla h_2)\cdot(\mu_1 \nabla h_1 - \mu_2 \nabla h_2) \\
&= -2\mathsf{c_d} \int_{\mathbb{R}^d} |\nabla(h_1 - h_2)|^2 \mu_1 \\
&\quad - 2\mathsf{c_d} \int_{\mathbb{R}^d} \nabla h_2 \cdot \nabla(h_1 - h_2)(\mu_1 - \mu_2).
\end{aligned} \tag{6.2.9}$$

In the right-hand side, we recognize from (6.1.6) the divergence of the stress-energy tensor $[\nabla(h_1 - h_2), \nabla(h_1 - h_2)]$, hence

$$\partial_t \int_{\mathbb{R}^d} |\nabla(h_1 - h_2)|^2 \le 2 \int_{\mathbb{R}^d} \nabla h_2 \cdot \mathrm{div}\,[\nabla(h_1 - h_2), \nabla(h_1 - h_2)]$$

so, if $\nabla^2 h_2$ is bounded, we may integrate by parts the right-hand side and bound it pointwise, thanks to (6.1.7), by

$$2\|\nabla^2 h_2\|_{L^\infty} \int_{\mathbb{R}^d} |[\nabla(h_1 - h_2), \nabla(h_1 - h_2)]| \le 2\|\nabla^2 h_2\|_{L^\infty} \int_{\mathbb{R}^d} |\nabla(h_1 - h_2)|^2.$$

By Gronwall's lemma, we conclude that

$$\int_{\mathbb{R}^d} |\nabla(h^{\mu_1^t} - h^{\mu_2^t})|^2 \le \exp\left(C\int_0^t \|\nabla^2(\mathsf{g} * \mu_2^s)\|_{L^\infty} ds\right) \int_{\mathbb{R}^d} |\nabla(h^{\mu_1^0} - h^{\mu_2^0})|^2. \tag{6.2.10}$$

Let us recall that $\frac{1}{2\mathsf{c_d}} \int_{\mathbb{R}^d} |\nabla(h^{\mu_1^t} - h^{\mu_2^t})|^2$ can also be rewritten as

$$\iint_{\mathbb{R}^d \times \mathbb{R}^d} \mathsf{g}(x - y) d\left(\mu_1^t - \mu_2^t\right)(x) d\left(\mu_1^t - \mu_2^t\right)(y)$$

and is the (square of the) $\dot{H}^{-1}(\mathbb{R}^d)$ seminorm of $\mu_1^t - \mu_2^t$. (6.2.10) constitutes a weak-strong uniqueness principle since it states that any solution (μ_1 here) that starts close to a strong solution (μ_2 here, for which $\|\nabla^2(\mathsf{g} * \mu_2^s)\|_{L^\infty}$ is to be controlled), remains close to it on fixed time intervals.

6.2.3. Time-derivative of the modulated energy. In the case of the empirical measure, the computation above is to be performed on the modulated energy, which takes care of the Dirac singularities.

To be able to use the electric formulation for $\mathsf{F}_N(X_N^t, \mu^t)$ and use Theorem 6.1, we need (4.1.1) to be satisfied by μ^t. This is the reason for the assumption (6.2.7). Indeed, if $\mathsf{s} > 0$, the assumption of boundedness of μ^t in L^∞ in (6.2.8) and the decay of g suffice to ensure that (4.1.1) holds. In the case $\mathsf{s} \le 0$, we claim that the assumption (6.2.7) is propagated in time by (6.2.5), i.e.,

$$\int_{\mathbb{R}^d} (-\mathsf{g}_-) d\mu^t < \infty, \tag{6.2.11}$$

where we recall that $\mathsf{g}_- = \min(\mathsf{g}, 0)$. Then we conclude the desired condition as in Step 2 of the proof of Lemma 2.20. To justify (6.2.11), it suffices to use (6.2.5), integration by parts and the bounds (6.2.8) to find

$$\partial_t \int_{\mathbb{R}^\mathsf{d}} (-\mathsf{g}_-)\mu^t = \int_{\mathbb{R}^\mathsf{d}} \nabla \mathsf{g}_- \cdot \mathbb{M}\nabla h^{\mu^t}\mu^t \le \|\nabla h^{\mu^t}\|_{L^\infty} \int_{\mathbb{R}^\mathsf{d}} |\nabla \mathsf{g}_-|\mu^t \le C_t \int_{\mathbb{R}^\mathsf{d}} (1-\mathsf{g}_-)\mu^t$$

by observing that $|\nabla \mathsf{g}_-| \le C(1-\mathsf{g}_-)$. Using Gronwall's lemma then allows to deduce from (6.2.7) that (6.2.11) holds for each $t \ge 0$.

We next turn to the computation of the time derivative of the modulated energy, taking advantage of the laws of evolution.

Lemma 6.7. *If X_N^t is a solution of (6.2.2) and μ^t solves (6.2.5), then letting μ_N^t be as in (6.2.1), we have*

(6.2.12)
$$\partial_t \mathsf{F}_N(X_N^t, \mu^t) \le -\frac{N^2}{2} \iint_{\mathbb{R}^\mathsf{d}\times\mathbb{R}^\mathsf{d}\setminus\triangle} \mathbb{M}\left(\nabla h^{\mu^t}(x) - \nabla h^{\mu^t}(y)\right) \cdot \nabla \mathsf{g}(x-y) d(\mu_N^t - \mu^t)(x) d(\mu_N^t-\mu^t)(y).$$

Proof. We note that if $\mathsf{s} \ge \mathsf{d}-1$, $\nabla \mathsf{g}$ is not integrable near 0, so $\nabla\mathsf{g} * \mu$ should be understood in a distributional sense and $\mu\nabla(\mathsf{g}*\mu) = \mu\mathsf{g} * \nabla\mu$ as well, assuming that μ is regular enough. We may also check that this distributional definition is equivalent to defining ∇h^μ in principal value:

$$\nabla h^\mu(x) = P.V. \int_{\mathbb{R}^\mathsf{d}\setminus\{x\}} \nabla \mathsf{g}(x-y) d\mu(y).$$

Returning to the definition (4.1.2) and using (6.2.2) and (6.2.5) and the symmetry of the problem, we have

$$\begin{aligned}
&\partial_t \mathsf{F}_N(X_N^t, \mu^t)\\
&= N^2 \partial_t \iint \frac{1}{2}\mathsf{g}(x-y) d\mu^t(x) d\mu^t(y) + \partial_t \sum_{i\neq j} \frac{1}{2}\mathsf{g}(x_i^t - x_j^t) - N\partial_t \sum_{i=1}^N \int_{\mathbb{R}^\mathsf{d}} \mathsf{g}(x_i^t - y) d\mu^t(y)\\
&= -N^2 \int_{\mathbb{R}^\mathsf{d}} \nabla h^{\mu^t} \cdot \mathbb{M}\nabla h^{\mu^t}(x) d\mu^t(x) - \frac{1}{N}\sum_{i=1}^N \sum_{j\neq i} \nabla\mathsf{g}(x_i^t - x_j^t) \cdot \mathbb{M} \sum_{k\neq i} \nabla \mathsf{g}(x_i^t - x_k^t)\\
&\quad + \sum_{j \neq i} \nabla h^{\mu^t}(x_i^t) \cdot \mathbb{M}\nabla\mathsf{g}(x_i^t - x_j^t)\\
&\quad + N\sum_{i=1}^N P.V. \int_{\mathbb{R}^\mathsf{d}\setminus\{x_i^t\}} \mathbb{M}\nabla h^{\mu^t}(x) \cdot \nabla \mathsf{g}(x - x_i^t) d\mu^t(x).
\end{aligned}$$

We then rewrite this as

$$\begin{aligned}\partial_t \mathsf{F}_N(X_N^t,\mu^t) = & - N^2 \int_{\mathbb{R}^{\mathsf{d}}} \nabla h^{\mu^t} \cdot \mathbb{M}\nabla h^{\mu^t}(x) d\mu^t(x) \\ & - \frac{1}{N}\sum_{i=1}^N \sum_{j\neq i} \nabla \mathsf{g}(x_i^t - x_j^t)\cdot \mathbb{M} \sum_{k\neq i} \nabla \mathsf{g}(x_i^t - x_k^t) \\ & + N^2 \int_{\mathbb{R}^{\mathsf{d}}} \nabla h^{\mu^t}(x) \cdot \int_{\mathbb{R}^{\mathsf{d}}\setminus\{x\}} \mathbb{M}\nabla \mathsf{g}(x-y) d\mu_N^t(y) d\mu_N^t(x) \\ & + N^2 \int_{\mathbb{R}^{\mathsf{d}}} P.V. \int_{\mathbb{R}^{\mathsf{d}}\setminus\{y\}} \mathbb{M}\nabla h^{\mu^t}(x) \cdot \nabla \mathsf{g}(x-y) d\mu^t(x) d\mu_N^t(y).\end{aligned}$$

We recognize that the right-hand side can be recombined and symmetrized into

$$\begin{aligned}-N^2 \int_{\mathbb{R}^{\mathsf{d}}} P.V. \int_{\mathbb{R}^{\mathsf{d}}\setminus\{x\}} & \nabla \mathsf{g}(x-y) d(\mu_N^t - \mu^t)(y) \\ & \cdot \mathbb{M}\, P.V. \int_{\mathbb{R}^{\mathsf{d}}\setminus\{x\}} \nabla \mathsf{g}(x-y) d(\mu_N^t-\mu^t)(y) d\mu_N^t(x) \\ - \frac{N^2}{2} \iint_{\triangle^c} \mathbb{M} & \left(\nabla h^{\mu^t}(x) - \nabla h^{\mu^t}(y)\right) \cdot \nabla \mathsf{g}(x-y) d(\mu_N^t - \mu^t)(x) d(\mu_N^t - \mu^t)(y),\end{aligned}$$

and since the first term is nonpositive by property of $\mathbb{M}$, we obtain the result. □

6.2.4. Conclusion. Combining (6.2.12) with the commutator estimate (6.1.4) applied to $v = \mathbb{M}\nabla h^{\mu^t}$, we obtain

(6.2.13)
$$\begin{aligned}&\partial_t \mathsf{F}_N(X_N^t,\mu^t) \\ &\quad \le C\|\nabla^{\otimes 2} h^{\mu^t}\|_{L^\infty} \left(\mathsf{F}_N(X_N^t,\mu^t) + N\frac{\log(N\|\mu^t\|_{L^\infty})}{2\mathsf{d}} \mathbf{1}_{\mathsf{s}=0} + C\|\mu^t\|_{L^\infty}^{\frac{\mathsf{s}}{\mathsf{d}}} N^{1+\frac{\mathsf{s}}{\mathsf{d}}}\right)\end{aligned}$$

where C depends only on s and d.

We may now apply Gronwall's inequality to the quantity

$$\begin{aligned}\Xi(t) := \mathsf{F}_N(X_N^t,\mu^t) & + \left(\frac{N}{2\mathsf{d}}\log N + \frac{N}{2\mathsf{d}} \log \sup_{s\in[0,t]} \|\mu^s\|_{L^\infty}\right) \mathbf{1}_{\mathsf{s}=0} \\ & + C_0 \sup_{s\in[0,t]} \|\mu^s\|_{L^\infty}^{\frac{\mathsf{s}}{\mathsf{d}}} N^{1+\frac{\mathsf{s}}{\mathsf{d}}},\end{aligned}$$

where C_0 is large enough that this quantity is nonnegative in view of (4.2.16). If that quantity is initially small (here $\ll N^2$), it will remain small on any time interval on which μ^t and ∇h^{μ^t} are bounded. More precisely, we may write

$$\Xi(t) \le C \exp\left(\int_0^t \|\nabla^{\otimes 2} h^{\mu^s}\|_{L^\infty} ds\right) \Xi(0). \tag{6.2.14}$$

In view of the assumptions, dividing by N^2, using that $\frac{\mathsf{s}}{\mathsf{d}} < 1$, (4.2.16) and Corollary 4.22, we obtain Theorem 6.6.

6.2.5. The case with noise: The modulated free energy method. If we now consider the dissipative or conservative case with noise (6.2.3) or (6.2.4), the modulated energy method, as generalized in [**NRS22, HCRS25**] to sub-Coulomb interactions, works in the case $\mathsf{s} < \mathsf{d} - 2$; see [**RS23**]: instead of considering

$\mathsf{F}_N(X_N^t, \mu^t)$, one simply considers $\mathbb{E}(\mathsf{F}_N(X_N^t, \mu^t))$ and differentiates it in time. One can also bound higher order moments.

However, in the case $\mathsf{s} \geq \mathsf{d} - 2$, this does not work. Bresch-Jabin-Wang introduced instead in [**BJW23**] the *modulated free energy* which works for the dissipative case $\mathbb{M} = I$ and marries the relative entropy method [**JW18**] and the modulated energy method. The point of view is then to consider $f_N^t \in \mathcal{P}((\mathbb{R}^{\mathsf{d}})^N)$, the joint law of the solution to the SDE system (6.2.3) at time t. By Itô's lemma, it satisfies the Liouville or forward Kolmogorov equation

$$\partial_t f_N^t = \sum_{i=1}^N \operatorname{div}_{x_i}\Big(\frac{1}{N}\sum_{j\neq i} \nabla \mathsf{g}(x_i - x_j) f_N\Big) + \frac{1}{\theta}\sum_{i=1}^N \Delta_{x_i} f_N. \tag{6.2.15}$$

The modulated free energy is then defined as

$$\mathcal{F}_N^\theta(f_N, \mu) = \frac{1}{\theta} H_N(f_N|\mu^{\otimes N}) + \int \mathsf{F}_N(X_N, \mu) df_N(X_N), \tag{6.2.16}$$

where the normalized relative entropy H_N is defined by

$$H_N(f|g) = N \int_{(\mathbb{R}^{\mathsf{d}})^N} f \log \frac{f}{g} dX_N = N\mathrm{ent}(f|g). \tag{6.2.17}$$

Note that in units of N^2, F_N is "almost positive" by (4.2.12), while the relative entropy is nonnegative, and controls the convergence of the k-point marginals of f_N to those of φ_N in total variation (TV) distance via the Csiszàr-Kullback-Pinsker inequality

$$\|\mu - \nu\|_{TV}^2 \leq 2 \int \mu \log \frac{\mu}{\nu}$$

combined with the subadditivity of entropy which yields

$$\int f_N \log \frac{f_N}{\mu^{\otimes N}} \geq \lfloor \frac{N}{k} \rfloor \int f_{N,k} \log \frac{f_{N,k}}{\mu^{\otimes k}},$$

where $\lfloor \cdot \rfloor$ denotes the integer part, and $f_{N,k}$ is the k-point marginal of f_N. Thus $\frac{1}{N^2}\mathcal{F}_N^\theta$ controls both relative entropy and modulated energy and metrizes the convergence of marginals of f_N to $\mu^{\otimes k}$.

It is very important that (6.2.16) has the structure of a free energy, i.e., an energy plus temperature times entropy. As seen from [**BJW23**] and reformulated in [**RS25b**], another way of rewriting (6.2.16) is as a relative entropy with respect to the *modulated Gibbs measure*

$$\mathbb{Q}_{N,\beta}(\mu) = \frac{1}{\mathsf{K}_{N,\beta}(\mu)} \exp\left(-\beta N^{-\frac{\mathsf{s}}{\mathsf{d}}} \mathsf{F}_N(X_N, \mu)\right) d\mu(x_1)\dots d\mu(x_N), \tag{6.2.18}$$

with $\theta = \beta N^{1-\frac{\mathsf{s}}{\mathsf{d}}}$, which is exactly as defined in (5.1.13). With this definition, a direct computation allows to check that

$$\mathcal{F}_N^\theta(f_N, \mu) = \frac{1}{\theta}\left(H_N(f_N|\mathbb{Q}_{N,\beta}(\mu)) - N \log \mathsf{K}_{N,\beta}(\mu)\right), \tag{6.2.19}$$

while in view of (5.2.47) and (5.2.12), we can bound

$$\frac{N}{\theta} |\log \mathsf{K}_{N,\beta}(\mu)| \leq C N^{1+\frac{\mathsf{s}}{\mathsf{d}}} \chi(\theta N^{\frac{\mathsf{s}}{\mathsf{d}}-1}) + \left(\frac{N}{2\mathsf{d}} \log N\right) \mathbf{1}_{\mathsf{s}=0} \ll N^2,$$

thus $\frac{1}{N^2}\mathcal{F}_N^\theta$ is equivalent to the suitably normalized relative entropy with respect to the modulated Gibbs measure, up to a small and constant perturbation. Convergence in modulated free energy will thus yield convergence in relative entropy.

As realized in [**BJW23**] (and rephrased in [**RS25b**]) thanks to its particular free energy structure, when differentiating in time $\mathcal{F}_N^\theta(f_N^t,\mu^t)$ with μ^t the mean-field limit solving (6.2.6), singular terms coming from $\frac{d}{dt}(\theta^{-1}H_N)$ and from $\frac{d}{dt}\int \mathsf{F}_N(X_N,\mu)df_N(X_N)$ exactly cancel and reveal only a nonpositive term and a commutator type term, as follows.

Lemma 6.8. *Consider* $\mathbb{M}=I$. *Assume that* f_N *solves* (6.2.15) *and* μ^t *solves* (6.2.6). *Then, we have*

$$\begin{aligned}(6.2.20)\quad &\frac{d}{dt}\mathcal{F}_N^\theta(f_N^t,\mu^t)=-\frac{N}{\theta^2}\mathcal{I}(f_N^t|\mathbb{Q}_{N,\beta}(\mu^t))\\ &-\frac12\int_{(\mathbb{R}^\mathsf{d})^N}\int_{(\mathbb{R}^\mathsf{d})^2\setminus\triangle}(u^t(x)-u^t(y))\cdot\nabla\mathsf{g}(x-y)d\left(\sum_{i=1}^N\delta_{x_i}-N\mu^t\right)^{\otimes 2}(x,y)df_N^t\end{aligned}$$

where

$$u^t=\frac1\theta\nabla\log\mu^t+\nabla h^{\mu^t}=\frac1\theta\nabla\log\mu^t+\nabla\mathsf{g}*\mu^t, \tag{6.2.21}$$

and $\mathcal{I}$ *is the Fisher information*

$$\mathcal{I}(f|g):=\int_{(\mathbb{R}^\mathsf{d})^N}\left|\nabla\log\frac fg\right|^2df. \tag{6.2.22}$$

PROOF. Thanks to the rewriting (6.2.19), we can provide a shorter proof of this, for the convenience of the reader. This can be interpreted as an application of De Bruijn's identity, once one finds the approximate Fokker-Planck equation solved by f_N and $\mathbb{Q}_{N,\beta}(\mu^t)$.

We will use that μ and f_N satisfy

$$\partial_t\mu^t=\operatorname{div}(u^t\mu) \tag{6.2.23}$$

$$\partial_t f_N=\operatorname{div}(bf_N)+\frac1\theta\Delta f_N,\qquad b:=\sum_{i=1}^N\frac1N\sum_{j\neq i}\nabla\mathsf{g}(x_i-x_j)e_i \tag{6.2.24}$$

where e_i is the i-th vector of the canonical basis of $(\mathbb{R}^\mathsf{d})^N$, and u^t is as in (6.2.21). We also note that in view of (6.2.18) and the definition of F_N, we have

$$\begin{aligned}\nabla\log\mathbb{Q}_{N,\beta}(\mu)&=-\frac\theta N\nabla\mathsf{F}_N(\cdot,\mu)+\sum_{i=1}^N\nabla\log\mu(x_i)\\ &=-\frac\theta N\left(\sum_{i=1}^N\int\nabla\mathsf{g}(x_i-y)d\Big(\sum_{j\neq i}\delta_{x_j}-N\mu\Big)(y)e_i\right)+\sum_{i=1}^N\nabla\log\mu(x_i)\\ (6.2.25)\quad&=-\theta b+\theta\sum_{i=1}^N u^t(x_i)e_i,\end{aligned}$$

and, computing from (6.2.23),

$$\begin{aligned}(6.2.26)\quad \partial_t \log \mathbb{Q}_{N,\beta}(\mu^t) &= -\partial_t \log \mathsf{K}_{N,\beta}(\mu^t) - \frac{\theta}{N}\partial_t \mathsf{F}_N(X_N,\mu^t) + \sum_{i=1}^N \partial_t \log \mu^t(x_i)\\ &= -\partial_t \log \mathsf{K}_{N,\beta}(\mu^t) - \theta \int_{(\mathbb{R}^\mathsf{d})^2} \nabla \mathsf{g}(x-y) d\left(\sum_{i=1}^N \delta_{x_i} - N\mu\right)(y)\cdot u(x)\mu(x)\\ &\quad + \sum_{i=1}^N \frac{(\operatorname{div}(u\mu))(x_i)}{\mu(x_i)}.\end{aligned}$$

We may now compute, using (6.2.24) and (6.2.26) together with the fact that f_N is a probability density,

$$\begin{aligned}(6.2.27)\quad & \partial_t\left(\frac{1}{N} H_N(f_N^t|\mathbb{Q}_{N,\beta}(\mu^t)) - \log \mathsf{K}_{N,\beta}(\mu^t)\right)\\ &= \partial_t \int_{(\mathbb{R}^\mathsf{d})^N} f_N^t \log \frac{f_N^t}{\mathbb{Q}_{N,\beta}(\mu^t)} - \partial_t \log \mathsf{K}_{N,\beta}(\mu^t)\\ &= \int_{(\mathbb{R}^\mathsf{d})^N} \operatorname{div}\left(b f_N + \frac{1}{\theta}\nabla f_N\right) \log \frac{f_N}{\mathbb{Q}_{N,\beta}(\mu)}\\ &\quad + \theta \int_{(\mathbb{R}^\mathsf{d})^N} f_N \int_{(\mathbb{R}^\mathsf{d})^2} \nabla \mathsf{g}(x-y) d\left(\sum_{i=1}^N \delta_{x_i} - N\mu\right)(y)\cdot u(x)\mu(x)\\ &\quad - \int_{(\mathbb{R}^\mathsf{d})^N} f_N \left(\sum_{i=1}^N \frac{(\operatorname{div}(u\mu))(x_i)}{\mu(x_i)}\right).\end{aligned}$$

We integrate by parts the first integral to obtain

$$\begin{aligned}&\int_{(\mathbb{R}^\mathsf{d})^N} \operatorname{div}\left(b f_N + \frac{1}{\theta}\nabla f_N\right) \log \frac{f_N}{\mathbb{Q}_{N,\beta}(\mu)}\\ &= \int_{(\mathbb{R}^\mathsf{d})^N} -b\cdot \nabla f_N + b f_N \cdot \nabla \log \mathbb{Q}_{N,\beta}(\mu) + \frac{1}{\theta}\nabla f_N \cdot \nabla \log \mathbb{Q}_{N,\beta}(\mu) - \frac{1}{\theta}\frac{|\nabla f_N|^2}{f_N}\end{aligned}$$

Inserting (6.2.25) to reexpress b, we obtain

$$\begin{aligned}&\int_{(\mathbb{R}^\mathsf{d})^N} \operatorname{div}\left(b f_N + \frac{1}{\theta}\nabla f_N\right) \log \frac{f_N}{\mathbb{Q}_{N,\beta}(\mu)}\\ &= \int_{(\mathbb{R}^\mathsf{d})^N} \frac{2}{\theta}\nabla \log \mathbb{Q}_{N,\beta}(\mu)\cdot \nabla f_N - \sum_{i=1}^N u(x_i)e_i \cdot \nabla f_N - \frac{1}{\theta}|\nabla \log \mathbb{Q}_{N,\beta}(\mu)|^2 f_N\\ &\quad + \int_{(\mathbb{R}^\mathsf{d})^N} \sum_{i=1}^N u(x_i)e_i\cdot \nabla \log \mathbb{Q}_{N,\beta}(\mu) f_N - \frac{1}{\theta}\frac{|\nabla f_N|^2}{f_N}\\ &= -\frac{1}{\theta}\int_{(\mathbb{R}^\mathsf{d})^N} |\nabla \log f_N - \nabla \log \mathbb{Q}_{N,\beta}(\mu)|^2 f_N\\ &\quad + \int_{(\mathbb{R}^\mathsf{d})^N} \sum_{i=1}^N u(x_i)e_i\cdot(-\nabla f_N + \nabla \log \mathbb{Q}_{N,\beta}(\mu) f_N).\end{aligned}$$

Recognizing the Fisher information and inserting this and (6.2.25) into (6.2.27), we are led to

$$
\begin{aligned}
(6.2.28)\quad \partial_t \left(\frac{1}{N} H(f_N^t|\mathbb{Q}_{N,\beta}(\mu^t)) - \log \mathsf{K}_{N,\beta}(\mu^t)\right)
&= -\frac{1}{\theta}\mathcal{I}(f_N|\mathbb{Q}_{N,\beta}(\mu)) \\
&+ \int_{(\mathbb{R}^\mathsf{d})^N} \sum_{i=1}^N u(x_i)e_i \cdot \left(-\nabla f_N + (-\theta b + \theta \sum_{i=1}^N u(x_i)e_i) f_N\right) \\
&+ \theta \int_{(\mathbb{R}^\mathsf{d})^N} f_N \int_{(\mathbb{R}^\mathsf{d})^2} \nabla \mathsf{g}(x-y) d\left(\sum_{i=1}^N \delta_{x_i} - N\mu\right)(y) \cdot u(x)\mu(x) \\
&- \int_{(\mathbb{R}^\mathsf{d})^N} \left(\sum_{i=1}^N \frac{(\operatorname{div}(u\mu))(x_i)}{\mu(x_i)}\right) f_N.
\end{aligned}
$$

We use an integration by parts to rewrite

$$
-\int_{(\mathbb{R}^\mathsf{d})^N} \sum_{i=1}^N u(x_i)e_i \cdot \nabla f_N = \int_{(\mathbb{R}^\mathsf{d})^N} \sum_{i=1}^N \operatorname{div} u(x_i) f_N.
$$

Thus, observing some cancellations,

$$
\begin{aligned}
&\int_{(\mathbb{R}^\mathsf{d})^N} \sum_{i=1}^N u(x_i)e_i \cdot \left(-\nabla f_N + (-\theta b + \theta \sum_{i=1}^N u(x_i)e_i) f_N\right) \\
&\quad - \int_{(\mathbb{R}^\mathsf{d})^N} \left(\sum_{i=1}^N \frac{(\operatorname{div}(u\mu))(x_i)}{\mu(x_i)}\right) f_N \\
&= \int_{(\mathbb{R}^\mathsf{d})^N} \sum_{i=1}^N u(x_i)e_i \cdot \left(-\theta b + \theta \sum_{i=1}^N u(x_i)e_i\right) f_N - \int_{(\mathbb{R}^\mathsf{d})^N} \sum_{i=1}^N u(x_i) \cdot \nabla \log \mu(x_i) f_N \\
&= \int_{(\mathbb{R}^\mathsf{d})^N} \theta \sum_{i=1}^N |u(x_i)|^2 f_N - \frac{\theta}{N}\left(\sum_{i=1}^N u(x_i) \cdot \sum_{j\neq i} \nabla \mathsf{g}(x_i - x_j)\right) f_N \\
&\quad - \int_{(\mathbb{R}^\mathsf{d})^N} \sum_{i=1}^N u(x_i) \cdot \nabla \log \mu(x_i) f_N.
\end{aligned}
$$

Inserting this into (6.2.28), we are led to

$$\partial_t \left(\frac{1}{N} H_N(f_N^t|\mathbb{Q}_{N,\beta}(\mu^t)) - \log \mathsf{K}_{N,\beta}(\mu) \right) = -\frac{1}{\theta} \mathcal{I}(f_N|\mathbb{Q}_{N,\beta}(\mu)) \tag{6.2.29}$$

$$+\theta \int_{(\mathbb{R}^\mathsf{d})^N} f_N \int_{(\mathbb{R}^\mathsf{d})^2} \nabla \mathsf{g}(x-y) d\left(\sum_{i=1}^N \delta_{x_i} - N\mu\right)(y) \cdot u(x)\mu(x)$$

$$+ \int_{(\mathbb{R}^\mathsf{d})^N} \theta \sum_{i=1}^N |u(x_i)|^2 f_N - \frac{\theta}{N} \left(\sum_{i=1}^N u(x_i) \cdot \sum_{j\neq i} \nabla \mathsf{g}(x_i - x_j) \right) f_N$$

$$- \int_{(\mathbb{R}^\mathsf{d})^N} \sum_{i=1}^N u(x_i) \cdot \nabla \log \mu(x_i) f_N$$

$$= -\frac{1}{\theta} \mathcal{I}(f_N|\mathbb{Q}_{N,\beta}(\mu))$$

$$+\theta \int_{(\mathbb{R}^\mathsf{d})^N} f_N \int_{(\mathbb{R}^\mathsf{d})^2} \nabla \mathsf{g}(x-y) d\left(\sum_{i=1}^N \delta_{x_i} - N\mu\right)(y) \cdot u(x)\mu(x)$$

$$- \frac{\theta}{N} \int_{(\mathbb{R}^\mathsf{d})^N} f_N \int_{(\mathbb{R}^\mathsf{d})^2 \setminus \triangle} \nabla \mathsf{g}(x-y) \cdot u(x) d\left(\sum_{i=1}^N \delta_{x_i}\right)(y) d\left(\sum_{i=1}^N \delta_{x_i}\right)(x)$$

$$+ \int_{(\mathbb{R}^\mathsf{d})^N} \theta \sum_{i=1}^N |u(x_i)|^2 f_N - \int_{(\mathbb{R}^\mathsf{d})^N} \sum_{i=1}^N u(x_i) \cdot \nabla \log \mu(x_i) f_N.$$

We may then use (6.2.21) to rewrite the terms on the last line as

$$\theta \int_{(\mathbb{R}^\mathsf{d})^N} \sum_{i=1}^N u(x_i) \cdot \nabla \mathsf{g} * \mu(x_i) f_N$$

$$= \theta \int_{(\mathbb{R}^\mathsf{d})^N} f_N \int_{(\mathbb{R}^\mathsf{d})^2} u(x) \cdot \nabla \mathsf{g}(x-y)\mu(y) d\left(\sum_{i=1}^N \delta_{x_i}\right)(x).$$

By definition (6.2.19), we may then group everything as

$$\partial_t \left(\frac{\theta}{N} \mathcal{F}_N^\theta \right) = -\frac{1}{\theta} \mathcal{I}(f_N|\mathbb{Q}_{N,\beta}(\mu))$$

$$- \frac{\theta}{N} \int_{(\mathbb{R}^\mathsf{d})^N} f_N \int_{(\mathbb{R}^\mathsf{d})^2 \setminus \triangle} \nabla \mathsf{g}(x-y) \cdot u(x) d\left(\sum_{i=1}^N \delta_{x_i} - N\mu\right)(x) d\left(\sum_{i=1}^N \delta_{x_i} - N\mu\right)(y).$$

Symmetrizing the last term with respect to x and y, we obtain the desired result. □

Discarding the nonpositive Fisher information term and using just the commutator estimate of Theorem 6.1, together with the fact that $\int \mathsf{F}_N df_N \le \mathcal{F}_N^\theta$ by nonnegativity of the entropy, this yields

$$\frac{d}{dt} \mathcal{F}_N^\theta(f_N^t, \mu^t) \le C \mathcal{F}_N^\theta(f_N^t, \mu^t) + o(N^2),$$

hence by Gronwall's lemma and the above remarks, we directly obtain local-in-time mean field convergence (in relative entropy) to the solution of (6.2.6). Using the commutator estimate requires a Lipschitz bound on u^t which is delicate when working in the whole space, due to the necessary decay of μ^t, however a self-similar transformation allows to reduce to a confined situation and obtain the needed bounds, which then even allows to conclude that global-in-time convergence holds [**RS24a**]. Global-in-time convergence in the two-dimensional log case was also obtained in [**CFGW24**].

The negative term in the right-hand side of (6.2.20) is the opposite of the relative Fisher information with respect to the modulated Gibbs measure and, as observed in [**RS25b**], in case of existence of a uniform-in-time "modulated" Logarithmic Sobolev Inequality for $\mathbb{Q}_{N,\beta}(\mu^t)$, it may be used to obtain an exponential decay of the modulated free energy, and what has been called *generation of chaos* (i.e., the probability becomes tensorized in large time, even if it is not initially), a term first coined by J. Lukkarinen. What we mean by uniform-in-time modulated Logarithmic Sobolev Inequality is that there exists a constant $C_{LS} > 0$ such that for all $N \geq 1$, all $t \geq 0$ and all $f \in C^1((\mathbb{R}^\mathsf{d})^N)$ we have

$$\int_{(\mathbb{R}^\mathsf{d})^N} f^2 \log \frac{f^2}{\int f^2 d\mathbb{Q}_{N,\beta}(\mu^t)} d\mathbb{Q}_{N,\beta}(\mu^t) \leq C_{LS} \int_{(\mathbb{R}^\mathsf{d})^N} |\nabla f|^2 d\mathbb{Q}_{N,\beta}(\mu^t), \tag{6.2.30}$$

which allows to compare the modulated Fisher information with the relative entropy appearing in (6.2.19), hence with $\mathcal{F}_N^\theta(f_N^t, \mu^t)$. Unfortunately, it seems very hard to prove such an inequality, except in the setting of $\mathsf{d} = 1$ with some convexity assumptions (we refer the reader to [**RS25b**]).

Even discarding the relative Fisher information term in (6.2.20), one may still obtain uniform-in-time convergence, or uniform-in-time propagation of chaos and even generation of chaos, by taking advantage of the decay rate of the vector field u^t in (6.2.20). This is easier to do in the setting of the torus, and exponential decay of u^t was proven, and used to conclude the uniform-in-time convergence in that context in [**CdCRS25**], where the commutator estimate is extended to the setting of the torus.

Part 3

Mesoscopic behavior

CHAPTER 7

The two energy quantities and the screening procedure

In previous chapters we have reduced the study of the energy $\mathcal{H}_N$ to that of the next order energy F_N, and then given an electric formulation for F_N which replaces pair interactions by an energy which is local in space in terms of the (opposite of the) electric field ∇h_N. The question we now turn to is how to show that this energy (and similarly the free energy) is *almost additive* in space so that disjoint regions of space can be considered independently. We can explain the idea of how to do so by using two energy quantities (in the spirit of Dirichlet-Neumann bracketing or the work of Armstrong-Smart for homogenization [**AS16**]), one subadditive and one superadditive, which converge to each other on large scales. This allows to quantify the defect of additivity of the energy or free energy over cubes. To quantify this we will need to control the difference between these two energy quantities, which differ only by the boundary condition which is imposed. This is where we use a *screening procedure* that allows to show that the effect of a specific boundary condition decays rapidly enough away from the boundary to be negligible.

In this whole Part 3, we restrict to the Coulomb case $\mathsf{s} = \mathsf{d} - 2$ and work in blown-up scale; see Section 5.2.2 for definitions. The adaptation of Part 3 to the other Riesz cases (1.1.2) can be found in [**Pei24**] for the one-dimensional logarithmic case, and in [**PS**] for the general case. The material in this chapter and the next originates in [**AS21**] but with new simplifications.

7.1. Dirichlet and Neumann problems

Let us now introduce the two local sub and super additive energy approximations, from [**AS21**]. Let us consider U a closed subset of $\mathbb{R}^{\mathsf{d}}$ with piecewise C^1 boundary. Most often, U will be $\mathbb{R}^{\mathsf{d}}$, a hyperrectangle or the complement of a hyperrectangle. Although N originally denoted the number of points in $\mathbb{R}^{\mathsf{d}}$ and defined the blown-up scale at which we are working, we will also use the notation N to denote the total number of points a system has in a generic set U which may not be the whole space.

7.1.1. Informal description. Given a nonnegative, bounded and integrable density μ over U, the first energy quantity is obtained by solving for

$$\begin{cases} -\Delta u = \mathsf{c_d}\Big(\sum_{i=1}^N \delta_{x_i} - \mu\Big) & \text{in } U \\ \frac{\partial u}{\partial \nu} = 0 & \text{on } \partial U \\ \nabla u \to 0 & \text{at } \infty. \end{cases} \tag{7.1.1}$$

Note that this equation is solvable (and the solution is unique up to addition of a constant) if and only if $\mu(U) = N$, which means if and only if the system in U is *neutral.* If U is bounded, there is no condition "at infinity". If U is unbounded, we can find a solution by considering first $\mathsf{g} * (\sum_{i=1}^N \delta_{x_i} - \mu \mathbf{1}_U)$ over $\mathbb{R}^\mathsf{d}$, and then subtracting off a harmonic function w with prescribed Neumann boundary condition on ∂U and such that $\nabla w \to 0$ at ∞. We will also later call *screened system* such a system with equation (7.1.1) solved.

Neglecting the issue of renormalization, the first energy quantity is defined informally as

$$\mathsf{F}(X_N, \mu, U) = \frac{1}{2\mathsf{c_d}} \int_U |\nabla u|^2.$$

Unless ambiguous, we omit the dependence in μ in the notation and simply write $\mathsf{F}(X_N, U)$ instead of $\mathsf{F}(X_N, \mu, U)$. We note that $\mathsf{F}(\cdot, \mathbb{R}^\mathsf{d})$ coincides with F defined in (5.2.9).

The second quantity is obtained by minimizing the energy with respect to all possible functions u compatible with the points in the sense of satisfying $-\Delta u = \mathsf{c_d}(\sum_{i=1}^N \delta_{x_i} - \mu)$, it naturally leads to a superadditive energy and to solving a Dirichlet problem:

$$\begin{cases} -\Delta v = \mathsf{c_d}\Big(\sum_{i=1}^N \delta_{x_i} - \mu \Big) & \text{in } U \\ v = 0 & \text{on } \partial U \text{ and at } \infty. \end{cases} \tag{7.1.2}$$

Neglecting again the issue of renormalization, the second energy quantity G is defined informally as

$$\mathsf{G}(X_N, \mu, U) = \frac{1}{2\mathsf{c_d}} \int_U |\nabla v|^2.$$

Again, we will often simply write $\mathsf{G}(X_N, U)$.

One may check that for any given distribution f, the solution of the variational problem

$$\min \left\{ \frac{1}{2\mathsf{c_d}} \int_U |\nabla w|^2, \quad -\Delta w = f \text{ in } U \right\} \tag{7.1.3}$$

is achieved by the solution of

$$\begin{cases} -\Delta v = f & \text{in } U \\ v = 0 & \text{on } \partial U \text{ and at } \infty. \end{cases} \tag{7.1.4}$$

Indeed, it suffices to consider competitors of the form $w + th$ with h harmonic, solving $\partial h / \partial \nu = g$ on ∂U for any given g of integral 0 on ∂U. Writing that $\int_U |\nabla(w + th)|^2 \geq \int_U |\nabla w|^2$, integrating by parts and letting $t \to 0$, we find that $\int_{\partial U} wg = 0$, and this being true for any g of integral 0, w must be constant on ∂U. Since the problem is unchanged by addition of a constant, we may assume that $w = v$, thus v minimizes the energy among all solutions of $-\Delta v = f$.

Applying this to $f = \mathsf{c_d}\Big(\sum_{i=1}^N \delta_{x_i} - \mu\Big)$, we immediately deduce that for any X_N such that F is defined, we have

$$\mathsf{F}(X_N, \mu, U) \geq \mathsf{G}(X_N, \mu, U). \tag{7.1.5}$$

We will now see that F is good for pasting while G is good for restricting, leading to the property that F is subadditive and G superadditive.

The property is easy for G. Assume U is the union of two sets U_1, U_2 with disjoint interiors and piecewise C^1 boundaries. If X_N is a configuration in U_1, v_1 the corresponding solution of (7.1.2) and $Y_{N'}$ a configuration in U_2, v_2 the corresponding solution of (7.1.2), and v the solution of (7.1.2) in U for the configuration $X_N \cup Y_{N'}$, then

$$\begin{aligned}\mathsf{G}(X_N \cup Y_{N'}, U) &= \frac{1}{2\mathsf{c_d}} \int_U |\nabla v|^2 = \frac{1}{2\mathsf{c_d}} \int_{U_1} |\nabla v|^2 + \frac{1}{2\mathsf{c_d}} \int_{U_2} |\nabla v|^2 \\ &\geq \frac{1}{2\mathsf{c_d}} \int_{U_1} |\nabla v_1|^2 + \frac{1}{2\mathsf{c_d}} \int_{U_2} |\nabla v_2|^2 = \mathsf{G}(X_N, U_1) + \mathsf{G}(Y_{N'}, U_2),\end{aligned} \tag{7.1.6}$$

where the inequality follows from the minimality property of v_1, resp. v_2, seen above. So G is superadditive as claimed.

It is convenient to also work with "electric fields" E which are gradients of electric potentials[1] and thus satisfy relations of the form

$$-\operatorname{div} E = \mathsf{c_d} \left(\sum_{i=1}^{N} \delta_{x_i} - \mu \right). \tag{7.1.7}$$

Any vector field, not necessarily a gradient, which satisfies such a relation, will be said to be compatible with (X_N, μ).

The subadditivity of F relies on the following lemma which exploits the fact that the Neumann electric field (that is, the electric field with zero normal component) is the L^2 projection of any compatible electric field onto gradients.

Lemma 7.1 (Projection lemma). *Assume that U is a bounded open of $\mathbb{R}^\mathsf{d}$ with piecewise C^1 boundary. Assume E is a vector-field and u a function satisfying*

$$\begin{cases} \operatorname{div} E = \Delta u & \text{in } U \\ (E - \nabla u) \cdot \nu = 0 & \text{on } \partial U \\ E - \nabla u \to 0 & \text{at } \infty. \end{cases} \tag{7.1.8}$$

Then

$$\int_U |\nabla u|^2 \leq \int_U |E|^2.$$

PROOF. It suffices to write

$$\begin{aligned}\int_U |E|^2 = \int_U |\nabla u + E - \nabla u|^2 &= \int_U |\nabla u|^2 + |E - \nabla u|^2 + 2\nabla u \cdot (E - \nabla u) \\ &\geq \int_U |\nabla u|^2 - 2 \int_U u \operatorname{div}(E - \nabla u) = \int_U |\nabla u|^2\end{aligned} \tag{7.1.9}$$

where we used (7.1.8) and Green's formula. □

This way, the energy $\mathsf{F}(X_N, U)$ can be estimated from above by that of any vector field with same divergence, i.e., after *relaxing* the condition of being a gradient, an idea already used in [**ACO09**] and [**SS12**].

Let us now see how the projection lemma implies subadditivity. Assume U is the union of two sets U_1, U_2 with disjoint interiors and piecewise C^1 boundaries. If X_N is a configuration in U_1 and $Y_{N'}$ a configuration in U_2 with $\mu(U_1) = N$, $\mu(U_2) = N'$, then

$$\mathsf{F}(X_N \cup Y_{N'}, U) \leq \mathsf{F}(X_N, U_1) + \mathsf{F}(Y_{N'}, U_2). \tag{7.1.10}$$

[1] What we define here is in truth the opposite of the physical electric field.

Indeed, let u_1 and u_2 be the solutions to the Neumann problems and set $E_1 = \nabla u_1$, $E_2 = \nabla u_2$. We have

$$(7.1.11)\quad -\mathrm{div}\, E_1 = \mathsf{c_d}\Big(\sum_{i=1}^{N} \delta_{x_i} - \mu\Big), \text{ in } U_1 \qquad -\mathrm{div}\, E_2 = \mathsf{c_d}\Big(\sum_{i=1}^{N'} \delta_{y_i} - \mu\Big) \text{ in } U_2.$$

We may now define $E = E_1 \mathbf{1}_{U_1} + E_2 \mathbf{1}_{U_2}$ and note that it satisfies

$$(7.1.12)\qquad \begin{cases} -\mathrm{div}\, E = \mathsf{c_d}\Big(\sum_{p\in X_N \cup Y_{N'}} \delta_p - \mu\Big) & \text{in } U \\ E\cdot \nu = 0 & \text{on } \partial U \\ E \to 0 & \text{at } \infty. \end{cases}$$

Indeed, no divergence is created across $\partial U_1 \cap \partial U_2$ thanks to the vanishing normal components on both sides: it suffices to consider a test function ϕ and use Green's formula twice together with the vanishing Neumann boundary conditions to obtain that

$$\begin{aligned} -\int_U \phi\, \mathrm{div}\, E = \int_U \nabla\phi\cdot E &= \int_{U_1} \nabla\phi \cdot E_1 + \int_{U_2} \nabla\phi\cdot E_2 \\ &= -\int_{U_1} \phi\, \mathrm{div}\, E_1 - \int_{U_2} \phi\, \mathrm{div}\, E_2 \\ &= \mathsf{c_d} \int \phi\Big(\sum_{p\in X_N\cup Y_{N'}} \delta_p - \mu\Big). \end{aligned}$$

Thus, E is compatible with $(X_N \cup Y_{N'}, \mu)$. It then follows from Lemma 7.1 that

$$\begin{aligned} \mathsf{F}(X_N \cup Y_{N'}, U) &\le \frac{1}{2\mathsf{c_d}} \int_U |E|^2 \\ &= \frac{1}{2\mathsf{c_d}} \int_{U_1} |\nabla u_1|^2 + \frac{1}{2\mathsf{c_d}} \int_{U_2} |\nabla u_2|^2 = \mathsf{F}(X_N, U_1) + \mathsf{F}(Y_{N'}, U_2), \end{aligned}$$

hence the claimed subadditivity.

We define corresponding partition functions.

Definition 7.2 (Neumann Gibbs measure and partition function). The Neumann partition function relative to U, is defined if $\mu(U) = N$ by

$$(7.1.13)\qquad \mathsf{K}_\beta(\mu, U) := N^{-N} \int_{U^N} \exp\left(-\beta \mathsf{F}(X_N, \mu, U)\right) d\mu^{\otimes N}(X_N).$$

The Neumann Gibbs measure is defined by

$$(7.1.14)\qquad d\mathbb{Q}_\beta(\mu, U) := \frac{1}{N^N \mathsf{K}_\beta(\mu, U)} \exp\left(-\beta \mathsf{F}(X_N, \mu, U)\right) d\mu^{\otimes N}(X_N).$$

The notation agrees with (5.2.10) and (5.2.11): when taking $U = \mathbb{R}^{\mathsf{d}}$ the definitions coincide, adopting the convention $\mathsf{K}_\beta(\mu, \mathbb{R}^{\mathsf{d}}) = \mathsf{K}_\beta(\mu)$ and $\mathbb{Q}_\beta(\mu, \mathbb{R}^{\mathsf{d}}) = \mathbb{Q}_\beta(\mu)$. Again we omit the N in the notation since it can be recovered from $\mu(U)$.

We may also consider in the same way the "Dirichlet partition function"

$$(7.1.15)\qquad \mathsf{L}_{N,\beta}(\mu, U) := N^{-N} \int_{U^N} \exp\left(-\beta \mathsf{G}(X_N, \mu, U)\right) d\mu^{\otimes N}(X_N).$$

In view of (7.1.5) we have a comparison for the free energies as well: whenever $\mu(U) = N$, we have

$$(7.1.16)\qquad \mathsf{L}_{N,\beta}(\mu, U) \le \mathsf{K}_\beta(\mu, U).$$

The subadditivity property has the following counterpart for the partition functions.

Lemma 7.3 ("Superadditivity" of Neumann partition functions). *Assume U is partitioned into p disjoint sets Q_i, $i \in [1,p]$ which are such that $\mu(Q_i) = N_i$ with N_i integer. We have*

$$\mathsf{K}_\beta(\mu, U) \geq \frac{N! N^{-N}}{N_1! \dots N_p! N_1^{-N_1} \dots N_p^{-N_p}} \prod_{i=1}^{p} \mathsf{K}_\beta(\mu, Q_i). \tag{7.1.17}$$

PROOF. It suffices to partition the phase space into sets of the form $\{x_{i_1}, \dots, x_{i_{N_j}} \in Q_j\}$ for each $j = 1, \dots, p$, then to use the subadditivity (7.1.10), or rigorously (7.1.28), noting that the number of ways to distribute N points in the p sets with N_i points in each set is $\frac{N!}{N_1! \dots N_p!}$. □

When we consider a uniform density μ, say $\mu = 1$ without loss of generality, then the sub and superadditivity property will imply that

$$-\frac{\log \mathsf{K}_\beta(1, \square_R)}{\beta R^{\mathsf{d}}} \tag{7.1.18}$$

is decreasing in R while

$$-\frac{\log \mathsf{L}_\beta(1, \square_R)}{\beta R^{\mathsf{d}}} \tag{7.1.19}$$

is increasing in R, when $\square_R$ are cubes of quantized volume. Thus each of them will have a limit, that we denote $f_{\mathsf{d}}(\beta)$, as it is a function of β and d only. Since on the other hand $\mathsf{L}_\beta \leq \mathsf{K}_\beta$, it means that if we can show that these limits are the same, then $\frac{1}{\beta R^{\mathsf{d}}} |\log \mathsf{K}_\beta(1, \square_R) - \log \mathsf{L}_\beta(1, \square_R)|$ will bound from above the distance of each of them to $f_{\mathsf{d}}(\beta)$ and allow to get a rate of convergence for the two limits. The goal is thus to show that

$$\frac{|\log \mathsf{K}_\beta(1, \square_R) - \log \mathsf{L}_\beta(1, \square_R)|}{\beta R^{\mathsf{d}}} \to 0 \quad \text{as } R \to \infty, \tag{7.1.20}$$

with a quantitative rate, which will bound the rate of convergence to the limit $f_{\mathsf{d}}(\beta)$.

The reason why (7.1.20) is true is that the effect of the boundary condition decays sufficiently fast away from the boundary that it only brings in a surface contribution of order $R^{\mathsf{d}-1}$, which is negligible compared to the volume R^{d}. In other words, we expect that (7.1.20) happens with rate $1/R$ (except that β comes into play when it is small). Moreover, since (7.1.18) and (7.1.19) converge rapidly to the same limit, it means that as R gets large they become almost constant, in particular it means that the free energies $\log \mathsf{K}_\beta(1, \square_R)$, superadditive, and $\log \mathsf{L}_\beta(1, \square_R)$, subadditive, must both be almost additive. This property of *almost additivity of the (free) energy* will be crucial.

Proving (7.1.20) requires work, and will rely on the *screening procedure*: that procedure consists in modifying arbitrary configurations and their potential (say v), into neutral configurations with the Neumann problem solved as in (7.1.1), without having added too much energy or entropy. Such modified configurations are called screened configurations. The screening procedure will create energy and volume errors that need to be precisely quantified.

7.1.2. Rigorous definitions. In order to deal with the renormalization properly, we need to introduce a new modified version of the minimal distance (5.2.18) that is relative to the domain U, and that makes the energy subadditive. In this part, we do not keep track of $\|\mu\|_{L^\infty}$ dependence so we replace λ by 1 in (5.2.18) and we let

$$\hat{\mathsf{r}}_i := \frac{1}{4}\min\left(\min_{x_j\in U, j\neq i}|x_i - x_j|, \mathrm{dist}(x_i, \partial U), 1\right). \tag{7.1.21}$$

This shrinks the radius of the balls when they approach ∂U, ensuring that all $B(x_i, \hat{\mathsf{r}}_i)$'s remain included in U if $x_i \in U$.

Definition 7.4 (Neumann electric energy in a domain U). If $\mu(U) = N$, for a configuration X_N of points in U and u as in (7.1.1) we define[2]

$$\mathsf{F}(X_N, \mu, U) := \frac{1}{2\mathsf{c_d}}\left(\int_U |\nabla u_{\hat{\mathsf{r}}}|^2 - \mathsf{c_d}\sum_{i=1}^N \mathsf{g}(\hat{\mathsf{r}}_i)\right) - \sum_{i=1}^N \int_U \mathsf{f}_{\hat{\mathsf{r}}_i}(x - x_i)d\mu(x), \tag{7.1.22}$$

where $u_{\vec{\eta}}$ is defined as in (4.1.9) and f_η as in (4.1.7).

Analogously to (4.5.2) we define a localized version of this energy in a measurable subset Ω

(7.1.23)

$$\mathsf{F}^\Omega(X_N, \mu, U) := \frac{1}{2\mathsf{c_d}}\left(\int_{\Omega\cap U} |\nabla u_{\hat{\mathsf{r}}}|^2 - \mathsf{c_d}\sum_{i\in I_\Omega} \mathsf{g}(\hat{\mathsf{r}}_i)\right) - \sum_{i\in I_\Omega} \int_U \mathsf{f}_{\hat{\mathsf{r}}_i}(x - x_i)d\mu(x)$$

where we let $I_\Omega = \{i, x_i \in \Omega\}$, and where this time

(7.1.24)

$$\hat{\mathsf{r}}_i := \frac{1}{4}\begin{cases} \min\left(\min_{j\neq i}|x_i - x_j|, \mathrm{dist}(x_i, \partial U), 1\right) & \text{if } \mathrm{dist}(x_i, \partial\Omega) \geq 2 \\ \min(1, \mathrm{dist}(x_i, \partial U)) & \text{if } \mathrm{dist}(x_i, \partial\Omega) \leq 1 \\ t\min\left(\min_{j\neq i}|x_i - x_j|, \mathrm{dist}(x_i, \partial U), 1\right) & \\ +(1-t)\min(\mathrm{dist}(x_i, \partial U), 1) & \text{if } \mathrm{dist}(x_i, \partial\Omega) = 1 + t, t \in [0, 1]. \end{cases}$$

We note that the definition (7.1.24) coincides with (7.1.21) when taking $\Omega = \mathbb{R}^{\mathsf{d}}$. Also, the definition has been made so that the radii are continuous with respect to the location of the points, which will be useful in Chapter 12. Note that the points in Ω^c influence the value of F^Ω via their truncated charge in $u_{\hat{\mathsf{r}}}$ but have no effect on the value of $\hat{\mathsf{r}}_i$ for $x_i \in \Omega$. This will be important in Definition 7.16 below. Here, the balls are enlarged to their largest possible values for points that approach the boundary of Ω (except for the part included in ∂U). This way, balls can potentially overlap the boundary of Ω and not be disjoint. This is the right choice to have the restriction property analogous to (4.5.3):

$$\mathsf{F}(X_N, \mu, U) \geq \mathsf{F}^\Omega(X_N, \mu, U) + \mathsf{F}^{\Omega^c}(X_N, \mu, U),$$

which we will justify below in Lemma 7.8.

[2]This definition is simpler than that in [**AS21**].

Finally, we will also use another definition, similar to (4.5.1) but in blown-up scale, which ignores ∂U (hence the balls may overlap ∂U):

(7.1.25)

$$\tilde{\mathsf{r}}_i := \frac{1}{4}\begin{cases} \min\left(\min_{j\neq i}|x_i - x_j|, 1\right) & \text{if } \operatorname{dist}(x_i, \partial\Omega) \geq 2 \\ 1 & \text{if } \operatorname{dist}(x_i, \partial\Omega) \leq 1 \\ t\min\left(\min_{j\neq i}|x_i - x_j|, 1\right) + (1-t) & \text{if } \operatorname{dist}(x_i, \partial\Omega) = 1+t, t\in[0,1]. \end{cases}$$

In the rigorous treatment, it is not useful to define the Dirichlet (or minimal energy) yet, it will need to be defined in conjunction with a screenability property below.

7.1.3. Projection lemma and consequences. A truncated version of an electric field E can be defined just as for electric potentials: for any E satisfying a relation of the form (7.1.7), for any $\vec{\eta}$, we let

$$E_{\vec{\eta}} = E - \sum_{i=1}^{n} \nabla \mathsf{f}_{\eta_i}(x - x_i). \tag{7.1.26}$$

Next we state a projection lemma with renormalized vector fields. The proof is identical to that of Lemma 7.1.

Lemma 7.5 (Projection lemma). *Assume that U is an open subset of $\mathbb{R}^{\mathsf{d}}$ with piecewise C^1 boundary. Assume E is a vector-field satisfying a relation of the form*

$$\begin{cases} -\operatorname{div} E = \mathsf{c_d}\left(\sum_{i=1}^N \delta_{x_i} - \mu\right) & \text{in } U \\ E\cdot\nu = 0 & \text{on } \partial U \\ E \to 0 & \text{as } |x|\to\infty, \end{cases} \tag{7.1.27}$$

and u solves

$$\begin{cases} -\Delta u = \mathsf{c_d}\left(\sum_{i=1}^N \delta_{x_i} - \mu\right) & \text{in } U \\ \frac{\partial u}{\partial \nu} = 0 & \text{on } \partial U \\ \nabla u \to 0 & \text{as } |x|\to\infty, \end{cases}$$

and

$$u\left(\frac{\partial u}{\partial \nu} - E\cdot\nu\right) \to 0 \quad \text{as } |x|\to\infty,\ x\in U.$$

Then

$$\int_U |\nabla u_{\hat{\mathsf{r}}}|^2 \leq \int_U |E_{\hat{\mathsf{r}}}|^2.$$

Corollary 7.6 (Subadditivity of F). *Assume U is the union of two sets U_1, U_2 with disjoint interiors and piecewise C^1 boundaries. If X_N is a configuration in U_1 and $Y_{N'}$ a configuration in U_2 with $\mu(U_1) = N$, $\mu(U_2) = N'$, then*

$$\mathsf{F}(X_N \cup Y_{N'}, \mu, U) \leq \mathsf{F}(X_N, \mu, U_1) + \mathsf{F}(Y_{N'}, \mu, U_2). \tag{7.1.28}$$

PROOF. For (7.1.28), let u and u' be the solutions to the Neumann problems (7.1.1) associated with the definition of F in (7.1.22) and set $E = \nabla u$, $E' = \nabla u'$. We have (7.1.11), and we may then define $E^0 = E\mathbf{1}_{U_1} + E'\mathbf{1}_{U_2}$, which satisfies (7.1.12). The result then follows from Lemma 7.5, noting that the renormalization does not interfere because the balls $B(x_i, \hat{\mathsf{r}}_i)$ remain included in U_1, resp. U_2. □

This proof tells us that screened configurations or screened electric fields can effectively be pasted together without creating any additional divergence, and their energies can be simply added to produce an upper bound on the true energy. We will use that property repeatedly.

Thanks to the subadditivity of the Neumann energy, letting $\mathsf{K}_\beta(\mu, U)$ and $\mathbb{Q}_\beta(\mu, U)$ be defined via (7.1.13) and (7.1.14), we have that (7.1.17) holds.

The same argument as in Corollary 7.6 allows to obtain the following.

Corollary 7.7 (The Neumann electric energy is larger than the regular one). *For any μ, U as above, and any configuration X_N in U, we have*

$$\mathsf{F}(X_N, \mu\mathbf{1}_U) = \mathsf{F}(X_N, \mu\mathbf{1}_U, \mathbb{R}^\mathsf{d}) \le \mathsf{F}(X_N, \mu, U), \tag{7.1.29}$$

in particular it follows from (5.2.19) *that*

$$\mathsf{F}(X_N, \mu, U) \ge -CN\mathbf{1}_{\mathsf{s}\ge 0}, \tag{7.1.30}$$

with C depending only on d, s and $\|\mu\|_{L^\infty}$.

PROOF. Let $E_{\hat{\mathsf{r}}}$ be as in (7.1.26) where $E = \nabla u$, with u solution to (7.1.1) arising in the definition of $\mathsf{F}(X_N, \mu, U)$. Let us extend E and $E_{\hat{\mathsf{r}}}$ by 0 in U^c. Since the balls $B(x_i, \hat{\mathsf{r}}_i)$ do not intersect ∂U by definition (7.1.24), both E and $E_{\hat{\mathsf{r}}}$ are normal to ∂U and thus no divergence is created at the boundary and the extended fields satisfy

(7.1.31)
$$-\mathrm{div}\, E = \mathsf{c_d}\left(\sum_{i=1}^N \delta_{x_i} - \mu\mathbf{1}_U\right) \text{ in } \mathbb{R}^\mathsf{d}, \qquad -\mathrm{div}\, E_{\hat{\mathsf{r}}} = \mathsf{c_d}\left(\sum_{i=1}^N \delta_{x_i}^{(\hat{\mathsf{r}}_i)} - \mu\mathbf{1}_U\right) \text{ in } \mathbb{R}^\mathsf{d}.$$

Let h_N be the electric potential associated to X_N and μ in $\mathbb{R}^\mathsf{d}$, as defined in (5.2.15). We have $\mathrm{div}\, E_{\hat{\mathsf{r}}} = \Delta h_{N,\hat{\mathsf{r}}}$, and $E_{\hat{\mathsf{r}}}$ and $h_{N,\hat{\mathsf{r}}}$ have the same decay at infinity. Hence, applying the projection lemma, Lemma 7.5, over U we find that

$$\int_{\mathbb{R}^\mathsf{d}} |\nabla h_{N,\hat{\mathsf{r}}}|^2 \le \int_{\mathbb{R}^\mathsf{d}} |E_{\hat{\mathsf{r}}}|^2 = \int_U |\nabla u_{\hat{\mathsf{r}}}|^2$$

and the claim (7.1.29) follows immediately in view of the definition (7.1.22). □

7.1.4. Local energy controls. We will repeatedly need the following lemma which is a variant in this context of Lemma 4.13, Propositions 4.19 and 4.28, and Lemma 4.20.

Lemma 7.8 (Local energy controls). *Let $\Omega \subset \mathbb{R}^\mathsf{d}$. First, let E be a vector field solving*

$$-\mathrm{div}\, E = \mathsf{c_{d,s}}\left(\sum_{i=1}^N \delta_{x_i} - \mu\mathbf{1}_U\right) \quad \text{in } \Omega \tag{7.1.32}$$

and let $E_{\vec{\alpha}}$, $E_{\vec{\eta}}$ be as in (7.1.26) *for $\vec{\alpha}$, $\vec{\eta}$ such that $\alpha_i \le \eta_i$ for all i. Then, letting $I_N = \{i, \alpha_i \ne \eta_i\}$, if for each $i \in I_N$, $B(x_i, \eta_i) \subset \Omega$, we have*

$$\begin{aligned}&\int_\Omega |E_{\vec{\eta}}|^2 - \mathsf{c_{d,s}}\sum_{i\in I_N} \mathsf{g}(\eta_i) - 2\mathsf{c_{d,s}}\sum_{i\in I_N}\int_\Omega \mathsf{f}_{\eta_i}(x - x_i)d\mu \\ &\quad - \left(\int_\Omega |E_{\vec{\alpha}}|^2 - \mathsf{c_{d,s}}\sum_{i\in I_N} \mathsf{g}(\alpha_i) - 2\mathsf{c_{d,s}}\sum_{i\in I_N}\int_\Omega \mathsf{f}_{\alpha_i}(x - x_i)d\mu\right) \le 0,\end{aligned} \tag{7.1.33}$$

with equality if the $B(x_i, \eta_i)$'s are disjoint from all the other $B(x_j, \eta_j)$'s for each $i \in I_N$.

Secondly, there exist $C_0, C_1, C > 0$ depending only on d *and $\|\mu\|_{L^\infty}$ such that for any configuration X_N in U and u corresponding via (7.1.1), and for any $\Omega \subset \mathbb{R}^{\mathsf{d}}$,*

$$\sum_{i \in I_{\Omega \cap U}} \mathsf{g}(\tilde{\mathsf{r}}_i) \le C_1 \left(\mathsf{F}^\Omega(X_N, \mu, U) + \# I_{\Omega \cap U} \right), \tag{7.1.34}$$

$$\int_{\Omega \cap U} |\nabla u_{\tilde{\mathsf{r}}}|^2 \le C \left(\mathsf{F}^\Omega(X_N, \mu, U) + C_0 \# I_{\Omega \cap U} \right), \tag{7.1.35}$$

with $C_1 = 2$ in the case $\mathsf{s} = 0$, $\tilde{\mathsf{r}}$ as in (7.1.25); and, if $\alpha_i = \tilde{\mathsf{r}}_i$ for all i's such that $\mathrm{dist}(x_i, \partial\Omega) \le \alpha_i$, *we have*

$$\mathsf{F}^\Omega(X_N, \mu, U) - \mathcal{F}^{\vec{\alpha}} \ge \frac{1}{2} \sum_{\substack{i,j \in I_\Omega, i \ne j \\ \mathrm{dist}(x_i, \partial\Omega) \ge \alpha_i}} (\mathsf{g}(x_i - x_j) - \mathsf{g}(\alpha_i))_+, \tag{7.1.36}$$

where

$$\mathcal{F}^{\vec{\alpha}} = \frac{1}{2\mathsf{c}_\mathsf{d}} \left(\int_{\Omega \cap U} |\nabla u_{\vec{\alpha}}|^2 - \mathsf{c}_\mathsf{d} \sum_{i \in I_{\Omega \cap U}} \mathsf{g}(\alpha_i) - 2\mathsf{c}_\mathsf{d} \sum_{i \in I_{\Omega \cap U}} \int_{\mathbb{R}^\mathsf{d} \cap U} \mathsf{f}_{\alpha_i}(x - x_i) d\mu(x) \right).$$

Moreover, we have

$$\mathsf{F}(X_N, \mu, U) \ge \mathsf{F}^\Omega(X_N, \mu, U) + \mathsf{F}^{\Omega^c}(X_N, \mu, U). \tag{7.1.37}$$

Let φ be a Lipschitz function in U with bounded support. Let Ω be an open set containing a 1-neighborhood of the support of φ in U. For any configuration X_N in U, letting u be defined as in (7.1.1) (resp. v as in (7.1.2)), we have

(7.1.38)
$$\left| \int_{\mathbb{R}^\mathsf{d}} \varphi \left(\sum_{i=1}^N \delta_{x_i} - d\mu \right) \right| \le \frac{1}{\mathsf{c}_\mathsf{d}} \|\nabla \varphi\|_{L^2(\Omega)} \|\nabla u_{\tilde{\mathsf{r}}}\|_{L^2(\Omega \cap U)} + C \|\nabla \varphi\|_{L^\infty(\Omega)} \# I_\Omega,$$

(and resp. the same with $v_{\tilde{\mathsf{r}}}$ in place of $u_{\tilde{\mathsf{r}}}$ if $\Omega \subset U$), where C depends only on d *and $\tilde{\mathsf{r}}$ is computed with respect to any set containing Ω.*

Remark 7.9. In the case where U is a hyperrectangle, we may reflect the configuration through the boundary, and then periodize it if we would like. Reflecting and periodizing the electric potential ∇u where u solves (7.1.1) provides an electric field E which satisfies (7.1.27) on $\mathbb{R}^\mathsf{d}$. Thanks to the projection lemma, Lemma 7.5, the energy per period of the configuration is then bounded by $\mathsf{F}(X_N, \mu, U)$. We may then apply the results of Lemma 7.8 to the extended configuration. The distance of a point of the original configuration to the boundary of ∂U is then half the distance of that point to its reflection through the boundary, hence is bounded below by (half) the minimal distance in the extended configuration. We deduce in particular that

$$\sum_{i=1}^N \mathsf{g}(\hat{\mathsf{r}}_i) \le C \left(\mathsf{F}(X_N, \mu, U) + CN \right), \tag{7.1.39}$$

and

$$\int_U |\nabla u_{\hat{\mathsf{r}}}|^2 \le C \left(\mathsf{F}(X_N, \mu, U) + CN \right). \tag{7.1.40}$$

PROOF OF LEMMA 7.8. The proof is an adaptation of those of Chapter 4, the main idea being to work with the electric field ∇u extended by 0 which brings us back to a full space situation.

Step 1: Monotonicity. The proof of (7.1.33) is the same as the proof Lemma 4.13 in terms of electric fields (and at the blown-up scale) with $\nabla u_{\vec\eta}$ replaced by $E_{\vec\eta}$ and $\nabla u_{\vec\alpha}$ by $E_{\vec\alpha}$. We note that the proof of that lemma only relies on the fact that $\nabla u_{\vec\eta} - \nabla u_{\vec\alpha} = \sum_{i\in I_N} \nabla \mathsf{f}_{\alpha_i,\eta_i}(z-x_i)$ which is still true for $E_{\vec\eta} - E_{\vec\alpha}$ and on the fact that $-\mathrm{div}\,\nabla u_{\vec\alpha} = \mathsf{c_d}\left(\sum_{i=1}^N \delta_{x_i}^{(\alpha)} - N\mu\right)$ (and the same with $\vec\eta$) which is replaced by $-\mathrm{div}\, E_{\vec\alpha} = \mathsf{c_d}\left(\sum_{i=1}^N \delta_{x_i}^{(\alpha)} - \mu\right)$ from (7.1.32), (7.1.26) and (4.1.22). We thus arrive at (7.1.33) without other changes than removing the N factors in front of μ.

Step 2: Rewriting of $\mathsf{F}^\Omega(X_N,\mu,U)$. As in the proof of Corollary 7.7, let us denote $E = (\nabla u)\mathbf{1}_U$, where u is the solution of (7.1.1) used in the definition of $\mathsf{F}(X_N,\mu,U)$ in (7.1.22). Again the crucial point is that thanks to the Neumann boundary condition, E solves (7.1.31) in the full space. From the definition (7.1.23), since $\mathsf{f}_{\hat{\mathsf{r}}_i}(x-x_i)$ is supported in $B(x_i,\hat{\mathsf{r}}_i)\subset U$, we may rewrite

$$\mathsf{F}^\Omega(X_N,\mu,U) = \frac{1}{2\mathsf{c_d}}\left(\int_{\Omega\cap U} |E_{\hat{\mathsf{r}}}|^2 - \mathsf{c_d}\sum_{i\in I_\Omega} \mathsf{g}(\hat{\mathsf{r}}_i)\right) - \sum_{i\in I_\Omega}\int_{\mathbb{R}^\mathsf{d}} \mathsf{f}_{\hat{\mathsf{r}}_i}(x-x_i)d\mu(x).$$

By the result (7.1.33), equality case, applied to E in Ω, if $\vec\alpha$ is such that $\alpha_i \le \hat{\mathsf{r}}_i$ with equality if $\mathrm{dist}(x_i,\partial\Omega)\le \alpha_i$, we have

(7.1.41)
$$\mathsf{F}^\Omega(X_N,\mu,U) = \frac{1}{2\mathsf{c_d}}\left(\int_{\Omega\cap U} |E_{\vec\alpha}|^2 - \mathsf{c_d}\sum_{i\in I_\Omega} \mathsf{g}(\alpha_i)\right) - \sum_{i\in I_\Omega}\int_{\mathbb{R}^\mathsf{d}} \mathsf{f}_{\alpha_i}(x-x_i)d\mu(x).$$

Step 3: Control of small scale interactions and of minimal distances. We can then copy without change the proof of Lemma 4.29 and Proposition 4.28, with $\mathcal{F}^{\vec\alpha}$ the quantity in the right-hand side of (7.1.41). The main differences are that we replace ∇h_N by E, remove the N factors in front of μ and replace λ by 1. The crucial point is that we may increase the balls from the radii $\hat{\mathsf{r}}_i$ to $\tilde{\mathsf{r}}_i$ and possibly overlap ∂U, as long as we do not overlap $\partial\Omega$, and then reinsert the controls of $\mathsf{g}(\tilde{\mathsf{r}}_i)$ for points near $\partial\Omega$ by absorbing them into the additive errors. We thus obtain (7.1.36), (7.1.34), and (7.1.35).

Step 4: Superadditivity for restriction: proof of (7.1.37)**.** In view of (7.1.33), changing the radii from $\hat{\mathsf{r}}_i$ of (7.1.21) to $\hat{\mathsf{r}}_i$ relative to Ω in (7.1.24) can only decrease the computed value of F. Splitting then $\int_U |\nabla u_{\hat{\mathsf{r}}}|^2$ into $\int_{U\cap\Omega} |\nabla u_{\hat{\mathsf{r}}}|^2 + \int_{U\cap\Omega^c} |\nabla u_{\hat{\mathsf{r}}}|^2$, we deduce the result in view of the definition (7.1.23).

Step 5: Proof of (7.1.38)**.** The proof is similar to that of (4.3.1). Integrating (7.1.1) against φ and using Green's formula, we have

$$\left|\int_\Omega \varphi\, d\Big(\sum_{i=1}^N \delta_{x_i} - \mu\Big)\right| = \frac{1}{\mathsf{c_d}}\left|\int_{\Omega\cap U} \nabla u\cdot\nabla\varphi\right|. \tag{7.1.42}$$

We may also bound, using (4.1.27) and $\tilde{\mathsf{r}}_i \le 1$,

$$\left|\int_\Omega \nabla\varphi\cdot\nabla(u-u_{\tilde{\mathsf{r}}})\right| \le \|\nabla\varphi\|_{L^\infty(\Omega)}\sum_{i\in I_\Omega}\|\nabla\mathsf{f}_{\tilde{\mathsf{r}}_i}\|_{L^1} \le C\|\nabla\varphi\|_{L^\infty(\Omega)}\#I_\Omega.$$

Meanwhile, by Cauchy-Schwarz,

$$\frac{1}{\mathsf{c_d}}\left|\int_{\Omega\cap U}\nabla u_{\tilde{\mathsf{r}}}\cdot\nabla\varphi\right|\le\frac{1}{\mathsf{c_d}}\|\nabla\varphi\|_{L^2(\Omega\cap U)}\|\nabla u_{\tilde{\mathsf{r}}}\|_{L^2(\Omega\cap U)}.$$

Combining the above we obtain the result. The proof of the relation in terms of v is the same. □

7.1.5. Free energy bounds. In this subsection, we show bounds on the free energy for the Neumann energy analogous to those obtained in Section 5.2.3. The upper bounds follow easily from (7.1.30) and the lower bounds will follow analogously to those of Section 5.2.3 from the subadditivity combined with the following lemma. This is a much simpler proof than that of [**AS21**], which allows to get rid of technical deterioration of ρ_β for $\mathsf{d}\ge 5$ in that paper.

Lemma 7.10 (Green's function representation of the Neumann electric energy). *Let U be an open subset of $\mathbb{R}^{\mathsf{d}}$ with bounded and piecewise C^1 boundary and μ a bounded nonnegative density such that $\mu(U)=N$ is an integer. Let G_U solve*

$$\begin{cases}-\Delta_x G_U(x,y)=\mathsf{c_d}(\delta_y(x)-\frac{1}{\mu(U)}\mu(x)) & \text{in } U\\ \frac{\partial G_U}{\partial\nu}=0 & \text{on } \partial U\\ \nabla G_U\to 0 & \text{as } |x|\to\infty,\end{cases}\tag{7.1.43}$$

and let

$$H_U(x,y)=G_U(x,y)-\mathsf{g}(x-y).\tag{7.1.44}$$

Then for any configuration X_N of points in U, we have

(7.1.45)

$$\begin{aligned}\mathsf{F}(X_N,\mu,U)=\frac12\iint_{(\mathbb{R}^{\mathsf{d}})^2\setminus\triangle}&\mathsf{g}(x-y)d\left(\sum_{i=1}^N\delta_{x_i}-\mu\mathbf{1}_U\right)(x)d\left(\sum_{i=1}^N\delta_{x_i}-\mu\mathbf{1}_U\right)(y)\\&+\frac12\iint_{U\times U}H_U(x,y)d\left(\sum_{i=1}^N\delta_{x_i}-\mu\right)(x)d\left(\sum_{i=1}^N\delta_{x_i}-\mu\right)(y).\end{aligned}$$

Proof. We check that $G_U(\cdot,y)$ and $H_U(\cdot,y)$ are well-defined up to additive constants, and that H_U is continuous. Arguing as in (2.2.13), we have that $\mathsf{g}*\mu$ is well-defined as an L^1_{loc} function, thus for a.e. y, we may consider $v=\mathsf{g}*(\delta_y-\frac{1}{\mu(U)}\mu)$ and solve for $w=G_U-v$ which satisfies

$$\begin{cases}\Delta w=0 & \text{in } U\\ \frac{\partial w}{\partial\nu}=-\frac{\partial v}{\partial\nu} & \text{on } \partial U\\ \nabla w\to 0 & \text{as } |x|\to\infty.\end{cases}\tag{7.1.46}$$

This can be done variationally since ∂U is bounded. Since G_U is defined up to an additive constant, we may in addition require that

$$\forall y\in U,\ \int_U G_U(x,y)d\mu(x)=0.\tag{7.1.47}$$

We may prove in the standard way that G_U is symmetric, hence so is H_U, by writing

$$G_U(y_0,x_0)-G_U(x_0,y_0) = \int_U G_U(x,x_0)\left(-\frac{1}{\mathsf{c_d}}\Delta G_U(x,y_0)+\frac{\mu(x)}{\mu(U)}\right) - G_U(x,y_0)\left(-\frac{1}{\mathsf{c_d}}\Delta G_U(x,x_0)+\frac{\mu(x)}{\mu(U)}\right)dx=0 \tag{7.1.48}$$

after using Green's theorem, the Neumann boundary condition and (7.1.47).

We may then observe that the function u of (7.1.1) satisfies

$$u(x)=\int_U G_U(x,y)d\left(\sum_{i=1}^N \delta_{x_i}-\mu\right)(y). \tag{7.1.49}$$

Inserting this into (7.1.22), using Green's theorem, the boundary condition in (7.1.1) and the fact that the balls $B(x_i,\hat{\mathsf{r}}_i)$ are included in U, we find

$$\begin{aligned}\mathsf{F}(X_N,\mu,U) &= -\frac{1}{2\mathsf{c_d}}\int_U u_{\hat{\mathsf{r}}}\Delta u_{\hat{\mathsf{r}}} - \frac{1}{2}\sum_{i=1}^N \mathsf{g}(\hat{\mathsf{r}}_i) - \sum_{i=1}^N\int_U \mathsf{f}_{\hat{\mathsf{r}}_i}(x-x_i)d\mu(x)\\ &= \frac{1}{2}\iint_{U\times U} G_U(x,y)d\Big(\sum_{i=1}^N \delta_{x_i}^{(\hat{\mathsf{r}}_i)}-\mu\Big)(x)d\Big(\sum_{i=1}^N \delta_{x_i}^{(\hat{\mathsf{r}}_i)}-\mu\Big)(y)\\ &\quad -\frac{1}{2}\sum_{i=1}^N \mathsf{g}(\hat{\mathsf{r}}_i) - \sum_{i=1}^N\int_U \mathsf{f}_{\hat{\mathsf{r}}_i}(x-x_i)d\mu(x).\end{aligned} \tag{7.1.50}$$

Splitting G_U as $\mathsf{g}(x-y)+H_U(x,y)$, we then obtain

$$\begin{aligned}\mathsf{F}(X_N,\mu,U) &= \frac{1}{2}\iint_{U\times U} \mathsf{g}(x-y)d\Big(\sum_{i=1}^N \delta_{x_i}^{(\hat{\mathsf{r}}_i)}-\mu\Big)(x)d\Big(\sum_{i=1}^N \delta_{x_i}^{(\hat{\mathsf{r}}_i)}-\mu\Big)(y)-\frac{1}{2}\sum_{i=1}^N \mathsf{g}(\hat{\mathsf{r}}_i)\\ &+\frac{1}{2}\iint_{U\times U} H_U(x,y)d\Big(\sum_{i=1}^N \delta_{x_i}^{(\hat{\mathsf{r}}_i)}-\mu\Big)(x)d\Big(\sum_{i=1}^N \delta_{x_i}^{(\hat{\mathsf{r}}_i)}-\mu\Big)(y)-\sum_{i=1}^N\int_U \mathsf{f}_{\hat{\mathsf{r}}_i}(x-x_i)d\mu(x).\end{aligned} \tag{7.1.51}$$

For the first line we use the fact that the balls $B(x_i,\hat{\mathsf{r}}_i)$ are disjoint, the harmonicity of g away from 0, and the fact that $\mathsf{g}*\delta_0^{(\eta)}=\mathsf{g}_\eta$ as seen in Section 4.1.2, to write

$$\int \mathsf{g}(x-y)d\delta_{x_i}^{(\hat{\mathsf{r}}_i)}(x)d\delta_{x_i}^{(\hat{\mathsf{r}}_i)}(y)=\mathsf{g}(\hat{\mathsf{r}}_i),$$

$$\int \mathsf{g}(x-y)d\delta_{x_i}^{(\hat{\mathsf{r}}_i)}(x)d\delta_{x_j}^{(\hat{\mathsf{r}}_j)}(y)=\mathsf{g}(x_i-x_j) \text{ if } i\neq j.$$

We thus obtain that

$$(7.1.52)\quad \frac{1}{2}\iint_{U\times U} \mathsf{g}(x-y)d\Big(\sum_{i=1}^N \delta_{x_i}^{(\hat{\mathsf{r}}_i)}-\mu\Big)(x)d\Big(\sum_{i=1}^N \delta_{x_i}^{(\hat{\mathsf{r}}_i)}-\mu\Big)(y)-\frac{1}{2}\sum_{i=1}^N \mathsf{g}(\hat{\mathsf{r}}_i)$$
$$=\frac{1}{2}\iint_{U\times U\setminus\triangle} \mathsf{g}(x-y)d\Big(\sum_{i=1}^N \delta_{x_i}-\mu\Big)(x)d\Big(\sum_{i=1}^N \delta_{x_i}-\mu\Big)(y)$$
$$-\iint_{U\times U\setminus\triangle} \mathsf{g}(x-y)d\Big(\sum_{i=1}^N \delta_{x_i}^{(\hat{\mathsf{r}}_i)}-\delta_{x_i}\Big)(x)d\mu(y).$$

Recalling (see (4.1.12)) that $\mathsf{g}*\delta_0^{(\eta)}-\mathsf{g}=\mathsf{g}_\eta-\mathsf{g}=-\mathsf{f}_\eta$, we may rewrite the second line in the right-hand side as $\sum_{i=1}^N\int_U \mathsf{f}_{\hat{\mathsf{r}}_i}(x-x_i)d\mu(x)$, and we have thus established that

$$(7.1.53)\quad \frac{1}{2}\iint_{U\times U} \mathsf{g}(x-y)d\Big(\sum_{i=1}^N \delta_{x_i}^{(\hat{\mathsf{r}}_i)}-\mu\Big)(x)d\Big(\sum_{i=1}^N \delta_{x_i}^{(\hat{\mathsf{r}}_i)}-\mu\Big)(y)$$
$$-\frac{1}{2}\sum_{i=1}^N \mathsf{g}(\hat{\mathsf{r}}_i)-\sum_{i=1}^N\int_U \mathsf{f}_{\hat{\mathsf{r}}_i}(x-x_i)d\mu(x)$$
$$=\frac{1}{2}\iint_{U\times U\setminus\triangle} \mathsf{g}(x-y)d\Big(\sum_{i=1}^N \delta_{x_i}-\mu\Big)(x)d\Big(\sum_{i=1}^N \delta_{x_i}-\mu\Big)(y).$$

We next turn to the second line of (7.1.51) and rewrite it as

$$(7.1.54)\quad \frac{1}{2}\iint_{U\times U} H_U(x,y)d\Big(\sum_{i=1}^N \delta_{x_i}^{(\hat{\mathsf{r}}_i)}-\mu\Big)(x)d\Big(\sum_{i=1}^N \delta_{x_i}^{(\hat{\mathsf{r}}_i)}-\mu\Big)(y)$$
$$=\frac{1}{2}\iint_{U\times U} H_U(x,y)d\Big(\sum_{i=1}^N \delta_{x_i}-\mu\Big)(x)d\Big(\sum_{i=1}^N \delta_{x_i}-\mu\Big)(y)$$
$$+\frac{1}{2}\iint_{U\times U} H_U(x,y)d\Big(\sum_{i=1}^N \delta_{x_i}^{(\hat{\mathsf{r}}_i)}-\delta_{x_i}\Big)(x)d\Big(\sum_{i=1}^N \delta_{x_i}^{(\hat{\mathsf{r}}_i)}+\delta_{x_i}-2\mu\Big)(y).$$

Recognizing $\delta_{x_i}^{(\hat{\mathsf{r}}_i)}-\delta_{x_i}$ as the Laplacian of $\mathsf{f}_{\hat{\mathsf{r}}_i}(x-x_i)$ from (4.1.13) and integrating by parts in x, using that $-\Delta H_U(x,y)=-\mathsf{c_d}\frac{\mu(x)}{\mu(U)}$ and that $\mathsf{f}_{\hat{\mathsf{r}}_i}(x-x_i)=0$ on ∂U by definition of $\hat{\mathsf{r}}_i$, we may rewrite the last line as

$$-\iint_{U\times U}\frac{\mu(x)}{\mu(U)}\left(\sum_{i=1}^N \mathsf{f}_{\hat{\mathsf{r}}_i}(x-x_i)\right)d\Big(\sum_{i=1}^N \delta_{x_i}^{(\hat{\mathsf{r}}_i)}+\delta_{x_i}-2\mu\Big)(y)=0,$$

by separation of variables, since $\mu(U)=N$. Inserting into (7.1.54) and combining with (7.1.53) and (7.1.51), we have obtained the result. □

We next turn to the lower bound of $\log \mathsf{K}_\beta(\mu,U)$, defined in (7.1.13).

Proposition 7.11 (Neumann free energy bound). *Let U be an open subset of $\mathbb{R}^{\mathsf{d}}$ with bounded and piecewise C^1 boundary and μ a bounded nonnegative density such that $\mu(U)=N$ is an integer. If $\mathsf{s}\le 0$ and U is unbounded, assume in addition that* (5.2.28), (5.2.29) *and* (5.2.37) *hold. Then*

$$(7.1.55)\qquad |\log \mathsf{K}_\beta(\mu,U)|\le C\beta\chi(\beta)N$$

where χ *is as in* (5.2.27), *and* $C > 0$ *depends only on* $\mathsf{d}, \mathsf{s}, \|\mu\|_{L^\infty}$ *and the constants in the assumptions.*

PROOF. The upper bound is straightforward from the definition (7.1.13) and (7.1.30). For the lower bound, we follow the steps of the proofs of Lemma 5.12 and Proposition 5.14. Assume first that $\mathsf{s} > 0$. Starting from (7.1.13) and using Jensen's inequality, we may then write

$$\log \mathsf{K}_\beta(\mu, U) \geq -\frac{\beta}{(\mu(U))^N} \int_{U^N} \mathsf{F}(X_N, \mu, U) d\mu^{\otimes N}(X_N).$$

We next insert (7.1.45) and argue as in (5.2.25) to obtain

$$\int_{U^N} \mathsf{F}(X_N, \mu, U) d\mu^{\otimes N}(X_N) = -\frac{N^{N-1}}{2} \iint_{U^2} \mathsf{g}(x-y) d\mu(x) d\mu(y), \tag{7.1.56}$$

after observing that the terms corresponding to the integrals of $H_U(x, y)$ cancel out. When $\mathsf{s} > 0$, this yields $\log \mathsf{K}_\beta(U, \mu) \geq 0$ hence the desired result.

In the situation where $\mathsf{s} \leq 0$, we may argue exactly as in the proof of Proposition 5.14: first we partition Λ into cells Q_i of size $O(R)$ with $\int_{Q_i} \mu = n_i$, then we replace the use of (5.2.39) by (7.1.17). Finally when integrating $\mathsf{F}(X_N, \mu, Q_i)$ within each cell, we notice that the contribution of the terms containing H_{Q_i} in (7.1.45) cancel after integration, so we are left with simply the terms in $\mathsf{g}(x-y)$, and

$$\log \mathsf{K}_\beta(\mu, U) \geq \frac{\beta}{2} \sum_{i=0}^{p} \frac{1}{n_i} \iint \mathsf{g}(x-y) d\mu_i(x) d\mu_i(y) + O(\log n_i)$$

and we finish exactly as in the proof of Proposition 5.14. □

Using (7.1.30), we have a lower bound for $\min \mathsf{F}(\cdot, \mu, U)$, and from (7.1.55) applied to, say, $\beta = 1$ we get that the exists a configuration such that a converse inequality holds, thus we have the following.

Corollary 7.12 (A priori bounds for minimizers). *Let* U *and* μ *be as in Proposition* 7.11, *under the same assumptions, defining*

$$\mathsf{E}_\infty(\mu, U) := \min_{X_N} \mathsf{F}(X_N, \mu, U), \tag{7.1.57}$$

we have

$$|\mathsf{E}_\infty(\mu, U)| \leq CN, \tag{7.1.58}$$

where $C > 0$ *depends only on* $\mathsf{d}, \mathsf{s}, \|\mu\|_{L^\infty}$ *and the constants in the assumptions.*

In view of (7.1.28) we also have the subbaditivity property: if U is the disjoint union of U_1 and U_2,

$$\mathsf{E}_\infty(\mu, U) \leq \mathsf{E}_\infty(\mu, U_1) + \mathsf{E}_\infty(\mu, U_2). \tag{7.1.59}$$

7.2. The screening procedure

7.2.1. Motivation and heuristics. Since Neumann boundary conditions allow to compute the energy subadditively over disjoint sets, this motivates the screening procedure, first introduced in [**SS12**] using ideas of [**ACO09**]. It consists in taking an arbitrary configuration in a cube (or rectangle) or its complement, with reasonably well controlled energy, and producing from it a configuration with Neumann boundary condition solved. The issue is to do so without changing the configuration too much—it will be possible to modify the configuration only in a boundary layer—and without adding too much energy. During the procedure, points will be deleted, the total number of points will change so as to achieve neutrality, and this modifies the phase-space volume of the family of configurations. We will also show that this change of volume can be well-controlled, which allows to give a probabilistic screening (or screening at the level of Gibbs measures): given a family of configurations we will produce a whole family of configurations, which will allow, roughly, to bound L_β in terms of K_β.

Remark 7.13. The two main situations we need to treat are:

(1) the case where $U = \mathbb{R}^{\mathsf{d}}$ and μ is a positive density with compact support or with convergent tails and $\int_{\mathbb{R}^{\mathsf{d}}} \mu = N$. We will denote Λ a set where $\mu \geq m > 0$ for some positive constant m (either the support of μ, or its "essential support" if it has tails).
(2) the case where U is a hyperrectangle (or disjoint union of hyperrectangles), we consider the Neumann partition function on U, and μ is a uniform density, say 1. Then we take $\Lambda = \mathbb{R}^{\mathsf{d}}$.

To perform the screening, we need to be in a region where the background density μ is not too small, because it is the negative background charge $-\mu$ which is used to neutralize and screen possible charge excesses. Thus, in the first case above, screening will only be possible in "the bulk," i.e., a region where μ is bounded from below by some constant $m > 0$ (and the error estimates will depend on m). Truly, we will have to be a positive (small) distance away from the set $\{\mu < m\}$ for results to hold. In the second case $\mu = 1$ up to the boundary, and the Neumann boundary condition holds there, the screening will preserve the Neumann boundary condition and there will be no difficulty in applying the screening procedure up to the boundary.

We have to perform two variants of the screening: an "inner screening" when $\Omega = Q_R$ and an "outer screening" when $\Omega = U \backslash Q_R$. Both are entirely parallel, so we present below the inner screening. Here Q_R is a hyperrectangle with sidelengths in $[R, 2R]$, and for any $t \in \mathbb{R}$ we will denote by Q_{R+t} the hyperrectangle of same center as Q_R and sidelengths increased by t.

The set Ω needs to be "quantized" in the sense that $\mu(\Omega)$ is an integer, equal to $\bar{\mathrm{n}}$, which is also the optimal number of points. Assume we are given a configuration X_n of n points in $\Omega = Q_R \cap U$, and let u be an associated potential satisfying a relation of the form

$$-\Delta u = \mathsf{c}_{\mathsf{d}} \left(\sum_{i=1}^{n} \delta_{x_i} - \mu \right) \quad \text{in } \Omega$$

for instance u is the function solving (7.1.2), and $E = \nabla u$.

The goal of the (inner) screening is to produce a family of configurations $Y_{\bar{\mathrm{n}}}$ having $\bar{\mathrm{n}} = \mu(\Omega)$ points in Ω, coinciding with X_n except in a boundary layer near $\partial\Omega$, for which we can solve

$$\begin{cases} -\operatorname{div} E^0 = \mathsf{c}_{\mathsf{d}} (\sum_{i=1}^{\bar{\mathrm{n}}} \delta_{y_i} - \mu) & \text{in } \Omega \\ E^0 \cdot \nu = 0 & \text{on } \partial\Omega. \end{cases}$$

The electric field E^0 will also coincide with E except in a boundary layer near $\partial\Omega$. We call $\mathcal{O}$ (like old) the interior set where the configurations X_n and $Y_{\bar{\mathrm{n}}}$ and their electric fields coincide, and by $\mathcal{N}$ (like new) the boundary layer where X_n and E are to be deleted and replaced by configurations that have the correct number of points.

The setup is represented in Figure 7.2.1. A recent feature of the construction, which appeared in [**AS21**], is to sample these new points according to a Coulomb Gibbs measure in $\mathcal{N}$.

We say that the configuration X_n has been screened because the resulting electric field E^0 can be extended to 0 outside Ω without any jump in its normal component, as if the system of point charges plus background charge μ was not generating any field outside Ω. The screened electric field E^0 may not be a gradient, however thanks to Lemma 7.5 its energy provides an upper bound for computing $\mathsf{F}(Y_{\bar{n}}, \Omega)$. The goal of the construction is to show that we can build E^0 and $Y_{\bar{n}}$ without adding too much energy to the original one.

Not all configurations are screenable: configurations that have for instance too many points very near the boundary, and no points inside, cannot be transformed into neutral configurations by changing the configuration in a boundary layer only. A reasonable energy bound rules out such pathological behavior and will be the "screenability condition" (7.2.7).

The boundary layer size depends on two lengthscales of choice ℓ and $\tilde{\ell}$, both $< R$. The lengthscale $\tilde{\ell}$ represents the distance over which one needs to look for a good contour near $\partial\Omega$ by a mean-value argument. For outer screening, this implies that we need a bit of buffer space to perform the screening argument. We will denote by $\bar{S}(X_n)$ the energy of the best screenable electric field $\int |E_{\mathsf{r}}|^2$ in a $\tilde{\ell}$-neighborhood of $\partial\Omega$. By a mean-value argument we find some contour Γ at distance $\leq \tilde{\ell}$ from $\partial\Omega$ so that the trace of $E \cdot \nu$ on Γ, denoted g, satisfies $\int_\Gamma |g|^2 \leq C\bar{S}(X_n)\tilde{\ell}^{-1}$. For the given configuration X_n, we denote by $\mathcal{O}$ the set enclosed by Γ and by $\mathcal{N}$ its complement. We let $n_{\mathcal{O}}$ be the number of points of X_n that belong to $\mathcal{O}$. The part of the configuration X_n which belongs to $\mathcal{N}$ is discarded, to be replaced by a configuration $Z_{\bar{n}-n_{\mathcal{O}}}$, in such a way that $Y_{\bar{n}}$, union of X_n restricted to $\mathcal{O}$ and $Z_{\bar{n}-n_{\mathcal{O}}}$ has exactly $\bar{n}$ points, as desired.

The other lengthscale $\ell \leq \tilde{\ell}$, represents the distance to Γ needed to "absorb" the possibly nonzero boundary condition g and replace it by a vanishing one. That size needs to be large enough depending on $\bar{S}(X_n)\tilde{\ell}^{-1}$ and the screening will only be possible if $\bar{S}(X_n)\tilde{\ell}^{-1}$ is small enough compared to ℓ, this is the screenability condition.

The "absorption" of g and its replacement by a zero Neumann boundary data (or zero boundary normal component for E) is done by splitting $\mathcal{N}$ into cells H_k of sidelength ℓ where we solve appropriate elliptic problems and estimate the energies by elliptic regularity estimates. More precisely the building block is to solve

$$\begin{cases} -\Delta u_k = \mathsf{c_d}\left(\sum \delta_{z_i} - \mu\right) & \text{in } H_k \\ \frac{\partial u_k}{\partial \nu} = \tilde{g} & \text{on } \partial H_k, \end{cases} \tag{7.2.1}$$

where $\tilde{g}$ is equal to g on the parts of H_k which intersect Γ (if any) and 0 otherwise, and z_i are points to be sampled in $\mathcal{N}$, leaving a point-free layer of width η near the boundary of $\mathcal{O}$.

Pasting together the electric fields ∇u_k obtained in each H_k produces an electric field which is compatible with $Z_{\bar{n}-n_{\mathcal{O}}}$, allowing to evaluate the energy F in $\mathcal{N}$ via Lemma 7.5, i.e., bounding it by $\sum_k |\nabla u_k|^2$.

The evaluation of the energy in each cell H_k follows the idea of [**ACO09**] of doing it in two parts by decomposing u into $h_1 + h_2$ where

$$\begin{cases} -\Delta h_1 = \mathsf{c_d}\left(\sum \delta_{x_i} - \mu - m_k\right) & \text{in } H_k \\ \frac{\partial h_1}{\partial \nu} = 0 & \text{on } \partial H_k, \end{cases} \tag{7.2.2}$$

where m_k is a constant chosen so that this problem is solvable and won't be too large (in particular we can keep $\mu + m_k > 0$) as soon as the initial energy is not too large; and

$$\begin{cases} -\Delta h_2 = \mathsf{c_d} m_k & \text{in } H_k \\ \frac{\partial h_2}{\partial \nu} = \tilde{g} & \text{on } \partial H_k \end{cases} \tag{7.2.3}$$

which will absorb the boundary condition. This will amount to replacing in the boundary layer $\mathcal{N}$ the reference measure μ by a modified measure $\tilde{\mu}$. By elliptic estimates, the energy cost is directly related to $\int_\Gamma |g|^2$, itself controlled by $\bar{S}(X_n)\tilde{\ell}^{-1}$. Note that some complications arise due to points that are very close to Γ and whose balls $B(x_i, \mathsf{r}_i)$ cut through Γ.

The energy error term (the sum of the old energy with $\mathsf{F}(Z_{\bar{\mathsf{n}}-n_{\mathcal{O}}}, \mathcal{N})$) can then be bounded by the number of such cells plus $\bar{S}(X_n)\tilde{\ell}^{-1}$. This produces an error on the free energy of order

$$C\beta R^{\mathsf{d}-1}\ell + C\frac{R^{\mathsf{d}-1}}{\ell}.$$

Optimizing this error over $\ell < R$ leads to stopping at $\ell = \frac{1}{\sqrt{\beta}}$ and $R > C\max(1, \frac{1}{\sqrt{\beta}})$. This is how our *minimal lengthscale* ρ_β of order $\max(1, \frac{1}{\sqrt{\beta}})$ appears. When β is small this becomes large, and screening cannot be obtained at smaller lengthscale. In physics, ρ_β corresponds to the *Debye screening length*; see for instance [**AM25**].

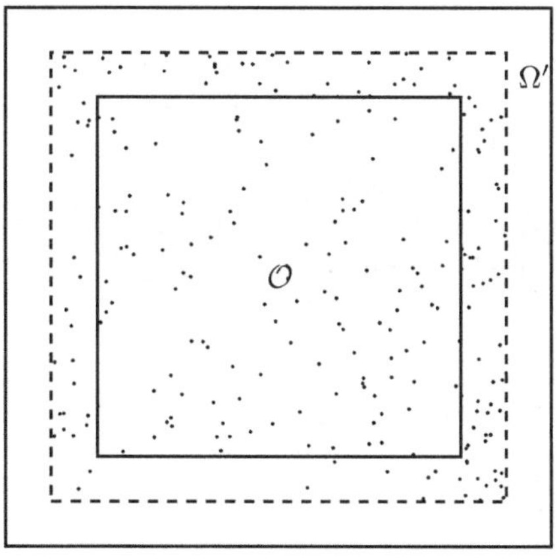

FIGURE 7.2.1. Setup for the screening

7.2.2. Screening statement. The following result can be found in [**AS21**], prior versions are found in [**SS12, RS15, PS17, LS17**].

Let us first start with the geometry and screenability condition.

Definition 7.14 (Quantized hyperrectangles of lengthscale R)**.** We define $\mathcal{Q}_R$ as the set of closed hyperrectangles Q in $\mathbb{R}^{\mathsf{d}}$ whose sidelengths belong to $[R, 2R]$ and are such that $\mu(Q)$ is an integer.

Given such a Q_R, we will abuse notation by denoting (for any $t \le R$) Q_t the hyperrectangle included in Q_R whose boundary is exactly at distance $R - t$ from ∂Q_R.

The presence of the set $\Omega' \subset \Omega$ below is meant to be able to take a configuration in a set Ω', screen it and extend it to a slightly larger set Ω. Since the configuration and its electric field will be completely discarded in a boundary layer near $\partial\Omega'$, it need only satisfy (7.2.4) in a possibly slightly smaller set than Ω', which will be useful later in conjunction with the best screenable energy G_U. By slightly smaller, we mean that the distance between the nested sets Ω'', Ω' and Ω is at most $2\tilde{\ell}$.

There are two variants of the construction: one, used to prove the local laws, starts from the control on the energy in Q_R only, uses the screenability condition (7.2.7) below, and finds a good boundary which is the boundary of a cube. The other, used once local laws are known, starts from the local control of the energy on cubes of size ρ_β to obtain improved error estimates, and uses a good boundary which is piecewise affine in order to have local controls on the boundary, and the screenability condition (7.2.8), applied only with ℓ and $\tilde{\ell}$ (approximately) equal to ρ_β.

Definition 7.15 (Screenability)**.** Assume μ is a bounded nonnegative density in U, which is either $\mathbb{R}^{\mathsf{d}}$, or a finite disjoint union of hyperrectangles with parallel sides, belonging to $\mathcal{Q}_R$ and all included in Λ, a set where $\mu \ge m > 0$; or the complement of such a set. Let ℓ and $\tilde{\ell}$ be such that $R \ge \tilde{\ell} \ge \ell \ge C$ for some $C > 0$ large enough depending only on d, m and $\|\mu\|_{L^\infty}$. Assume Q_R is a hyperrectangle of sidelengths in $[R, 2R]$ with sides parallel to those of U. Let $\Omega = Q_R \cap U$ (inner case) or $\Omega = U \backslash Q_R$ (outer case). In the inner case, assume $\mu \ge m > 0$ in Ω, in the outer case assume $\mu \ge m > 0$ in $Q_{R+\tilde{\ell}} \backslash Q_R$. In both cases, assume that $\mu(\Omega) = \bar{\mathrm{n}}$, an integer. In the case where U is a disjoint union of hyperrectangles with parallel sides, also assume that the faces of ∂Q_R are at distance $\ge 2\tilde{\ell}$ from their respective parallel faces of ∂U.

Let $\Omega'' \subset \Omega' \subset \Omega$ and $Q_{R-\tilde{\ell}} \cap U \subset \Omega''$ in the inner case, respectively $U \backslash Q_{R+\tilde{\ell}} \subset \Omega''$ in the outer case. Let X_n be a configuration of points in Ω' and assume that w solves

$$
\begin{cases} -\Delta w = \mathsf{c_d} \left(\sum_{i=1}^n \delta_{x_i} - \mu \right) & \text{in } \Omega'' \\ \frac{\partial w}{\partial \nu} = 0 & \text{on } \partial U \cap \Omega''. \end{cases} \tag{7.2.4}
$$

We denote if $\Omega = Q_R \cap U$,

$$
S(X_n, w) = \int_{(Q_{R-\tilde{\ell}} \backslash Q_{R-2\tilde{\ell}}) \cap U} |\nabla w_{\hat{\mathsf{r}}}|^2, \tag{7.2.5}
$$

$$
S'(X_n, w) = \sup_x \int_{(Q_{R-\tilde{\ell}} \backslash Q_{R-2\tilde{\ell}}) \cap \square_\ell(x) \cap U} |\nabla w_{\hat{\mathsf{r}}}|^2,
$$

respectively if $\Omega = U \backslash Q_R$,

$$S(X_n, w) = \int_{(Q_{R+2\tilde{\ell}} \backslash Q_{R+\tilde{\ell}}) \cap U} |\nabla w_{\hat{\mathsf{r}}}|^2, \tag{7.2.6}$$

$$S'(X_n, w) = \sup_x \int_{(Q_{R+2\tilde{\ell}} \backslash Q_{R+\tilde{\ell}}) \cap \square_\ell(x) \cap U} |\nabla w_{\hat{\mathsf{r}}}|^2,$$

where $\hat{\mathsf{r}}$ is defined as in (7.1.24). We say that w is inner screenable (resp. outer screenable) if either

$$\ell^{\mathsf{d}+1} \geq C \frac{S(X_n, w)}{\tilde{\ell}}, \tag{7.2.7}$$

or

$$\ell^{\mathsf{d}+1} \geq C S'(X_n, w), \tag{7.2.8}$$

for some $C > 0$ depending only on d, m and $\|\mu\|_{L^\infty}$.

We can now define the best screenable potential and its energy. There are two complications in the definition compared to the formal definition of the Dirichlet energy in Section 1.2. The first one is that we have to retain the Neumann boundary condition on ∂U when we treat the case of Neumann partition functions (as in item (2) of Remark 7.13). The second is that we have to add to the definition the condition of being screenable. During the bootstrap proof of the local laws, the bounds on the energy obtained at prior scales are sufficient to ensure inner screenability in small boxes, however we cannot know for certain that outer screenability also holds on the sole basis of these energy controls (because the needed screenability is in terms of the boundary size of $U \backslash Q_R$ which when R is relatively small is much smaller than the volume of $U \backslash Q_R$).

We may now define the appropriate notion of the quantity G.

Definition 7.16 (Best screenable potential and energy). With the same notation as above and $\hat{\mathsf{r}}$ as in (7.1.24), given a configuration X_n of points in Ω, we let

(7.2.9)

$$\mathsf{G}_U^{\mathrm{inn}}(X_n, \mu, \Omega) = \min \Bigg\{ \frac{1}{2\mathsf{c_d}} \left(\int_\Omega |\nabla w_{\hat{\mathsf{r}}}|^2 - \mathsf{c_d} \sum_{i=1}^n \mathsf{g}(\hat{\mathsf{r}}_i) \right) - \sum_{i=1}^n \int_\Omega \mathsf{f}_{\hat{\mathsf{r}}_i}(x - x_i) d\mu(x),$$

$$\text{s.t. } w \text{ is inner screenable satisfying a relation of the form}$$

$$\begin{cases} -\Delta w = \mathsf{c_d} \left(\sum_{i=1}^n \delta_{x_i} - \mu + \sum_j \delta_{x_j}^{(\eta_j)} \right) & \text{in } \Omega \cap U \\ \frac{\partial w}{\partial \nu} = 0 & \text{on } \partial U \cap \partial \Omega, \end{cases}$$

$$\text{where } x_j \notin \Omega,\ \eta_j \leq \frac{1}{4} \min(1, \mathrm{dist}(x_j, \partial U)),\ w_{\hat{\mathsf{r}}} = w - \sum_{i=1}^n \mathsf{f}_{\hat{\mathsf{r}}_i}(\cdot - x_i), \hat{\mathsf{r}} \text{ as in (7.1.24)} \Bigg\},$$

respectively,

(7.2.10)
$$\mathsf{G}_U^{\mathrm{out}}(X_n,\mu,\Omega)=\min\left\{\frac{1}{2\mathsf{c_d}}\left(\int_\Omega|\nabla w_{\hat{\mathsf{r}}}|^2-\mathsf{c_d}\sum_{i=1}^n\mathsf{g}(\hat{\mathsf{r}}_i)\right)-\sum_{i=1}^n\int_\Omega \mathsf{f}_{\hat{\mathsf{r}}_i}(x-x_i)d\mu(x),\right.$$
$$\left.\text{s.t. } w \text{ is outer screenable satisfying a relation of the same form in } \Omega\right\}.$$

By the direct method in the calculus of variations, one may check that the minima are achieved. We also define
(7.2.11)
$$\bar{S}(X_n)=\inf\{S(X_n,w), w \text{ achieving the min in } \mathsf{G}_U^{\mathrm{inn}}(X_n,\Omega), \text{ resp. } \mathsf{G}_U^{\mathrm{out}}(X_n,\mu,\Omega)\}$$
(7.2.12)
$$\bar{S}'(X_n)=\inf\{S'(X_n,w), w \text{ achieving the min in } \mathsf{G}_U^{\mathrm{inn}}(X_n,\Omega), \text{ resp. } \mathsf{G}_U^{\mathrm{out}}(X_n,\mu,\Omega)\}.$$

Remark 7.17. The introduction of $\sum_j \delta_{x_j}^{(\eta_j)}$ in the equation solved by w is meant to account for the possibility of points outside Ω whose smeared charges could overlap Ω. That additional charge can only be supported in a layer of distance $\le \frac14$ from $\partial\Omega$, so that w solves (7.2.4) in $\Omega''=\Omega\backslash\{x,\mathrm{dist}(x,U\backslash\Omega)\le\frac14\}$, which allows to apply the screening procedure to w in Ω. Note also that the radii $\hat{\mathsf{r}}_i$ depend only on the points in Ω, i.e., are determined by X_n only, so $\mathsf{G}_U^{\mathrm{inn/out}}$ is a well-defined function of X_n.

When u is inner/outer screenable, it is a competitor in the definition of $\mathsf{G}_U^{\mathrm{inn/out}}$, thus we have the following.

Lemma 7.18. *Let u be the solution of* (7.1.1) *used in the definition of* (7.1.23)*. If u is inner screenable, resp. outer screenable, then*

(7.2.13)
$$\mathsf{F}^\Omega(X_N,\mu,U)\ge \mathsf{G}_U^{\mathrm{inn/out}}(X_N|_\Omega,\mu,\Omega).$$

The following proposition is the main screening result, it is proved in the appendix.

Proposition 7.19 (Screening). *Let us use the same assumptions and notation as in Definition* 7.15*. There exists $C>5$ depending only on d, m, and $\|\mu\|_{L^\infty}$ such that the following holds. Let X_n be a configuration of points in Ω' and let w solve* (7.2.4) *in $\Omega''\subset\Omega'$ and be screenable via* (7.2.7) *or* (7.2.8)*, with $\ell\ge C$ large enough. There exists a set $\mathcal{O}$ such that $Q_{R-2\tilde{\ell}}\cap U\subset\mathcal{O}\subset Q_{R-\tilde{\ell}}\cap U$ in the case of inner screening (resp. $U\backslash Q_{R+2\tilde{\ell}}\subset\mathcal{O}\subset U\backslash Q_{R+\tilde{\ell}}$ for outer screening), a subset $I_\partial\subseteq\{1,\dots,n\}$ and a positive density $\tilde{\mu}$ supported in $\mathcal{N}:=\Omega\backslash\mathcal{O}$ (all depending on X_n) such that the following holds:*

- *$n_{\mathcal{O}}$ being the number of points of X_n such that $B(x_i,\hat{\mathsf{r}}_i)$ intersects $\mathcal{O}$, we have*

(7.2.14)
$$\tilde{\mu}(\mathcal{N})=\bar{\mathrm{n}}-n_{\mathcal{O}},\qquad |\mu(\mathcal{N})-\tilde{\mu}(\mathcal{N})|\le C\left(R^{\mathsf{d}-1}+\frac{S(X_n,w)}{\tilde{\ell}}\right),$$

(7.2.15)
$$\|\mu-\tilde{\mu}\|_{L^\infty(\mathcal{N})}\le\frac{m}{2},\qquad \int_{\mathcal{N}}(\tilde{\mu}-\mu)^2\le C\frac{S(X_n,w)}{\ell\tilde{\ell}}+C\ell^{-1}R^{\mathsf{d}-1};$$

- *we have* $\#I_\partial \le C\frac{S(X_n,w)}{\ell}$;
- *for any configuration* $Z_{\bar{\mathrm{n}}-n_{\mathcal{O}}}$ *of* $\bar{\mathrm{n}}-n_{\mathcal{O}}$ *points in* $\mathcal{N}$, *the configuration* $Y_{\bar{\mathrm{n}}}$ *in* Ω *equal to the union of the points* x_i *of* X_n *such that* $B(x_i,\hat{\mathrm{r}}_i)$ *intersects* $\mathcal{O}$ *and the points* z_i *of* $Z_{\bar{\mathrm{n}}-n_{\mathcal{O}}}$ *(which thus has* $\bar{\mathrm{n}}$ *points) satisfies*

$$\mathsf{F}(Y_{\bar{\mathrm{n}}},\mu,\Omega) \le \frac{1}{2\mathsf{c_d}}\left(\int_{\Omega'}|\nabla w_{\hat{\mathrm{r}}}|^2 - \mathsf{c_d}\sum_{i=1}^n \mathsf{g}(\hat{\mathrm{r}}_i)\right) - \sum_{i=1}^n\int_{\Omega'}\mathsf{f}_{\hat{\mathrm{r}}_i}(x-x_i)d\mu(x)$$
$$+C\left(\frac{\ell S(X_n,w)}{\tilde{\ell}} + R^{\mathsf{d}-1}\tilde{\ell} + \mathsf{F}(Z_{\bar{\mathrm{n}}-n_{\mathcal{O}}},\tilde{\mu},\mathcal{N}) + |\bar{\mathrm{n}}-n| + \sum_{(i,j)\in J}\mathsf{g}(x_i-z_j)\right), \tag{7.2.16}$$

where the index set $J=J(X_n)$ *in the sum is given by*

$$J := \{(i,j)\in I_\partial\times\{1,\dots,\bar{\mathrm{n}}-n_{\mathcal{O}}\} : |x_i-z_j|\le \hat{\mathrm{r}}_i\}.$$

Moreover, if (7.2.7) is used, we can take $\mathcal{O}$ *to be equal to* $Q_t\cap U$ *for some* $t\in[R-2\tilde{\ell},R-\tilde{\ell}]$ *(resp.* $U\backslash Q_t$ *for some* $t\in[R+\tilde{\ell},R+2\tilde{\ell}]$*) and we can make* t *a measurable function of* X_n, *while if (7.2.8) is used, the boundary of* $\mathcal{O}$ *is in general piecewise affine and made of facets parallel to the faces of* Q_R *of sidelengths bounded above and below by constants times* ℓ, *all included in some* $Q_{t+\ell}\backslash Q_t$ *for* $t\in[R-2\tilde{\ell},R-\tilde{\ell}-\ell]$, *resp.* $t\in[R+\tilde{\ell},R+2\tilde{\ell}-\ell]$.

Applying the proposition to w achieving the min in (7.2.9), resp. (7.2.10), we directly obtain the upper bound

$$\mathsf{F}(Y_{\bar{\mathrm{n}}},\mu,\Omega) - \mathsf{G}_U^{\mathrm{inn/out}}(X_n,\mu,\Omega')$$
$$\le C\left(\frac{\ell S(X_n,w)}{\tilde{\ell}} + R^{\mathsf{d}-1}\tilde{\ell} + \mathsf{F}(Z_{\bar{\mathrm{n}}-n_{\mathcal{O}}},\tilde{\mu},\mathcal{N}) + |n-\bar{\mathrm{n}}| + \sum_{(i,j)\in J}\mathsf{g}(x_i-z_j)\right) \tag{7.2.17}$$

with the right-hand side equal to the various screening errors (boundary energy, new added energy, interactions of points near the boundary of $\mathcal{O}$).

Once this result is established one may tune the parameters $\ell,\tilde{\ell}$ to obtain the best results. For instance, at the beginning we may only know that $\int_{Q_R}|\nabla w_{\hat{\mathrm{r}}}|^2$ is bounded by $O(R^{\mathsf{d}})$, we then bound $S(X_n,w)$ and $S'(X_n,w)$ by $O(R^{\mathsf{d}})$, optimize the right-hand side of (7.2.16) and choose $\ell\le\tilde{\ell}$ satisfying the constraints and obtain

$$\mathsf{F}(Y_{\bar{\mathrm{n}}},\Omega)\le \mathsf{G}_U(X_n,\Omega) + C(R^{\mathsf{d}-\sigma}+|n-\bar{\mathrm{n}}|),$$

for some $\sigma>0$, i.e., we get an error which is smaller than the order of the energy. The error $|n-\bar{\mathrm{n}}|$ can be controlled via the energy on a slightly larger domain, and shown to be negligible as well.

At the end of the bootstrap argument in the next chapter, we will know that the energy and points are well distributed down to scale C (ignoring the β-dependence). This means that we then know that (for most configurations) $\bar{S}'(X_n)$ is controlled by $\tilde{\ell}^{\mathsf{d}}$ and $\bar{S}(X_n)$ by $R^{\mathsf{d}-1}\tilde{\ell}$. The condition (7.2.8) is then automatically satisfied and we can thus take $\ell=\tilde{\ell}=C$, and we may also control $|n-\bar{\mathrm{n}}|$ by $O(R^{\mathsf{d}-1})$ to obtain a bound

$$\mathsf{F}(Y_{\bar{\mathrm{n}}},\Omega)\le \mathsf{G}_U(X_n,\Omega)+CR^{\mathsf{d}-1},$$

i.e., with an error only proportional to the surface, the best one can hope to achieve by this approach.

The screening is also possible in the Riesz cases $\mathsf{d} - 2 < \mathsf{s} < \mathsf{d}$ using the dimension extension procedure, it was first done in [**PS17**]. The extension involves additional technical difficulties in the construction. We refer to [**PS**] for the most up-to-date version.

The screening procedure is instrumental to show the local laws of Chapter 8, and also free energy expansions and CLT results for fluctuations of linear statistics in Chapters 9 and 10.

CHAPTER 8

Local laws and almost additivity of the free energy

In this chapter, we continue to focus on the Coulomb case. The goal is twofold:
(i) prove the almost additivity of the free energy as outlined in the previous chapter, thanks to the screening procedure.
(ii) provide local laws, which are analogues of the concentration bounds of (5.2.59) but localized at mesoscopic and microscopic scales, i.e., prove that

$$\left|\log \mathbb{E}_{\mathbb{Q}_\beta(\mu,U)}\left(\exp \frac{\beta}{2}\mathsf{F}^{\square_R}(X_N,\mu,U)\right)\right| \le C\beta\chi(\beta)R^{\mathsf{d}}, \tag{8.0.1}$$

where $\mathbb{Q}_\beta(\mu, U)$ is as in (7.1.14), $\mathsf{F}^{\square_R}$ as in (7.1.23) and $\chi(\beta)$ as in (5.2.27). This means that the share of the energy in a cube $\square_R$ of size R is controlled by the volume of that cube, as long as $\rho_\beta \le R \le N^{1/\mathsf{d}}$ with

$$\rho_\beta = C\max\left(1, \sqrt{\frac{\chi(\beta)}{\beta}}\right) \tag{8.0.2}$$

the minimal scale or Debye length scale already alluded to at the end of the previous chapter. Since we are discussing things in blown-up coordinates, this corresponds to a control down to the (temperature-dependent) microscale $\rho_\beta N^{-1/\mathsf{d}}$ in original coordinates.

This local law will allow to control the number of points that can fall in a small cube and provide discrepancy estimates, and the local energy control will also be crucial in conjunction with the use of Theorem 6.1 to obtain control of fluctuations in the next chapter. Such controls can be viewed as a manifestation of *rigidity* of the Coulomb gas down to the minimal scale.

The local law down to microscale allows to deduce that the energy in a boundary layer of microscopic size around $\partial\square_R$ is proportional to the surface $\rho_\beta R^{\mathsf{d}-1}$ which gives an optimal screening error of that order, and in view of the heuristics given in the previous chapter, yields almost additivity of the free energy up to such boundary terms (see Proposition 8.13) and in particular, as announced in the previous chapter, that $\frac{\log \mathsf{K}_\beta(\square_R,1)}{\beta R^{\mathsf{d}}}$ converges to a limit at speed $R^{\mathsf{d}-1}$ (see Theorem 8.14) with the correction due to ρ_β blowing up when β gets small.

These local laws, first obtained in [**AS21**], extend to Coulomb interactions in arbitrary dimension results obtained for the two-dimensional Coulomb case at fixed β and down to mesoscales in [**Leb17**] (for electric energy and point discrepancy) and [**BBNY17**] (for point discrepancy), it also extends them *down to the microscale* and to possibly N-dependent β. We recall that better local laws on the number of points (but not on the energy F) valid up to the boundary in all Coulomb cases, were proven by Eric Thoma in [**Tho24**]; see Section 5.3.2.

The extension of the local laws (8.0.1) to general Riesz cases (1.1.2) are to be found in [**PS**]. The one-dimensional logarithmic case, or β-ensembles case, has on the other hand been much studied. For this situation, local laws for the electric energy (down to microscale) can be found in [**Pei24**]. Local laws concerning the number of points, Stieltjes transform, as well as description of particle spacings, local statistics and their universality, have been the object of many papers, we refer to [**Joh98, Shc13, BG13, BG24, BMP22, BFG15, BEY14, BEY12, LLW19, BL18, BLS18**] and references therein.

8.1. Method and heuristics

The goals stated above are accomplished jointly by a procedure of bootstrap on scales, originally introduced in this context in the case $\mathsf{d} = 2$ in [**Leb17**], also taking inspiration from [**AS16**], which is akin to a renormalization argument in statistical physics. The result of [**BBNY17**] also relied on a bootstrap on scales, but using loop equations in place of the screening procedure.

At the largest scale $R = N^{1/\mathsf{d}}$, the law (8.0.1) is already known from (5.2.59). Assuming it is true down to some scale $2R$, we wish to show it is also true for scale R, without any deterioration of the constant $\mathcal{C}$ in (8.0.1). The control at scale $2R$ ensures that configurations have a well-controlled energy and that most of them are thus screenable, and it will allow to control the screening errors and show that they are $\ll R^{\mathsf{d}}$. This will allow to show that the two free energy quantities $\log \mathsf{K}_\beta(\square_R)$ and $\log \mathsf{L}_\beta(\square_R)$ of the previous chapter are very close, which in turn will imply the desired estimate (8.0.1) at scale R. The closeness of the free energy quantities will also naturally imply the almost additivity of the free energy, as hinted to in the previous chapter. The bootstrap procedure has to stop at scale ρ_β, the Debye length, that we call minimal scale (by analogy with [**AS16**]), below which the screening errors become as large as the volume, thus deteriorating $\mathcal{C}$.

We note that at the regular scale, ρ_β becomes $\rho_\beta N^{-1/\mathsf{d}}$, which is

$$\chi(\beta)^{1/2}\beta^{-1/2}N^{-1/\mathsf{d}} = \chi(\beta)^{1/2}\theta^{-1/2},$$

with θ as in (3.2.5). Thus the same lengthscale for rigidity as the macroscopic one in Section 2.5 appears (they are exactly equal if $\mathsf{d} \geq 3$ for which $\chi(\beta) = 1$). We believe this to probably be close to optimal, as formulated in the following.

Conjecture 8.1. *The lengthscale $\beta^{-1/2}N^{-1/\mathsf{d}}$ is sharp and (8.0.1) is optimal, in particular local laws do not hold below ρ_β and particle numbers behave in a Poissonian way below that scale.*

Again we are in the situation of Remark 7.13. In the first situation of that remark, the screening procedure requires a bit of "buffer space" to be executed and the local laws will only be valid in the bulk, at a distance (still much smaller than $N^{1/\mathsf{d}}$) from the boundary of the "essential support of μ" (a set where $\mu \geq m > 0$). In the second situation where we consider a hyperrectangle with Neumann boundary condition, results are valid up to the boundary.

Let us now get into more detail about the heuristics. We wish to obtain the local law (8.0.1), where $\square_R$ denotes a *closed* cube of radius R (not necessarily centered at the origin). We know from (5.2.59) that this holds at the largest scale $R = N^{\frac{1}{\mathsf{d}}}$. Assume we know it holds for all cubes of size $\geq 2R$ sufficiently far from the boundary, with the same constant $\mathcal{C}$, and let us try to show it holds for down to R with the same constant $\mathcal{C}$.

Recalling that $\mathbb{Q}_\beta(\mu, U)$ is defined in (7.1.14), let us try to estimate

$$\mathbb{E}_{\mathbb{Q}_\beta(\mu,U)}\left(\exp\left(\frac{\beta}{2}\mathsf{F}^{\square_R}(X_N,\mu)\right)\right) = \frac{\int_{(\mathbb{R}^\mathsf{d})^N} \exp\left(-\beta\mathsf{F}(X_N,\mu)+\frac{\beta}{2}\mathsf{F}^{\square_R}(X_N,\mu)\right)d\mu^{\otimes N}}{\int_{(\mathbb{R}^\mathsf{d})^N} \exp\left(-\beta\mathsf{F}(X_N,\mu)\right)d\mu^{\otimes N}}. \tag{8.1.1}$$

In the numerator we will bound below $\mathsf{F}^{\square_R}(X_N,\mu)$ by $\mathsf{G}^{\mathrm{inn}}(X_N|_{\square_R},\square_R)$ by simple comparison as in Lemma 7.18 and the same for $\mathsf{F}^{\square_R^c}(X_N,\mu)$ (we omit the μ-dependence in the notation for G). We bound from below the denominator by the integral over configurations obtained by gluing together configurations in $\square_R$ and $\square_R^c$ with Neumann boundary conditions. Let us denote by $\bar{\mathrm{n}}$ the optimal number of points in $\square_R$ that is $\bar{\mathrm{n}} = \int_{\square_R}\mu$ (assumed to be an integer), and for a configuration X_N in $\mathbb{R}^\mathsf{d}$, let us denote by n the actual number of points that fall in $\square_R$. Using for now the notation (7.1.15), we thus bound

$$\mathbb{E}_{\mathbb{Q}_\beta(\mu,U)}\left(\exp\left(\frac{\beta}{2}\mathsf{F}^{\square_R}(X_N,\mu)\right)\right) \tag{8.1.2}$$
$$\le \frac{\sum_{n=0}^{N}\binom{N}{n}\int_{(\mathbb{R}^\mathsf{d})^N}\exp\left(-\frac{1}{2}\beta\mathsf{G}^{\mathrm{inn}}(X_N|_{\square_R},\square_R)\right)\exp\left(-\beta\mathsf{G}^{\mathrm{out}}(X_N|_{\square_R^c},\square_R^c)\right)d\mu^{\otimes N}(X_N)}{\binom{N}{\bar{\mathrm{n}}}\int_{(\square_R)^{\bar{\mathrm{n}}}}\exp\left(-\beta\mathsf{F}(\cdot,\mu,\square_R)\right)d\mu^{\otimes\bar{\mathrm{n}}}\int_{\left(\square_R^c\right)^{N-\bar{\mathrm{n}}}}\exp\left(-\beta\mathsf{F}(\cdot,\mu,\square_R^c)\right)d\mu^{\otimes(N-\bar{\mathrm{n}})}}$$
$$\le \frac{\sum_{n=0}^{N}\binom{N}{n}\mathsf{L}_{n,\beta/2}(\mu,\square_R)\mathsf{L}_{N-n,\beta}(\mu,\square_R^c)}{\binom{N}{\bar{\mathrm{n}}}\mathsf{K}_\beta(\mu,\square_R)\mathsf{K}_\beta(\mu,\square_R^c)}.$$

Heuristically, the sum in the numerator concentrates near $n=\bar{\mathrm{n}}$, otherwise the cube is very unbalanced and the energy is too large by discrepancy estimates such as (5.2.20)–(5.2.21) in view of the law (8.0.1) at scale $2R$. We are left with having to evaluate

$$\frac{\mathsf{L}_{\bar{\mathrm{n}},\beta/2}(\mu,\square_R)}{\mathsf{K}_{\beta/2}(\mu,\square_R)}\,\frac{\mathsf{L}_{N-\bar{\mathrm{n}},\beta}(\mu,\square_R^c)}{\mathsf{K}_\beta(\mu,\square_R^c)}\,\frac{\mathsf{K}_{\beta/2}(\mu,\square_R)}{\mathsf{K}_\beta(\mu,\square_R)}.$$

To evaluate the last ratio, we may make use of the Neumann free energy bounds (7.1.55). On the other hand, the first two ratios can be shown to be close to 1 by screening since screening precisely allows to compare L_β and K_β. Moreover, the local law (8.0.1) at scale $2R$ provides the bounds needed for the screening (by providing good energy bounds for most configurations). We will then obtain

$$\log\mathbb{E}_{\mathbb{Q}_\beta(\mu,U)}\left(\exp\left(\frac{\beta}{2}\mathsf{F}^{\square_R}(X_N,\mu)\right)\right)\le C\beta\chi(\beta)R^\mathsf{d}+o(R^\mathsf{d}),$$

which yields the desired estimate if $\mathcal{C}$ was chosen large enough to start with. Once this is obtained, we know that local laws hold down to scale ρ_β and we can use this to find that the difference between $\log\mathsf{L}_\beta$ and $\log\mathsf{K}_\beta$ is only a surface error, hence the almost additivity of $\log\mathsf{K}_\beta$ up to this surface error.

8.2. The case of minimizers

Let us now present rigorous results, starting with the case of energy minimizers (or $\beta = \infty$) as an easier illustration of the method. We will treat in parallel the two situations described in Remark 7.13.

The idea is a simple comparison: let ∇h be the electric field for a minimizer X_N^0 of $\mathsf{F}(\cdot, \mu, \mathbb{R}^\mathsf{d})$, i.e., achieving $\mathsf{E}_\infty(\mu, \mathbb{R}^\mathsf{d})$ as in (7.1.57), and let $\square_R$ be a cube of size R included in the (blown-up) essential support of the measure μ (a bit far from the boundary) then if the energy in $\square_{2R}$ can be well bounded in terms of R, we can perform inner and outer screening relative to $\square_R$ and replace the electric field in $\square_R$ by the electric field associated to a minimizer of $\mathsf{F}(\cdot, \mu, \square_R)$, while adding only $O(R^{\mathsf{d}-1})$ to the energy. Comparison then allows to deduce a bound on $\mathsf{F}^{\square_R}(X_N^0, \mu, \mathbb{R}^\mathsf{d})$.

Thanks to the comparison between the sub and superadditive energies, we can get a convergence with a rate in the case of the uniform density $\mu = 1$, second case in Remark 7.13.

Theorem 8.2 (Uniform distribution of energy and points down to microscale for minimizers). *Assume* $\mathsf{d} \geq 1$ *and* $\mathsf{s} = \mathsf{d} - 2$.

(1) Neumann problems in cubes. *Let* $\square_R$ *be a closed cube of size* R *with* R^d *an integer. There exists* $C > 0$ *depending only on* d *such that*

$$\left| \frac{\mathsf{E}_\infty(1, \square_R)}{R^\mathsf{d}} - f_\mathsf{d}(\infty) \right| \leq \frac{C}{R}, \tag{8.2.1}$$

where $f_\mathsf{d}(\infty) \in \mathbb{R}$ *and moreover, if* X_{R^d} *is a minimizer for* $\mathsf{E}_\infty(1, \square_R)$, *for any cube* $\square_L(x) \subset \square_R$ *such that* $L \geq C$, *we have*

$$\left| \int_{\square_L(x)} \sum_{i=1}^{R^\mathsf{d}} \delta_{x_i} - |\square_L| \right| \leq C L^{\mathsf{d}-1} \tag{8.2.2}$$

and the energy is uniformly distributed in the sense that

$$\mathsf{F}^{\square_L(x)}(X_{R^\mathsf{d}}, 1, \square_R) = L^\mathsf{d} f_\mathsf{d}(\infty) + O(L^{\mathsf{d}-1}). \tag{8.2.3}$$

(2) Minimizers of the Coulomb energy at the usual scale. *Assume that* μ *is a bounded probability density supported in some compact set* Σ *and satisfies* $\mu \geq m > 0$ *in* Σ *(resp. let* $\mu = \mu_V$*). There exists* $C > 0$ *depending only on* $\mathsf{d}, m, \|\mu\|_{L^\infty}$ *such that the following holds.*

Let X_N^0 *minimize* $\mathsf{F}_N(X_N, \mu)$ *(resp. minimize* $\mathcal{H}_N$*). Let* $R \geq C N^{-1/\mathsf{d}}$ *and* Q_R *be a hyperrectangle of sidelengths in* $[R, 2R]$ *satisfying* $N \int_{Q_R} \mu \in \mathbb{N}$, *and*

$$\operatorname{dist}(Q_R, \partial\Sigma) \geq C N^{\frac{-2}{\mathsf{d}(\mathsf{d}+2)}} \tag{8.2.4}$$

we have

$$\left| \int_{Q_R} \sum_{i=1}^{N} \delta_{x_i} - N \int_{Q_R} d\mu \right| \leq C \left(N^{\frac{1}{\mathsf{d}}} R \right)^{\mathsf{d}-1}, \tag{8.2.5}$$

and

$$\left| \mathsf{F}^{Q_{RN^{1/\mathsf{d}}}}((X_N^0)', \mu') - \mathsf{E}_\infty(\mu, Q_{RN^{1/\mathsf{d}}}) \right| \leq C (N^{\frac{1}{\mathsf{d}}} R)^{\mathsf{d}-1}. \tag{8.2.6}$$

Remark 8.3. We will give in Corollary 12.7 a variational interpretation for $f_{\mathsf{d}}(\infty)$,

$$f_{\mathsf{d}}(\infty) = \min \mathbb{W}(\cdot, 1)$$

where $\mathbb{W}$ is the "jellium renormalized energy" defined in Chapter 11. We also give an expression for $\mathsf{E}_\infty(\mu, Q)$ in (9.4.3).

The explicit rate in (8.2.3) is an improvement compared to [**RNS15**, **PRN18**]. As in [**RNS15**], we can also prove with the same method the same results on minimizers and the minimum of the renormalized energy $\mathbb{W}(\cdot, 1)$ of [**SS12**, **RS15**, **PS17**]. For instance the limit as $R \to \infty$ that defines $\mathbb{W}(\cdot, 1)$ can be shown to be $f_{\mathsf{d}}(\infty)$ with rate $1/R$: the upper bound is by periodization of a minimizer for E_∞ while the lower bound is obtained as in (8.2.12) to be combined with (8.2.1).

PROOF. **Step 1: Bootstrap.** In order to accommodate both situations of the theorem, we work at the blown up scale, and consider that μ satisfies $0 < m \le \mu$ in Σ, and let X_N^0 be a minimizer of $\mathsf{F}(\cdot, \mu, U)$ among configurations with $N = \mu(U)$ points, where U is either a cube $\square_R$ or $\mathbb{R}^{\mathsf{d}}$. Let u be the electric potential satisfying (7.1.1). We claim that if $L \ge C$ then

$$\mathsf{F}^{\square_L(x)}(X_N^0, \mu, U) + 2C_0 \# I_{\square_L(x)} \le C L^{\mathsf{d}} \tag{8.2.7}$$

for some $C > 0$ depending only on d and μ, and C_0 the constant in (7.1.35).

This is proven by a bootstrap. First we may reduce to proving this for hypercubes with faces parallel to the boundary of U in instance (1). We know that (8.2.7) is true for the maximal scale $L = R$ in instance (1) or $L = N^{1/\mathsf{d}}$ in instance (2) by (7.1.58). Assume this is true for some $\tilde{L}$, i.e., assume

$$\mathsf{F}^{\square_{\tilde{L}}(x)}(X_N^0, \mu, U) + 2C_0 \# I_{\square_{\tilde{L}}(x)} \le \mathcal{C} \tilde{L}^{\mathsf{d}}, \tag{8.2.8}$$

we need to show it is true for $\frac{3}{4}\tilde{L} \ge L \ge \tilde{L}/2$. We have emphasized the constant $\mathcal{C}$ which must be independent of L, so we need to prove that the same result holds for L with the same constant $\mathcal{C}$. Let us reduce to Q_L such that $\bar{\mathrm{n}} := \int_{Q_L \cap U} \mu$ is an integer, and denote by $n = \# I_{Q_L}$.

First by (8.2.8), we have from (7.1.35), (5.2.20)–(5.2.21) that

$$|n - \bar{\mathrm{n}}| \le C L^{\mathsf{d}-1} + C\sqrt{\mathcal{C}} L^{\mathsf{d}-\frac{1}{2}}, \tag{8.2.9}$$

and moreover that

$$\int_{\square_{\tilde{L}}} |\nabla u_{\tilde{\mathsf{r}}}|^2 + C C_0 \# I_{\square_L(x)} \le C \left(\mathsf{F}^{\square_{\tilde{L}}(x)}(X_N^0, U) + C_0 \# I_{\square_{\tilde{L}}(x)} \right) \le C \mathcal{C} \tilde{L}^{\mathsf{d}}.$$

We then apply the outer screening (with respect to Q_L) of Proposition 7.19 with $S(X_N^0, u) \le \mathcal{C}\tilde{L}^{\mathsf{d}}$, and we make the choices of $\ell \le \tilde{\ell} \le L$ that are the smallest possible to satisfy (7.2.7)[1] and obtain small errors in (7.2.16). For instance we choose $\ell = L^{\frac{\mathsf{d}}{\mathsf{d}+2}}$, $\tilde{\ell} = M L^{\frac{\mathsf{d}}{\mathsf{d}+2}}$, with $M \ge 1$ large enough. The condition $\ell \le \tilde{\ell} \le L$ is satisfied as long as L is larger than some constant depending on M. The separation from the boundary conditions will be satisfied by assumption (see below). The screenability condition (7.2.7) is satisfied for the potential u since $\tilde{\ell}\ell^{\mathsf{d}+1} = M\ell^{\mathsf{d}+2} \ge S(X_N^0, u)$ in view of (8.2.8) and $\tilde{L} \le 2L$, if M is chosen sufficiently large in terms of $\mathcal{C}$. This implies that the best inner and outer screenable potentials have an even smaller energy (7.2.9) and (7.2.10), and are thus screenable as well.

[1] Note that in the definition of $S(X_n, w)$, $w_{\hat{\mathsf{r}}}$ can be changed into $w_{\tilde{\mathsf{r}}}$ as in (7.1.25) since $\hat{\mathsf{r}}$ and $\tilde{\mathsf{r}}$ coincide in the region of integration.

Let x_i be the points of X_N^0, let us apply the screening procedure of Proposition 7.19, which yields a set $\mathcal{N}$ and a number of points $n_{\mathcal{O}}$. Using the notation of (7.2.17), arguing as in Proposition 7.11 we may obtain the bound

$$\begin{aligned}\int_{\mathcal{N}^{\bar{\mathsf{n}}-n_{\mathcal{O}}}} \Big(\mathsf{F}(Z_{\bar{\mathsf{n}}-n_{\mathcal{O}}}, \tilde{\mu}, \mathcal{N}) + \sum_{(i,j)\in J} \mathsf{g}(x_i - z_j)\Big) d\mu^{\otimes \bar{\mathsf{n}}-n_{\mathcal{O}}}(Z_{\bar{\mathsf{n}}-n_{\mathcal{O}}}) \\ \le C\tilde{\ell}L^{\mathsf{d}-1} + \sum_{i\in I_\partial} \int_{|z-x_i|\le \tilde{\mathsf{r}}_i} \mathsf{g}(x_i - z)\, d\tilde{\mu}(z) \\ \le C\tilde{\ell}L^{\mathsf{d}-1} + \#I_\partial.\end{aligned}$$

We may thus find a configuration $Z_{\bar{\mathsf{n}}-n_{\mathcal{O}}}$ in $\mathcal{N}$ such that

$$\mathsf{F}(Z_{\bar{\mathsf{n}}-n_{\mathcal{O}}}, \tilde{\mu}, \mathcal{N}) + \sum_{(i,j)\in J} \mathsf{g}(x_i - z_j) \le C\tilde{\ell}L^{\mathsf{d}-1} + \#I_\partial.$$

Let us thus apply Proposition 7.19 to minimizers of (7.2.9), respectively (7.2.10), with that particular $Z_{\bar{\mathsf{n}}-n_{\mathcal{O}}}$. The relation (7.2.17) then yields in view of (8.2.9) and (8.2.8)

$$\mathsf{E}_\infty(\mu, Q_L^c) \le \mathsf{G}_U^{\mathrm{out}}(X_N^0|_{Q_L^c}, Q_L^c) + C\left(\frac{\mathcal{C}}{M}L^{\mathsf{d}} + ML^{\mathsf{d}-1}L^{\frac{\mathsf{d}}{\mathsf{d}+2}} + \sqrt{\mathcal{C}}L^{\mathsf{d}-\frac{1}{2}}\right). \tag{8.2.10}$$

Choosing M large enough, the error in the right-hand side can be made $\le \frac{1}{2}\mathcal{C}L^{\mathsf{d}}$ when L is large enough, and thus combining (7.1.59), and (8.2.10), it follows that

$$\begin{aligned}\mathsf{F}(X_N^0, U) = \mathsf{E}_\infty(\mu, U) \le \mathsf{E}_\infty(\mu, Q_L) + \mathsf{E}_\infty(\mu, Q_L^c) \\ \le \mathsf{E}_\infty(\mu, Q_L) + \mathsf{G}_U^{\mathrm{out}}(X_N^0|_{Q_L^c}, Q_L^c) + \frac{1}{2}\mathcal{C}L^{\mathsf{d}}.\end{aligned}$$

On the other hand, by (7.1.37) and Lemma 7.18 (which applies since u is screenable), we have

$$\mathsf{F}^{Q_L}(X_N^0, U) + \mathsf{G}_U^{\mathrm{out}}(X_N^0|_{Q_L^c}, Q_L^c) \le \mathsf{F}(X_N^0, U).$$

Hence, combining the two, we deduce that if L is large enough (depending on $\mathcal{C}$)

$$\mathsf{F}^{Q_L}(X_N^0, \mu, U) \le \mathsf{E}_\infty(\mu, Q_L) + \frac{1}{2}\mathcal{C}L^{\mathsf{d}}.$$

In view of (8.2.9) we have as well $n \le \frac{1}{2}\mathcal{C}L^{\mathsf{d}}$ if L is large enough. With (7.1.58) applied in Q_L, this concludes the proof of (8.2.7).

Let us examine the distance condition to the boundary which the screening proposition requires. If Σ is a proper subset of U, at each step we need a safety distance $\tilde{\ell}$. But we chose $\tilde{\ell} = ML^{\frac{\mathsf{d}}{\mathsf{d}+2}}$. At the first iteration L is of order R, respectively $N^{1/\mathsf{d}}$, so $\tilde{\ell} = MR^{\frac{\mathsf{d}}{\mathsf{d}+2}}$. At further iterations, we need $Q_{L+\tilde{\ell}}$ to be included in $Q_{\tilde{L}}$ so we need a further distance $CL^{\frac{\mathsf{d}}{\mathsf{d}+2}}$, and since L gets multiplied by a factor $\in [\frac{1}{2}, \frac{3}{4}]$ at each step, summing the distances over the iteration still gives a condition of the form dist $\ge M(N^{1/\mathsf{d}})^{\frac{\mathsf{d}}{\mathsf{d}+2}}$ at the blow-up scale, hence condition (8.2.4) at the original scale.

In the case where $U = \square_R$ and we can take $\Sigma = \mathbb{R}^{\mathsf{d}}$, we need that $Q_L(x)$ has sides parallel to those of $\square_R$, satisfying that the faces of $\partial(\square_L(x) \cap \square_R)$ not belonging to $\partial\square_R$ are at a distance $\ge CR^{\frac{\mathsf{d}}{\mathsf{d}+2}}$ from the parallel faces of $\partial\square_R$ for some $C > 0$. This condition is however not really restrictive, as in the first step we can always include a given Q_L into one satisfying the required condition and

$L \in (\frac{1}{8}, \frac{1}{4})$. We can then repeat the argument during the bootstrap and obtain the result up the boundary.

Step 2: Local laws. Now that we know (8.2.7) down to scale C, we can use it to obtain, via (7.1.35), (5.2.20), (5.2.21),

$$|n - \bar{\mathrm{n}}| \le CL^{\mathsf{d}-1}. \tag{8.2.11}$$

Following the relations after (8.2.10), we find the improvement

$$\mathsf{F}^{Q_L}(X_N^0, \mu, U) \le \mathsf{E}_\infty(\mu, Q_L) + CL^{\mathsf{d}-1}$$

and also that (8.2.11) holds. By screening, i.e., by Proposition 7.19 applied in Q_L to $X_N^0|_{Q_L}$ and u, we also have

$$\mathsf{E}_\infty(\mu, Q_L) \le \mathsf{F}^{Q_L}(X_N^0, \mu, U) + CL^{\mathsf{d}-1}, \tag{8.2.12}$$

so we conclude that

$$\mathsf{F}^{Q_L}(X_N^0, \mu, U) = \mathsf{E}_\infty(\mu, Q_L) + O(L^{\mathsf{d}-1}), \tag{8.2.13}$$

with the O depending only on d, m and $\|\mu\|_{L^\infty}$.

Step 3: Conclusion. Let $U = \square_{2R}$ with R^{d} an integer, and $\mu = 1$, and X_N^0 be a minimizer of $\mathsf{F}(\cdot, 1, \square_{2R})$, splitting U into 2^{d} cubes of size R, denoted $\square_R^i$, and using (7.1.37), we have

$$\mathsf{E}_\infty(1, \square_{2R}) = \mathsf{F}(X_N^0, 1, \square_{2R}) \ge \sum_i \mathsf{F}^{\square_R^i}(X_N^0, 1, \square_{2R}).$$

Inserting the result (8.2.13), it follows that

$$\mathsf{E}_\infty(1, \square_{2R}) \ge 2^{\mathsf{d}} \mathsf{E}_\infty(1, \square_R) + O(R^{\mathsf{d}-1}) \tag{8.2.14}$$

and combining with (7.1.59), we arrive at

$$\frac{\mathsf{E}_\infty(1, \square_{2R})}{|\square_{2R}|} = \frac{\mathsf{E}_\infty(1, \square_R)}{|\square_R|} + O(R^{-1}). \tag{8.2.15}$$

Applying this iteratively to $2^k R$ and summing the series, we deduce that

$$f_{\mathsf{d}}(\infty) := \lim_{R \to \infty} \frac{\mathsf{E}_\infty(1, \square_R)}{|\square_R|}$$

exists and that (8.2.1) holds. We also then obtain (8.2.3) from (8.2.13) and (8.2.2) from (8.2.11).

For the case of (2), we apply the results of the previous steps with $U = \mathbb{R}^{\mathsf{d}}$, $\mu = \mu'$, and then a blow-down, using (5.2.8). The relation (8.2.6) is a rewriting of (8.2.13) and (8.2.5) follows from (8.2.11). □

8.3. Local laws with temperature

8.3.1. Statement. We now turn to the question of proving (8.0.1) with temperature $\beta > 0$. This is more difficult since we have to deal with many configurations at once. The following statement is meant again to either be applied with $U = \mathbb{R}^{\mathsf{d}}$, giving information on the Gibbs measure $\mathbb{Q}_\beta(\mu, \mathbb{R}^{\mathsf{d}})$, or to U a hyperrectangle and $\Lambda = \mathbb{R}^{\mathsf{d}}$ (in which case condition (8.3.1) is empty) giving the local law for the Neumann Gibbs measure $\mathbb{Q}_\beta(\mu, U)$ in a hyperrectangle. Since we have made some simplifications compared to [**AS21**], we obtain a better result than in that paper, in particular some assumption is removed and ρ_β (defined in (8.0.2)) is improved, and the result is valid for $\mathsf{d} = 1$.

In the theorem, we place the same assumptions as in Proposition 7.11 to ensure the a priori bound (7.1.55). We recall that by (5.1.11) the original Coulomb gas Gibbs measure is equal to $\mathbb{Q}_\beta(\mu'_\theta, \mathbb{R}^\mathsf{d})$ for μ'_θ the blown-up thermal equilibrium measure, and that as discussed after (5.2.28)–(5.2.29), these assumptions are satisfied for μ'_θ in $\Lambda = \Sigma'$ provided (A5) holds and $\theta \ge \theta_0 > 0$. Thus the theorem below applies to (1.1.5) as soon as $\theta > \theta_0$ and (A1)–(A5) hold.

Theorem 8.4 (Local laws). *Assume $\mathsf{d} \ge 1$ and $\mathsf{s} = \mathsf{d} - 2$. Let U be an open subset of $\mathbb{R}^\mathsf{d}$ with bounded and piecewise C^1 boundary and μ be a bounded nonnegative density such that $\mu(U) = N$ is an integer. Assume that $\mu \ge m > 0$ in a set Λ. If $\mathsf{s} \le 0$ and U is unbounded, assume in addition that* (5.2.28), (5.2.29), *and* (5.2.37) *hold. There exists a constant $C > 0$ depending only on $\mathsf{d}, m, \|\mu\|_{L^\infty}$ and the constants in the assumptions, such that the following holds.*

Let $\square_R(x)$ be a closed cube of size $R \ge \rho_\beta$ centered at x (where ρ_β as is in (8.0.2)*) with*

$$\mathrm{dist}(\square_R(x), \partial\Lambda \cap U) \ge d_0 \tag{8.3.1}$$

where

$$d_0 := C \max\left(\left(\frac{N^{\frac{1}{\mathsf{d}}}}{\max(1, \beta^{-\frac{1}{2}}\chi(\beta)^{\frac{1}{2}})} \right)^{-\frac{2}{3}} N^{\frac{1}{\mathsf{d}}}, N^{\frac{1}{\mathsf{d}+2}} \right), \tag{8.3.2}$$

we have, for C_0 the constant in (7.1.35),

(8.3.3)
$$\left| \log \mathbb{E}_{\mathbb{Q}_\beta(\mu,U)} \left(\exp\left(\frac{1}{2}\beta \left(\mathsf{F}^{\square_R(x)}(\cdot, \mu, U) + 2C_0 \# I_{\square_R(x)} \right) \right) \right) \right| \le C\beta\chi(\beta) R^\mathsf{d}.$$

8.3.2. Proof of the local laws with temperature. When studying Gibbs measures, we need to perform a "probabilistic screening", i.e., show that given a set of configurations with well-controlled energy, we may screen them and sample new points in $\mathcal{N}$ to obtain a set with large enough volume in which (7.2.16) holds. This is possible and yields comparison of partition functions (reduced to screenable configurations) as stated in the following proposition (essentially following [**AS21**]).

We are going to condition on n, the number of points of the configuration that fall in Ω, the set as in Definition 7.15. Below, the quantity ε_e corresponds to the energy error while ε_v corresponds to the volume error. We want the volume errors to be bounded by $O(\beta)$ times the volume, which is more difficult to obtain when β is small.

Proposition 8.5. *With the same assumptions and notation as in Proposition 7.19, assume in addition that $\tilde{\ell} \ge \max(\beta^{-\frac{1}{\mathsf{d}-\mathsf{s}}} \mathbf{1}_{\mathsf{s} \le 0}, 1)$. Let us define the set*

$$\mathcal{D}_{s,z} = \left\{ X_n \in \Omega^n, \bar{S}(X_n) \le s, \bar{S}'(X_n) \le z \right\}, \tag{8.3.4}$$

where $\bar{S}, \bar{S}'$ are as in (7.2.11), *resp.* (7.2.12), *relative to Ω. For any numbers s, z such that*

$$\ell^{\mathsf{d}+1} \ge C\frac{s}{\tilde{\ell}} \quad \text{or } \ell^{\mathsf{d}+1} \ge Cz, \tag{8.3.5}$$

and

$$s < c\tilde{\ell}^2 R^{\mathsf{d}-1} \tag{8.3.6}$$

for some $c > 0$ small enough (depending only on $\mathsf{d}, m, \|\mu\|_{L^\infty}$), there exist α, α' satisfying $\alpha + n - \bar{\mathrm{n}} \ge 1$, and

$$\left|\frac{\alpha'}{\alpha} - 1\right| \le C\left(\frac{1}{\tilde{\ell}} + \frac{s}{\tilde{\ell}^2 R^{\mathsf{d}-1}}\right), \qquad \frac{1}{C}\tilde{\ell}R^{\mathsf{d}-1} \le \alpha \le C\tilde{\ell}R^{\mathsf{d}-1}, \tag{8.3.7}$$

such that letting

$$\varepsilon_e := C\left(\frac{s\ell}{\tilde{\ell}} + R^{\mathsf{d}-1}\tilde{\ell}\chi(\beta) + |n - \bar{\mathrm{n}}|\right) \tag{8.3.8}$$

and

- *if $\ell^{\mathsf{d}+1} \ge C\frac{s}{\tilde{\ell}}$,*

$$\varepsilon_v := C\left(\frac{s}{\ell\tilde{\ell}} + \ell^{-1}R^{\mathsf{d}-1} + \log R\right) \tag{8.3.9}$$
$$+ n - \bar{\mathrm{n}} + \alpha - \alpha' - (n - \bar{\mathrm{n}} + \alpha)\log\frac{\alpha}{\alpha'} - (n - \bar{\mathrm{n}} + \alpha + \frac{1}{2})\log\left(1 + \frac{n - \bar{\mathrm{n}}}{\alpha}\right) - \frac{1}{2}\log\frac{n}{\bar{\mathrm{n}}},$$

- *respectively, if $\ell^{\mathsf{d}+1} \ge Cz$,*

$$\varepsilon_v := C\left(\frac{s}{\ell\tilde{\ell}} + R^{\mathsf{d}-1}(\ell^{-1} + \ell^{1-\mathsf{d}}\log\ell) + \log R\right) \tag{8.3.10}$$
$$+ n - \bar{\mathrm{n}} + \alpha - \alpha' + (\bar{\mathrm{n}} - n - \alpha)\log\frac{\alpha}{\alpha'} - (\alpha + n - \bar{\mathrm{n}} + \frac{1}{2})\log\left(1 + \frac{n - \bar{\mathrm{n}}}{\alpha}\right) - \frac{1}{2}\log\frac{n}{\bar{\mathrm{n}}}$$

in both cases we have

$$\begin{aligned} &\frac{1}{n!}\int_{\mathcal{D}_{s,z}} \exp\left(-\beta \mathsf{G}_U^{\mathrm{inn/out}}(X_n, \mu, \Omega)\right) d\mu^{\otimes n}(X_n) \\ &\quad \le C\exp\left(\beta\varepsilon_e + \varepsilon_v\right)\frac{\bar{\mathrm{n}}^{\bar{\mathrm{n}}}}{\bar{\mathrm{n}}!}\mathsf{K}_\beta(\mu, \Omega)\int_{\Omega^{\bar{\mathrm{n}}}} \exp\left(-\beta\mathsf{F}(Y_{\bar{\mathrm{n}}}, \mu, \Omega)\right) d\mu^{\otimes\bar{\mathrm{n}}}(Y_{\bar{\mathrm{n}}}), \end{aligned} \tag{8.3.11}$$

where $\mathsf{G}_U^{\mathrm{inn/out}}$ is as in Definition 7.16 and K_β as in (7.1.13). Here $C > 0$ depends only on $\mathsf{d}, m, \|\mu\|_{L^\infty}$.

For the proof, we will need the following.

Lemma 8.6. *For each $X_n \in \mathcal{D}_{s,z}$, we have*

$$\begin{aligned} &\int_{\mathcal{N}^{\bar{\mathrm{n}} - n_{\mathcal{O}}}} \exp\Bigg(-C\beta\Big(\mathsf{F}(Z_{\bar{\mathrm{n}}-n_{\mathcal{O}}}, \tilde{\mu}, \mathcal{N}) + \sum_{(i,j)\in J}\mathsf{g}(x_i - z_j)\Big)\Bigg) d\mu|_{\mathcal{N}}^{\otimes(\bar{\mathrm{n}}-n_{\mathcal{O}})}(Z_{\bar{\mathrm{n}}-n_{\mathcal{O}}}) \\ &\quad \ge (\bar{\mathrm{n}} - n_{\mathcal{O}})^{\bar{\mathrm{n}}-n_{\mathcal{O}}}\exp\left(\mu(\mathcal{N}) - \tilde{\mu}(\mathcal{N}) - C\left(\beta\chi(\beta)R^{\mathsf{d}-1}\tilde{\ell} + \frac{s}{\ell\tilde{\ell}} + \ell^{-1}R^{\mathsf{d}-1}\right)\right). \end{aligned} \tag{8.3.12}$$

PROOF. Applying Jensen's inequality, we find

$$\begin{aligned}
&\int_{\mathcal{N}^{\bar{n}-n_{\mathcal{O}}}} \exp\Big(-C\beta\Big(\mathsf{F}(Z_{\bar{n}-n_{\mathcal{O}}},\tilde{\mu},\mathcal{N})+\sum_{(i,j)\in J}\mathsf{g}(x_i-z_j)\Big)\Big)\,d\mu|_{\mathcal{N}}^{\otimes(\bar{n}-n_{\mathcal{O}})}(Z_{\bar{n}-n_{\mathcal{O}}})\\
&=\int_{\mathcal{N}^{\bar{n}-n_{\mathcal{O}}}} \exp\Big[-C\beta\Big(\mathsf{F}(Z_{\bar{n}-n_{\mathcal{O}}},\tilde{\mu},\mathcal{N})+\sum_{(i,j)\in J}\mathsf{g}(x_i-z_j)\Big)\\
&\qquad\qquad+\sum_{i=1}^{\bar{n}-n_{\mathcal{O}}}\log\frac{\mu}{\tilde{\mu}}(z_i)\Big]\,d\tilde{\mu}^{\otimes(\bar{n}-n_{\mathcal{O}})}(Z_{\bar{n}-n_{\mathcal{O}}})\\
&\geq \tilde{\mu}(\mathcal{N})^{\bar{n}-n_{\mathcal{O}}}\exp\Big[\tilde{\mu}(\mathcal{N})^{n_{\mathcal{O}}-\bar{n}}\int_{\mathcal{N}^{\bar{n}-n_{\mathcal{O}}}}\Big(-C\beta\Big(\mathsf{F}(Z_{\bar{n}-n_{\mathcal{O}}},\tilde{\mu},\mathcal{N})+\sum_{(i,j)\in J}\mathsf{g}(x_i-z_j)\Big)\\
&\qquad\qquad+\sum_{i=1}^{\bar{n}-n_{\mathcal{O}}}\log\frac{\mu}{\tilde{\mu}}(z_i)\Big)\,d\tilde{\mu}^{\otimes(\bar{n}-n_{\mathcal{O}})}(Z_{\bar{n}-n_{\mathcal{O}}})\Big],
\end{aligned}$$

where we recall that $\tilde{\mu}(\mathcal{N})=\bar{n}-n_{\mathcal{O}}$. We then use the same proof as in Lemma 5.12, Proposition 5.14 and Proposition 7.11. The term $\sum_{(i,j)\in J}\mathsf{g}(x_i-z_j)$ adds a contribution

$$-C\beta(\bar{n}-n_{\mathcal{O}})^{\bar{n}-n_{\mathcal{O}}}\sum_{i\in I_\partial}\int_{|z-x_i|\leq\tilde{\mathsf{r}}_i}\mathsf{g}(x_i-z)\,d\tilde{\mu}(z)\geq -C\beta(\bar{n}-n_{\mathcal{O}})^{\bar{n}-n_{\mathcal{O}}}\#I_\partial$$

and, by $\#I_\partial\leq Cs/\tilde{\ell}$ and (8.3.6), we conclude that

$$\begin{aligned}
&\int_{\mathcal{N}^{\bar{n}-n_{\mathcal{O}}}} \exp\Big(-C\beta\Big(\mathsf{F}(Z_{\bar{n}-n_{\mathcal{O}}},\tilde{\mu},\mathcal{N})+\sum_{(i,j)\in J}\mathsf{g}(x_i-z_j)\Big)\Big)\,d\mu|_{\mathcal{N}}^{\otimes(\bar{n}-n_{\mathcal{O}})}(Z_{\bar{n}-n_{\mathcal{O}}})\\
&\qquad\geq(\bar{n}-n_{\mathcal{O}})^{\bar{n}-n_{\mathcal{O}}}\exp\Big(\int_{\mathcal{N}}\tilde{\mu}\log\frac{\mu}{\tilde{\mu}}-C\beta R^{\mathsf{d}-1}\tilde{\ell}(1+|\mathsf{g}(R)|\mathbf{1}_{\mathsf{s}\leq 0})\Big).
\end{aligned}$$

In the case $\mathsf{s}\leq 0$, i.e., $\mathsf{d}=1,2$, in view of the fact that $\tilde{\ell}\geq\beta^{-\frac{1}{\mathsf{d}-\mathsf{s}}}$, we see from its construction that $\mathcal{N}$ can be partitioned into disjoint nondegenerate cells of size $\max(1,\beta^{-\frac{1}{\mathsf{d}-\mathsf{s}}})$ in which $\tilde{\mu}$ integrates to an integer. Using superadditivity as in the proof of Propositions 5.14 and 7.11, we can improve this into

(8.3.13)

$$\begin{aligned}
&\int_{\mathcal{N}^{\bar{n}-n_{\mathcal{O}}}} \exp\Big(-C\beta\Big(\mathsf{F}(Z_{\bar{n}-n_{\mathcal{O}}},\tilde{\mu},\mathcal{N})+\sum_{(i,j)\in J}\mathsf{g}(x_i-z_j)\Big)\Big)\,d\mu|_{\mathcal{N}}^{\otimes(\bar{n}-n_{\mathcal{O}})}(Z_{\bar{n}-n_{\mathcal{O}}})\\
&\qquad\geq(\bar{n}-n_{\mathcal{O}})^{\bar{n}-n_{\mathcal{O}}}\exp\Big(\int_{\mathcal{N}}\tilde{\mu}\log\frac{\mu}{\tilde{\mu}}-C\beta\chi(\beta)R^{\mathsf{d}-1}\tilde{\ell}\Big).
\end{aligned}$$

Using then (7.2.15), (8.3.6), and (7.2.14) we have $|\frac{\mu}{\tilde{\mu}}-1|<\frac{1}{2}$ in $\mathcal{N}$ if c is chosen small enough and thus by Taylor expansion

(8.3.14)

$$\int_{\mathcal{N}}\tilde{\mu}\log\frac{\mu}{\tilde{\mu}}=\int_{\mathcal{N}}\mu-\tilde{\mu}+O\Big(\int_{\mathcal{N}}\frac{|\mu-\tilde{\mu}|^2}{\tilde{\mu}}\Big)=\mu(\mathcal{N})-\tilde{\mu}(\mathcal{N})+O\Big(\frac{s}{\ell\tilde{\ell}}+\ell^{-1}R^{\mathsf{d}-1}\Big).$$

Together with (8.3.13) this concludes the proof. □

PROOF OF PROPOSITION 8.5. Again, we consider the inner screening case, the outer case being parallel. The set $\mathcal{D}_{s,z}$ consists of configurations with controlled energy, which satisfy the screenability condition (7.2.7) or (7.2.8) as soon as (8.3.5) holds. We apply the screening result, Proposition 7.19 with $\Omega = \Omega' = \Omega''$. Let us first consider the situation in which (7.2.7) holds, in which case, for any configuration X_n, the screening contour Γ is the boundary of a hyperrectangle $Q_t = \mathcal{O}$ where we will denote $t := t(X_n)$. We will also denote $n_{\mathcal{O}}(X_n)$ the number of points of X_n in $\mathcal{O}$, as in Proposition 7.19. Let us also denote by $\Phi(X_n)$ the set of all configurations $Y_{\bar{\mathrm{n}}}$ produced by screening the configuration X_n, i.e., all those $Y_{\bar{\mathrm{n}}}$'s coinciding with X_n in $\mathcal{O} = \mathcal{O}(X_n)$ and with an arbitrary configuration $Z_{\bar{\mathrm{n}}-n_{\mathcal{O}}(X_n)}$ in $\mathcal{N} = \mathcal{N}(X_n)$, with all possible relabellings. We wish to integrate the screening estimates over all possible $Y_{\bar{\mathrm{n}}}$'s but we have to do it with care as a given configuration could be produced multiple times by the screening. We will also condition on $n_{\mathcal{O}}(X_n)$ being equal to a given integer $k \le \bar{\mathrm{n}}$.

The parameter $0 \le \eta \le \ell$ being given, we split the possibly values of t into intervals $t \in I_p := [R - \tilde{\ell} - (p+1)\eta, R - \tilde{\ell} - p\eta)$, p integer. In view of Proposition 7.19, the integer p ranges from 0 to $C\tilde{\ell}/\eta$. We note that p and k are constrained, indeed in view of (7.2.14), we must have on the one hand $\tilde{\mu}(\mathcal{N}(X_n)) = \bar{\mathrm{n}} - n_{\mathcal{O}}(X_n)$ and on the other hand

$$
\begin{aligned}
(8.3.15)\quad & \mu(Q_R \backslash Q_{R-(p+1)\eta}) = \mu(\mathcal{N}(X_n)) + O(\eta R^{\mathsf{d}-1}) \\
&= \tilde{\mu}(\mathcal{N}(X_n)) + O(R^{\mathsf{d}-1} + \frac{s}{\tilde{\ell}}) + O(\eta R^{\mathsf{d}-1}) = \tilde{\mu}(\mathcal{N}(X_n)) \left(1 + O\left(\frac{1+\eta}{\tilde{\ell}} + \frac{s}{\tilde{\ell}^2 R^{\mathsf{d}-1}}\right)\right) \\
&= (\bar{\mathrm{n}} - n_{\mathcal{O}}(X_n)) \left(1 + O\left(\frac{1+\eta}{\tilde{\ell}} + \frac{s}{\tilde{\ell}^2 R^{\mathsf{d}-1}}\right)\right)
\end{aligned}
$$

where we used the upper and lower bounds on μ and the fact that $Q_R \backslash Q_{R-\tilde{\ell}} \subset \mathcal{N} \subset Q_R \backslash Q_{R-2\tilde{\ell}}$.

We denote by $\mathcal{A}$ the set of admissible integer p's and k's, it is included in $[0, \lfloor \frac{C\tilde{\ell}}{\eta} \rfloor] \times [0, \bar{\mathrm{n}}]$, and satisfies the constraint that

$$
(8.3.16)\qquad \mu(Q_R \backslash Q_{R-(p+1)\eta}) = (\bar{\mathrm{n}} - k) \left(1 + O\left(\frac{1+\eta}{\tilde{\ell}} + \frac{s}{\tilde{\ell}^2 R^{\mathsf{d}-1}}\right)\right).
$$

We next let $\nu = (t(X_n), n_{\mathcal{O}}(X_n)) \# \mu^{\otimes n}$, i.e., the "law" on $\mathbb{N} \times \mathbb{N}$ equal to the "law" of $(t(X_n), n_{\mathcal{O}}(X_n))$ for $X_n \sim \mu^{\otimes n}$. We have a natural disintegration

$$
(8.3.17)\qquad d\mu^{\otimes n}(X_n) = \sum_{(p,k) \in \mathcal{A}} d(\mu^{\otimes n})_{p,k}(X_n) \nu(p,k).
$$

Here, the conditional measure $(\mu^{\otimes n})_{p,k}$ is supported on $\{X_n, t(X_n) \in I_p, n_{\mathcal{O}}(X_n) = k\}$, and given by

$$
\begin{aligned}
(8.3.18)\qquad (\mu^{\otimes n})_{p,k}(A) &= \frac{\mu^{\otimes n}(A \cap \{t(X_n) \in I_p, n_{\mathcal{O}}(X_n) = k\})}{\mu^{\otimes n}(\{t(X_n) \in I_p, n_{\mathcal{O}}(X_n) = k\})} \\
&= \frac{\mu^{\otimes n}(A \cap \{t(X_n) \in I_p, n_{\mathcal{O}}(X_n) = k\})}{\nu(p,k)}
\end{aligned}
$$

Then, for given p, k, we define the multiplicity measure

$$
(8.3.19)\qquad d\mathsf{m}(Y_{\bar{\mathrm{n}}}, p, k) = \int_{X_n | Y_{\bar{\mathrm{n}}} \in \Phi(X_n)} d\mu^{\otimes(\bar{\mathrm{n}}-k)}((Y_{\bar{\mathrm{n}}})_{S_k^c}) d(\mu^{\otimes n})_{p,k}(X_n),
$$

where for each $Y_{\bar{n}} \in \Phi(X_n)$, S_k is a set of indices of cardinality k such that the configuration $(Y_{\bar{n}})_{S_k}$ coincides with $\{X_n\} \cap Q_{t(X_n)}$ (here $(Y_{\bar{n}})_S$ denotes the restriction of the $\bar{n}$-tuple to the indices in S). By Fubini's theorem, we can write

$$\begin{aligned}(8.3.20)\quad &\int_{\Omega^{\bar{n}}} \exp\left(-\beta \mathsf{F}(Y_{\bar{n}}, \mu, \Omega)\right) d\mathsf{m}(Y_{\bar{n}}, p, k)\\ &= \int_{\Omega^n} \left(\int_{\Phi(X_n)} \exp\left(-\beta \mathsf{F}(Y_{\bar{n}}, \mu, \Omega)\right) d\mu^{\otimes(\bar{n}-k)}((Y_{\bar{n}})_{S_k^c}) \right) d(\mu^{\otimes n})_{p,k}(X_n).\end{aligned}$$

We next rewrite the right-hand side. In view of (7.2.17), we have

$$\begin{aligned}\mathsf{F}(Y_{\bar{n}}, \mu, \Omega) \le\ &\mathsf{G}_U^{\mathrm{inn}}(X_n, \Omega)\\ &+ C\left(\frac{\ell s}{\tilde{\ell}} + R^{\mathsf{d}-1}\tilde{\ell} + \mathsf{F}(Z_{\bar{n}-n_{\mathcal{O}}}, \tilde{\mu}, \mathcal{N}) + |n-\bar{n}| + \sum_{(i,j)\in J} \mathsf{g}(x_i - z_j) \right).\end{aligned}$$

Inserting into the above, we obtain

$$\begin{aligned}(8.3.21)\quad &\int_{\Omega^{\bar{n}}} \exp\left(-\beta \mathsf{F}(Y_{\bar{n}}, \mu, \Omega)\right) d\mathsf{m}(Y_{\bar{n}}, p, k)\\ &\ge \int_{\Omega^n} \left(\int_{\Phi(X_n)} e^{-\beta \mathsf{G}_U^{\mathrm{inn}}(X_n,\Omega) - C\beta \mathsf{F}(Z_{\bar{n}-n_{\mathcal{O}}}, \tilde{\mu}, \mathcal{N}) - C\beta\left(\frac{\ell s}{\tilde{\ell}} + R^{\mathsf{d}-1}\tilde{\ell} + |n-\bar{n}| + \sum_{(i,j)\in J} \mathsf{g}(x_i - z_j)\right)} \right.\\ &\qquad\qquad \left. \times\, d\mu^{\otimes(\bar{n}-k)}((Y_{\bar{n}})_{S_k^c}) \right) d(\mu^{\otimes n})_{p,k}(X_n)\\ &= \binom{\bar{n}}{k} \int_{\Omega^n} \left(\int e^{-C\beta \mathsf{F}(Z_{\bar{n}-n_{\mathcal{O}}}, \tilde{\mu}, \mathcal{N}) - C\beta \sum_{(i,j)\in J} \mathsf{g}(x_i - z_j)} d\mu^{\otimes(\bar{n}-k)}|_{Q_R \setminus Q_{t(X_n)}}(Z_{\bar{n}-k}) \right)\\ &\quad \times \exp\left(-\beta \mathsf{G}_U^{\mathrm{inn}}(X_n, \mu, \Omega) - C\beta\left(\frac{\ell s}{\tilde{\ell}} + R^{\mathsf{d}-1}\tilde{\ell}\right) - C\beta|n-\bar{n}| \right) d(\mu^{\otimes n})_{p,k}(X_n).\end{aligned}$$

Lemma 8.6 above provides the bound (8.3.12), inserting it into the above relation, we are led to

$$\begin{aligned}(8.3.22)\quad &\int_{\Omega^{\bar{n}}} \exp\left(-\beta \mathsf{F}(Y_{\bar{n}}, \mu, \Omega)\right) d\mathsf{m}(Y_{\bar{n}}, p, k)\\ &\ge \exp\left(-C\left(\beta\frac{\ell s}{\tilde{\ell}} + \beta\chi(\beta) R^{\mathsf{d}-1}\tilde{\ell} + \frac{s}{\ell\tilde{\ell}} + \ell^{-1} R^{\mathsf{d}-1} \right) - C\beta|n-\bar{n}| \right)\\ &\times \binom{\bar{n}}{k} (\bar{n}-k)^{\bar{n}-k} \int_{\Omega^n} \exp\left(\mu(\mathcal{N}) - \tilde{\mu}(\mathcal{N})\right) \exp\left(-\beta \mathsf{G}_U^{\mathrm{inn}}(X_n, \mu, \Omega)\right) d(\mu^{\otimes n})_{p,k}(X_n).\end{aligned}$$

As in (8.3.15), in view of (7.2.14), we have $\tilde{\mu}(\mathcal{N}) = \bar{\mathrm{n}} - k$, on the other hand we have $\mu(\mathcal{N}) = \mu(Q_R \backslash Q_{R-(p+1)\eta}) + O(\eta R^{\mathsf{d}-1})$. Inserting this into (8.3.22), passing factors onto the left-hand side and summing over p and k after multiplying by $\nu(p,k)$, in view of (8.3.17) we obtain

$$\begin{aligned}
&\int_{\Omega^n} \exp\left(-\beta \mathsf{G}_U^{\mathrm{inn}}(X_n,\mu,\Omega)\right) d\mu^{\otimes n}(X_n) \\
&\quad \le \exp\left(C\left(\beta \frac{\ell s}{\tilde{\ell}} + \beta\chi(\beta) R^{\mathsf{d}-1}\tilde{\ell} + \frac{s}{\ell\tilde{\ell}} + \ell^{-1}R^{\mathsf{d}-1}\right) - C\beta|n-\bar{\mathrm{n}}|\right) \\
&\quad \times \sum_{(p,k)\in\mathcal{A}} \nu(p,k) \frac{e^{(\bar{\mathrm{n}}-k)-\mu(Q_R\backslash Q_{R-(p+1)\eta}) + O(\eta R^{\mathsf{d}-1})}}{\binom{\bar{\mathrm{n}}}{k}(\bar{\mathrm{n}}-k)^{\bar{\mathrm{n}}-k}} \\
&\quad \times \int_{\Omega^{\bar{\mathrm{n}}}} \exp\left(-\beta \mathsf{F}(Y_{\bar{\mathrm{n}}},\mu,\Omega)\right) d\mathsf{m}(Y_{\bar{\mathrm{n}}},p,k).
\end{aligned} \tag{8.3.23}$$

We may rewrite this as

$$\begin{aligned}
&\int_{\Omega^n} \exp\left(-\beta \mathsf{G}_U^{\mathrm{inn}}(X_n,\mu,\Omega)\right) d\mu^{\otimes n}(X_n) \\
&\le \exp\left(C\left(\beta \frac{\ell s}{\tilde{\ell}} + \beta\chi(\beta) R^{\mathsf{d}-1}\tilde{\ell} + \frac{s}{\ell\tilde{\ell}} + \ell^{-1}R^{\mathsf{d}-1} + \eta R^{\mathsf{d}-1}\right) - C\beta|n-\bar{\mathrm{n}}|\right) \\
&\quad \times \max_{(k,p)\in\mathcal{A}} \left(e^{(\bar{\mathrm{n}}-k)-\mu(Q_R\backslash Q_{R-(p+1)\eta})} \frac{\binom{n}{k}\mu(Q_R\backslash Q_{R-(p+1)\eta})^{n-k}}{\binom{\bar{\mathrm{n}}}{k}(\bar{\mathrm{n}}-k)^{\bar{\mathrm{n}}-k}} \right) \\
&\quad \times \sum_{(p,k)\in\mathcal{A}} \frac{\nu(p,k)}{\binom{n}{k}\mu(Q_R\backslash Q_{R-(p+1)\eta})^{n-k}} \int_{\Omega^{\bar{\mathrm{n}}}} \exp\left(-\beta \mathsf{F}(Y_{\bar{\mathrm{n}}},\mu,\Omega)\right) d\mathsf{m}(Y_{\bar{\mathrm{n}}},p,k).
\end{aligned} \tag{8.3.24}$$

But by definition of m (8.3.19), we have

$$\begin{aligned}
&\sum_{(p,k)\in\mathcal{A}} \frac{\nu(p,k)}{\binom{n}{k}\mu(Q_R\backslash Q_{R-(p+1)\eta})^{n-k}} \int_{\Omega^{\bar{\mathrm{n}}}} \exp\left(-\beta \mathsf{F}(Y_{\bar{\mathrm{n}}},\mu,\Omega)\right) d\mathsf{m}(Y_{\bar{\mathrm{n}}},p,k) \\
&= \sum_{(p,k)\in\mathcal{A}} \int_{\Omega^{\bar{\mathrm{n}}}} \int_{X_n | Y_{\bar{\mathrm{n}}}\in\Phi(X_n)} d\mu^{\otimes(\bar{\mathrm{n}}-k)}((Y_{\bar{\mathrm{n}}})_{S_k^c}) d(\mu^{\otimes n})_{p,k}(X_n)\nu(p,k) \\
&\qquad\qquad \times \frac{e^{-\beta\mathsf{F}(Y_{\bar{\mathrm{n}}},\mu,\Omega)}}{\binom{n}{k}\mu(Q_R\backslash Q_{R-(p+1)\eta})^{n-k}}.
\end{aligned}$$

Inserting (8.3.18), letting S'_k denote a set of indices of cardinality $n-k$ such that the points of $(X_n)_{S'_k}$ coincide with $\{X_n\} \cap \mathcal{O}(X_n)$, we obtain

$$\begin{aligned}
&\sum_{(p,k)\in\mathcal{A}} \frac{\nu(p,k)}{\binom{n}{k}\mu(Q_R\backslash Q_{R-(p+1)\eta})^{n-k}} \int_{\Omega^{\bar{n}}} e^{-\beta \mathsf{F}(Y_{\bar{n}},\mu,\Omega)} d\mathsf{m}(Y_{\bar{n}},p,k) \\
&= \sum_{(p,k)\in\mathcal{A}} \int_{\Omega^{\bar{n}}} \int_{X_n|Y_{\bar{n}}\in\Phi(X_n)} d\mu^{\otimes(\bar{n}-k)}((Y_{\bar{n}})_{S_k^c}) d\mu^{\otimes n}(X_n) \mathbf{1}_{n_{\mathcal{O}}(X_n)=k, t(X_n)\in I_p} \\
&\quad \times \frac{e^{-\beta \mathsf{F}(Y_{\bar{n}},\mu,\Omega)}}{\binom{n}{k}\mu(Q_R\backslash Q_{R-(p+1)\eta})^{n-k}} \\
&= \sum_{(p,k)\in\mathcal{A}} \int_{\Omega^{\bar{n}}} \int_{X_n|Y_{\bar{n}}\in\Phi(X_n)} d\mu^{\otimes(\bar{n}-k)}((Y_{\bar{n}})_{S_k^c}) d\mu^{\otimes k}((X_n)_{S'_k}) d\mu^{\otimes(n-k)}((X_n)_{(S'_k)^c}) \\
&\quad \times \mathbf{1}_{n_{\mathcal{O}}(X_n)=k, t(X_n)\in I_p} \frac{e^{-\beta \mathsf{F}(Y_{\bar{n}},\mu,\Omega)}}{\binom{n}{k}\mu(Q_R\backslash Q_{R-(p+1)\eta})^{n-k}} \\
&= \sum_{(p,k)\in\mathcal{A}} \int_{\Omega^{\bar{n}}} \int_{X_n|Y_{\bar{n}}\in\Phi(X_n)} d\mu^{\otimes\bar{n}}(Y_{\bar{n}}) d\mu^{\otimes(n-k)}((X_n)_{(S'_k)^c}) \mathbf{1}_{n_{\mathcal{O}}(X_n)=k, t(X_n)\in I_p} \\
&\quad \times \frac{e^{-\beta \mathsf{F}(Y_{\bar{n}},\mu,\Omega)}}{\binom{n}{k}\mu(Q_R\backslash Q_{R-(p+1)\eta})^{n-k}} \\
&= \int_{\Omega^{\bar{n}}} e^{-\beta \mathsf{F}(Y_{\bar{n}},\mu,\Omega)} d\mu^{\otimes\bar{n}}(Y_{\bar{n}}) \sum_{(p,k)\in\mathcal{A}} \int_{X_n|Y_{\bar{n}}\in\Phi(X_n)} d\mu^{\otimes(n-k)}((X_n)_{(S'_k)^c}) \\
&\quad \times \frac{\mathbf{1}_{n_{\mathcal{O}}(X_n)=k, t(X_n)\in I_p}}{\binom{n}{k}\mu(Q_R\backslash Q_{R-(p+1)\eta})^{n-k}} \\
&\le \int_{\Omega^{\bar{n}}} e^{-\beta \mathsf{F}(Y_{\bar{n}},\mu,\Omega)} d\mu^{\otimes\bar{n}}(Y_{\bar{n}}) \sum_{(p,k)\in\mathcal{A}} 1.
\end{aligned}$$

Inserting that $\#\mathcal{A} \le C\frac{\bar{n}\tilde{\ell}}{\eta}$, we conclude that

$$\begin{aligned}
(8.3.25)\quad &\int_{\Omega^n} \exp\left(-\beta \mathsf{G}_U^{\mathrm{inn}}(X_n,\mu,\Omega)\right) d\mu^{\otimes n}(X_n) \\
&\le C\frac{\bar{n}\tilde{\ell}}{\eta} e^{C\left(\beta\frac{\ell_s}{\ell}+\beta\chi(\beta)R^{\mathsf{d}-1}\tilde{\ell}+\frac{s}{\ell\tilde{\ell}}+\ell^{-1}R^{\mathsf{d}-1}+\eta R^{\mathsf{d}-1}\right)-C\beta|n-\bar{n}|} \\
&\quad \times \max_{(k,p)\in\mathcal{A}} \left(e^{\bar{n}-k-\mu(Q_R\backslash Q_{R-(p+1)\eta})} \frac{\binom{n}{k}\mu(Q_R\backslash Q_{R-(p+1)\eta})^{n-k}}{\binom{\bar{n}}{k}(\bar{n}-k)^{\bar{n}-k}} \right) \\
&\quad \times \int_{\Omega^{\bar{n}}} \exp\left(-\beta \mathsf{F}(Y_{\bar{n}},\mu,\Omega)\right) d\mu^{\otimes\bar{n}}(Y_{\bar{n}}).
\end{aligned}$$

We now let $(k_0,p_0)\in\mathcal{A}$ be an integer couple where the max is achieved, which must satisfy (8.3.16). Denoting for shortcut $\alpha=\bar{n}-k_0$ and $\alpha'=\mu(Q_R\backslash Q_{R-(p_0+1)\eta})$, the relation (8.3.7) is then satisfied, and we next study the factor

$$\frac{\binom{n}{k_0}e^{\bar{n}-k_0-\alpha'}\mu(Q_R\backslash Q_{R-(p_0+1)\eta})^{n-k_0}}{\binom{\bar{n}}{k_0}(\bar{n}-k_0)^{\bar{n}-k_0}} = \frac{n!}{\bar{n}!}\left(\frac{\bar{n}!}{n!}\frac{\binom{n}{k_0}e^{\alpha-\alpha'}(\alpha')^{n-k_0}}{\binom{\bar{n}}{k_0}(\bar{n}-k_0)^{\bar{n}-k_0}}\right).$$

By Stirling's formula, substituting $k_0 = \bar{n} - \alpha$, we have

$$\log\left(\frac{\bar{n}!}{n!}\frac{\binom{n}{k_0}e^{\alpha-\alpha'}(\alpha')^{n-k_0}}{\binom{\bar{n}}{k_0}(\bar{n}-k_0)^{\bar{n}-k_0}}\right)$$

$$= \alpha-\alpha'-\bar{n}+n+(n-k_0)\log\alpha'-(n-k_0)\log(n-k_0)-\frac{1}{2}\log\frac{\bar{n}(n-k_0)}{(\bar{n}-k_0)n}-O(1)$$

$$= \alpha-\alpha'+n-\bar{n}-(n-\bar{n}+\alpha)\log\frac{\alpha}{\alpha'}+(n-\bar{n}+\alpha)\log\alpha$$

$$-(n-\bar{n}+\alpha)\log(n-\bar{n}+\alpha)-\frac{1}{2}\log\frac{\bar{n}(n-\bar{n}+\alpha)}{n\alpha}+O(1)$$

$$= \alpha-\alpha'+n-\bar{n}-(n-\bar{n}+\alpha)\log\frac{\alpha}{\alpha'}$$

$$-(n-\bar{n}+\alpha+\frac{1}{2})\log(1+\frac{n-\bar{n}}{\alpha})-\frac{1}{2}\log\frac{\bar{n}}{n}+O(1).$$

Choosing $\eta = \ell^{-1}$ and inserting into (8.3.25) we obtain the result (the outer case is completely analogous).

We next turn to the second case, where $\ell^{\mathsf{d}+1} \geq Cz$ and (7.2.8) is used in Proposition 7.19. Then, for each configuration X_n, Γ, the boundary of $\mathcal{O}$ is a piecewise affine boundary with facets parallel to the faces of Q_R, included in some $Q_{t+\ell}\backslash Q_\ell$. We need to split the phase-space into regions where we know this boundary location up to a $O(\eta)$ error. There are $O(\frac{\tilde{\ell}}{\ell})$ strips of the form $Q_{t+\ell}\backslash Q_\ell$ in which Γ can lie. For each of them there are $O(\frac{R^{\mathsf{d}-1}}{\ell^{\mathsf{d}-1}})$ facets forming Γ, and $O(\ell/\eta)$ choices for each so that we know their location up to an error η. This allows to partition the phase-space into $C_1\frac{\tilde{\ell}}{\ell}\left(C_2\frac{\ell}{\eta}\right)^{C_3\frac{R^{\mathsf{d}-1}}{\ell^{\mathsf{d}-1}}}$ sets such that each Γ in one of these sets is at distance $\leq \eta$ of a reference boundary. Up to using translations and increasing the factor C_2 in the multiplicity, we may ensure that each Γ in entirely included in one such set. This allows to run the same proof as in the first case, except with the multiplicity $\frac{\bar{n}\tilde{\ell}}{\eta}$ at the end replaced by $\bar{n}\frac{\tilde{\ell}}{\ell}\left(C\frac{\ell}{\eta}\right)^{\frac{R^{\mathsf{d}-1}}{\ell^{\mathsf{d}-1}}}$, which adds a term $O\left(\frac{R^{\mathsf{d}-1}}{\ell^{\mathsf{d}-1}}\log\frac{\ell}{\eta}\right)$ to the volume errors. The choice $\eta = \ell^{1-\mathsf{d}}$ yields the announced estimate. □

Remark 8.7. When summing the contributions over Ω where n points fall and $U\backslash\Omega$ where $N-n$ points fall, the errors of (8.3.9) compensate and add up to a well bounded error. More precisely, if α, α', respectively γ, γ' satisfy (8.3.7), then for every n we have

(8.3.26)

$$\alpha-\alpha'-(n-\bar{n}+\alpha)\log\frac{\alpha}{\alpha'}-(n-\bar{n}+\alpha+\frac{1}{2})\log\left(1+\frac{n-\bar{n}}{\alpha}\right)-\frac{1}{2}\log\frac{n}{\bar{n}}$$

$$+\gamma-\gamma'-(\bar{n}-n+\gamma)\log\frac{\gamma}{\gamma'}-(\bar{n}-n+\gamma+\frac{1}{2})\log\left(1+\frac{\bar{n}-n}{\gamma}\right)-\frac{1}{2}\log\frac{N-n}{N-\bar{n}}$$

$$\leq C\left(\frac{R^{\mathsf{d}-1}}{\tilde{\ell}}+\frac{s^2}{\tilde{\ell}^3R^{\mathsf{d}-1}}\right).$$

PROOF. First we notice that since the expressions arising here originate in Stirling's formula, they can be restricted to the case of $\alpha+n-\bar{n}\geq 1$, $\gamma+\bar{n}-n\geq 1$, $n\geq 1$ and $N-n\geq 1$ (all the quantities involved are integers).

We then study the expression in the left-hand side of (8.3.26) as a function of the real variable n (with the above constraints). Differentiating in n, we find that it achieves its maximum when

$$\log\frac{\gamma\alpha'}{\gamma'\alpha} - \log\left(1+\frac{n-\bar{\mathrm{n}}}{\alpha}\right) + \frac{1}{2(\alpha+n-\bar{\mathrm{n}})} + \log\left(1+\frac{\bar{\mathrm{n}}-n}{\gamma}\right) - \frac{1}{2(\gamma+\bar{\mathrm{n}}-n)} + \frac{1}{2n} - \frac{1}{2(N-n)} = 0.$$

Using $\alpha+n-\bar{\mathrm{n}}\geq 1$, $\gamma+\bar{\mathrm{n}}-n\geq 1$, $n\geq 1$, $N-n\geq 1$ and (8.3.7) we deduce that

$$\left|\log\left(1+\frac{\bar{\mathrm{n}}-n}{\gamma}\right) - \log\left(1+\frac{n-\bar{\mathrm{n}}}{\alpha}\right)\right| \leq C$$

and thus

$$\frac{1+\frac{\bar{\mathrm{n}}-n}{\gamma}}{1+\frac{n-\bar{\mathrm{n}}}{\alpha}} \text{ is bounded above and below}$$

and it follows easily in view of (8.3.7) that $|n-\bar{\mathrm{n}}|\leq C\tilde{\ell}R^{\mathsf{d}-1}$. To find the maximum of (8.3.26) it thus suffices to maximize it for such n's. But for such n's we may check that $\frac12\log\left(1+\frac{n-\bar{\mathrm{n}}}{\alpha}\right)$, $\frac12\log\left(1+\frac{\bar{\mathrm{n}}-n}{\gamma}\right)$, $\log\frac{n}{\bar{\mathrm{n}}}$ and $\log\frac{N-n}{N-\bar{\mathrm{n}}}$ are all bounded by a constant depending only on $\mathsf{d}, m, \|\mu\|_{L^\infty}$, hence it suffices to obtain a bound for

$$\alpha-\alpha'+(\bar{\mathrm{n}}-n-\alpha)\log\frac{\alpha}{\alpha'} - (\alpha+n-\bar{\mathrm{n}})\log\left(1+\frac{n-\bar{\mathrm{n}}}{\alpha}\right) + \gamma'-\gamma+(n-\bar{\mathrm{n}}-\gamma)\log\frac{\gamma}{\gamma'} - (\gamma+\bar{\mathrm{n}}-n)\log\left(1+\frac{\bar{\mathrm{n}}-n}{\gamma}\right). \tag{8.3.27}$$

Differentiating in n, we find that this expression is maximal exactly for

$$1+\frac{\bar{\mathrm{n}}-n}{\gamma} = \frac{\gamma'\alpha}{\gamma\alpha'}\left(1+\frac{n-\bar{\mathrm{n}}}{\alpha}\right) \Leftrightarrow n = \bar{\mathrm{n}} + \frac{\frac{\gamma}{\gamma'}-\frac{\alpha}{\alpha'}}{\frac{1}{\gamma'}+\frac{1}{\alpha'}},$$

and we observe that this ensures that

$$(\bar{\mathrm{n}}-n)\log\left(\frac{\alpha}{\alpha'}\frac{\gamma'}{\gamma}\frac{1+\frac{n-\bar{\mathrm{n}}}{\alpha}}{1+\frac{\bar{\mathrm{n}}-n}{\gamma}}\right) = 0.$$

Inserting this into (8.3.27) we find that the maximum of the expression is then equal to

$$\alpha-\alpha'-\alpha\log\frac{\alpha}{\alpha'}+\gamma-\gamma'-\gamma\log\frac{\gamma}{\gamma'} - \alpha\log\left(1+\frac{n-\bar{\mathrm{n}}}{\alpha}\right) - \gamma\log\left(1+\frac{\bar{\mathrm{n}}-n}{\gamma}\right) = O\left(\frac{R^{\mathsf{d}-1}}{\tilde{\ell}} + \frac{s^2}{\tilde{\ell}^3R^{\mathsf{d}-1}}\right),$$

where we used Taylor expansion and (8.3.7). □

The next corollary allows to compare the Neumann partition function to the "Dirichlet" partition function reduced to good (and screenable) configurations. It consists in applying the screening procedure with the condition (7.2.8), after local laws have been proved.

Corollary 8.8. *With the same assumptions and notation as in the previous proposition and Definition 7.15, there exists $C > 0$ depending only on $\mathsf{d}, m, \|\mu\|_{L^\infty}$ such that the following holds. Let $M \geq 1$ and*

$$\mathcal{B}_n = \left\{ X_n \in \Omega^n, \sup_x \int_{(\partial\Omega)_{-2\tilde{\ell}} \cap \square_L(x)} |\nabla w_{\tilde{\mathrm{r}}}|^2 \leq M\chi(\beta) L^{\mathsf{d}} \right\}$$

where w is a minimizer for (7.2.9), resp. (7.2.10), relative to Ω, and $(\partial\Omega)_{-2\tilde{\ell}}$ denotes $Q_{R-\tilde{\ell}} \backslash Q_{R-2\tilde{\ell}} \cap U$ if $\Omega = Q_R \cap U$ and $(Q_{R+2\tilde{\ell}} \setminus Q_{R+\tilde{\ell}}) \cap U$ if $\Omega = U \setminus Q_R$. If

$$R > L > CM \max(\chi(\beta), \beta^{-\frac{1}{\mathsf{d}-\mathsf{s}}} \mathbf{1}_{\mathsf{s} \leq 0}), \tag{8.3.28}$$

and $\mathrm{dist}(Q_R, \partial\Lambda \cap U) \geq L$, *we have*

$$\begin{aligned} &\frac{1}{n!} \int_{\mathcal{B}_n} \exp\left(-\beta \mathsf{G}_U^{\mathrm{inn/out}}(X_n, \mu, \Omega)\right) d\mu^{\otimes n}(X_n) \\ &\leq C \frac{\bar{\mathrm{n}}^{\bar{\mathrm{n}}}}{\bar{\mathrm{n}}!} \mathsf{K}_\beta(\Omega, \mu) \exp\Bigg[\beta \left(C R^{\mathsf{d}-1} L \chi(\beta) M + |n - \bar{\mathrm{n}}|\right) \\ &\qquad + C\left(\frac{M\chi(\beta) R^{\mathsf{d}-1}}{L} + \frac{R^{\mathsf{d}-1}}{L^{\mathsf{d}-1}} \log L + \log R\right) \\ &+ n - \bar{\mathrm{n}} + \alpha - \alpha' - (n - \bar{\mathrm{n}} + \alpha) \log \frac{\alpha}{\alpha'} - (n - \bar{\mathrm{n}} + \alpha + \frac{1}{2}) \log\left(1 + \frac{n - \bar{\mathrm{n}}}{\alpha}\right) - \frac{1}{2} \log \frac{n}{\bar{\mathrm{n}}}\Bigg], \end{aligned} \tag{8.3.29}$$

with α, α' satisfying

$$\left|\frac{\alpha'}{\alpha} - 1\right| \leq C \frac{\chi(\beta)}{L}, \qquad \frac{1}{C} L R^{\mathsf{d}-1} \leq \alpha \leq C L R^{\mathsf{d}-1}.$$

PROOF. If X_n in $\mathcal{B}_n$ then

$$\bar{S}(X_n) \leq \frac{R^{\mathsf{d}-1}}{L^{\mathsf{d}-1}} M\chi(\beta) L^{\mathsf{d}}, \quad \bar{S}'(X_n) \leq M\chi(\beta) L^{\mathsf{d}}.$$

using the definition (7.2.11) or (7.2.12). We check that setting $\ell = \tilde{\ell} = L$, $s = M\frac{R^{\mathsf{d}-1}}{L^{\mathsf{d}-1}}\chi(\beta)L^{\mathsf{d}}$ and $z = M\chi(\beta)L^{\mathsf{d}}$, we have that if (8.3.28) holds, then, up to making the constant larger in (8.3.28), in view of (5.2.27), (8.3.5) and (8.3.6) hold. The result follows by applying the result of Proposition 8.5. □

The next goal is to select $s, \ell, \tilde{\ell}$ to optimize the errors made in Proposition 8.5. This way we obtain the main result of this section, which implements the heuristic idea described in the first section of this chapter, and shows that one can run the bootstrap procedure and show that if one has good energy controls at some scale, one can deduce control at half scales.

In the rest of the paper, we will denote the event that X_N has n points in Ω by

$$\mathcal{A}_n := \{X_N \in U^N, \#I_\Omega = n\}. \tag{8.3.30}$$

The next proposition allows to leverage on the screening for a given number of points n falling in Ω. The work in the main proof will then consist in conditioning on n.

Proposition 8.9 (Bootstrap). *Assume U is either $\mathbb{R}^{\mathsf{d}}$ or a finite union of disjoint hyperrectangles all included in Λ with parallel sides belonging to $\mathcal{Q}_\rho$ for some $\rho \geq \max(1, \beta^{-\frac{1}{\mathsf{d}-\mathsf{s}}}\mathbf{1}_{\mathsf{d}\leq 2})$, or the complement of such a set. Let μ be a density such that $\mu \geq m > 0$ in the set Λ and $\mu(U) = N$ is an integer. Let C_0 be the constant of* (7.1.35).

There exists a constant $C > 0$ depending only on d, m and $\|\mu\|_{L^\infty}$ such that the following holds. Assume that Q_R is a hyperrectangle of sidelengths in $[R, 2R]$ with sides parallel to those of U, that $\mu(Q_R \cap U) = \bar{\mathrm{n}}$ and $Q_R \cap U \subset \Lambda$. Assume that there exists a cube $\square_L$ of size L such that

$$\left|\log \mathbb{E}_{\mathbb{Q}_\beta(\mu,U)}\left(\exp\left(\frac{\beta}{2}\left(\mathsf{F}^{\square_L}(\cdot,\mu,U) + 2C_0\#I_{\square_L}\right)\right)\right)\right| \leq \mathcal{C}\beta\chi(\beta)L^{\mathsf{d}} \tag{8.3.31}$$

with $\mathcal{C} > 1$, and assume that $\square_L$ contains $Q_{R+2\tilde{\ell}} \cap U$ with

$$L \geq R \geq \frac{1}{2}L,$$

$$R \geq C' \max(1, \beta^{-\frac{1}{2}}\chi(\beta)^{\frac{1}{2}}) := \rho_\beta \tag{8.3.32}$$

and

$$\tilde{\ell} = C'' \max\left(\left(\frac{R}{\rho_\beta}\right)^{-\frac{2}{3}} R, R^{1-\frac{2}{\mathsf{d}+2}}\right) \tag{8.3.33}$$

for some C', C'' large enough, both depending only on d, m, $\|\mu\|_{L^\infty}$ and $\mathcal{C}$. Assume in addition that

$$\mathrm{dist}(Q_R \cap U, \partial\Lambda \cap U) \geq \tilde{\ell}. \tag{8.3.34}$$

Then there exists a sequence γ_n satisfying

$$\sum_{n=0}^{N} \gamma_n \leq \exp\left(-\mathcal{C}\beta\chi(\beta)R^{\mathsf{d}}\right) \tag{8.3.35}$$

such that we have

$$\begin{aligned} \mathbb{E}_{\mathbb{Q}_\beta(\mu,U)}&\left(\exp\left(\frac{\beta}{2}\left(\mathsf{F}^{Q_{R-2\tilde{\ell}}}(X_N,\mu,U)\right)\mathbf{1}_{\mathcal{A}_n}\right)\right) \\ &\leq \gamma_n + \frac{\mathsf{K}_{\beta/2}(\mu, Q_R)}{\mathsf{K}_\beta(\mu, Q_R)} \exp\left(\beta\left(\frac{\mathcal{C}}{4}\chi(\beta)R^{\mathsf{d}} + C|n - \bar{\mathrm{n}}| + \frac{C_0}{2}n\right)\right). \end{aligned} \tag{8.3.36}$$

Once one has obtained local laws down to the minimal scale ρ_β, Corollary 8.8 will allow to improve the error term and bound it by $R^{\mathsf{d}-1}$.

Proof. **Step 1: The case of excess energy**. We denote $\Omega = Q_R \cap U$ and $\mathring{\Omega} = Q_{R-2\tilde{\ell}} \cap U$. Recalling the definition of $\mathcal{A}_n$ in (8.3.30), and letting $M \geq 1$ be a constant to be determined below, we define

$$\mathcal{B}_n := \left\{X_N \in \mathcal{A}_n, \int_{\square_L \setminus \mathring{\Omega}} |\nabla u_{\tilde{\mathsf{r}}}|^2 \leq M\mathcal{C}\chi(\beta)L^{\mathsf{d}}\right\},$$

where for each configuration, u is defined as solving (7.1.1) over U and $\tilde{\mathsf{r}}$ is as in (7.1.25) relative to $\square_L$. If $X_N \in \mathcal{B}_n^c \cap \mathcal{A}_n$ then

$$\int_{\square_L \setminus \mathring{\Omega}} |\nabla u_{\tilde{\mathsf{r}}}|^2 > M\mathcal{C}\chi(\beta)L^{\mathsf{d}}, \tag{8.3.37}$$

and view of the definition (7.1.23) and (7.1.35) we then have that

$$\begin{aligned}\mathsf{F}^{\square_L}(X_N,\mu,U)+C_0\#I_{\square_L} &\geq \mathsf{F}^{\mathring{\Omega}}(X_N,\mu,U)+\frac{1}{C}\int_{\square_L\setminus\mathring{\Omega}}|\nabla u_{\tilde{\mathsf{r}}}|^2\\ &\geq \mathsf{F}^{\mathring{\Omega}}(X_N,\mu,U)+\frac{M\mathcal{C}\chi(\beta)L^{\mathsf{d}}}{C}.\end{aligned}$$

We deduce that

$$\begin{aligned}&\mathbb{E}_{\mathbb{Q}_\beta(\mu,U)}\left(\exp\left(\frac{\beta}{2}\left(\mathsf{F}^{\square_L}(\cdot,\mu,U)+C_0\#I_{\square_L}\right)\right)\mathbf{1}_{\mathcal{B}_n^c\cap\mathcal{A}_n}\right)\\ &\quad\geq \exp\left(\frac{\beta}{2}\frac{M\mathcal{C}\chi(\beta)L^{\mathsf{d}}}{C}\right)\mathbb{E}_{\mathbb{Q}_\beta(\mu,U)}\left(\exp\left(\frac{\beta}{2}\mathsf{F}^{\mathring{\Omega}}(\cdot,\mu,U)\right)\mathbf{1}_{\mathcal{B}_n^c\cap\mathcal{A}_n}\right).\end{aligned}$$

It follows from (8.3.31) that

$$\mathbb{E}_{\mathbb{Q}_\beta(\mu,U)}\left(\exp\left(\frac{\beta}{2}\mathsf{F}^{\mathring{\Omega}}(\cdot,\mu,U)\right)\mathbf{1}_{\mathcal{B}_n^c\cap\mathcal{A}_n}\right)\leq\gamma_n, \tag{8.3.38}$$

with $\sum_{n=0}^N\gamma_n\leq\exp\left(-\mathcal{C}\beta\chi(\beta)R^{\mathsf{d}}\right)$, provided M is chosen large enough, depending only on $\mathsf{d},m,\|\mu\|_{L^\infty}$. We henceforth fix M.

Step 2: The case of good energy bounds.

We now wish to estimate the same quantity in the event $\mathcal{B}_n$. First we note that being in $\mathcal{B}_n$ implies a control of $S(X_n,u)$ defined as in (7.2.5) or (7.2.6) by $M\mathcal{C}\chi(\beta)L^{\mathsf{d}}$.

We next wish to choose $\tilde{\ell}=\tilde{\epsilon}R$ with $\tilde{\epsilon}<\frac{1}{4}$ to be determined later, and $\ell=\epsilon\tilde{\ell}$ with $0\leq\epsilon\leq 1$, satisfying

$$\ell^{\mathsf{d}+1}\geq\frac{CM\mathcal{C}\chi(\beta)L^{\mathsf{d}}}{\tilde{\ell}} \tag{8.3.39}$$

with C as in (8.3.5). This way, choosing $s=M\mathcal{C}\chi(\beta)L^{\mathsf{d}}$, the screenability condition (8.3.5) is verified for u. To apply Proposition 8.5, we also need

$$\max\left(\beta^{-\frac{1}{\mathsf{d}-\mathsf{s}}}\mathbf{1}_{\mathsf{d}\leq 2},1\right)\leq\tilde{\epsilon}R \tag{8.3.40}$$

and

$$M\mathcal{C}\chi(\beta)L^{\mathsf{d}}<c\tilde{\ell}^2R^{\mathsf{d}-1}. \tag{8.3.41}$$

Using (7.1.37), Lemma 7.18 (which applies since u is screenable) and (7.1.35) which yields some nonnegativity, we have

$$\begin{aligned}&\frac{\beta}{2}\mathsf{F}^{\mathring{\Omega}}(X_N,U)-\beta\mathsf{F}(X_N,U)\\ &\quad\leq\frac{\beta}{2}\mathsf{F}^{\mathring{\Omega}}(X_N,U)-\beta\mathsf{F}^{\Omega}(X_N,U)-\beta\mathsf{F}^{U\setminus\Omega}(X_N,U)\\ &\quad\leq\frac{\beta}{2}\mathsf{F}^{\mathring{\Omega}}(X_N,U)-\frac{\beta}{2}\mathsf{F}^{\Omega}(X_N,U)-\frac{\beta}{2}\mathsf{F}^{\Omega}(X_N,U)-\beta\mathsf{F}^{U\setminus\Omega}(X_N,U)\\ &\quad\leq-\frac{\beta}{2}\mathsf{F}^{\Omega\setminus\mathring{\Omega}}(X_N,U)-\frac{\beta}{2}\mathsf{F}^{\Omega}(X_N,U)-\beta\mathsf{F}^{U\setminus\Omega}(X_N,U)\\ &\quad\leq-\frac{\beta}{2}\mathsf{F}^{\Omega}(X_N,U)-\beta\mathsf{F}^{U\setminus\Omega}(X_N,U)+\frac{\beta}{2}C_0n\\ &\quad\leq-\frac{\beta}{2}\mathsf{G}_U^{\mathrm{inn}}(X_N|_\Omega,\Omega)-\beta\mathsf{G}_U^{\mathrm{out}}(X_N|_{U\setminus\Omega},U\setminus\Omega)+\frac{\beta}{2}C_0n.\end{aligned}$$

Thus,

$$\begin{aligned}
&\mathbb{E}_{\mathbb{Q}_\beta(\mu,U)}\left(\exp\left(\frac{\beta}{2}\mathsf{F}^{\mathring{\Omega}}(\cdot,U)\right)\mathbf{1}_{\mathcal{B}_n}\right)\\
&= \frac{1}{N^N\mathsf{K}_\beta(\mu,U)}\int_{\mathcal{B}_n}\exp\left(\frac{\beta}{2}\mathsf{F}^{\mathring{\Omega}}(X_N,U)-\beta\mathsf{F}(X_N,U)\right)d\mu^{\otimes N}\\
&\le \frac{1}{N^N\mathsf{K}_\beta(\mu,U)}\frac{N!}{n!(N-n)!}\int_{\Omega^n\cap\mathcal{B}_n}\exp\left(-\frac{\beta}{2}\mathsf{G}_U^{\mathrm{inn}}(\cdot,\Omega)+\frac{\beta}{2}C_{\mathrm{O}}n\right)d\mu^{\otimes n}\\
&\quad\times\int_{(U\backslash\Omega)^{N-n}\cap\mathcal{B}_n}\exp\left(-\beta\mathsf{G}_U^{\mathrm{out}}(\cdot,U\backslash\Omega)\right)d\mu^{\otimes(N-n)}.
\end{aligned}$$

Inserting (8.3.11) applied in Ω (with $\beta/2$ instead of β) and in $U\backslash\Omega$ and using Remark 8.7, we deduce that

$$\begin{aligned}
&\mathbb{E}_{\mathbb{Q}_\beta(\mu,U)}\left(\exp\left(\frac{\beta}{2}\left(\mathsf{F}^{\mathring{\Omega}}(\cdot,U)\right)\right)\mathbf{1}_{\mathcal{B}_n}\right)\\
&\le \frac{1}{N^N\mathsf{K}_\beta(\mu,U)}\\
&\quad\times\frac{N!}{\bar{\mathrm{n}}!(N-\bar{\mathrm{n}})!}C\bar{\mathrm{n}}^{\bar{\mathrm{n}}}(N-\bar{\mathrm{n}})^{N-\bar{\mathrm{n}}}\mathsf{K}_{\beta/2}(\mu,\Omega)\mathsf{K}_\beta(\mu,U\backslash\Omega)\exp\left(\beta\varepsilon_e+\varepsilon_v+\frac{\beta}{2}C_{\mathrm{O}}n\right)
\end{aligned}$$

with

$$\varepsilon_e := C\left(\ell\frac{MC\chi(\beta)L^{\mathsf{d}}}{\tilde{\ell}}+R^{\mathsf{d}-1}\tilde{\ell}\chi(\beta)+|n-\bar{\mathrm{n}}|\right) \tag{8.3.42}$$

and

$$\varepsilon_v := C\left(\frac{MC\chi(\beta)L^{\mathsf{d}}}{\ell\tilde{\ell}}+\frac{R^{\mathsf{d}-1}}{\ell}+\frac{(MC\chi(\beta)L^{\mathsf{d}})^2}{\tilde{\ell}^3R^{\mathsf{d}-1}}+\log R\right), \tag{8.3.43}$$

where we used the choice $s := MC\chi(\beta)L^{\mathsf{d}}$ and $\ell \le \tilde{\ell}$. We may also bound from below $\mathsf{K}_\beta(\mu,U)$ using (7.1.17) applied with the sets Ω and $U\backslash\Omega$, which yields

$$\frac{N!\bar{\mathrm{n}}^{\bar{\mathrm{n}}}(N-\bar{\mathrm{n}})^{N-\bar{\mathrm{n}}}}{N^N\mathsf{K}_\beta(\mu,U)\bar{\mathrm{n}}!(N-\bar{\mathrm{n}})!}\mathsf{K}_{\beta/2}(\mu,\Omega)\mathsf{K}_\beta(\mu,U\backslash\Omega)\le\frac{\mathsf{K}_{\beta/2}(\mu,\Omega)}{\mathsf{K}_\beta(\mu,\Omega)}.$$

Inserting into the above, we obtain that

$$\mathbb{E}_{\mathbb{Q}_\beta(\mu,U)}\left(\exp\left(\frac{\beta}{2}\mathsf{F}^{\mathring{\Omega}}(\cdot,U)\right)\mathbf{1}_{\mathcal{B}_n}\right)\le C\frac{\mathsf{K}_{\beta/2}(\mu,\Omega)}{\mathsf{K}_\beta(\mu,\Omega)}\exp\left(\beta\varepsilon_e+\varepsilon_v+\frac{\beta}{2}C_{\mathrm{O}}n\right). \tag{8.3.44}$$

We now search for the smallest $\tilde{\ell}$ such that the terms of $\beta\varepsilon_e+\varepsilon_v$ (except those involving n and $\bar{\mathrm{n}}$) are $\le \beta\chi(\beta)\frac{\mathcal{C}}{8}R^{\mathsf{d}}$ that is

$$\begin{cases}
C\frac{MC\chi(\beta)L^{\mathsf{d}}\ell}{\tilde{\ell}}\le\frac{\mathcal{C}}{8}\chi(\beta)R^{\mathsf{d}},\\
CR^{\mathsf{d}-1}\tilde{\ell}\chi(\beta)\le\frac{\mathcal{C}}{8}\chi(\beta)R^{\mathsf{d}},\\
C\frac{MC\chi(\beta)L^{\mathsf{d}}}{\ell\tilde{\ell}}\le\frac{\mathcal{C}}{8}\beta\chi(\beta)R^{\mathsf{d}},\\
C\frac{R^{\mathsf{d}-1}}{\ell}\le\frac{\mathcal{C}}{8}\beta\chi(\beta)R^{\mathsf{d}},\\
C\frac{(MC\chi(\beta)L^{\mathsf{d}})^2}{\tilde{\ell}^3R^{\mathsf{d}-1}}\le\frac{\mathcal{C}}{8}\beta\chi(\beta)R^{\mathsf{d}}\\
\log R\le\frac{\mathcal{C}}{8}\beta\chi(\beta)R^{\mathsf{d}},
\end{cases}$$

and also (8.3.39), (8.3.40), and (8.3.41) are satisfied. With our choice $R \le L \le 2R$ and $\ell = \epsilon\tilde{\ell}$, $\tilde{\ell} = \tilde{\epsilon}R$, after direct computations we find that these reduce to

$$\begin{cases} C\epsilon M \le \frac{1}{8}, \\ C\tilde{\epsilon} \le \frac{\mathcal{C}}{8}, \\ \frac{CM}{\epsilon\tilde{\epsilon}^2} \le \frac{\beta}{8}R^2, \\ \frac{C}{\epsilon\tilde{\epsilon}} \le \frac{\mathcal{C}}{8}\beta\chi(\beta)R^2, \\ CM^2\mathcal{C}\chi(\beta) \le \frac{\beta}{8}R^2\tilde{\epsilon}^3, \\ C\log R \le \frac{\mathcal{C}}{8}\beta\chi(\beta)R^{\mathsf{d}}, \\ \epsilon^{\mathsf{d}+1}\tilde{\epsilon}^{\mathsf{d}+2}R^2 \ge CM\mathcal{C}\chi(\beta), \\ \tilde{\epsilon}R \ge \max\left(\beta^{-\frac{1}{\mathsf{d}-\mathsf{s}}}\mathbf{1}_{\mathsf{d}\le 2}, 1\right), \\ CM\mathcal{C}\chi(\beta) < \tilde{\epsilon}^2 R, \end{cases}$$

for some constant $C > 0$ large enough, and depending only on d, m and $\|\mu\|_{L^\infty}$. We can take ϵ small enough that the first condition is realized. Inspecting the definition (5.2.27), we find that in all dimensions we can then make the other conditions realized by requiring

$$\begin{cases} R^2\min(\beta\chi(\beta)^{-1}, 1) \ge C' \\ \tilde{\epsilon} = C''\max\left((R^2\beta\chi(\beta)^{-1})^{-1/3}, R^{-\frac{2}{\mathsf{d}+2}}\right), \end{cases}$$

where C', C'' are large enough constants depending on the other parameters.

Combining (8.3.44) with (8.3.38), we obtain the result. □

We now can complete the proof of the local laws.

PROOF OF THEOREM 8.4. We first note that it is enough to prove the local law result in hyperrectangles $Q_R \in \mathcal{Q}_R$ (as in Definition 7.14), with sides parallel to those of U and even more generally in $Q_{R-2\tilde{\ell}}$ if $\tilde{\ell} < \frac{1}{4}R$, with $R \ge \rho_\beta$. Also, if U is a hyperrectangle, thanks to a similar reasoning, we may ignore the condition in Definition 7.15 about the faces of $\partial(Q_R \cap U)$ being sufficiently far from the parallel faces of U (see also the discussion in the proof of Theorem 8.2) .

Indeed, thanks to the lower bound on μ, general cubes of size $R \ge \rho_\beta$ can be covered by a finite number of such hyperrectangles. The proof then proceeds by a bootstrap on the scales: we wish to show that if

$$\log \mathbb{E}_{\mathbb{Q}_\beta(\mu,U)}\left(\exp\left(\frac{\beta}{2}\left(\mathsf{F}^{\square_L(x)}(\cdot,U) + 2C_0\#I_{\square_L(x)}\right)\right)\right) \le \mathcal{C}\beta\chi(\beta)L^{\mathsf{d}}, \tag{8.3.45}$$

for any $\square_L(x)$ sufficiently far from $\partial\Lambda$, then if $\frac{3}{4}L \ge R \ge \frac{1}{2}L$, and as long as R is large enough, we have

$$\log \mathbb{E}_{\mathbb{Q}_\beta(\mu,U)}\left(\exp\left(\frac{\beta}{2}\left(\mathsf{F}^{Q_{R-2\tilde{\ell}}}(\cdot,U) + 2C_0\#I_{Q_R}\right)\right)\right) \le \mathcal{C}\beta\chi(\beta)R^{\mathsf{d}}. \tag{8.3.46}$$

By iteration, this will clearly imply the result: indeed rewriting

$$\mathbb{E}_{\mathbb{Q}_\beta(\mu,U)}\left(\exp\left(\frac{\beta}{2}(\mathsf{F}(\cdot,U))\right)\right)$$

as $\frac{\mathsf{K}_{\beta/2}(\mu,U)}{\mathsf{K}_\beta(\mu,U)}$ as in the proof of Corollary 5.20, and using (7.1.55), we have that (8.3.45) holds for $L \geq \frac{1}{2}N^{\frac{1}{\mathsf{d}}}$ up to changing $\mathcal{C}$ if necessary. Without loss of generality, we may now assume for the rest of the proof that $L \leq \frac{1}{2}N^{\frac{1}{\mathsf{d}}}$.

To make sure that the constants are independent of β and R, we have used the notation $\mathcal{C}$, and we wish to prove (8.3.46) with the same constant $\mathcal{C}$ as in (8.3.45). In the sequel, unless specified, *all constants* $C>0$ *will be independent of* $\mathcal{C}$, i.e., they may depend only on d, m and $\|\mu\|_{L^\infty}$.

Let us now consider $Q_R \in \mathcal{Q}_R$, denote $\bar{\mathrm{n}} = \mu(Q_R \cap U)$ and as previously, denote by $\mathcal{A}_n$ the event that X_N, a configuration of N points in U, has n points in $Q_R \cap U$. We wish to control

$$\mathbb{E}_{\mathbb{Q}_\beta(\mu,U)}\left(\exp\left(\frac{\beta}{2}(\mathsf{F}^{Q_{R-2\tilde{\ell}}}(\cdot,U)+2C_0 n)\right)\right)$$
$$= \sum_{n=0}^{N} \exp(\beta C_0 n)\, \mathbb{E}_{\mathbb{Q}_\beta(\mu,U)}\left(\exp\left(\frac{\beta}{2}(\mathsf{F}^{Q_{R-2\tilde{\ell}}}(\cdot,U))\mathbf{1}_{\mathcal{A}_n}\right)\right).$$

The terms in the sum for which n is close to $\bar{\mathrm{n}}$, more precisely $|n-\bar{\mathrm{n}}| \leq KR^{\mathsf{d}-\frac{1}{2}}$ are easily treated using (8.3.36). The terms for which $|n-\bar{\mathrm{n}}| > KR^{\mathsf{d}-\frac{1}{2}}$ will be handled separately and controlled by energy-excess considerations.

To apply Proposition 8.9 we need $Q_{R+\tilde{\ell}}$ to be included in a cube $\square_L$ in which the local laws hold and at distance $\geq \tilde{\ell}$ as in (8.3.33) from $\partial\Lambda$. At the first iteration, L is of order $N^{\frac{1}{\mathsf{d}}}$ and $R \geq \frac{1}{2}L$ so we need

$$\mathrm{dist}(Q_R, \partial\Lambda) \geq C'' \max\left(\left(\frac{N^{\frac{1}{\mathsf{d}}}}{\max(1, \beta^{-\frac{1}{2}}\chi(\beta)^{\frac{1}{2}})}\right)^{-\frac{2}{3}} N^{\frac{1}{\mathsf{d}}}, N^{\frac{1}{\mathsf{d}+2}}\right).$$

At further iterations, to have $Q_{R+\tilde{\ell}}$ be included in $\square_L$, we need a further distance of

$$C'' \max\left(\left(\frac{R}{\max(1, \beta^{-\frac{1}{2}}\chi(\beta)^{\frac{1}{2}})}\right)^{-\frac{2}{3}} R, R^{1-\frac{2}{\mathsf{d}+2}}\right).$$

Since R is multiplied by a factor in $[\frac{1}{2}, \frac{3}{4}]$ at each step, summing the series over the iterations gives a total distance

$$C''' \max\left(\left(\frac{N^{\frac{1}{\mathsf{d}}}}{\max(1, \beta^{-\frac{1}{2}}\chi(\beta)^{\frac{1}{2}})}\right)^{-\frac{2}{3}} N^{\frac{1}{\mathsf{d}}}, N^{\frac{1}{\mathsf{d}+2}}\right)$$

hence a condition of the form (8.3.1) suffices.

Step 1: The bad event. We claim that in the bad event $|\bar{\mathrm{n}} - n| > KR^{\mathsf{d}-\frac{1}{2}}$, we have

$$\mathsf{F}^{Q_{R+3}}(X_N,U) - \mathsf{F}^{Q_R}(X_N,U) \geq CR^{1-\mathsf{d}}|\bar{\mathrm{n}}-n|^2 - C\#I_{Q_{R+3}\setminus Q_R}, \tag{8.3.47}$$

where $C > 0$ depends only on $\|\mu\|_{L^\infty}$ and d. Assuming this, and changing C_0 to the larger constant in (8.3.47) if necessary, we then write

$$\begin{aligned}
&\mathbb{E}_{\mathbb{Q}_\beta(\mu,U)}\left(\exp\left(\frac{\beta}{2}(\mathsf{F}^{Q_R}(\cdot,U)+2C_0 n)\right)\mathbf{1}_{\mathcal{A}_n}\right)\\
&\le \mathbb{E}_{\mathbb{Q}_\beta(\mu,U)}\left(\exp\left(\frac{\beta}{2}(\mathsf{F}^{Q_{R+3}}(\cdot,U)+2C_0\#I_{Q_{R+3}\setminus Q_R}+C_0 n)\right)\mathbf{1}_{\mathcal{A}_n}\right)\\
&\quad\times\exp\left(-\frac{\beta}{2}CR^{1-\mathsf{d}}|\bar{\mathrm{n}}-n|^2\right)\\
&\le \mathbb{E}_{\mathbb{Q}_\beta(\mu,U)}\left(\exp\left(\frac{\beta}{2}(\mathsf{F}^{Q_{R+3}}(\cdot,U)+2C_0\#I_{Q_{R+3}})\right)\mathbf{1}_{\mathcal{A}_n}\right)\\
&\quad\times\exp\left(-\frac{\beta}{2}CR^{1-\mathsf{d}}|\bar{\mathrm{n}}-n|^2\right).
\end{aligned}\tag{8.3.48}$$

Since $L \le 2R$ and $|\bar{\mathrm{n}}-n| > KR^{\mathsf{d}-\frac{1}{2}}$, we now see that if we choose $K = C\sqrt{\mathcal{C}\chi(\beta)}$ where $C > 0$ is large enough and depends only on C, C_0 and d, the exponent in the second term in the right-hand side is at most $-\mathcal{C}\beta\chi(\beta)L^{\mathsf{d}}$.

On the other hand, using Lemma 7.18, (7.1.37) and (7.1.35), we may check that

$$\mathsf{F}^{Q_{R+3}}(\cdot,U)+2C_0\#I_{Q_{R+3}} \le \mathsf{F}^{\square_L}(\cdot,U)+2C_0\#I_{\square_L},$$

hence in view of (8.3.48) and the assumption that (8.3.45) satisfied in a cube $\square_L$ containing Q_{R+3}, we may then bound

$$\begin{aligned}
&\sum_{n,|\bar{\mathrm{n}}-n|>KR^{\mathsf{d}-\frac{1}{2}}}\mathbb{E}_{\mathbb{Q}_\beta(\mu,U)}\left(\exp\left(\frac{\beta}{2}(\mathsf{F}^{Q_R}(\cdot,U)+2C_0 n)\right)\mathbf{1}_{\mathcal{A}_n}\right)\\
&\le \exp\left(-\mathcal{C}\beta\chi(\beta)L^{\mathsf{d}}\right)\sum_{n=0}^{N}\mathbb{E}_{\mathbb{Q}_\beta(\mu,U)}\left(\exp\left(\frac{\beta}{2}(\mathsf{F}^{\square_L}(\cdot,U)+2\beta C_0\#I_{\square_L})\right)\mathbf{1}_{\mathcal{A}_n}\right)\le 1.
\end{aligned}\tag{8.3.49}$$

To prove the claim, in view of (5.2.20) we may write

$$C\int_{Q_{R+2}\setminus Q_{R+1}}|\nabla u_{\tilde{\mathrm{r}}}|^2 \ge CR^{1-\mathsf{d}}\left(|n-\bar{\mathrm{n}}|-C\|\mu\|_{L^\infty}R^{\mathsf{d}-1}\right)^2 \ge cR^{1-\mathsf{d}}|\bar{\mathrm{n}}-n|^2 \tag{8.3.50}$$

if K is chosen large enough (depending on d and $\|\mu\|_{L^\infty}$), where $c > 0$ is a constant depending only on d, m and $\|\mu\|_{L^\infty}$. In view of (7.1.37) we have

$$\mathsf{F}^{Q_{R+3}}(X_N,U)-\mathsf{F}^{Q_R}(X_N,U)\ge \mathsf{F}^{Q_{R+3}\setminus Q_R}(X_N,U).\tag{8.3.51}$$

By (7.1.35), and performing first a reflection through the boundary of U in the Neumann case as in Remark 7.9 if necessary, we may write that

$$C\mathsf{F}^{Q_{R+3}\setminus Q_R}(X_N,U)\ge\int_{(Q_{R+3}\setminus Q_R)\cap U}|\nabla u_{\tilde{\mathrm{r}}}|^2 - C\#I_{Q_{R+3}\setminus Q_R},$$

where $u_{\tilde{\mathrm{r}}}$ is computed with respect to $Q_{R+3}\setminus Q_R$. But by definition $\int_{Q_{R+3}\setminus Q_R}|\nabla u_{\tilde{\mathrm{r}}}|^2$ is larger than $\int_{Q_{R+2}\setminus Q_{R+1}}|\nabla u_{\hat{\mathrm{r}}}|^2$ with this time $\hat{\mathrm{r}}$ computed with respect to U, which is bounded below by (8.3.50). Inserting into (8.3.51) we thus conclude (8.3.47).

Step 2: The good event. We next consider the terms for which $|\bar{\mathrm{n}}-n| \le KR^{\mathsf{d}-1/2}$. For those, we may apply Proposition 8.9 (at least if $R > C$ with C made large enough). We need to assume (8.3.32). In view of (8.3.36) we may thus write

$$\sum_{|n-\bar{\mathrm{n}}|\le KR^{\mathsf{d}-1/2}} \mathbb{E}_{\mathbb{Q}_\beta(\mu,U)}\left(\exp\left(\frac{\beta}{2}\left(\mathsf{F}^{Q_{R-2\tilde{\ell}}}(\cdot,U)+2C_0 n\right)\right)\mathbf{1}_{\mathcal{A}_n}\right)$$
$$\le \sum_{n=\bar{\mathrm{n}}-KR^{\mathsf{d}-1/2}}^{\bar{\mathrm{n}}+KR^{\mathsf{d}-1/2}} \exp\left(\beta\left(\frac{\mathcal{C}}{4}\chi(\beta)R^{\mathsf{d}}+C|n-\bar{\mathrm{n}}|+2C_0 n\right)\right)\frac{\mathsf{K}_{\beta/2}(\mu,Q_R)}{\mathsf{K}_\beta(\mu,Q_R)}$$
$$+\gamma_n\exp\left(\beta C_0 n\right).$$

Recalling the choice of K as $C\sqrt{\mathcal{C}\chi(\beta)}$ and using that $\bar{\mathrm{n}}=\mu(Q_R)\le\|\mu\|_{L^\infty}R^{\mathsf{d}}$, we have that if $|n-\bar{\mathrm{n}}|\le KR^{\mathsf{d}-1/2}$ and $R\ge\mathcal{C}\chi(\beta)$, then $KR^{\mathsf{d}-1/2}\le CR^{\mathsf{d}}$ and $n\le CR^{\mathsf{d}}$, with C depending only on $\mathsf{d},m,\|\mu\|_{L^\infty}$. But $R\ge\mathcal{C}\chi(\beta)$ is satisfied as soon as (8.3.32) holds (up to changing C' if necessary).

Using (8.3.35) and using (7.1.55) to bound the ratio of partition functions, we deduce that, for every R satisfying (8.3.32),

$$\sum_{n=\bar{\mathrm{n}}-KR^{\mathsf{d}-1/2}}^{\bar{\mathrm{n}}+KR^{\mathsf{d}-1/2}} \mathbb{E}_{\mathbb{Q}_\beta(\mu,U)}\left(\exp\left(\frac{\beta}{2}\left(\mathsf{F}^{Q_{R-2\tilde{\ell}}}(\cdot,U)+2C_0 n\right)\right)\mathbf{1}_{\mathcal{A}_n}\right)$$
$$\le CR^{\mathsf{d}}\exp\left(\beta\left(\frac{3\mathcal{C}}{8}\chi(\beta)R^{\mathsf{d}}+C_0CR^{\mathsf{d}}\right)\right)\exp\left(C\beta\chi(\beta)R^{\mathsf{d}}\right)$$
$$+\exp\left(\beta C_0CR^{\mathsf{d}}-\mathcal{C}\beta\chi(\beta)R^{\mathsf{d}}\right).$$

Making $\mathcal{C}$ larger if necessary (compared to the constants C_0, C appearing here) we deduce

$$\sum_{n=\bar{\mathrm{n}}-KR^{\mathsf{d}-1/2}}^{\bar{\mathrm{n}}+KR^{\mathsf{d}-1/2}} \mathbb{E}_{\mathbb{Q}_\beta(\mu,U)}\left(\exp\left(\frac{\beta}{2}\left(\mathsf{F}^{Q_{R-2\tilde{\ell}}}(\cdot,U)+2C_0 n\right)\right)\mathbf{1}_{\mathcal{A}_n}\right)$$
$$\le\exp\left(\beta\frac{\mathcal{C}}{2}\chi(\beta)R^{\mathsf{d}}+C\log R\right). \tag{8.3.52}$$

The logarithmic term can then be absorbed using that $R\ge\rho_\beta$ and a careful examination of the definition (5.2.27).

Combining (8.3.49) and (8.3.52), we conclude that (8.3.46) holds. This yields that for any $\square_R(x)$ satisfying (8.3.1), the estimate (8.3.3) holds. □

8.4. Consequences of the local laws and almost additivity

8.4.1. Direct corollaries. Since the electric energy F controls discrepancies and linear statistics, we can immediately deduce controls on such quantities. These discrepancy controls are not expected to be sharp, but they already provide a form of rigidity of the particles numbers. We refer to Section 5.3.2 for the better discrepancy estimates due to Thoma and obtained by isotropic averaging, and to Section 10.3 for the best-to-date two-dimensional hyperuniformity (or variance of the number of points) result of Leblé [**Leb**].

Corollary 8.10 (Discrepancy controls). *Assume the hypotheses of Theorem 8.4 for $\square_R(x)$ with $R\ge\rho_\beta$ and let B be a ball such that $2B\subseteq\square_R(x)\cap U$. There* *exists $C>0$ depending only on d, m and $\|\mu\|_{L^\infty}$ such that letting*

$$D := \int_B \left(\sum_{i=1}^N \delta_{x_i} - d\mu\right),$$

we have

$$\log \mathbb{E}_{\mathbb{Q}_\beta(\mu,U)}\left(\exp\left(\frac{\beta}{C}\chi(\beta)^{1/3}R^{2(1-\mathsf{d})}\rho_\beta^{\mathsf{d}-2/3}D^2\right)\right) \le C\beta\chi(\beta)\rho_\beta^{\mathsf{d}}, \tag{8.4.1}$$

and

$$\log \mathbb{E}_{\mathbb{Q}_\beta(\mu,U)}\left(\exp\left(\frac{\beta}{C}\frac{D^2}{R^{\mathsf{d}-2}}\min\left(1,\frac{|D|}{R^{\mathsf{d}}}\right)\right)\right) \le C\beta\chi(\beta)R^{\mathsf{d}}. \tag{8.4.2}$$

PROOF. We may suppose $x=0$. First, we observe that by choice of C_0 and (7.1.35) we have for any $R \ge \rho_\beta$,

$$\log \mathbb{E}_{\mathbb{Q}_\beta(\mu,U)}\left(\exp\left(\frac{1}{2C}\beta\int_{\square_R\cap U}|\nabla u_{\tilde{\mathsf{r}}}|^2\right)\right) \le C\beta\chi(\beta)R^{\mathsf{d}}, \tag{8.4.3}$$

where $\tilde{\mathsf{r}}$ is computed with respect to $\partial\square_R$. We next may use either first (5.2.20)–(5.2.21) or second (4.4.8)–(4.4.7) (after suitable blow-up) to deduce from this a control of the discrepancy, after noting that these inequalities apply as well to u instead of h_N.

In the first way we cover $\square_{R+2}\backslash\square_{R-2}$ by at most $O((R/\rho_\beta)^{\mathsf{d}-1})$ cubes Q_k of size ρ_β. Applying (8.4.3) for the cubes Q_k and using the generalized Hölder inequality

$$\mathbb{E}(f_1\dots f_k) \le \prod_{i=1}^k \mathbb{E}(f_i^k)^{\frac{1}{k}}, \tag{8.4.4}$$

which can be proved by induction, we find

$$\log \mathbb{E}_{\mathbb{Q}_\beta(\mu,U)}\left(\exp\left(C^{-1}\beta(\frac{R}{\rho_\beta})^{1-\mathsf{d}}\int_{\square_{R+\rho_\beta}\backslash\square_{R-\rho_\beta}}|\nabla u_{\tilde{\mathsf{r}}}|^2\right)\right) \le C\beta\chi(\beta)\rho_\beta^{\mathsf{d}}, \tag{8.4.5}$$

for some constants C depending only on d, m and $\|\mu\|_{L^\infty}$. In view of (5.2.20)–(5.2.21), we then have that for all $1 \le \delta \le \rho_\beta$,

$$\left|\int_{\square_R}\sum_{i=1}^N \delta_{x_i} - d\mu\right|^2 \le C\|\mu\|_{L^\infty}^2 R^{2(\mathsf{d}-1)}\delta^2 + C\frac{R^{\mathsf{d}-1}}{\delta}\int_{\square_{R+\rho_\beta}\backslash\square_{R-\rho_\beta}}|\nabla u_{\tilde{\mathsf{r}}}|^2.$$

Choosing $\delta = (\chi(\beta)\rho_\beta)^{1/3}$ and inserting into (8.4.5), we find (8.4.1).

In the second way, we simply bound $\int_{B_{2R}}|\nabla u_{\tilde{\mathsf{r}}}|^2$ using (8.4.3). Inserting into (4.4.8)–(4.4.7) directly yields (8.4.2). □

The second corollary is a control on linear statistics for Lipschitz functions. It follows from the combination of (8.4.3) and (7.1.38) applied in $\mathbb{R}^{\mathsf{d}}$. In Chapter 9, we will see how to obtain a better control by the transport method, but under a stronger regularity assumption on the test function.

Corollary 8.11 (Linear statistics control). *Under the same assumptions as Theorem 8.4, if φ is a 1-Lipschitz function supported in $\square_R(x)\cap U$, then*

$$\left|\log \mathbb{E}_{\mathbb{Q}_\beta(\mu,U)}\left(\exp\frac{\beta}{CR^{\mathsf{d}}}\left(\int_{\mathbb{R}^{\mathsf{d}}}\varphi\, d(\sum_{i=1}^N\delta_{x_i}-\mu)\right)^2\right)\right| \le C\beta\chi(\beta)R^{\mathsf{d}}, \tag{8.4.6}$$

for some $C>0$ depending only on d, m and $\|\mu\|_{L^\infty}$.

The final corollary is a control of minimal distances, direct consequence of (7.1.34) and (8.3.3) applied with $R = \rho_\beta$.

Corollary 8.12 (Minimal distance control). *Under the same assumptions as Theorem 8.4, for any point x_i of the (blown-up) configuration at distance $\ge d_0$ from $\partial\Lambda \cap U$ and at distance ≥ 1 from ∂U, denoting*

$$\mathsf{r}_i = \min\left(\min_{j\neq i}|x_i - x_j|, \frac{1}{4}\right)$$

we have

$$\left|\log \mathbb{E}_{\mathbb{Q}_\beta(\mu,U)}\left(\exp\left(\frac{\beta}{2}\mathsf{g}(\mathsf{r}_i)\right)\right)\right| \le C\beta\chi(\beta)\rho_\beta^{\mathsf{d}}. \tag{8.4.7}$$

The fact that (8.4.1) gives a bound on all the moments of the number of points in a compact set centered at x satisfying (8.3.1) yields that if x is far enough from $\partial\Lambda$, the law of the point configuration $\{x_1 - x, \dots, x_N - x\}_N$ converges as $N \to \infty$, up to extraction of a subsequence, to a limiting point process with simple points and finite correlation functions of all order. As mentioned in Section 5.3.2, the overcrowding estimates of Theorem 5.30 from [**Tho24**] give an easier proof of subsequential convergence to a limiting point process, which in additions works up to the boundary.

8.4.2. Almost additivity of the free energy.

As explained in Section 8.1, the reasoning of the bootstrap proof of the local laws also yields the almost additivity of the free energy up to surface errors.

Proposition 8.13 (Almost additivity of the free energy). *Assume $\mathsf{d} \ge 1$ and $\mathsf{s} = \mathsf{d} - 2$. Let U be an open subset of $\mathbb{R}^{\mathsf{d}}$ with bounded and piecewise C^1 boundary and μ be a bounded nonnegative density such that $\mu(U) = N$ is an integer. Assume that $\mu \ge m > 0$ in a set Λ. If $\mathsf{s} \le 0$ and U is unbounded, assume in addition that (5.2.28), (5.2.29), and (5.2.37) hold. Assume $\hat{U}$ is a subset of Λ at distance $\ge d_0$ from $\partial\Lambda$ with d_0 as in (8.3.2), and is a disjoint union of p hyperrectangles Q_i belonging to $\mathcal{Q}_R$, with $R \ge \rho_\beta$ satisfying*

$$R \ge \rho_\beta + \left(\frac{1}{\beta\chi(\beta)}\log\frac{R^{\mathsf{d}-1}}{\rho_\beta^{\mathsf{d}-1}}\right)^{\frac{1}{\mathsf{d}}}. \tag{8.4.8}$$

Then there exists $C > 0$, depending only on d, m and $\|\mu\|_{L^\infty}$, such that

$$\left|\log \mathsf{K}_\beta(\mu, \mathbb{R}^{\mathsf{d}}) - \left(\log \mathsf{K}_\beta(\mu, \mathbb{R}^{\mathsf{d}}\backslash\hat{U}) + \sum_{i=1}^{p}\log \mathsf{K}_\beta(\mu, Q_i)\right)\right| \le Cp\left(\beta R^{\mathsf{d}-1}\rho_\beta\chi(\beta) + \beta^{1-\frac{1}{\mathsf{d}}}\chi(\beta)^{1-\frac{1}{\mathsf{d}}}\left(\log\frac{R}{\rho_\beta}\right)^{\frac{1}{\mathsf{d}}}R^{\mathsf{d}-1}\right). \tag{8.4.9}$$

If U is a subset of Λ equal to a disjoint union of p hyperrectangles Q_i belonging to $\mathcal{Q}_R$, with $R \geq \rho_\beta$ satisfying (8.4.8), $N_i = \mu(Q_i)$, then we have,

$$\left|\log \mathsf{K}_\beta(\mu, U) - \sum_{i=1}^{p} \log \mathsf{K}_\beta(\mu, Q_i)\right| \leq Cp\left(\beta R^{\mathsf{d}-1}\chi(\beta)\rho_\beta + \beta^{1-\frac{1}{\mathsf{d}}}\chi(\beta)^{1-\frac{1}{\mathsf{d}}}\left(\log \frac{R}{\rho_\beta}\right)^{\frac{1}{\mathsf{d}}} R^{\mathsf{d}-1}\right), \tag{8.4.10}$$

with C as above.

PROOF. We only need to prove upper bounds for $\log \mathsf{K}_\beta(\mathbb{R}^\mathsf{d})$ and $\log \mathsf{K}_\beta(U)$, since the matching lower bounds are direct consequences of (7.1.17), Stirling's formula and the control (8.4.15) below.

We recall that by Theorem 8.4 the local laws hold down to scale ρ_β in $U = \mathbb{R}^\mathsf{d}$. In particular, for any cube $\square$ in $\hat{U}$ of size $r \geq \rho_\beta$, we have

$$\log \mathbb{E}_{\mathbb{Q}_\beta(\mu,\mathbb{R}^\mathsf{d})}\left(\exp\left(\frac{1}{2C}\beta\int_\square |\nabla u_{\vec{\mathsf{r}}}|^2\right)\right) \leq C r^\mathsf{d}\beta\chi(\beta). \tag{8.4.11}$$

Let Q_1 be the first rectangle in the list, and let us denote by n the number of points a configuration has in Q_1 and by $\bar{\mathrm{n}} = \mu(Q_1)$. Let us also define

$$\hat{Q}_1 := \{x \in Q_1, \mathrm{dist}(x, \partial Q_1) \leq r\}$$

and

$$\mathcal{B} := \left\{X_N \in (\mathbb{R}^\mathsf{d})^N \,:\, |n - \bar{\mathrm{n}}| \leq \varepsilon, \quad \sup_x \int_{\hat{Q}_1 \cap \square_r(x)} |\nabla u_{\vec{\mathsf{r}}}|^2 \leq M\chi(\beta)r^\mathsf{d}\right\}$$

where we let

$$\varepsilon := M\left(R^{\mathsf{d}-1}\sqrt{\chi(\beta)\rho_\beta}\right)$$

and $M > 0$ is to be selected below. The first condition $|n - \bar{\mathrm{n}}| \leq \varepsilon$ in the definition of $\mathcal{B}$ has large probability in view of (8.4.1). For the second condition, by a covering argument we have $O(\frac{R^{\mathsf{d}-1}}{r^{\mathsf{d}-1}})$ conditions to satisfy and each of them has probability of the complement bounded by $\exp\left(-\frac{M}{C}\beta\chi(\beta)r^\mathsf{d}\right)$ if M is large enough, in view of (8.4.11). Using a union bound we thus have

$$\mathbb{Q}_\beta(\mu, \mathbb{R}^\mathsf{d})[\mathcal{B}^c] \leq C\frac{R^{\mathsf{d}-1}}{r^{\mathsf{d}-1}}\exp\left(-\frac{M}{C}\beta\chi(\beta)r^\mathsf{d}\right)$$

and this is $\leq \frac{1}{2}$ if

$$C\frac{R^{\mathsf{d}-1}}{r^{\mathsf{d}-1}}\exp\left(-\frac{M}{C}\beta\chi(\beta)r^\mathsf{d}\right) \leq \frac{1}{2}$$

so we choose

$$r = \rho_\beta + \left(\frac{1}{\beta\chi(\beta)}\log\frac{R^{\mathsf{d}-1}}{\rho_\beta^{\mathsf{d}-1}}\right)^{\frac{1}{\mathsf{d}}}, \tag{8.4.12}$$

which satisfies the condition if M is large enough. In view of the definitions (5.2.10) and (5.2.11), it follows that

$$N^{-N}\int_{\mathcal{B}^c} \exp\left(-\beta \mathsf{F}(\cdot)\right) d\mu^{\otimes N} = \mathsf{K}_\beta(\mu, \mathbb{R}^\mathsf{d})\mathbb{Q}_\beta(\mu)[\mathcal{B}^c] \leq \frac{1}{2}\mathsf{K}_\beta(\mu).$$

We thus have

$$\begin{aligned}\frac{N^N}{2}\mathsf{K}_\beta(\mu,\mathbb{R}^\mathsf{d}) &\le \int_{\mathcal{B}} \exp\left(-\beta\mathsf{F}(\cdot,\mu)\right) d\mu^{\otimes N}\\ &\le \sum_{n=\bar{\mathrm{n}}-\varepsilon}^{\bar{\mathrm{n}}+\varepsilon} \frac{N!}{n!(N-n)!}\int_{\mathcal{B}} \exp\left(-\beta \mathsf{G}^{\mathrm{inn}}_{\mathbb{R}^\mathsf{d}}(\cdot,Q_1)\right) d\mu^{\otimes n}\\ &\qquad \times \int_{\mathcal{B}} \exp\left(-\beta \mathsf{G}^{\mathrm{out}}_{\mathbb{R}^\mathsf{d}}(\cdot,\mathbb{R}^\mathsf{d}\backslash Q_1)\right) d\mu^{\otimes(N-n)},\end{aligned}$$

where for the second line we subdivided the event over the possible values of n and applied (7.1.37) and Lemma 7.18.

We now apply the results of Corollary 8.8 with $L = r$ to Q_1 and $\mathbb{R}^\mathsf{d}\backslash Q_1$, combined with Remark 8.7. For that we check that (8.3.28) is satisfied since $r \ge \rho_\beta$, and obtain

$$\begin{aligned}\mathsf{K}_\beta(\mu,\mathbb{R}^\mathsf{d}) &\le 2\mathsf{K}_\beta(\mu,Q_1)\mathsf{K}_\beta(\mu,\mathbb{R}^\mathsf{d}\backslash Q_1) \sum_{n=\bar{\mathrm{n}}-\varepsilon}^{\bar{\mathrm{n}}+\varepsilon} \frac{N!N^{-N}}{\bar{\mathrm{n}}!(N-\bar{\mathrm{n}})!}\bar{\mathrm{n}}^{\bar{\mathrm{n}}}(N-\bar{\mathrm{n}})^{N-\bar{\mathrm{n}}}\\ &\quad \times \exp\Bigg(C\beta\left(R^{\mathsf{d}-1}r\chi(\beta)M+\varepsilon\right)\\ &\qquad + C\left(\log R + \frac{M^2\chi(\beta)^2R^{\mathsf{d}-1}}{r} + \frac{R^{\mathsf{d}-1}}{r^{\mathsf{d}-1}}\log r\right)\Bigg).\end{aligned}$$

Next, using Stirling's formula we have

$$\frac{N!N^{-N}\bar{\mathrm{n}}^{\bar{\mathrm{n}}}(N-\bar{\mathrm{n}})^{N-\bar{\mathrm{n}}}}{\bar{\mathrm{n}}!(N-\bar{\mathrm{n}})!} \le C\sqrt{\frac{N}{2\pi\bar{\mathrm{n}}(N-\bar{\mathrm{n}})}} \le C$$

and we deduce

$$\begin{aligned}(8.4.13)\quad &\log \mathsf{K}_\beta(\mu,\mathbb{R}^\mathsf{d}) \le \log \mathsf{K}_\beta(\mu,Q_1) + \log \mathsf{K}_\beta(\mu,\mathbb{R}^\mathsf{d}\backslash Q_1)\\ &+C\left(1+\log\varepsilon + \beta\left(MR^{\mathsf{d}-1}r\chi(\beta)+\varepsilon\right) + \log R + \frac{M^2\chi(\beta)^2R^{\mathsf{d}-1}}{r} + \frac{R^{\mathsf{d}-1}}{r^{\mathsf{d}-1}}\log r\right).\end{aligned}$$

Since

$$(8.4.14)\qquad r \ge \rho_\beta \ge \max(1, \chi(\beta)^{\frac12}\beta^{-\frac12}) \ge 1$$

we have $\frac{\chi(\beta)}{r} \le \beta r$ so we may absorb the last term into $\beta MR^{\mathsf{d}-1}r\chi(\beta)$. Inspecting the definition (5.2.27) in all dimensions, we check that we may also absorb the last term of the bound (8.4.13) into the previous one. Also, since $r \ge \rho_\beta \ge 1$ and $\chi(\beta) \ge 1$, by definition of ε we may absorb ε into $MR^{\mathsf{d}-1}r\chi(\beta)$. Since $R \ge \rho_\beta \ge \sqrt{\chi(\beta)}$, we have $\varepsilon = MR^{\mathsf{d}-1}\sqrt{\chi(\beta)\rho_\beta} \le CR^\mathsf{d}$, which allows to absorb $\log\varepsilon$ into $\log R$. Finally, since $R \ge \rho_\beta \ge C\chi(\beta)^{\frac12}\beta^{-\frac12}$ we have that, for every $R \ge \rho_\beta$,

$$(8.4.15)\qquad \log R \le C\beta\chi(\beta)\rho_\beta R^{\mathsf{d}-1},$$

which allows us to absorb the $\log R$ term into the others.

Inserting the definition of r, we have obtained

$$\begin{aligned}\log \mathsf{K}_\beta(\mu) &\le \log \mathsf{K}_\beta(\mu,Q_1) + \log \mathsf{K}_\beta(\mu,U\backslash Q_1)\\ &\qquad + C\left(\beta R^{\mathsf{d}-1}\rho_\beta\chi(\beta) + \beta^{1-\frac{1}{\mathsf{d}}}\chi(\beta)^{1-\frac{1}{\mathsf{d}}}R^{\mathsf{d}-1}\left(\log\frac{R}{\rho_\beta}\right)^{\frac{1}{\mathsf{d}}}\right).\end{aligned}$$

We may now iterate this by bounding $\log \mathsf{K}_\beta(\mu, \mathbb{R}^{\mathsf{d}} \setminus \cup_{i=1}^{j} Q_i)$ in the same way thanks to the local laws up to the boundary of Theorem 8.4. This yields (8.4.9).

The proof of (8.4.10) is analogous, using that the local laws hold up to the boundary for $\mathbb{Q}_\beta(\mu, U)$. □

Fixing μ to be the constant 1 and restricting to R such that R^{d} is an integer, we can then assert that $-\frac{\log \mathsf{K}_\beta(1,\square_R)}{\beta R^{\mathsf{d}}}$ has a finite limit as $R \to \infty$, which we denote $f_{\mathsf{d}}(\beta)$. Moreover, the convergence is at speed R^{-1}. This is the analogue with temperature of Theorem 8.2(1). It proves the existence of a thermodynamic limit for this uniform gas in a box.

Theorem 8.14 (Free energy expansion, Neumann jellium case). *Assume* $\mathsf{d} \geq 1$ *and* $\mathsf{s} = \mathsf{d} - 2$. *There exists a function* $f_{\mathsf{d}} : (0, \infty) \to \mathbb{R}$ *and a constant* $C > 0$ *depending only on* d *such that*

$$-C \leq f_{\mathsf{d}}(\beta) \leq C\chi(\beta) \tag{8.4.16}$$

$$f_{\mathsf{d}} \text{ is locally Lipschitz in } (0,\infty) \text{ with } |f_{\mathsf{d}}'(\beta)| \leq \frac{C\chi(\beta)}{\beta}, \tag{8.4.17}$$

and such that if R^{d} *is an integer we have*

$$\left| -\frac{\log \mathsf{K}_\beta(1,\square_R)}{\beta R^{\mathsf{d}}} - f_{\mathsf{d}}(\beta) \right| \leq C \left(\chi(\beta)\frac{\rho_\beta}{R} + \frac{\beta^{-\frac{1}{\mathsf{d}}}\chi(\beta)^{1-\frac{1}{\mathsf{d}}}}{R} \log^{\frac{1}{\mathsf{d}}} \frac{R}{\rho_\beta} \right) \tag{8.4.18}$$

where ρ_β *is as in* (8.0.2).

The function $f_{\mathsf{d}}(\beta)$ equal to the limit free energy per unit volume, is called the *pressure* in physics. Note that in dimension 1 an explicit formula for $f_{\mathsf{d}}(\beta)$ in terms of the first eigenvalue of a Fröbenius operator was provided in [**Kun74**]. In particular, that formula implies that $f_{\mathsf{d}}(\beta)$ is analytic in β and thus the one-dimensional jellium has no phase transitions. We will use this fact for the proof of the CLT in Chapter 10.

By scaling, if instead $\mu \equiv m$, a constant, we obtain that letting $Q' = m^{\frac{1}{\mathsf{d}}} Q$, where Q is a hyperrectangle such that $m|Q|$ is an integer,

$$\mathsf{F}(X_N, m, Q) = \begin{cases} m^{\frac{\mathsf{s}}{\mathsf{d}}} \mathsf{F}(m^{\frac{1}{\mathsf{d}}} X_N, 1, Q') & \text{if } \mathsf{s} \neq 0, \\ \mathsf{F}(m^{\frac{1}{\mathsf{d}}} X_N, 1, Q') - \frac{m|Q|}{2\mathsf{d}} \log m & \text{if } \mathsf{s} = 0. \end{cases}$$

Thus, we have that

$$\mathsf{K}_\beta(m, Q) = m^{-m|Q|} \mathsf{K}_{\beta m^{\mathsf{s}/\mathsf{d}}}(1, Q') \exp\left(\frac{\beta}{2\mathsf{d}} |Q| m \log m \mathbf{1}_{\mathsf{s}=0} \right),$$

and we deduce from (8.4.18) that

$$\lim_{R \to \infty} -\frac{\log \mathsf{K}_\beta(m, \square_R)}{\beta R^{\mathsf{d}}} = m^{1+\frac{\mathsf{s}}{\mathsf{d}}} f_{\mathsf{d}}(\beta m^{\frac{\mathsf{s}}{\mathsf{d}}}) + \left(\frac{1}{\beta} - \frac{1}{2\mathsf{d}} \mathbf{1}_{\mathsf{s}=0} \right) m \log m. \tag{8.4.19}$$

This is also valid for $\beta = \infty$ hence for E_∞. This then allows to deduce the free energy for a general μ by using the almost additivity after splitting the support of μ into hypercubes small enough that μ is almost constant in each.

PROOF OF THEOREM 8.14. Let us first start by treating the case of a cube $\square_R$ with R^{d} integer. In view of (7.1.17) and Stirling's formula, we have

$$\frac{1}{\beta}\log \mathsf{K}_\beta(1,\square_{2R}) \geq O\left(\frac{\log N}{\beta}\right) + \frac{2^{\mathsf{d}}}{\beta}\log \mathsf{K}_\beta(1,\square_R).$$

Thus, denoting $\phi(R) = \frac{\log \mathsf{K}_\beta(1,\square_R)}{\beta R^{\mathsf{d}}}$, this means that

$$\phi(2R) \geq \phi(R) + O\left(\frac{\log R}{\beta R^{\mathsf{d}}}\right)$$

and summing these relations we have

$$\phi(\infty) \geq \phi(R) + O\left(\sum_{k=1}^{\infty} \frac{\log R}{\beta 2^k R^{\mathsf{d}}}\right),$$

that is,

$$\phi(R) \leq \phi(\infty) + O\left(\frac{\log R}{\beta R^{\mathsf{d}}}\right). \tag{8.4.20}$$

On the other hand, in view of (8.4.10), we have

$$\log \mathsf{K}_\beta(1,\square_{2R}) \leq 2^{\mathsf{d}}\log \mathsf{K}_\beta(1,\square_R) + C\beta R^{\mathsf{d}}\left(\frac{\chi(\beta)}{R}\left(\rho_\beta + \beta^{-\frac{1}{\mathsf{d}}}\chi(\beta)^{-\frac{1}{\mathsf{d}}}\log^{\frac{1}{\mathsf{d}}}\frac{R}{\rho_\beta}\right)\right), \tag{8.4.21}$$

that is,

$$\phi(2R) \leq \phi(R) + C\left(\frac{\chi(\beta)}{R}\left(\rho_\beta + \beta^{-\frac{1}{\mathsf{d}}}\chi(\beta)^{-\frac{1}{\mathsf{d}}}\log^{\frac{1}{\mathsf{d}}}\frac{R}{\rho_\beta}\right)\right).$$

Summing these relations, we conclude just as above that

$$\phi(\infty) \leq \phi(R) + O\left(\frac{\chi(\beta)}{R}\left(\rho_\beta + \beta^{-\frac{1}{\mathsf{d}}}\chi(\beta)^{-\frac{1}{\mathsf{d}}}\log^{\frac{1}{\mathsf{d}}}\frac{R}{\rho_\beta}\right)\right). \tag{8.4.22}$$

Denoting by $-f_{\mathsf{d}}(\beta)$ the value $\phi(\infty)$ and recalling (8.4.15), we have the desired bound by combining (8.4.20) and (8.4.22).

In view of (5.2.3) and (5.2.38) applied with $\mu = 1$ and $U = \square_R$, we also have $-C\chi(\beta) \leq \phi(R) \leq C$ with C independent of β, which implies that $-C \leq f_{\mathsf{d}}(\beta) \leq C\chi(\beta)$.

To check that f_{d} is locally Lipschitz, let us observe that

$$\begin{aligned}\log \frac{\mathsf{K}_{\beta+\delta}(1,\square_R)}{\mathsf{K}_\beta(1,\square_R)} &= \log \mathbb{E}_{\mathbb{Q}_\beta(1,\square_R)}\left(\exp\left(-\delta \mathsf{F}(\cdot,1,\square_R)\right)\right)\\ &\leq \frac{2|\delta|}{\beta}\log \mathbb{E}_{\mathbb{Q}_\beta(1,\square_R)}\left(\exp\left(\frac{1}{2}\beta \mathsf{F}(\cdot,1,\square_R)\right)\right)\\ &\leq C|\delta|\chi(\beta)R^{\mathsf{d}},\end{aligned}$$

using Hölder's inequality and (8.3.3). Dividing by βR^{d} and sending $R \to \infty$ yields (8.4.17). □

CHAPTER 9

The transport method and free energy expansions

In this chapter, we return to the normal scale, we continue to specialize in the Coulomb case, and we wish to analyze fluctuations of linear statistics, i.e., quantities of the form

$$\mathrm{Fluct}_\mu(\varphi) := \sum_{i=1}^N \varphi(x_i) - N \int_{\mathbb{R}^{\mathsf{d}}} \varphi\, d\mu, \tag{9.0.1}$$

for φ a regular enough function, and where μ is the reference measure, typically the equilibrium or thermal equilibrium measure, satisfying (4.1.1). We recall that a first, nonoptimal, bound on fluctuations was given in (5.2.64). We will now see how to obtain a finer bound.

The starting point is to reexpress the Laplace transform (or moment generating function) of $\mathrm{Fluct}_\mu(\varphi)$. This approach was pioneered by Johansson [**Joh98**]. Indeed, to show that a random variable is often bounded, it suffices to show that its Laplace transform is, and to show that a random variable converges to a Gaussian, it suffices to show that its Laplace transform converges to that of a Gaussian.

By definition of the Gibbs measure (1.1.5), letting $V_t := V + t\varphi$ and inserting the V-dependence in the notation for $Z_{N,\beta}$, we have

(9.0.2)

$$\begin{aligned}
&\mathbb{E}_{\mathbb{P}_{N,\beta}}\left(e^{-\beta t N^{1-\frac{\mathsf{s}}{\mathsf{d}}}\sum_{i=1}^N \varphi(x_i)}\right)\\
&\quad= \frac{1}{Z_{N,\beta}(V)} \int_{(\mathbb{R}^{\mathsf{d}})^N} \exp\left(-\beta N^{-\frac{\mathsf{s}}{\mathsf{d}}}\Big(\frac{1}{2}\sum_{i\neq j} \mathsf{g}(x_i-x_j) + N\sum_{i=1}^N (V+t\varphi)(x_i)\Big)\right) dX_N\\
&\quad= \frac{Z_{N,\beta}(V_t)}{Z_{N,\beta}(V)},
\end{aligned}$$

thus, understanding fluctuations via the Laplace transform leads to understanding ratios of partition functions. To do so, we may first use the splitting with respect to the equilibrium measure or the thermal equilibrium measure, as in (5.1.7) or (5.1.14), and we are thus led to comparing reduced partition functions $\mathsf{K}_{N,\beta}$. This brings us to the question of estimating the change in reduced partition function, or in free energy, under a perturbation of the external potential V, or equivalently under a perturbation of the reference measure.

The evaluation of the ratio of partition functions will be done in two different ways:
(i) the first way is to view the perturbed measure as the push-forward of the reference measure by a transport map. For that we introduce a transport method, which

allows to evaluate the change in partition function when making a perturbation of the equilibrium measure.
(ii) the second way is to estimate the partition function for general equilibrium measures by using the almost additivity of Chapter 8, partitioning the domain into small cubes in which the equilibrium measure is almost constant, and using (i) to estimate the error made in each cube.
Comparing the two ways of estimating the partition function then allows to deduce an improved estimate.

In this chapter, we start by introducing the transport method, which allows to evaluate the change in partition function and other quantities under the variation induced by a transport, which we call a "transport calculus". The quantities from Chapter 6 will naturally appear in the form

$$\frac{d}{dt}\log \mathsf{K}_{N,\beta}(\mu_t) = -\beta N^{-\frac{\mathsf{s}}{\mathsf{d}}}\mathbb{E}_{\mathbb{Q}_{N,\beta}(\mu_t)}\left(\mathsf{A}_1(X_N,\mu_t,\psi_t)\right),$$

where μ_t is the transport of μ_0 along the flow of ψ_t. The term A_1, which is defined in (6.1.3) involves a singular kernel, except in very special cases in $\mathsf{d}=1$ such as log gases that we will discuss below, and the commutator estimate of Theorem 6.1 will play a major role in estimating it. This part is not restricted to the Coulomb case, but works in the Riesz case as well.

We will then show how to deduce free energy expansions for general (nonhomogeneous) equilibrium measures, taking advantage of the almost additivity of the energy. This is where we restrict to the Coulomb case. The Riesz case $\mathsf{d}-2<\mathsf{s}<\mathsf{d}$ is treated in [**PS**].

In the next chapter, we will leverage the two expansions of the free energy to study fluctuations and obtain a central limit theorem (CLT) in one and two dimensions.

The transport method to obtain the CLT was pioneered in [**LS18**]. It has also been used in [**BLS18**] for the one-dimensional log case, [**Leb21**] for the sine-β process, in [**LZ21**] to obtain a local CLT for the two-dimensional Coulomb gas, and in [**LLZ24, Pei25**] to analyze the maximum of the log-gas potential. Finding the appropriate transport map is equivalent to the inversion of the "master operator" in work on one-dimensional log gases such as [**BG13, BG24**]. The transport method and the terms A_1 that appear in it are also a substitute for the Dyson-Schwinger equations in these contexts.

We present here the method with an improved version over the more recent presentation of [**Ser23**], including adding the treatment of the Coulomb $\mathsf{d}=1$ case.

9.1. Transport calculus

We recall the definition of general reduced Gibbs measures (5.1.13).

We wish to understand, with the help of the quantities of (6.1.3) in Chapter 6, the variations of energy, free energy, and other quantities along a transport. For that we will exploit the Eulerian/Lagrangian correspondence. We first recall a few basic facts and definitions.

Definition 9.1 (Push-forward of a measure). If X and Y are measurable spaces, Φ a measurable map from X to Y, and μ a measure on X, the push forward of μ by Φ is defined as the measure on Y such that $\Phi\#\mu(A)=\mu(\Phi^{-1}(A))$ for any measurable set A in Y. In particular, if f is a measurable function on Y, we have

$$\int_Y f(y)\, d(\Phi\#\mu)(y) = \int_X f(\Phi(x)) d\mu(x). \tag{9.1.1}$$

Let $\psi_t : \mathbb{R}^\mathsf{d} \to \mathbb{R}^\mathsf{d}$ be a Lipschitz vector field depending continuously on a "time" parameter $t \in [-1, 1]$. Let us define the flow $\Phi_t : \mathbb{R}^\mathsf{d} \to \mathbb{R}^\mathsf{d}$ to be the solution to

$$\begin{cases} \frac{d\Phi_t}{dt}(x) = \psi_t(\Phi_t(x)) \\ \Phi_0(x) = x. \end{cases} \tag{9.1.2}$$

This flow is well-defined for $t \in [-1, 1]$ by standard ODE theory. Moreover, it is standard and easy to check that if μ is a probability density on $\mathbb{R}^\mathsf{d}$, then the push-forward

$$\mu_t := \Phi_t\#\mu \tag{9.1.3}$$

solves

$$\partial_t \mu_t + \operatorname{div}(\psi_t \mu_t) = 0. \tag{9.1.4}$$

Indeed, let φ be a test function, by definition of the push-forward and by (9.1.2), we have

$$\begin{aligned} \int_{\mathbb{R}^\mathsf{d}} \varphi\, \partial_t \mu_t = \partial_t \int_{\mathbb{R}^\mathsf{d}} \varphi\, d\mu_t = \partial_t \int_{\mathbb{R}^\mathsf{d}} \varphi(\Phi_t(x)) d\mu(x) &= \int_{\mathbb{R}^\mathsf{d}} \nabla\varphi(\Phi_t(x)) \cdot \psi_t(\Phi_t(x)) d\mu(x) \\ &= \int_{\mathbb{R}^\mathsf{d}} \nabla\varphi \cdot \psi_t d\mu_t. \end{aligned}$$

This is the Eulerian formulation, as opposed to the Lagrangian formulation (9.1.2) which follows "particle trajectories."

The next proposition provides the main results of differentiation of energy and free energy along a transport in terms of the quantities of Chapter 6. We recall the notation A_n from (6.1.2), in particular that

$$\begin{aligned} \frac{d^n}{dt^n}\Big|_{t=0} \mathsf{F}_N((I + tv)^{\oplus N}(X_N), (I + tv)_{\#}\mu) &= \mathsf{A}_n(X_N, \mu, v) \\ &= \frac{1}{2} \int_{(\mathbb{R}^\mathsf{d})^2 \setminus \triangle} \nabla^{\otimes n} \mathsf{g}(x - y) : (v(x) - v(y))^{\otimes n} d\Big(\sum_{i=1}^N \delta_{x_i} - N\mu\Big)^{\otimes 2}(x, y), \end{aligned} \tag{9.1.5}$$

and $\mathsf{K}_{N,\beta}$ and $\mathbb{Q}_{N,\beta}$ from (5.1.12), (5.1.13). We will also make use of the Hölder seminorms as in (A.0.3)

$$|\varphi|_{C^\alpha} = \sup_{x,y} \frac{|\varphi(x) - \varphi(y)|}{|x - y|^\alpha},$$

and for $0 < \alpha \le 1$ and k integer, $|\varphi|_{C^{k+\alpha}}$ denotes the C^α seminorm of $\nabla^{\otimes k}\varphi$.

Proposition 9.2 (Transport calculus). *Let μ be a probability density on $\mathbb{R}^\mathsf{d}$ satisfying* (4.1.1). *Let ψ_t, $t \in [-1, 1]$, be a Lipschitz vector field, Φ_t solve* (9.1.2), *and let μ_t be as in* (9.1.3). *Let us abuse notation by also denoting $\Phi_t(X_N) = (\Phi_t(x_1), \dots, \Phi_t(x_N))$ for any $X_N \in (\mathbb{R}^\mathsf{d})^N$. For any $t \in [-1, 1]$, we have*

$$\frac{d}{dt} \mathsf{F}_N(\Phi_t(X_N), \Phi_t\#\mu) = \mathsf{A}_1(\Phi_t(X_N), \mu_t, \psi_t), \tag{9.1.6}$$

$$\frac{d}{dt} \mathsf{A}_1(\Phi_t(X_N), \mu_t, \psi_t) = \mathsf{A}_2(\Phi_t(X_N), \mu_t, \psi_t), \tag{9.1.7}$$

or more generally, for any $n \geq 0$,

$$\frac{d^n}{dt^n}\mathsf{F}_N(\Phi_t(X_N), \Phi_t\#\mu) = \mathsf{A}_n(\Phi_t(X_N), \mu_t, \psi_t). \tag{9.1.8}$$

Moreover,

$$\frac{d}{dt}\log \mathsf{K}_{N,\beta}(\mu_t) = -\beta N^{-\frac{\mathsf{s}}{\mathsf{d}}}\mathbb{E}_{\mathbb{Q}_{N,\beta}(\mu_t)}\left(\mathsf{A}_1(X_N, \mu_t, \psi_t)\right). \tag{9.1.9}$$

Letting $G(X_N, t)$ be a general function of X_N and t, we have

$$\begin{aligned}\frac{d}{dt}\mathbb{E}_{\mathbb{Q}_{N,\beta}(\mu_t)}\left(G(X_N,t)\right) = &- \beta N^{-\frac{\mathsf{s}}{\mathsf{d}}}\operatorname{Cov}_{\mathbb{Q}_{N,\beta}(\mu_t)}\left[G(X_N,t), \mathsf{A}_1(X_N,\mu_t,\psi_t)\right]\\ &+ \mathbb{E}_{\mathbb{Q}_{N,\beta}(\mu_t)}\left(\nabla_{X_N}G(X_N,t)\cdot(\psi_t)^{\otimes N} + \partial_t G(X_N,t)\right),\end{aligned} \tag{9.1.10}$$

where Cov *denotes the covariance.*

As an example of application, combining (9.1.10) and (9.1.9), we may find that

$$\begin{aligned}&\frac{d^2}{dt^2}\log \mathsf{K}_{N,\beta}(\mu_t)\\ &= (\beta N^{-\frac{\mathsf{s}}{\mathsf{d}}})^2 \operatorname{Var}_{\mathbb{Q}_{N,\beta}(\mu_t)}\left(\mathsf{A}_1(X_N,\mu_t,\psi_t)\right) - \beta N^{-\frac{\mathsf{s}}{\mathsf{d}}}\mathbb{E}_{\mathbb{Q}_{N,\beta}(\mu_t)}\left(\mathsf{A}_2(X_N,\mu_t,\psi_t)\right),\end{aligned} \tag{9.1.11}$$

where Var denotes the variance.

Here, the fact that we have reduced (thanks to the use of the thermal equilibrium measure) to partition functions expressed relative to $\mu^{\otimes N}$ and not the Lebesgue measure, is very convenient for the proof; see in particular the crucial computation (9.1.14) below.

Proof. By semigroup property of the flow Φ_t, we have

$$\Phi_{t+h}(x) = (I + h\psi_t)(\Phi_t(x)) + o(h), \quad \text{as } h \to 0, \tag{9.1.12}$$

and also

$$\mu_{t+h} = (I + h\psi_t)\#\mu_t + o(h), \quad \text{as } h \to 0. \tag{9.1.13}$$

Thus by definition (9.1.5), we find that (9.1.6) holds. The proof of (9.1.7) and (9.1.8) is analogous.

Let ϕ_h be a diffeomorphism from $\mathbb{R}^\mathsf{d}$ to $\mathbb{R}^\mathsf{d}$ (extended into one from $(\mathbb{R}^\mathsf{d})^N$ to itself). By (5.1.12) and the change of variables $y = \phi_h(x)$ (or by definition of the push-forward) we have

$$\begin{aligned}\frac{\mathsf{K}_{N,\beta}(\phi_h\#\mu)}{\mathsf{K}_{N,\beta}(\mu)} &= \frac{1}{\mathsf{K}_{N,\beta}(\mu)}\int_{(\mathbb{R}^\mathsf{d})^N}\exp\left(-\beta N^{-\frac{\mathsf{s}}{\mathsf{d}}}\mathsf{F}_N(X_N,\phi_h\#\mu)\right)d(\phi_h\#\mu)^{\otimes N}(X_N)\\ &= \frac{1}{\mathsf{K}_{N,\beta}(\mu)}\int_{(\mathbb{R}^\mathsf{d})^N}\exp\left(-\beta N^{-\frac{\mathsf{s}}{\mathsf{d}}}\mathsf{F}_N(\phi_h(X_N),\phi_h\#\mu)\right)d\mu^{\otimes N}(X_N)\\ &= \mathbb{E}_{\mathbb{Q}_{N,\beta}(\mu)}\left(\exp\left(-\beta N^{-\frac{\mathsf{s}}{\mathsf{d}}}(\mathsf{F}_N(\phi_h(X_N),\phi_h\#\mu) - \mathsf{F}_N(X_N,\mu))\right)\right).\end{aligned} \tag{9.1.14}$$

Applying to $\mu = \mu_t$ and $\phi_h = (I + h\psi_t)$, letting $h \to 0$, since (9.1.13) holds we obtain in view of (9.1.6) the relation (9.1.9).

Let us now turn to (9.1.10). We have

(9.1.15)
$$
\begin{aligned}
&\frac{d}{dt}\mathbb{E}_{\mathbb{Q}_{N,\beta}(\mu_t)}\left(G(X_N,t)\right)\\
&= -\frac{1}{\mathsf{K}_{N,\beta}(\mu_t)^2}\frac{d}{dt}\mathsf{K}_{N,\beta}(\mu_t)\int_{(\mathbb{R}^{\mathsf{d}})^N} e^{-\beta N^{-\frac{\mathsf{s}}{\mathsf{d}}}\mathsf{F}_N(X_N,\mu_t)}G(X_N,t)d\mu_t^{\otimes N}(X_N)\\
&\quad + \frac{1}{\mathsf{K}_{N,\beta}(\mu_t)}\frac{d}{dt}\int_{(\mathbb{R}^{\mathsf{d}})^N} e^{-\beta N^{-\frac{\mathsf{s}}{\mathsf{d}}}\mathsf{F}_N(X_N,\mu_t)}G(X_N,t)d\mu_t^{\otimes N}(X_N).
\end{aligned}
$$

The first term in the right-hand side is equal to

$$\left(-\frac{d}{dt}\log \mathsf{K}_{N,\beta}(\mu_t)\right)\mathbb{E}_{\mathbb{Q}_{N,\beta}(\mu_t)}(G(X_N,t))$$

for which we insert (9.1.9) to find it is equal to

$$\beta N^{-\frac{\mathsf{s}}{\mathsf{d}}}\mathbb{E}_{\mathbb{Q}_{N,\beta}(\dot{\mu}_t)}\left(\mathsf{A}_1(X_N,\mu_t,\psi_t)\right)\mathbb{E}_{\mathbb{Q}_{N,\beta}(\mu_t)}(G(X_N,t)).$$

By a change of variables as above, the second term in (9.1.15) is equal to

$$\frac{1}{\mathsf{K}_{N,\beta}(\mu_t)}\frac{d}{dh}_{|h=0}\int_{(\mathbb{R}^{\mathsf{d}})^N} e^{-\beta N^{-\frac{\mathsf{s}}{\mathsf{d}}}\mathsf{F}_N(\phi_h(X_N),\phi_h\#\mu_t)}G(\phi_h(X_N),t+h)d\mu_t^{\otimes N}(X_N)$$

and in view of (9.1.6), this is

$$
\begin{aligned}
&\mathbb{E}_{\mathbb{Q}_{N,\beta}(\mu_t)}\left(-\beta N^{-\frac{\mathsf{s}}{\mathsf{d}}}\mathsf{A}_1(X_N,\mu_t,\psi_t)G(X_N,t)\right)\\
&\quad + \mathbb{E}_{\mathbb{Q}_{N,\beta}(\mu_t)}\left(\nabla_{X_N}G(X_N,t)\cdot(\psi_t)^{\otimes N}+\partial_t G\right).
\end{aligned}
$$

Combining all these relations, we find (9.1.10). □

With this result in hand, we can take advantage of the commutator estimate of Theorem 6.1 which, in view of (6.1.3), states that[1]

(9.1.16)
$$|\mathsf{A}_1(X_N,\mu,\psi)| \le C\|\nabla\psi\|_{L^\infty}\left(\mathsf{F}_N(X_N,\mu)+\left(\frac{N\log N}{2\mathsf{d}}\right)\mathbf{1}_{\mathsf{s}=0}+CN^{1+\frac{\mathsf{s}}{\mathsf{d}}}\right)$$

with C depending only on d,s and $\|\mu\|_{L^\infty}$. In view of Proposition 9.2 this allows to bound the variation of the energy (or free energy) along a transport by the energy itself. Moreover, since the commutator estimate can be localized, we can combine this with the local laws to obtain localized bounds.

Here is an example of use where we deduce a control on the variation of energy along a transport via an ODE argument.

Corollary 9.3. *Assume μ is a bounded probability density satisfying* (4.1.1). *Let ψ_t, $t\in[-1,1]$ be a Lipschitz vector field, Φ_t solve* (9.1.2), *and μ_t be as in* (9.1.3). *Let C_0 be chosen large enough in view of* (4.5.5) *so that*

(9.1.17)
$$\Xi(t) := \mathsf{F}_N(\Phi_t(X_N),\mu_t)+\left(\frac{\#I_N}{2\mathsf{d}}\log N\right)\mathbf{1}_{\mathsf{s}=0}+C_0\#I_\Omega N^{\frac{\mathsf{s}}{\mathsf{d}}}\ge 0.$$

Then we have, for all $|\tau|\le 1$,

(9.1.18)
$$\forall t\in[-\tau,\tau],\qquad \Xi(t)\le \exp\left(C\int_0^t|\psi_s|_{C^1}ds\right)\Xi(0),$$

where C depends only on s,d.

[1] We note that λ in that theorem and its proof can simply be replaced by $N^{-1/\mathsf{d}}$.

PROOF. In view of (9.1.6), we have

$$\Xi'(t) = \frac{d}{dt}\mathsf{F}_N(\Phi_t(X_N), \mu_t) = \mathsf{A}_1(\Phi_t(X_N), \mu_t, \psi_t).$$

Combining with (9.1.16), we deduce

$$\Xi'(t) \le C|\psi_t|_{C^1}\Xi(t),$$

with $C > 0$ depending only of d and s, after making C_0 larger if necessary. Applying Gronwall's lemma to the function Ξ, we deduce the result. □

We finish with stating the adaptation of the transport calculus when working instead with the standard equilibrium measure and the definitions (5.1.5), (5.1.6). The main difference is that a Jacobian term appears.

Lemma 9.4 (Transport calculus, usual equilibrium measure). *Let μ be a probability density satisfying* (4.1.1). *Let ψ_t, $t \in [-1, 1]$ be a Lipschitz vector field, Φ_t solve* (9.1.2), *and μ_t be as in* (9.1.3). *Let $\zeta_t : \mathbb{R}^{\mathsf{d}} \to \mathbb{R}$ be space-time Lipschitz. We have, for any $t \in [-1, 1]$,*

(9.1.19)

$$\frac{d}{dt}\log \tilde{\mathsf{K}}_{N,\beta}(\mu_t, \zeta_t) = \mathbb{E}_{\mathbb{Q}_{N,\beta}(\mu_t,\zeta_t)}\Bigg(-\beta N^{-\frac{\mathsf{s}}{\mathsf{d}}}\left(\mathsf{A}_1(X_N, \mu_t, \psi_t) + N\sum_{i=1}^N \partial_t\zeta_t(x_i) + \nabla\zeta_t \cdot \psi_t(x_i)\right) + \sum_{i=1}^N \operatorname{div}\psi_t(x_i)\Bigg).$$

PROOF. Let ϕ_h be a diffeomorphism from $\mathbb{R}^{\mathsf{d}}$ to $\mathbb{R}^{\mathsf{d}}$ (extended into one from $(\mathbb{R}^{\mathsf{d}})^N$ to itself) with nonnegative Jacobian. By the change of variables $y_i = \phi_h(x_i)$, we have

(9.1.20)

$$\begin{aligned}
&\frac{\tilde{\mathsf{K}}_{N,\beta}(\phi_h\#\mu_t, \zeta_{t+h})}{\tilde{\mathsf{K}}_{N,\beta}(\mu_t, \zeta_t)}\\
&= \frac{1}{\tilde{\mathsf{K}}_{N,\beta}(\mu_t, \zeta_t)}\int_{(\mathbb{R}^{\mathsf{d}})^N} e^{-\beta N^{-\frac{\mathsf{s}}{\mathsf{d}}}\left(\mathsf{F}_N(\phi_h(X_N), \phi_h\#\mu_t) + N\sum_{i=1}^N \zeta_{t+h}(\phi_h(x_i))\right)} \det D\phi_h(X_N)dX_N\\
&= \mathbb{E}_{\mathbb{Q}_{N,\beta}(\mu_t,\zeta_t)}\\
&\times\left(e^{-\beta N^{-\frac{\mathsf{s}}{\mathsf{d}}}\left(\mathsf{F}_N(\phi_h(X_N),\phi_h\#\mu_t) - \mathsf{F}_N(X_N,\mu_t) + N\sum_{i=1}^N \zeta_{t+h}(\phi_h(x_i)) - \zeta_t(x_i)\right) + \sum_{i=1}^N \log\det D\phi_h(x_i)}\right).
\end{aligned}$$

Applying to $\phi_h = I + h\psi_t$, letting $h \to 0$ and using (9.1.6) we deduce (9.1.19). □

9.2. Transport calculus for the Neumann energy

Let us start by defining a blown-down version of the Neumann partition function of (7.1.13). Let U be an open subset of $\mathbb{R}^{\mathsf{d}}$ with piecewise C^1 boundary. Let μ be a nonnegative density on U with $N\mu(U) = \bar{\mathrm{n}}$, an integer. Following the rescaling in (5.2.6)–(5.2.8) we let

$$\mathsf{F}_N(X_{\bar{\mathrm{n}}}, \mu, U) := N^{\frac{\mathsf{s}}{\mathsf{d}}}\mathsf{F}(X'_{\bar{\mathrm{n}}}, \mu', U') - \left(\frac{\bar{\mathrm{n}}}{2\mathsf{d}}\log N\right)\mathbf{1}_{\mathsf{s}=0} \tag{9.2.1}$$

with F as in (7.1.22) and $(X'_{\bar{n}}, \mu', U')$ the blown-up quantities $X'_{\bar{n}} = N^{1/\mathsf{d}} X_{\bar{n}}$, $\mu'(x) = \mu(xN^{-1/\mathsf{d}})$ and $U' = N^{1/\mathsf{d}} U$. We then let

$$d\mathbb{Q}_{N,\beta}(\mu, U) := \frac{1}{\mathsf{K}_{N,\beta}(\mu, U)} \exp\left(-\beta N^{-\frac{\mathsf{s}}{\mathsf{d}}} \mathsf{F}_N(X_{\bar{n}}, \mu, U)\right) d\mu^{\otimes \bar{n}}(X_{\bar{n}}), \tag{9.2.2}$$

$$\mathsf{K}_{N,\beta}(\mu, U) := \int_{U^{\bar{n}}} \exp\left(-\beta N^{-\frac{\mathsf{s}}{\mathsf{d}}} \mathsf{F}_N(X_{\bar{n}}, \mu, U)\right) d\mu^{\otimes \bar{n}}(X_{\bar{n}}), \tag{9.2.3}$$

and note that these coincide with (5.1.13) and (5.1.12) when $U = \mathbb{R}^{\mathsf{d}}$ and μ is a probability density, so that in that case $\mathbb{Q}_{N,\beta}(\mu) = \mathbb{Q}_{N,\beta}(\mu, \mathbb{R}^{\mathsf{d}})$ and $\mathsf{K}_{N,\beta}(\mu) = \mathsf{K}_{N,\beta}(\mu, \mathbb{R}^{\mathsf{d}})$.

We can phrase a transport calculus and commutator estimates adapted to this situation.

Lemma 9.5 (First variation and commutator estimate for Neumann energy). *Let U be a hyperrectangle. The Neumann energy being defined in (9.2.1), letting ψ_t, $t \in [-1, 1]$, be a Lipschitz vector field in U with $\psi_t \cdot \nu = 0$ on ∂U, Φ_t solve*

$$\begin{cases} \frac{d\Phi_t}{dt}(x) = \psi_t(\Phi_t(x)) & \text{in } U \\ \Phi_0(x) = x & \text{in } U, \end{cases} \tag{9.2.4}$$

and $\mu_t := \Phi_t \# \mu$, there exists $\mathsf{A}_1(X_{\bar{n}}, \mu, \psi, U)$, linear in its third entry, such that we have

$$\frac{d}{dt} \mathsf{F}_N(\Phi_t(X_{\bar{n}}), \Phi_t \# \mu, U) = \mathsf{A}_1(\Phi_t(X_{\bar{n}}), \mu_t, \psi_t, U), \tag{9.2.5}$$

$$\frac{d}{dt} \log \mathsf{K}_{N,\beta}(\mu_t, U) = -\beta N^{-\mathsf{s}/\mathsf{d}} \mathbb{E}_{\mathbb{Q}_{N,\beta}(\mu_t, U)}(\mathsf{A}_1(\Phi_t(X_{\bar{n}}), \mu_t, \psi_t, U)), \tag{9.2.6}$$

and for any Lipschitz vector field $\psi : U \to U$ such that $\psi \cdot \nu = 0$ on ∂U, we have

$$|\mathsf{A}_1(X_{\bar{n}}, \mu, \psi, U)| \le C \|\nabla \psi\|_{L^\infty} \left(\mathsf{F}_N(X_{\bar{n}}, \mu, U) - \bar{n}\left(\frac{\log N}{2\mathsf{d}}\right) \mathbf{1}_{\mathsf{s}=0} + C \bar{n} N^{\frac{\mathsf{s}}{\mathsf{d}}} \right), \tag{9.2.7}$$

for some $C > 0$ depending only on d and s.

Proof. We have the analogue of Lemma 7.10 in rescaled coordinates: let G_U solve

$$\begin{cases} -\Delta_x G_U(x, y) = \mathsf{c}_{\mathsf{d}}(\delta_y(x) - \frac{1}{\mu(U)} \mu(x)) & \text{in } U \\ \frac{\partial G_U}{\partial \nu} = 0 & \text{on } \partial U, \end{cases} \tag{9.2.8}$$

and let

$$H_U(x, y) = G_U(x, y) - \mathsf{g}(x - y), \tag{9.2.9}$$

then for any configuration $X_{\bar{n}}$ of points in U, the function u solving the analogue of (7.1.1)

$$\begin{cases} -\Delta u = \mathsf{c}_{\mathsf{d}}\left(\sum_{i=1}^{\bar{n}} \delta_{x_i} - N\mu\right) & \text{in } U \\ \frac{\partial u}{\partial \nu} = 0 & \text{on } \partial U, \end{cases} \tag{9.2.10}$$

satisfies

$$u(x) = \int_U G_U(x, y) d\left(\sum_{i=1}^{\bar{n}} \delta_{x_i} - N\mu\right)(y) \tag{9.2.11}$$

and we have

(9.2.12)
$$\begin{aligned}&\mathsf{F}_N(X_{\bar{\mathrm{n}}},\mu,U)\\&=\frac{1}{2}\iint_{(\mathbb{R}^{\mathsf{d}})^2\setminus\triangle}\mathsf{g}(x-y)d\left(\sum_{i=1}^{\bar{\mathrm{n}}}\delta_{x_i}-N\mu\mathbf{1}_U\right)(x)d\left(\sum_{i=1}^{\bar{\mathrm{n}}}\delta_{x_i}-N\mu\mathbf{1}_U\right)(y)\\&+\frac{1}{2}\iint_{U\times U}H_U(x,y)d\left(\sum_{i=1}^{\bar{\mathrm{n}}}\delta_{x_i}-N\mu\right)(x)d\left(\sum_{i=1}^{\bar{\mathrm{n}}}\delta_{x_i}-N\mu\right)(y).\end{aligned}$$

Arguing as in (9.1.12)–(9.1.13), a direct computation using the definition of the push-forward yields that

(9.2.13)
$$\begin{aligned}&\frac{d}{dt}\mathsf{F}_N(\Phi_t(X_{\bar{\mathrm{n}}}),\Phi_t\#\mu,U)\\&=\frac{1}{2}\iint_{(\mathbb{R}^{\mathsf{d}})^2\setminus\triangle}(\psi_t(x)-\psi_t(y))\cdot\nabla\mathsf{g}(x-y)\\&\qquad\times d\left(\sum_{i=1}^{\bar{\mathrm{n}}}\delta_{x_i^t}-N\mu_t\mathbf{1}_U\right)(x)d\left(\sum_{i=1}^{\bar{\mathrm{n}}}\delta_{x_i^t}-N\mu_t\mathbf{1}_U\right)(y)\\&+\frac{1}{2}\iint_{U\times U}(\psi_t(x)\cdot\partial_xH_U(x,y)+\psi_t(y)\cdot\partial_yH_U(x,y))\\&\qquad\times d\left(\sum_{i=1}^{\bar{\mathrm{n}}}\delta_{x_i^t}-N\mu_t\right)(x)d\left(\sum_{i=1}^{\bar{\mathrm{n}}}\delta_{x_i^t}-N\mu_t\right)(y).\end{aligned}$$

Setting

(9.2.14)
$$\begin{aligned}&\mathsf{A}_1(X_{\bar{\mathrm{n}}},\mu,\psi,U)\\&:=\frac{1}{2}\iint_{(\mathbb{R}^{\mathsf{d}})^2\setminus\triangle}(\psi(x)-\psi(y))\cdot\nabla\mathsf{g}(x-y)\\&\qquad\times d\left(\sum_{i=1}^{\bar{\mathrm{n}}}\delta_{x_i}-N\mu\mathbf{1}_U\right)(x)d\left(\sum_{i=1}^{\bar{\mathrm{n}}}\delta_{x_i}-N\mu\mathbf{1}_U\right)(y)\\&+\frac{1}{2}\iint_{U\times U}(\psi(x)\cdot\partial_xH_U(x,y)+\psi(y)\cdot\partial_yH_U(x,y))\\&\qquad\times d\left(\sum_{i=1}^{\bar{\mathrm{n}}}\delta_{x_i}-N\mu\right)(x)d\left(\sum_{i=1}^{\bar{\mathrm{n}}}\delta_{x_i}-N\mu\right)(y),\end{aligned}$$

the relation (9.2.5) is thus proved and (9.2.6) follows as in the proof of (9.1.9).

Let us now justify (9.2.7). Let us consider balls $B(x_i,\eta_i)$ with $\eta_i\in[\frac{1}{2}\hat{\mathsf{r}}_i,\hat{\mathsf{r}}_i]$ (where $\hat{\mathsf{r}}_i$ is defined as in (7.1.24) except maxed at scale $N^{-1/\mathsf{d}}$ instead of 1), so that the balls are disjoint and included in U. Let $u_{\vec{\eta}}$ denote the truncated potential and $u^i(x)=u(x)-\mathsf{g}(x-x_i)$ for each i. By symmetry of G_U hence H_U, desymmetrizing both integrals, we may write

$$\mathsf{A}_1(X_{\bar{n}}, \mu, \psi, U) := \sum_{i=1}^{\bar{n}} \int_U \psi(x_i) \cdot (\nabla \mathsf{g}(x_i - x) + \nabla_x H_U(x_i, x)) d\left(\sum_{j \neq i}^{\bar{n}} \delta_{x_j} - N\mu\right)(x)$$

$$+ \sum_{i=1}^{\bar{n}} \psi(x_i) \cdot \nabla_x H_U(x_i, x_i)$$

$$- N \iint_{U \times U} \psi(x) \cdot (\nabla \mathsf{g}(x-y) + \partial_x H_U(x,y)) d\mu(x) d\left(\sum_{i=1}^{\bar{n}} \delta_{x_i} - N\mu\right)(y),$$

hence in view of (9.2.11) and the definition of u^i, we may rewrite this as

$$\mathsf{A}_1(X_{\bar{n}}, \mu, \psi, U) = \sum_{i=1}^{\bar{n}} \int_U \psi(x_i) \cdot \nabla u^i(x_i) - N \int_U \psi \cdot \nabla u \, d\mu.$$

The rest of the proof is the same as that of Theorem 6.1, making use of Lemma 7.8 and Remark 7.9. An important point is the integration by parts

$$\int_U \psi \cdot \operatorname{div} [\nabla u_{\vec{\eta}}, \nabla u_{\vec{\eta}}]$$

that yields a boundary term $\int_{\partial U} \sum_{i,j} \psi_i [\nabla u_{\vec{\eta}}, \nabla u_{\vec{\eta}}]_{ij} \nu_j$ which vanishes thanks to the condition $\nabla u \cdot \nu = 0$. □

9.3. Application: Lipschitzness of the free energy

We will denote

$$\bar{d}_0 := C N^{-\frac{1}{\mathsf{d}}} \max\left(\left(\frac{N^{\frac{1}{\mathsf{d}}}}{\max(1, \beta^{-\frac{1}{2}} \chi(\beta)^{\frac{1}{2}})}\right)^{-\frac{2}{3}} N^{\frac{1}{\mathsf{d}}}, N^{\frac{1}{\mathsf{d}+2}}\right) \tag{9.3.1}$$

the blown-down of the distance d_0 of (8.3.2).

In Theorem 8.14, we proved the convergence of the Neumann partition function for growing cubes in the case where the background measure μ is constant. In order to understand the free energy for nonhomogeneous background, we want to proceed by approximation, reducing to the case of a constant one after showing that $\log \mathsf{K}_{N,\beta}(Q, \mu)$ varies little if μ varies little.

The method consists in using an explicit transport ψ_t that transports μ_0 into μ_1 as in (9.1.3) and using the transport calculus. We start by an elementary lemma that allows us to work with linear interpolants within this framework.

Lemma 9.6. *Let μ_0 and μ_1 be two probability densities on $\mathbb{R}^{\mathsf{d}}$ and let $\mu_s = (1-s)\mu_0 + s\mu_1$ be their linear interpolant. Assume that for $s \in [-1, 1]$, ψ_s is a Lipschitz vector field on $\mathbb{R}^{\mathsf{d}}$, depending continuously on s, and satisfying*

$$-\operatorname{div}(\psi_s \mu_s) = \mu_1 - \mu_0. \tag{9.3.2}$$

Then defining Φ_s as in (9.1.2), we have that $\mu_s = \Phi_s \# \mu_0$.

PROOF. Let $\bar{\mu}_s = \Phi_s \# \mu_0$ as defined in (9.1.3). The argument consists in showing that by uniqueness, we must have $\mu_s = \bar{\mu}_s$. As seen in (9.1.4), $\bar{\mu}_s$ solves

$$\partial_s \bar{\mu}_s + \operatorname{div}(\psi_s \bar{\mu}_s) = 0.$$

On the other hand, μ_s solves

$$\partial_s \mu_s = \mu_1 - \mu_0 = -\mathrm{div}\,(\psi_s \mu_s).$$

So μ_s and $\bar{\mu}_s$ solve the same linear continuity equation. Since ψ_s is Lipschitz, this equation enjoys uniqueness (for instance take $u_s = \mu_s - \bar{\mu}_s$ and check that $\frac{d}{ds}\int u_s(x)\varphi(\Phi_s(x))dx = 0$ for all test functions φ), hence $\bar{\mu}_s = \mu_s$, as desired. □

We use this to evaluate the change in $\log \mathsf{K}_{N,\beta}(\mathbb{R}^{\mathsf{d}}, \mu)$ when μ varies only in a hyperrectangle Q_ℓ. The change is controlled by the energy via the commutator estimate, which is then combined with the local laws to control the energy when on mesoscales. For that reason, we restrict to the Coulomb case and place the same assumptions as for the local laws.

Proposition 9.7 (Variation of free energy—global case). *Assume* $\mathsf{d} \geq 1$ *and* $\mathsf{s} = \mathsf{d}-2$. *Assume* $\rho_\beta N^{-1/\mathsf{d}} \leq \ell \leq C$. *Let* μ_0, μ_1 *be two nonnegative Lipschitz densities bounded above and below by positive constants in a set* Λ. *If* $\mathsf{s} \leq 0$, *assume in addition that* (5.2.28), (5.2.29) *and* (5.2.37) *hold for* μ_0 *and* μ_1. *Assume that* $\varphi \in C^3$, *that* $\square_\ell$ *is a hypercube of sidelength* ℓ *included in* Λ *and containing a* $2N^{-1/\mathsf{d}}$-*neighborhood of* $\operatorname{supp}\varphi$. *If* $\ell \ll 1$ *as* $N \to \infty$, *assume also that* $\mathrm{dist}(\square_\ell, \partial\Lambda) \geq \bar{d}_0$ *as in* (9.3.1). *Assume that*

$$\mu_1 - \mu_0 = -\Delta\varphi.$$

Then

$$\begin{aligned}(9.3.3)\qquad &|\log \mathsf{K}_{N,\beta}(\mu_1, \mathbb{R}^{\mathsf{d}}) - \log \mathsf{K}_{N,\beta}(\mu_0, \mathbb{R}^{\mathsf{d}})| \\ &\leq C\beta\chi(\beta)N\ell^{\mathsf{d}}\left((|\mu_0|_{C^1} + |\varphi|_{C^3})|\varphi|_{C^1} + |\varphi|_{C^2}\right).\end{aligned}$$

where C *depends only on* d *and the upper and lower bounds for* μ_0 *and* μ_1 *and the constants in the assumptions.*

Proof. Consider the linear interpolant $\mu_s = (1-s)\mu_0 + s\mu_1$, and

$$\psi_s := \frac{\nabla\varphi}{\mu_s}.$$

By definition, ψ_s is Lipschitz, $\square_\ell$ contains a $2N^{-1/\mathsf{d}}$-neighborhood of its support, and ψ_s solves (9.3.2), thus, by Lemma 9.6, $\mu_s = \Phi_s \# \mu_0$ where Φ_s solves (9.1.2). Moreover, we can estimate

$$\begin{aligned}(9.3.4)\qquad |\psi_s|_{C^1} &\leq C\left(\left\|\frac{1}{\mu_s}\right\|_{L^\infty}^2 |\mu_s|_{C^1}|\varphi|_{C^1} + \left\|\frac{1}{\mu_s}\right\|_{L^\infty}|\varphi|_{C^2}\right) \\ &\leq C\left((|\mu_0|_{C^1} + |\varphi|_{C^3})|\varphi|_{C^1} + |\varphi|_{C^2}\right)\end{aligned}$$

where C depends on d and the upper and lower bounds for μ_0 and μ_1. By (9.1.9) in Proposition 9.2, we obtain

$$(9.3.5)\qquad \frac{d}{ds}\log \mathsf{K}_{N,\beta}(\mathbb{R}^{\mathsf{d}}, \mu_s) = \mathbb{E}_{\mathbb{Q}_{N,\beta}(\mu_s)}\left(-\beta N^{-\frac{\mathsf{s}}{\mathsf{d}}}\mathsf{A}_1(X_N, \mu_s, \psi_s)\right).$$

Inserting the commutator estimate of Theorem 6.1 recalled in (9.1.16), we obtain

$$\left|\frac{d}{ds}\log \mathsf{K}_{N,\beta}(\mathbb{R}^{\mathsf{d}},\mu_s)\right| \\ \le C|\psi_s|_{C^1}\mathbb{E}_{\mathbb{Q}_{N,\beta}(\mu_s)}\Big(-\beta N^{-\frac{s}{\mathsf{d}}}\Big(\mathsf{F}_N^{\square_\ell}(X_N,\mu_s) \\ +\left(\frac{\#I_{\square_\ell}\log N}{2\mathsf{d}}\right)\mathbf{1}_{\mathsf{s}=0}+C\#I_{\square_\ell}N^{\frac{s}{\mathsf{d}}}\Big)\Big).$$

Inserting (9.3.4), the rescaling formula as in (5.2.8) and using the local laws (8.3.3) which hold in $\square_\ell$ (if ℓ is not small they hold automatically by (5.2.62)), we deduce that

$$\left|\frac{d}{ds}\log \mathsf{K}_{N,\beta}(\mathbb{R}^{\mathsf{d}},\mu_s)\right|\le C\beta\chi(\beta)N\ell^{\mathsf{d}}\left((|\mu_0|_{C^1}+|\varphi|_{C^3})|\varphi|_{C^1}+|\varphi|_{C^2}\right).$$

Integrating between 0 and 1 gives the result. □

Proposition 9.8 (Variation of free energy—Neumann cube case). *Assume $\rho_\beta N^{-1/\mathsf{d}}\le \ell\le C$. Let μ_0,μ_1 be two Lipschitz densities bounded above and below by positive constants in Q_ℓ, a hyperrectangle of sidelengths in $[\ell,2\ell]$ with $N\mu_0(Q_\ell)=N\mu_1(Q_\ell)=\bar{\mathrm{n}}$ an integer. Then*

$$(9.3.6)\quad |\log \mathsf{K}_{N,\beta}(Q_\ell,\mu_1)-\log \mathsf{K}_{N,\beta}(Q_\ell,\mu_0)| \\ \le C\beta\chi(\beta)N\ell^{\mathsf{d}}\left(\ell^2(|\mu_0|_{C^1}+|\mu_1-\mu_0|_{C^1})|\mu_1-\mu_0|_{C^1}+\ell|\mu_1-\mu_0|_{C^1}\right),$$

where C depends only on d and a lower bound for μ_0 and μ_1.

PROOF. Since we are working with the Neumann energy in a cube, we need to find a transport that preserves the cube and solves (9.3.2). For that we let φ solve

$$(9.3.7)\quad \begin{cases}-\Delta\varphi=\mu_1-\mu_0 & \text{in } Q_\ell\\ \dfrac{\partial\varphi}{\partial\nu}=0 & \text{on } \partial Q_\ell.\end{cases}$$

By elliptic regularity (applied after reflection through the boundary) and scaling we have

$$|\varphi|_{C^1}\le C\ell^2|\mu_1-\mu_0|_{C^1},\quad |\varphi|_{C^2}\le C\ell|\mu_1-\mu_0|_{C^1}.$$

Consider, for $0\le s\le 1$, the linear interpolant $\mu_s=(1-s)\mu_0+s\mu_1$. Setting

$$\psi_s:=\frac{\nabla\varphi}{\mu_s},$$

we have

$$\begin{cases}-\mathrm{div}\,(\psi_s\mu_s)=\mu_1-\mu_0 & \text{in } Q_\ell\\ \psi_s\cdot\nu=0 & \text{on } \partial Q_\ell.\end{cases}$$

An analogous reasoning to that of Lemma 9.6 yields that $\mu_s=\Phi_s\#\mu_0$ where Φ_t solves

$$\begin{cases}\frac{d\Phi_t}{dt}(x)=\psi_t(\Phi_t(x))\\ \Phi_0(x)=x.\end{cases}\quad \text{in } Q_\ell$$

Moreover,

$$(9.3.8) \qquad |\psi_s|_{C^1} \le C\left(\left\|\frac{1}{\mu_s}\right\|_{L^\infty}^2 |\mu_s|_{C^1}|\varphi|_{C^1} + \left\|\frac{1}{\mu_s}\right\|_{L^\infty}|\varphi|_{C^2}\right)$$
$$\le C\left(\ell^2\left(|\mu_0|_{C^1}+|\mu_1-\mu_0|_{C^1}\right)|\mu_1-\mu_0|_{C^1}+\ell|\mu_1-\mu_0|_{C^1}\right),$$

where C depends only on d and a lower bound for μ_0 and μ_1. The rest of the proof is identical to the previous one, using Lemma 9.5. □

9.4. Free energy expansions for inhomogeneous density

The results of Proposition 8.13 allowed to evaluate $\log \mathsf{K}_{N,\beta}(\mu)$ by almost additivity, cutting the region into cells in which μ is almost constant. Since the only available formula (8.4.19) is for uniform densities (provided by Theorem 8.14 where the function $f_\mathsf{d}(\beta)$ was introduced), in each cell we need to compare $\log \mathsf{K}_{N,\beta}(\mu, Q_\ell)$ with $\log \mathsf{K}_{N,\beta}(\fint_{Q_\ell}\mu, Q_\ell)$ where $\fint \mu$ is the average of μ. This can be done via Proposition 9.8. In order to have a good precision, we need the cells of the partition to be small, while in order to have small additivity errors in Proposition 8.13, we need them to be large. Optimizing the effect of these two types of error (additivity error and approximation error) over the size of the partitioning cells leads to choosing cells of size $N^{\frac{1}{2\mathsf{d}}}$ in blown up scale, and we get after some direct but tedious computations (for which we refer to [**Ser23**, Section 6]) the following result.

Proposition 9.9 (Free energy expansion for general density in a hyperrectangle). *Assume $\mathsf{s}=\mathsf{d}-2$. Let R be such that $RN^{1/\mathsf{d}}$ satisfies (8.4.8). Let Q be a hyperrectangle of sidelengths in $(R,2R)$. Let μ be a Lipschitz density bounded above and below by positive constants in Q, and assume $N\int_Q \mu = \bar{\mathrm{n}}$ is an integer. Then,*

$$(9.4.1) \qquad \log \mathsf{K}_{N,\beta}(\mu, Q) = -\beta N\int_Q \mu^{1+\frac{\mathsf{s}}{\mathsf{d}}} f_\mathsf{d}(\beta\mu^{\frac{\mathsf{s}}{\mathsf{d}}})$$
$$-\frac{\beta}{2\mathsf{d}}N\left(\int_Q \mu\log\mu\right)\mathbf{1}_{\mathsf{s}=0}+\left(\frac{\beta}{2\mathsf{d}}\bar{\mathrm{n}}\log N\right)\mathbf{1}_{\mathsf{s}=0}$$
$$+O\left(\beta\chi(\beta)NR^\mathsf{d}\mathcal{R}(N,R,\mu)\right)$$

where we let

$$(9.4.2) \qquad \mathcal{R}(N,R,\mu) := \max\left(x(1+|\log x|),(y^{\frac{1}{2}}+y)(1+|\log y|^{\frac{1}{\mathsf{d}}})\right)$$

after setting

$$x := \frac{\rho_\beta}{RN^{\frac{1}{\mathsf{d}}}}, \quad y := \frac{\rho_\beta|\mu|_{C^1}}{N^{\frac{1}{\mathsf{d}}}},$$

and the O depends only on d and the upper and lower bounds for μ.

What is important here is that we get an explicit error rate, valid for broad temperature regimes. The quantity x is small by (8.4.8), the estimate is interesting when y is small too. If β is fixed then so is ρ_β and the error rate obtained is $R^\mathsf{d}N^{1-\frac{1}{2\mathsf{d}}}\log^{\frac{1}{\mathsf{d}}}N$. Note that we may let $\beta\to\infty$ in the result and obtain the following expansion for the minimal energy (7.1.57):

$$(9.4.3)\qquad \mathsf{E}_\infty(\mu,Q)=Nf_\mathsf{d}(\infty)\int_Q\mu^{1+\frac{\mathsf{s}}{\mathsf{d}}}+\left(\frac{N}{2\mathsf{d}}\int_Q\mu\log\mu-\frac{\bar{\mathrm{n}}}{2\mathsf{d}}\log N\right)\mathbf{1}_{\mathsf{s}=0}+O(NR^\mathsf{d}\mathcal{R}(N,R,\mu)).$$

We deduce the following.

Corollary 9.10 (Relative expansion, local version). *Let μ_0 and μ_1 be two non-negative Lipschitz densities bounded above and below by positive constants in a set Λ, and coinciding outside Q_ℓ, a hyperrectangle included in Λ of sidelengths in $(\ell, 2\ell)$ with $C \ge \ell \ge \rho_\beta N^{-1/\mathsf{d}}$ that satisfies $\mathrm{dist}(Q_\ell, \partial\Lambda) \ge \bar{d}_0$. Assume $N\int_{Q_\ell}\mu_0 = N\int_{Q_\ell}\mu_1 = \bar{\mathrm{n}}$ is an integer. We have*

$$
\begin{aligned}
(9.4.4)\qquad & \log \mathsf{K}_{N,\beta}(\mu_1, \mathbb{R}^\mathsf{d}) - \log \mathsf{K}_{N,\beta}(\mu_0, \mathbb{R}^\mathsf{d}) \\
&= -\beta N \int_{Q_\ell} \mu_1^{1+\frac{\mathsf{s}}{\mathsf{d}}} f_\mathsf{d}(\beta \mu_1^{\frac{\mathsf{s}}{\mathsf{d}}}) - \frac{\beta}{2\mathsf{d}} N \left(\int_{Q_\ell} \mu_1 \log \mu_1\right) \mathbf{1}_{\mathsf{s}=0} \\
&\quad + \beta N \int_{Q_\ell} \mu_0^{1+\frac{\mathsf{s}}{\mathsf{d}}} f_\mathsf{d}(\beta \mu_0^{\frac{\mathsf{s}}{\mathsf{d}}}) + \frac{\beta}{2\mathsf{d}} N \left(\int_{Q_\ell} \mu_0 \log \mu_0\right) \mathbf{1}_{\mathsf{s}=0} \\
&\quad + O\left(\beta \chi(\beta) N \ell^\mathsf{d} \left(\mathcal{R}(N,\ell,\mu_0) + \mathcal{R}(N,\ell,\mu_1)\right)\right),
\end{aligned}
$$

where $\mathcal{R}$ is as in Proposition 9.9, and the O depends only on d and the upper and lower bounds for μ_1 and μ_0 in Q_ℓ.

Proof. We may apply (8.4.9) (in rescaled form) to both μ_0 and μ_1 and subtract the obtained relations to get that

$$
\begin{aligned}
\log \mathsf{K}_{N,\beta}(\mu_1, \mathbb{R}^\mathsf{d}) - \log \mathsf{K}_{N,\beta}(\mu_0, \mathbb{R}^\mathsf{d}) &= \log \mathsf{K}_{N,\beta}(\mu_1, Q_\ell) - \log \mathsf{K}_{N,\beta}(\mu_0, Q_\ell) \\
&+ O\left(N^{1-\frac{1}{\mathsf{d}}} \beta \ell^{\mathsf{d}-1} \rho_\beta \chi(\beta) + N^{1-\frac{1}{\mathsf{d}}} \beta^{1-\frac{1}{\mathsf{d}}} \chi(\beta)^{1-\frac{1}{\mathsf{d}}} \left(\log \frac{\ell N^{\frac{1}{\mathsf{d}}}}{\rho_\beta}\right)^{\frac{1}{\mathsf{d}}} \ell^{\mathsf{d}-1}\right).
\end{aligned}
$$

Inserting the result of (9.4.1) applied to μ_0 and μ_1, we deduce

$$
\begin{aligned}
&\log \mathsf{K}_{N,\beta}(\mu_1) - \log \mathsf{K}_{N,\beta}(\mu_0) \\
&= -\beta N \int_Q \mu_1^{1+\frac{\mathsf{s}}{\mathsf{d}}} f_\mathsf{d}(\beta \mu_1^{\frac{\mathsf{s}}{\mathsf{d}}}) - \frac{\beta}{2\mathsf{d}} \left(N \int_Q \mu_1 \log \mu_1\right) \mathbf{1}_{\mathsf{s}=0} \\
&\quad + \beta N \int_Q \mu_0^{1+\frac{\mathsf{s}}{\mathsf{d}}} f_\mathsf{d}(\beta \mu_0^{\frac{\mathsf{s}}{\mathsf{d}}}) + \frac{\beta}{2\mathsf{d}} N \left(\int_Q \mu_0 \log \mu_0\right) \mathbf{1}_{\mathsf{s}=0} \\
&\quad + O\Bigg(\beta \chi(\beta) N \ell^\mathsf{d} \Bigg(\mathcal{R}(N,\ell,\mu_0) + \mathcal{R}(N,\ell,\mu_1) \\
&\qquad + N^{-\frac{1}{\mathsf{d}}} \ell^{-1} \rho_\beta + \beta^{-\frac{1}{\mathsf{d}}} \chi(\beta)^{-\frac{1}{\mathsf{d}}} \left(\log \frac{\ell N^{\frac{1}{\mathsf{d}}}}{\rho_\beta}\right)^{\frac{1}{\mathsf{d}}} \ell^{-1} N^{-\frac{1}{\mathsf{d}}}\Bigg)\Bigg).
\end{aligned}
$$

Checking that $\beta^{-\frac{1}{\mathsf{d}}}\chi(\beta)^{-\frac{1}{\mathsf{d}}} \le \rho_\beta$ by (8.0.2) and (5.2.27), by definition of x we see that we may absorb the last error terms into $\mathcal{R}$. □

Remark 9.11. A crucial point here, that will be used in the proof of the CLT in Chapter 10, is that for $\mathsf{s} = 0$, the terms $-\beta N \int_Q \mu_1^{1+\frac{\mathsf{s}}{\mathsf{d}}} f_\mathsf{d}(\beta \mu_1^{\frac{\mathsf{s}}{\mathsf{d}}})$ and $\beta N \int_Q \mu_0^{1+\frac{\mathsf{s}}{\mathsf{d}}} f_\mathsf{d}(\beta \mu_0^{\frac{\mathsf{s}}{\mathsf{d}}})$ simply cancel out. This is a manifestation of the conformal invariance of the Laplacian (or the Coulomb interaction) in two dimensions. In contrast, for all other dimensions, a density-dependent effective temperature $\beta\mu^{\mathsf{s}/\mathsf{d}}$ appears.

Returning to the splitting formula (5.1.9), we may now obtain a general expansion for the free energy or minimal energy. For this, it suffices to partition the space into regions with quantized mass, use the almost additivity of the free energy of Proposition 8.13, the expansion just obtained, and control the contributions of the tails by the cruder bounds of (7.1.55).

The result is given in the Coulomb case, but let us emphasize that the same formula is proven for all Riesz cases $\mathsf{s} \in [\mathsf{d}-2, \mathsf{d})$, albeit with a nonexplicit error term in [**LS17**], and now in [**PS**] with an explicit error rate. This provides the precise next order term (order N term) in the (free) energy expansion, with an explicit power error rate. In the one-dimensional logarithmic case, free energy expansions formulae are known, even up to arbitrary order, for quadratic potential V thanks to Selberg's formula, or for regular enough V's, see in particular [**BG13, BG24, Shc13**].

Other than this one-dimensional setting, prior results are restricted to the $\mathsf{d} = 1$ Coulomb case and the $\beta = 2$ two-dimensional Coulomb case: the most recent results obtain expansions to all order in radial situations including the possibility of a multi-connected droplet [**BKS23, ACC23**]. Wiegmann and Zabrodin [**ZW06**] made a prediction for the order 1 term in the free energy expansion in the general β two-dimensional case, another derivation via formal free field (or path integral) computations can be found in [**Kle16**]. A more precise prediction is also made in [**Sha11, CFTW15**] for β equal to 2 and nearing 2, the next order term being of order $\sqrt{N}$, and predicted to be independent of μ_V. Recently, this has been proved in the $\beta = 2$ case for some specific non radially symmetric potential [**BSY25**], see also [**Kle14, KMMW17**] in the Riemannian setup. Finally [**TF99, JGP94**] provide finite effect corrections to the free energy. Assembling all the above predictions, the prediction for formula (1.3.5) in the two-dimensional Coulomb case is

$$\boxed{\log Z_{N,\beta} = A_\beta N^2 + \frac{\beta}{4} N \log N + B_\beta N + C_\beta \sqrt{N} + D_\beta \log N + E_\beta + o(1)}$$

with

$$\begin{aligned}
A_\beta &= -\beta \mathcal{E}(\mu_V)\\
B_\beta &= \beta f_{\mathsf{d}}(\beta) + (1 - \frac{\beta}{4}\mathbf{1}_{\mathsf{s}=0}) \int_\Sigma \mu_V \log \mu_V\\
C_\beta &= k \frac{4}{3\sqrt{\pi}} \log \frac{\beta}{2}\\
D_\beta &= \frac{1}{2} - \frac{\chi}{12}\\
E_\beta &= -\frac{1}{2} \log \left(\frac{\det_\zeta(-\Delta_{\mathbb{R}^2 \setminus \Sigma})}{\det_\zeta(-\Delta_{\mathbb{R}^2})} \right) + c(\beta) + d(\beta) \int_{\partial \Sigma} \frac{\partial \log \Delta V}{\partial \nu}\\
&\quad + \frac{(\frac{\beta}{4} - 1)^2}{\beta} \left(\frac{1}{4\pi} \int_{\mathbb{R}^2} |\nabla (\log \Delta V)^\Sigma|^2 \right)
\end{aligned}$$

where k is the number of connected components of Σ, χ is the Euler characteristic of Σ, B_β involves an entropy term called in this context *Mabuchi functional*, E_β from [**ZW06**] is based on the Zeta regularized determinant of the Laplacian and involves a *Liouville functional* from conformal field theory, where f^Σ denotes the harmonic extension of f outside Σ, and c and d are unknown functions.

We are here confirming the next-to-leading order, or order N, term B_β. Note that we are expressing it via the thermal equilibrium measure but we could as well express it via the usual equilibrium measure, as was done in [**LS17**]. The reader interested in this can refer to that paper or use the definition (2.0.4) and the results of Theorem 2.23.

In the general Riesz case, other than [**LS17**] there are few results as well, except results in [**BHS12**] on Riesz-energy minimizers on the torus. In contrast to all the results in the literature, we emphasize that the errors in the formulas below are independent of β as long as $\theta > \theta_0$ where θ is as in (3.2.5), which allows to reach very high-temperature regimes. The formula for the general Riesz case proven in [**LS17, PS**] takes the same form, but is expressed in terms of μ_V and the quantitative error rate is not as good.

Theorem 9.12 (Free energy/minimal energy expansion). *Assume* $\mathsf{s} = \mathsf{d} - 2$ *with* $\mathsf{d} \geq 2$. *Assume* (A1)–(A5), $\theta \geq \theta_0 > 0$, *and that the equilibrium measure* μ_V *is Lipschitz on its support (if* $\beta = \infty$*), resp. that the thermal equilibrium measure is Lipschitz. We have*

$$(9.4.5)\quad \min \mathcal{H}_N(X_N) = N^2 \mathcal{E}(\mu_V) - \frac{1}{2\mathsf{d}}(N \log N)\mathbf{1}_{\mathsf{s}=0} + \frac{N}{2\mathsf{d}}\left(\int \mu_V \log \mu_V\right)\mathbf{1}_{\mathsf{s}=0} + N^{1+\frac{\mathsf{s}}{\mathsf{d}}} f_{\mathsf{d}}(\infty) \int_{\mathbb{R}^{\mathsf{d}}} \mu_V^{1+\frac{\mathsf{s}}{\mathsf{d}}} + O(N^{1+\frac{\mathsf{s}}{\mathsf{d}}-\sigma})$$

and

(9.4.6)

$$\log Z_{N,\beta} = -\beta N^{2-\frac{\mathsf{s}}{\mathsf{d}}} \mathcal{E}_\theta(\mu_\theta) + \frac{\beta}{2\mathsf{d}}(N \log N)\mathbf{1}_{\mathsf{s}=0} - N\frac{\beta}{2\mathsf{d}}\left(\int_{\mathbb{R}^{\mathsf{d}}} \mu_\theta \log \mu_\theta\right)\mathbf{1}_{\mathsf{s}=0} - N\beta \int_{\mathbb{R}^{\mathsf{d}}} \mu_\theta^{1+\frac{\mathsf{s}}{\mathsf{d}}} f_{\mathsf{d}}(\beta \mu_\theta^{\frac{\mathsf{s}}{\mathsf{d}}}) + O(\beta\chi(\beta)N^{1-\sigma})$$

where $\sigma > 0$ *depends only on* d, *and the* O *depends on* d, *upper and lower bounds for* μ_V *in* Σ *and* $|\mu_V|_{C^1(\Sigma)}$, *respectively* θ_0, *an upper bound for* μ_θ, *and a lower bound for* μ_θ *and control for the Lipschitz norm of* μ_θ *in* $\{x \in \Sigma, \mathrm{dist}(x, \partial\Sigma) \geq \max(\theta^{\varepsilon-\frac{1}{2}}, \bar{d}_0)\}$ *(provided for instance by* (2.5.32)*), and additionally in dimension* 2 *a bound like* (2.5.28).

We also recall that it will be proven in Corollary 12.7, that

$$f_{\mathsf{d}}(\infty) = \min \mathbb{W}(\cdot, 1)$$

while a variational formula will be given for $f_{\mathsf{d}}(\beta)$ in Corollary 13.11

$$(9.4.7)\qquad \beta f_{\mathsf{d}}(\beta) = \min \mathcal{I}_\beta^1.$$

Proof. We give the proof for the case with temperature, the case of minimizers is similar, working with μ_V instead of μ_θ. First, as verified after (5.2.29), the thermal equilibrium measure μ_θ satisfies (5.2.28) and (5.2.29) for $\Lambda = \Sigma$, the support of the equilibrium measure μ_V and $\mu_\theta(\Sigma^c) \leq C\beta^{-1/2}N^{-1/\mathsf{d}}$ by Theorem 2.23. Let us define

$$(9.4.8)\qquad \hat{\Sigma} = \{x \in \Sigma, \mathrm{dist}(x, \partial\Sigma) \geq \bar{d}_0\}.$$

One may check that $\theta^{\varepsilon-\frac{1}{2}}$ can be made smaller than $\bar{d}_0$, thus by (2.5.32) we have uniform bounds for $|\mu_\theta|_{C^1(\hat{\Sigma})}$ and we also have that the local laws hold in $\hat{\Sigma}$. Using

Lemma 5.13, we then partition most of $\hat{\Sigma}$ into hyperrectangles Q_i of sidelengths of order $\rho_\beta N^{-1/\mathsf{d}} \le r \le \bar{d}_0$, such that $N \int_{Q_i} \mu_\theta = \bar{\mathrm{n}}_i$ is an integer. We keep only the hyperrectangles that are inside $\hat{\Sigma}$. This way the local laws are satisfied in $U := \cup_i Q_i$ and (8.4.9) applies. Moreover

$$\mu_\theta(\mathbb{R}^\mathsf{d} \backslash U) \le C(r + \beta^{-1/2} N^{-1/\mathsf{d}}). \tag{9.4.9}$$

We apply (8.4.9) to μ_θ and combine it with the result of Proposition 9.9 in each Q_i to obtain

$$\begin{aligned}\log \mathsf{K}_{N,\beta}(\mu_\theta) = &-\beta N \int_U \mu_\theta^{1+\frac{\mathsf{s}}{\mathsf{d}}} f_\mathsf{d}(\beta \mu_\theta^{\frac{\mathsf{s}}{\mathsf{d}}}) - \frac{\beta}{2\mathsf{d}} N \left(\int_U \mu_\theta \log \mu_\theta \right) \mathbf{1}_{\mathsf{s}=0} \\ &- \frac{\beta}{2\mathsf{d}} N \mu_\theta(U)(\log N) \mathbf{1}_{\mathsf{s}=0} + \log \mathsf{K}_{N,\beta}(\mu_\theta, \mathbb{R}^\mathsf{d} \backslash U) + O\left(\beta \chi(\beta) N |U| \mathcal{R}\left(N, r, \mu_\theta\right)\right)\end{aligned} \tag{9.4.10}$$

where we can absorb the errors in (8.4.9) into the $\mathcal{R}$. To bound $\log \mathsf{K}_{N,\beta}(\mathbb{R}^\mathsf{d} \backslash U, \mu_\theta)$ we use (7.1.55) and obtain (after rescaling)

$$\begin{aligned}&\left| \log \mathsf{K}_{N,\beta}(\mu_\theta, \mathbb{R}^\mathsf{d} \backslash U) + \frac{\beta}{4} N \mu_\theta(U^c)(\log N) \mathbf{1}_{\mathsf{d}=2} \right| \\ &\qquad \le C \beta \chi(\beta) N \mu_\theta(\mathbb{R}^\mathsf{d} \backslash U) \le C \beta \chi(\beta)(r + \beta^{-1/2} N^{-1/\mathsf{d}}).\end{aligned}$$

It remains to bound

$$-\beta N \int_{U^c} \mu_\theta^{1+\frac{\mathsf{s}}{\mathsf{d}}} f_\mathsf{d}(\beta \mu_\theta^{\frac{\mathsf{s}}{\mathsf{d}}}) - \frac{\beta}{4} N \left(\int_{U^c} \mu_\theta \log \mu_\theta \right) \mathbf{1}_{\mathsf{d}=2},$$

for this we may use the bound for f_d given in (8.4.16) and (9.4.9).

In dimension $\mathsf{d} = 2$ we bound $\int_{U^c} \mu_\theta \log \mu_\theta$ by a bound provided for instance by (2.5.28). We conclude that

$$\begin{aligned}&-\beta N \int_{\cup_i Q_i} \mu_\theta^{1+\frac{\mathsf{s}}{\mathsf{d}}} f_\mathsf{d}(\beta \mu_\theta^{\frac{\mathsf{s}}{\mathsf{d}}}) - \frac{\beta}{4} N \left(\int_{\cup_i Q_i} \mu_\theta \log \mu_\theta \right) \mathbf{1}_{\mathsf{d}=2} \\ &= -\beta N \int_{\mathbb{R}^\mathsf{d}} \mu_\theta^{1+\frac{\mathsf{s}}{\mathsf{d}}} f_\mathsf{d}(\beta \mu_\theta^{\frac{\mathsf{s}}{\mathsf{d}}}) \\ &\quad - \frac{\beta}{4} N \left(\int_{\mathbb{R}^\mathsf{d}} \mu_\theta \log \mu_\theta \right) \mathbf{1}_{\mathsf{d}=2} + O\left(N \beta \chi(\beta)(r + \beta^{-1/2} N^{-1/\mathsf{d}}) \right).\end{aligned}$$

Inserting this into (9.4.10), there only remains to optimize over r to obtain the result. □

CHAPTER 10

Analysis of fluctuations

In this chapter, we return to our initial purpose in the previous one which is to analyze the fluctuations via the Laplace transform as in (9.0.2). In the general Coulomb case, this will lead us to improved controls of fluctuations compared to Corollary 5.22, which are sharp in two dimensions. In the one- and two-dimensional (Coulomb) case, by comparing two ways of estimating the ratio of partition functions in (9.0.2), one via almost additivity and approximation, and one by transport, we are able to deduce a better error on the expansion, which allows to precisely evaluate the terms in (9.0.2). This leads to a full Central Limit Theorem for fluctuations in that case, where we show that the fluctuations of smooth test functions converge to a specific Gaussian, and the electric potential converges to the Gaussian free field.

Prior to this, such results were known in the extensively-studied one dimensional logarithmic case $\mathsf{d} = 1$, $\mathsf{s} = 0$, under various regularity assumptions on V, for test functions that live at the macroscale [**Joh98, BG13, BG24, Shc13, BLS18, LLW19**] and at mesoscales [**BL18, BMP22, Pei24**], and in [**Boub**] for the $\mathsf{d} = 1$ circular Riesz case. In dimension $\mathsf{d} = 2$, it was first proven in [**RV07**] in the determinantal case $\beta = 2$, with V quadratic, and in [**AHM11, AHM15**] still for $\beta = 2$ but for V analytic. This was extended to all β and all mesoscales $\ell \geq N^{-\alpha}$, $\alpha < \frac{1}{2}$ in [**LS18**] (which includes the boundary case) and in [**BBNY19**].

We will describe the result from [**Ser23**] valid for all β provided $\theta \gg 1$, i.e., $\beta \gg N^{-1+\frac{\mathsf{s}}{\mathsf{d}}}$, and valid down to the microscale, more precisely to the *minimal scale*, i.e., for $\ell \gg \rho_\beta$. Such very large (and also very small) temperature regimes were not treated before. An exception is the one-dimensional β-ensemble case in the borderline regime $\beta = cN^{-1}$ for which the limit point process is Poissonian [**AD14, NT18, NT20, HL21, Lam21**].

The result we will present for the one-dimensional Coulomb case $\mathsf{d} = 1$, $\mathsf{s} = -1$ is new.

10.1. Improved control of fluctuations

To deduce the convergence of the Laplace transform of (9.0.1) from (9.0.2), the correct scaling is to choose a t which depends on N and tends to 0 as $N \to \infty$ (for instance $t = 1/N$ if $\mathsf{s} = 0$), thus it will be important to think in terms of a linearization as $t \to 0$.

To understand the ratio of partition functions $Z_{N,\beta}(V_t)/Z_{N,\beta}(V)$ in (9.0.2), we proceed by using the splitting, either with respect to the thermal equilibrium measure as in (5.1.14) (this is what is done in [**Ser23**]), or with respect to the usual equilibrium measure as in (5.1.11) (this is what was originally done in [**LS18**]). The former has the advantage of working for a broad temperature regime, including for small β and of giving the simple expressions of Proposition 9.2, without volume change terms. The disadvantage compared to the latter is that the transformation of the thermal equilibrium measure under perturbation of V is less simple than that of the regular equilibrium measure, a problem which will be solved by making an approximation, and that it requires more regularity of V as well as appealing to the estimates for $|\mu_\theta|_{C^\sigma}$ provided by Theorem 2.23. The approach with the regular equilibrium measure has the disadvantage of the volume change terms in Proposition 9.4, but the advantage that the equilibrium measure change is completely explicit in the bulk and that there are no approximation errors related to it.

Here the Coulomb and Riesz cases differ a lot: indeed in the Coulomb case the perturbed equilibrium measure μ_V is easy to compute when φ is supported in Σ (the support of μ_V): it is $\mu_{V_t} = \mu_V + \frac{t}{\mathsf{c_d}}\Delta\varphi$ (see (2.1.17)), where $V_t = V + t\varphi$. In the Riesz case, the perturbed equilibrium measure depends on the perturbation φ in a nonlocal way, and thus the construction of an appropriate transport is much more delicate, this is done in [**PS**] where the whole study made in this chapter is extended to the Riesz $\mathsf{d}-2 < \mathsf{s} < \mathsf{d}$ case.

The Coulomb case where φ is not supported in Σ, i.e., has a support intersecting $\partial\Sigma$ is also much more delicate: one needs to understand how $\partial\Sigma$ is displaced under the perturbation. This is described precisely (in all dimensions) in [**SS18**]. The PDE approach used there replaces Sakai's theory used in [**AHM11**], which is restricted to the two-dimensional analytic setting. Doing so allowed us to treat the case where the support of φ may overlap $\partial\Sigma$ (the first instance was in [**AHM11**] in the case of V analytic and $\beta = 2$), we refer to [**LS18**]. To treat that situation, the use of the regular equilibrium measure is much more convenient.

We note the very interesting physics results of [**CSA21**] that show that the edge behavior of the Coulomb gas is different from the bulk one: they predict "freezing at the edge" and observe much stronger oscillation of the density near the edge, as if the effective temperature was larger there than in the bulk. Related are predictions of slow decay of correlations at the boundary made in [**FJ96**] and justified in the determinantal case $\beta = 2$ with Szegö kernel formulae in [**AC22, ACC25, ACC24**].

We start here by presenting the proof with the thermal equilibrium measure, and explain how to use the approach via the regular equilibrium measure in Section 10.2.3.

10.1.1. Assumptions. We assume $\mathsf{s} = \mathsf{d} - 2$, and continue to assume (A1)–(A5), $\theta > \theta_0$. We assume in addition that $V \in C^5$ or even C^8 (here we are not trying to minimize the regularity assumption) and (2.5.20) holds, so that the results of Theorem 2.23 hold.

The result will be valid down to the (temperature dependent) microscale, i.e., for test functions φ supported in a cube of size ℓ which may depend on N, such that

$$\rho_\beta N^{-\frac{1}{\mathsf{d}}} \ll \ell \le C \quad \text{as } N \to \infty \tag{10.1.1}$$

where ρ_β is as in (8.0.2). In view of (3.2.5) and since $\rho_\beta \ge C\beta^{-1/2}$ by (8.0.2), this implies in particular in view of (3.2.5) that $C \ge \ell \gg \theta^{-1/2}$ hence $\theta \gg 1$.

They will also be valid away from a boundary layer (we recall we treat here only the interior case and refer to [**LS18**] otherwise), i.e., where the local laws of Chapter 8 hold:

$$\hat{\Sigma} := \{x \in \Sigma, \mathrm{dist}(x, \partial\Sigma) \ge \bar{d}_0\} \tag{10.1.2}$$

with Σ the support of μ_V and $\bar{d}_0$ is as in (9.3.1).

We recall that from (2.5.32) we have uniform bounds on μ_θ in C^k in $\hat{\Sigma}$, provided $V \in C^{k+4}$, and μ_θ is bounded below in $\hat{\Sigma}$ by a constant $m > 0$ depending only on μ_V and $\theta_0 > 0$.

10.1.2. Preliminary results. As explained above, we assume that φ is supported in $\hat{\Sigma}$.

We do not have an exact formula for the perturbed thermal equilibrium measure $\mu_\theta^{V_t}$ with perturbed potential $V_t = V + t\varphi$, however [**AS22**] provides an expansion up to arbitrary inverse powers of θ (the large effective temperature of (3.2.5)) already stated in (2.5.31). It will suffice for our purposes to retain only the leading order term in $1/\theta$, however a more precise description, requiring more regularity of φ can be obtained by using next order terms, we refer to [**Ser23**] for that treatment.

Our first step is thus to replace $\mu_\theta^{V_t}$ by the approximation

$$\nu_\theta^t := \mu_\theta + \frac{t}{\mathsf{c_d}} \Delta\varphi. \tag{10.1.3}$$

Observe that if φ is supported in $\hat{\Sigma}$ where $\mu_\theta \geq m > 0$, ν_θ^t is also a probability density, as long as $|t||\varphi|_{C^2}$ is smaller than a constant depending only on d and m.

Instead of splitting the energy with respect to $\mu_\theta^{V_t}$ we can provide a splitting with respect to ν_θ^t. The proof of the following lemma is presented below in Section 10.2.4. We recall the definition of Fluct in (9.0.1). We will also highlight the V dependence in (2.0.3) as a superscript.

Lemma 10.1. *If $t|\varphi|_{C^2}$ is small enough (depending on $m > 0$ above), we have*

$$\begin{aligned} &\mathbb{E}_{\mathbb{P}_{N,\beta}}\left(e^{-t\beta N^{1-\frac{\mathsf{s}}{\mathsf{d}}}\sum_{i=1}^N \varphi(x_i)}\right) \\ &= e^{-\beta N^{2-\frac{\mathsf{s}}{\mathsf{d}}}(\mathcal{E}_\theta^{V_t}(\nu_\theta^t) - \mathcal{E}_\theta^V(\mu_\theta))} \frac{\mathsf{K}_{N,\beta}(\nu_\theta^t)}{\mathsf{K}_{N,\beta}(\mu_\theta)} \mathbb{E}_{\mathbb{Q}_{N,\beta}(\nu_\theta^t)}\left(\exp\left(-\theta \operatorname{Fluct}_{\nu_\theta^t}(\varepsilon_t)\right)\right) \end{aligned} \tag{10.1.4}$$

where

$$\varepsilon_t := \mathsf{g} * \nu_\theta^t + V + t\varphi + \frac{1}{\theta}\log \nu_\theta^t - c_\theta \tag{10.1.5}$$

with c_θ as in (5.1.8). We have that ε_t is supported in the support of φ and if $V \in C^4$,

$$\|\varepsilon_t\|_{L^\infty} \leq C\frac{t}{\theta}|\varphi|_{C^2}, \tag{10.1.6}$$

and if in addition $V \in C^5$,

$$|\varepsilon_t|_{C^1} \leq C\frac{t}{\theta}(|\varphi|_{C^2} + |\varphi|_{C^3}), \tag{10.1.7}$$

where $C > 0$ depends only on $\mathsf{d}, \mathsf{s}, m$ and the norms of V.

Examining (10.1.4), three things thus need to be done:

1) evaluate the limit as $t \to 0$ of $\exp\left(-\beta N^{2-\frac{\mathsf{s}}{\mathsf{d}}}(\mathcal{E}_\theta^{V_t}(\nu_\theta^t) - \mathcal{E}_\theta^V(\mu_\theta))\right)$, this is done in Lemma 10.2
2) control $\frac{\mathsf{K}_{N,\beta}(\nu_\theta^t)}{\mathsf{K}_{N,\beta}(\mu_\theta)}$ by Proposition 9.7
3) show that the last expectation, really the approximation error, is close to 1 by a priori rough estimates on fluctuations from Chapter 4.

For the first item we will prove in Section 10.2.4 the following, by explicit computations.

Lemma 10.2. *We have*

$$\mathcal{E}_\theta^{V_t}(\nu_\theta^t) - \mathcal{E}_\theta^V(\mu_\theta) - t\int_{\mathbb{R}^\mathsf{d}} \varphi d\mu_\theta = -\frac{t^2}{2\mathsf{c}_\mathsf{d}}\int_{\mathbb{R}^\mathsf{d}} |\nabla\varphi|^2 + \frac{t^2}{2\theta\mathsf{c}_\mathsf{d}^2}\int_{\mathbb{R}^\mathsf{d}} \frac{|\Delta\varphi|^2}{\mu_\theta} + O\left(\frac{t^3}{\theta}\int_{\mathbb{R}^\mathsf{d}} \frac{|\Delta\varphi|^3}{\mu_\theta^2}\right) \tag{10.1.8}$$

and

$$\left|\mathcal{E}_\theta(\nu_\theta^t) - \mathcal{E}_\theta(\mu_\theta) - t\int_{\mathbb{R}^\mathsf{d}} \varphi d\mu_\theta\right| \le Ct^2|\mathrm{supp}\nabla\varphi|\left(|\varphi|_{C^1}^2 + \frac{|\varphi|_{C^2}^2}{\theta}\right), \tag{10.1.9}$$

where O *and* C *are universal.*

For the second item we observe that by (10.1.3) and Proposition 9.7 we have

$$\left|\log\frac{\mathsf{K}_{N,\beta}(\nu_\theta^t)}{\mathsf{K}_{N,\beta}(\mu_\theta)}\right| \le C\beta\chi(\beta)N\ell^\mathsf{d}\left((|\mu_\theta|_{C^1(Q_\ell)} + |t||\varphi|_{C^3})|t||\varphi|_{C^1} + |t||\varphi|_{C^2}\right). \tag{10.1.10}$$

For the third item we have the following, whose proof relies on the local laws and its consequence (8.4.6), see proof in Section 10.2.4.

Lemma 10.3. *If* $t|\varphi|_{C^2}$ *is small enough (depending on* $m > 0$ *above), we have*

$$\left|\log\mathbb{E}_{\mathbb{Q}_{N,\beta}(\nu_\theta^t)}\left(\exp\left(-\theta\,\mathrm{Fluct}_{\nu_\theta^t}(\varepsilon_t)\right)\right)\right| \le C\sqrt{\chi(\beta)}\beta N^{1+\frac{1}{\mathsf{d}}}\ell^\mathsf{d}|\varepsilon_t|_{C^1} + C\theta N\ell^\mathsf{d}|\varepsilon_t|_{C^1}^2 \tag{10.1.11}$$

and

$$\left|\log\mathbb{E}_{\mathbb{Q}_{N,\beta}(\nu_\theta^t)}\left(\exp\left(-\theta\,\mathrm{Fluct}_{\nu_\theta^t}(\varepsilon_t)\right)\right)\right| \le C\|\varepsilon_t\|_{L^\infty}\theta N\ell^\mathsf{d} + C\|\varepsilon_t\|_{L^\infty}^2\theta N\ell^{\mathsf{d}-2}. \tag{10.1.12}$$

10.1.3. Main result on fluctuations control. Inserting the result of (10.1.6), (10.1.9), (10.1.10), and (10.1.12) into (10.1.4), we have proven the following bound on linear statistics fluctuations, originally found in [**Ser23**]. As mentioned in Section 5.3.2, [**CH25**] provides a short and elegant proof of this bound in the macroscopic case $\ell \propto 1$.

Theorem 10.4 (Control of fluctuations, general Coulomb case)**.** *Let* $\mathsf{s} = \mathsf{d} - 2$. *Assume* $V \in C^5$, (A1)–(A5), (2.5.20) *hold,* $\theta > \theta_0 > 0$, *and* $\varphi \in C^3$. *Assume that there exists a cube* $\square_\ell$ *(not necessarily centered at* 0*) such that* $\mathrm{supp}\,\varphi \subset \square_{\ell/2} \subset \square_\ell \subset \hat\Sigma$, *with* ℓ *satisfying* (10.1.1) *(if it does not hold that* $\ell \ll 1$, *assume only* $\square_\ell \subset \Sigma$*). Then, letting* $\mathrm{Fluct}_{\mu_\theta}(\varphi) = \sum_{i=1}^N \delta_{x_i} - N\mu_\theta$, *for any* $t \in \mathbb{R}$ *such that* $|t||\varphi|_{C^2}$ *is smaller than a positive constant depending only on* d *and* m, *we have*

$$\begin{aligned}\left|\log\mathbb{E}_{\mathbb{P}_{N,\beta}}\left(e^{\beta t N^{1-\frac{\mathsf{s}}{\mathsf{d}}}\mathrm{Fluct}_{\mu_\theta}(\varphi)}\right)\right| &\le C|t|N\ell^\mathsf{d}\left(|\varphi|_{C^2} + \beta\chi(\beta)(|\varphi|_{C^1} + |\varphi|_{C^2})\right) \\ &\quad + Ct^2N\ell^\mathsf{d}\left(\beta N^{\frac{2}{\mathsf{d}}}|\varphi|_{C^1}^2 + \beta\chi(\beta)|\varphi|_{C^3}|\varphi|_{C^1}\right),\end{aligned} \tag{10.1.13}$$

where the constants depend only on the constants in the assumptions and upper and lower bounds for μ_θ *in* $\square_\ell$, *but not on* N, β, φ *or* t.

PROOF. We insert all the announced elements into (10.1.4) to find

$$\begin{aligned}(10.1.14)\qquad &\left|\log \mathbb{E}_{\mathbb{P}_{N,\beta}}\left(\exp(-\beta t N^{1-\frac{\mathsf{s}}{\mathsf{d}}} \operatorname{Fluct}_{\mu_\theta}(\varphi)\right)\right| \\ &\le C\beta N^{2-\frac{\mathsf{s}}{\mathsf{d}}}\left(t^2\ell^{\mathsf{d}}\left(|\varphi|_{C^1}^2+\frac{|\varphi|_{C^2}^2}{\theta}\right)\right) \\ &+C\beta\chi(\beta)N\ell^{\mathsf{d}}\left((|\mu_\theta|_{C^1(Q_\ell)}+|t||\varphi|_{C^3})|t||\varphi|_{C^1}+|t||\varphi|_{C^2}\right) \\ &+C|t||\varphi|_{C^2}N\ell^{\mathsf{d}}+Ct^2|\varphi|_{C^2}^2\frac{N\ell^{\mathsf{d}-2}}{\theta}.\end{aligned}$$

Using that $|t||\varphi|_{C^2}$ is smaller than a constant and $\ell \le C$ to absorb some terms, using (3.2.5), $\mathsf{s}=\mathsf{d}-2$ and (10.1.1) to find $\theta\ell^2 \ge \chi(\beta)\max(1,\beta)\ge 1$, we obtain

$$\begin{aligned}(10.1.15)\qquad &\left|\log \mathbb{E}\left(\exp(-\beta|t|N^{1-\frac{\mathsf{s}}{\mathsf{d}}} \operatorname{Fluct}_{\mu_\theta}(\varphi)\right)\right| \\ &\le Ct^2\beta N^{2-\frac{\mathsf{s}}{\mathsf{d}}}\ell^{\mathsf{d}}|\varphi|_{C^1}^2+C|t|N\ell^{\mathsf{d}}|\varphi|_{C^2}+C\beta\chi(\beta)N\ell^{\mathsf{d}}\left((1+|t||\varphi|_{C^3})|t||\varphi|_{C^1}+|t||\varphi|_{C^2}\right)\end{aligned}$$

hence the result. □

We may then optimize over t and as an illustration, we give the following corollary in the macroscopic case $\ell = 1$, obtained with $t = -\min(N^{-1}, N^{-\frac{1}{2}-\frac{1}{\mathsf{d}}})$ after using that $e^{|X|} \le e^X + e^{-X}$. This allows to treat the case where $\varphi(x) = \varphi_0(x\ell^{-1})$ for some fixed φ_0.

Corollary 10.5 (Macroscopic scale). *Under the same assumptions, if* $\mathsf{d} \ge 2$, *we have*

$$(10.1.16)\quad \left|\log \mathbb{E}_{\mathbb{P}_{N,\beta}}\left(e^{\beta N^{-\frac{\mathsf{s}}{\mathsf{d}}}|\operatorname{Fluct}_{\mu_\theta}(\varphi)|}\right)\right| \le C\left(|\varphi|_{C^2}+|\varphi|_{C^1}+N^{\frac{2}{\mathsf{d}}-1}|\varphi|_{C^1}^2+N^{-1}|\varphi|_{C^1}|\varphi|_{C^3}\right),$$

while if $\mathsf{d}=1$, *we have*

$$(10.1.17)\quad \left|\log \mathbb{E}_{\mathbb{P}_{N,\beta}}\left(e^{\beta N^{\frac{1}{2}}|\operatorname{Fluct}_{\mu_\theta}(\varphi)|}\right)\right| \le C\left(N^{-\frac{1}{2}}(|\varphi|_{C^2}+|\varphi|_{C^1})+|\varphi|_{C^1}^2+N^{-2}|\varphi|_{C^1}|\varphi|_{C^3}\right),$$

where the constant depends on d, μ_θ *and* β, *but not on* N *or* φ.

In the meso/microscales, we apply instead to

$$t = -\ell^2\min\left((N\ell^{\mathsf{d}})^{-1}, (N\ell^{\mathsf{d}})^{-\frac{1}{2}-\frac{1}{\mathsf{d}}}\right)$$

and obtain

Corollary 10.6 (Mesoscopic and microscopic scale). *Let* $\mathsf{s}=\mathsf{d}-2$. *Assume* $V \in C^5$, (A1)–(A5), (2.5.20) *hold,* $\theta > \theta_0 > 0$. *Assume that* $\varphi \in C^3$ *is supported in a cube of sidelength* ℓ *included in* $\hat\Sigma$ *and* $|\varphi|_{C^k} \le M\ell^{-k}$ *for all* $k \le 3$. *If* $\mathsf{d}\ge 2$ *we have*

$$(10.1.18)\qquad \left|\log \mathbb{E}_{\mathbb{P}_{N,\beta}}\left(e^{\beta(N^{\frac{1}{\mathsf{d}}}\ell)^{-\mathsf{s}}|\operatorname{Fluct}_{\mu_\theta}(\varphi)|}\right)\right| \le C(M+M^2),$$

while if $\mathsf{d}=1$, *we have*

$$(10.1.19)\qquad \left|\log \mathbb{E}_{\mathbb{P}_{N,\beta}}\left(e^{\beta(N\ell)^{\frac{1}{2}}|\operatorname{Fluct}_{\mu_\theta}(\varphi)|}\right)\right| \le C(M+M^2),$$

where C *depends only on* d, μ_θ *and* β, *but not on* N *or* φ.

These results express a strong form of rigidity of the Coulomb gas: when β is fixed the fluctuations are much smaller than those of a Poisson point process (see Figure 1.3.3, left), since they do not need to be normalized by $\sqrt{N}$. Note that a control under the sole assumption that $\nabla\varphi \in L^2$, but valid only for $\beta \le 2$, is obtained in [**Ber19**].

The results are particularly simple in the two-dimensional Coulomb case for which $\mathsf{s} = 0$, they say that fluctuations of functions at any scale larger than the microscale are bounded. This turns out to be sharp as we will see in the next sections. We can already see that the order of fluctuations in other dimension is more subtle, and we do not claim that the bounds obtained here are optimal.

In the physics literature, the papers [**JLM93, Leb83**] (see also [**Mar88, MY80**]) contain a well-known prediction of an order $N^{1-1/\mathsf{d}}$ for the variance of the number of points in a domain (see Section 10.3 below), however there is no prediction for the order of fluctuations of smooth linear statistics. On the other hand [**Cha19, GS20**] obtained bounds on the number of points and linear statistics for the hierarchical Coulomb gas model, a simplified model introduced by Dyson, and the full Jancovici-Lebowitz-Manificat law was established for that model in [**NY24**]. In [**Cha19**] the order of fluctuations of smooth linear statistics was speculated upon ($N^{1/3}$ vs. $N^{1/6}$) with supporting arguments from the example of orthogonal polynomial ensemble treated in [**BH20**] in favor of $N^{1/3}$, and finally it was shown in [**GS20**] to be in $N^{1-2/\mathsf{d}}$, again still for the hierarchical model instead of the full model and for β of order $N^{1/3}$.

10.2. Central limit theorem in one- and two-dimensional Coulomb cases

To go beyond and really understand the limit of Laplace transform of fluctuations, we need to return to the identity obtained in (10.1.4). We wish to show that the log Laplace transform converges to that of a Gaussian random variable, i.e., to a function of the form $t\,\mathrm{mean} + \frac{t^2}{2}\mathrm{var}$. The variance term already appears explicitly in the quadratic term of the right-hand side of (10.1.4) as can be seen in (10.1.8). That term indicates that in order to have a finite limiting variance, one should take $t = \tau N^{-1+\frac{\mathsf{s}}{2\mathsf{d}}}$, with τ of order 1, thus a small t as $N \to \infty$. In the mesoscales, we will consider that $\varphi = \varphi_0(\frac{\cdot}{\ell})$ for some fixed φ_0. Then one should take $t = \tau N^{-1+\frac{\mathsf{s}}{2\mathsf{d}}}\ell^{-\frac{\mathsf{s}}{2}} = \tau \ell^{\mathsf{d}-\mathsf{s}}(N^{\frac{1}{\mathsf{d}}}\ell)^{\frac{\mathsf{s}}{2}-\mathsf{d}}$. The question is then to understand more precisely the term $\log \frac{\mathsf{K}_{N,\beta}(\nu_\theta^t)}{\mathsf{K}_{N,\beta}(\mu_\theta)}$ that appeared in (10.1.4), since the other terms appearing there have precise enough expansions or bounds. A first interpretation of this ratio is via (9.3.5), which already allowed in the previous chapter to bound the term with the help of the commutator estimate, but a second one is provided by the expansion (9.4.1) which was obtained by leveraging the almost additivity. Since neither of these estimates is sufficient on its own, the idea is to compare them to obtain a stronger estimate.

10.2.1. Statement of results. Let us now state the main results, starting with the two-dimensional case. The two-dimensional case is special as the order N term in the relative expansion of (9.4.4) does not involve the function f_d, as seen in Remark 9.11. This makes it so that the relatively unknown function f_d does not appear at all in the proof or in the limiting quantities.

Theorem 10.7 (CLT in dimension 2 for possibly small β). *Let* $\mathsf{d} = 2$, $\mathsf{s} = 0$. *Assume* $V \in C^8$, (A1)–(A5), $\theta > \theta_0 > 0$ *and* (2.5.20) *hold, assume that* $\varphi \in C^6$, *and* $\square_\ell$ *is a hypercube of sidelength* ℓ *such that* $\operatorname{supp}\varphi \subset \square_{\ell/2} \subset \square_\ell \subset \hat{\Sigma}$ *(if it does not hold that* $\ell \ll 1$ *then only assume* $\square_\ell \subset \Sigma$*). Assume also that* φ *takes the form* $\varphi_0(\frac{x-\bar{x}}{\ell})$ *for some fixed function* $\varphi_0 \in C^6(\mathbb{R}^{\mathsf{d}})$. *Assume* $N^{\frac{1}{\mathsf{d}}}\ell\rho_\beta^{-1} \to \infty$ *as* $N \to \infty$,[1] *and*[2]

$$\beta^{\frac{1}{2}} \ll (N^{\frac{1}{\mathsf{d}}}\ell)^{\frac{5}{4}-\frac{\mathsf{d}}{2}} \log^{-\frac{1}{2\mathsf{d}}}(N^{\frac{1}{\mathsf{d}}}\ell). \tag{10.2.1}$$

Then for any fixed τ, *we have*

(10.2.2)

$$\left|\log \mathbb{E}_{\mathbb{P}_{N,\beta}}\left(\exp\left(-\tau\beta^{\frac{1}{2}} \operatorname{Fluct}_{\mu_\theta}(\varphi)\right)\right) + \tau \operatorname{mean}(\varphi) - \frac{\tau^2}{2}\operatorname{var}(\varphi)\right| \to 0 \quad \text{as } \frac{N^{\frac{1}{\mathsf{d}}}\ell}{\rho_\beta} \to \infty$$

with

$$\operatorname{var}(\varphi) = \frac{1}{\mathsf{c}_{\mathsf{d}}}\int_{\mathbb{R}^{\mathsf{d}}} |\nabla\varphi_0|^2 \tag{10.2.3}$$

and

$$\operatorname{mean}(\varphi) = -\frac{\beta^{\frac{1}{2}}}{4\mathsf{c}_{\mathsf{d}}}\int_{\mathbb{R}^{\mathsf{d}}} \log\mu_\theta\, \Delta\varphi.$$

Note that the convergence rates are independent of β, which means that β can be dependent on N, the only condition being $N^{\frac{1}{\mathsf{d}}}\ell \gg \rho_\beta$ and (10.2.1).

The covariance structure found in (10.2.3) characterizes the convergence to a Gaussian free field, see [**KM13**] for definitions and reference on this conformal field theory context. More precisely we have shown the following.

Corollary 10.8. *Under the same assumptions,*

$$\beta^{1/2}\left(\operatorname{Fluct}_{\mu_\theta}(\varphi) + \frac{1}{4\mathsf{c}_{\mathsf{d}}}\int_{\mathbb{R}^{\mathsf{d}}} (\Delta\varphi)\log\mu_\theta\right)$$

converges[3] *as* $N \to \infty$ *to a Gaussian of mean* 0 *and variance* $\frac{1}{\mathsf{c}_{\mathsf{d}}}\int_{\mathbb{R}^2} |\nabla\varphi_0|^2$. *Moreover, letting* $h_N = \mathsf{g} * \left(\sum_{i=1}^N \delta_{x_i} - N\mu_\theta\right)$ *as in* (4.1.4), *we have that* $\beta^{\frac{1}{2}}h_N$ *converges to the Gaussian free field as* $N \to \infty$.

These results were first shown for fixed β (independent of N) and for mesoscales $\ell \geq N^{-\alpha}$, $\alpha < \frac{1}{2}$, in [**LS18, BBNY19**]. It is important in all such results that φ is supported in one connected component of the droplet Σ, for otherwise testing the fluctuations against φ would allow to count the number of particles in each component, while that number may fluctuate by an $O(1)$. The correct result in such a case must then include an extra discrete Gaussian in the limit, as seen in the one-dimensional "multi-cut" log gas case in [**Shc13, BG24**] as well as in the two dimensional Coulomb gas case [**ACC25**].

The case where Σ has one connected component (to avoid the subtlety mentioned just above) but the support of φ may intersect $\partial\Sigma$ was only treated in [**AHM11, AHM15**] in the case $\beta = 2$ and in [**LS18**] for all β (we refer the interested reader to that paper) and one gets instead a limiting variance $\frac{1}{\mathsf{c}_{\mathsf{d}}}\int_{\mathbb{R}^2} |\nabla\varphi^\Sigma|^2$

[1] This implies $\theta \gg 1$.

[2] The assumption allows β to tend to $+\infty$ as $N \to \infty$ but not at too fast of a rate.

[3] The convergence is in the sense of convergence of the Laplace transforms, which implies convergence in law but is in fact a bit stronger.

where φ^Σ denotes the harmonic extension of φ outside Σ. This was also recently extended to fractal contours in the context of determinantal point processes in [**Lin23**]. In the Riesz case treated in [**PS**], because of the nonlocal nature of the problem, a fractional harmonic extension φ^Σ appears in all cases (even if $\operatorname{supp}\varphi \subset \Sigma$) and the limit variance is proportional to $\|\varphi^\Sigma\|^2_{\dot H^{\frac{\mathsf{d}-\mathsf{s}}{2}}}$ in the macroscopic case (respectively $\|\varphi\|^2_{\dot H^{\frac{\mathsf{d}-\mathsf{s}}{2}}}$ in the case $\ell \to 0$) leading to the limit of the Riesz potential being a fractional Gaussian field.

The extension to all $\ell \gg N^{-1/2}\rho_\beta$ and to possibly small β was obtained in [**Ser23**]. Further corrections to the variance (10.2.3), in power of $1/\theta$, are provided there. The assumptions of regularity on V and especially on φ are certainly not optimal, it is an interesting open question to find what the minimal regularity on φ needed is for the result to hold. The application of the second order commutator estimates (6.1.40) is the main bottleneck. In [**Ser23**], we only required $V \in C^7$ and $\varphi \in C^4$ because we used a different version of (6.1.40), the new version allows for a slightly simpler proof and can better generalize to higher dimension.

In radial setups, one may hope to compute more things: for instance there are physics predictions on next order cumulants [**BDMS23**], and studies of fluctuations in the case $\beta = 2$ when Σ has more than one connected component [**AC22, ACC25, ACC24**]. This shows that the soft edge situation (V smooth) is quite different from the hard edge ($V = \infty$ outside Σ) one, in particular fluctuations may be non-Gaussian even though the droplet is connected.

When β is so large that (10.2.1) fails, it is likely that the CLT is not true as one expects crystallization and fluctuations near a lattice rather than GFF-like fluctuations, see Chapters 11 and 12. Instead we can normalize $\mathrm{Fluct}(\varphi)$ differently to obtain a convergence result.

Theorem 10.9 (Low temperature and minimizers in dimension 2)**.** *Let* $\mathsf{d} = 2$, $\mathsf{s} = 0$. *Assume the same as in the previous theorem on* V, φ, ℓ. *Assume that* $\beta \gg 1$ *and* $N^{\frac{1}{\mathsf{d}}}\ell \gg 1$ *as* $N \to \infty$. *Then we have (in the sense of convergence of the Laplace transforms)*

$$\lim_{N\to\infty}\left(\mathrm{Fluct}_{\mu_\theta}(\varphi) + \frac{1}{4\mathsf{c}_\mathsf{d}}\int_{\mathbb{R}^\mathsf{d}} (\log \mu_\theta)\Delta\varphi\right) = 0.$$

The case of minimizers of $\mathcal{H}_N$ corresponds to $\beta = \infty$ and can be obtained by simply letting $\beta \to \infty$ in the case with temperature since the constants are independent of β.

Corollary 10.10. *Under the same assumption, assume that* X_N *minimizes* $\mathcal{H}_N$ *and* $N^{\frac{1}{\mathsf{d}}}\ell \gg 1$ *as* $N \to \infty$. *Then we have*

$$\lim_{N\to\infty} \mathrm{Fluct}_{\mu_V}(\varphi) = -\frac{1}{4}\int_{\mathbb{R}^\mathsf{d}} \frac{\Delta\varphi}{\mathsf{c}_\mathsf{d}} \log \mu_V.$$

Note that this generalizes [**LS18**] and also complements the results on minimizers in [**AOC12, RNS15, PS17, PRN18**], see Theorem 8.2.

In the one-dimensional Coulomb case, the function $f_\mathsf{d}(\beta)$ appears in the relative free energy expansion (9.4.4). We are able to use the analyticity of f_d proven in [**Kun74**] to still deduce a similar result for fixed β. The fact that the bounds on derivatives of f_d may degenerate as $\beta \to 0$ is the reason why we do not treat the high-temperature regime in this setting.

Theorem 10.11 (CLT in dimension 1). *Let $\mathsf{d}=1$, $\mathsf{s}=-1$. Assume $V\in C^5$, (A1)–(A5), $\beta>\beta_0>0$ and (2.5.20) hold, assume that $\varphi\in C^3$, and $\square_\ell$ is a cube of sidelength ℓ such that $\operatorname{supp}\varphi\subset\square_{\ell/2}\subset\square_\ell\subset\hat\Sigma$ (if it does not hold that $\ell\ll 1$ then only assume $\square_\ell\subset\Sigma$). Assume also that φ takes the form $\varphi_0(\frac{x-\bar x}{\ell})$ for some fixed function $\varphi_0\in C^4(\mathbb{R}^{\mathsf{d}})$. Assume $N^{\frac{1}{\mathsf{d}}}\ell\to\infty$ as $N\to\infty$ and*

$$\beta^{\frac12}\ll (N^{\frac{1}{\mathsf{d}}}\ell)^{\frac54-\frac{\mathsf{d}}{2}}\log^{-\frac{1}{2\mathsf{d}}}(N^{\frac{1}{\mathsf{d}}}\ell). \tag{10.2.4}$$

Then for any fixed τ, we have

$$\lim_{N^{1/\mathsf{d}}\ell\to+\infty}\log\mathbb{E}_{\mathbb{P}_{N,\beta}}\left(\exp\left(-\tau\beta^{\frac12}(N\ell)^{\frac12}\operatorname{Fluct}_{\mu_\theta}(\varphi)\right)\right)=\frac{\tau^2}{2\mathsf{c}_{\mathsf{d}}}\int_{\mathbb{R}^{\mathsf{d}}}|\nabla\varphi_0|^2. \tag{10.2.5}$$

In other words, $\beta^{\frac12}(N\ell)^{\frac12}\operatorname{Fluct}_{\mu_\theta}(\varphi)$ converges to a Gaussian of mean zero and variance $\frac{1}{\mathsf{c}_{\mathsf{d}}}\int_{\mathbb{R}^{\mathsf{d}}}|\nabla\varphi_0|^2$.

Theorem 10.12 (Low temperature and minimizers in dimension 1). *Let $\mathsf{d}=1$, $\mathsf{s}=-1$. Assume $V\in C^5$, (A1)–(A5) and (2.5.20) hold. Assume that $\varphi\in C^3$ and $\operatorname{supp}\varphi\subset\square_{\ell/2}\subset\square_\ell\subset\hat\Sigma$, (if it does not hold that $\ell\to 0$ then only assume $\square_\ell\subset\Sigma$) with ℓ satisfying $N^{-1/\mathsf{d}}\le\ell\le C$, and that (10.2.20) holds. Assume $\beta\gg 1$ and $N^{\frac{1}{\mathsf{d}}}\ell\gg 1$ as $N\to\infty$. Then we have (in the sense of convergence of the Laplace transforms)*

$$\lim_{N\to\infty}(N\ell)^{\frac12}\operatorname{Fluct}_{\mu_\theta}(\varphi)=0.$$

10.2.2. Proof of the CLT results.

10.2.2.1. *Preliminary: Control of the variation of local energy.* In the mesoscopic case, we will need the following generalization of Corollary 9.3, which allows to control the local energy after a transport.

Lemma 10.13. *Assume μ is a bounded probability density satisfying (4.1.1). Let ψ_t, $t\in[-1,1]$ be a Lipschitz vector field, Φ_t solve (9.1.2), and let μ_t be as in (9.1.3). Assume that $\ell\ge 4N^{-1/\mathsf{d}}$ and Ω is a cube $\square_\ell$ such that $\operatorname{supp}\psi_t\subset\square_{\ell/2}\subset\square_\ell\subset\hat\Sigma$ and that Φ_t maps Ω to itself for every $t\in[0,\tau]$. Let X_N be a configuration and assume that $\vec\eta$ is such that for every i, $\eta_i\le\frac12\mathsf{r}_i$. Let*

$$\Xi^\Omega(t):=\int_\Omega|\nabla h^{\mu_t}_{N,\vec\eta}[\Phi_t(X_N)]|^2+\#I_\Omega N^{\frac{\mathsf{s}}{\mathsf{d}}}+N^{-\frac{2}{\mathsf{d}}}(\#I_\Omega)^2\ell^{-\mathsf{d}}, \tag{10.2.6}$$

where we reuse the notation of (4.1.4). Then, making τ smaller enough that $\sup_{|t|\le\tau}(t|\psi_t|_{C^1})<\frac12$, we have

$$\forall t\in[-\tau,\tau],\qquad \Xi^\Omega(t)\le\exp\left(C\int_0^t(\ell^{-1}\|\psi_s\|_{L^\infty}+|\psi_s|_{C^1})ds\right)\Xi^\Omega(0), \tag{10.2.7}$$

where C depends only on d.

Proof. The commutator estimate provides us an estimate on

$$\frac{d}{dt}\mathsf{F}_N(\Phi_t(X_N),\mu_t),$$

so we need to estimate the difference between that and $(\Xi^\Omega)'(t)$. For shortcut, let us denote h^t_N the electric potential associated to $\Phi_t(X_N)$ and μ_t, as in (4.1.4).

Since Φ_t coincides with the identity map outside Ω for each $t \in [-\tau, \tau]$, using the definition (4.2.2) (the equality case with $\eta_i = \mathsf{r}_i$), we may write that

$$\begin{aligned}(10.2.8)\qquad & \mathsf{F}_N(\Phi_t(X_N), \mu_t) - \mathsf{F}_N(X_N, \mu_0) \\ &= \frac{1}{2\mathsf{c_d}} \int_\Omega |\nabla h^t_{N,\vec\eta}|^2 - |\nabla h^0_{N,\vec\eta}|^2 \\ &\quad - N \sum_{i \in I_\Omega} \int \mathsf{f}_{\eta_i}(x - \Phi_t(x_i)) d\mu_t(x) + N \sum_{i \in I_\Omega} \int \mathsf{f}_{\eta_i}(x - x_i) d\mu_0(x) \\ &\quad + \frac{1}{2\mathsf{c_d}} \int_{\Omega^c} |\nabla h^t_{N,\vec\eta}|^2 - |\nabla h^0_{N,\vec\eta}|^2,\end{aligned}$$

where we observe that η_i is smaller than the minimum of the r_i for X_N and $\Phi_t(X_N)$—since $|t\psi|_{C^1} < \frac{1}{2}$, they are within a factor in $[\frac{1}{2}, 2]$ of each other. Let us first evaluate the last term. Since $\Phi_t = I$ in Ω^c, the charges coincide there and the function $h^t_{N,\vec\eta} - h^0_{N,\vec\eta} := u^t$ is harmonic in Ω^c and decaying at infinity, and its gradient as well, arguing as in the proof of Proposition 2.16. Hence, integrating by parts, we find

$$\begin{aligned}(10.2.9)\qquad \int_{\Omega^c} |\nabla h^t_{N,\vec\eta}|^2 - |\nabla h^0_{N,\vec\eta}|^2 &= \int_{\Omega^c} \nabla u^t \cdot (\nabla u^t + 2\nabla h^0_N) \\ &= \int_{\Omega^c} |\nabla u^t|^2 + 2 \int_{\partial\Omega} (h^0_{N,\vec\eta} - \bar h^0_{N,\vec\eta}) \frac{\partial u^t}{\partial \nu},\end{aligned}$$

where ν is the outer pointing unit normal to Ω on its boundary and $\bar h^0_{N,\vec\eta}$ is the average of $h^0_{N,\vec\eta}$ on $\partial\Omega$. By Cauchy-Schwarz and trace theorem, we may write that

$$(10.2.10)\qquad \left| \int_{\partial\Omega} (h^0_{N,\vec\eta} - \bar h^0_{N,\vec\eta}) \frac{\partial u^t}{\partial \nu} \right| \le C \ell^{\frac{1}{2}} \|\nabla h^0_{N,\vec\eta}\|_{L^2(\Omega)} \|\nabla u^t\|_{L^2(\partial\Omega)}.$$

We next estimate ∇u^t. By definition,

$$u^t(x) = \int_{\mathbb{R}^\mathsf{d}} \mathsf{g}(x-y) d\left(\sum_{i=1}^N \delta^{(\eta_i)}_{\Phi_t(x_i)} - N\mu_t\right)(y) - \int_{\mathbb{R}^\mathsf{d}} \mathsf{g}(x-y) d\left(\sum_{i=1}^N \delta^{(\eta_i)}_{x_i} - N\mu_0\right)(y).$$

By definition of $\delta^{(\eta_i)}_{x_i}$ in (4.1.10), we compute that

$$\begin{aligned}\partial_t u^t(x) &= \partial_t \left(\sum_{i=1}^N \fint_{\partial B(0,\eta_i)} \mathsf{g}(x - \Phi_t(x_i) - \cdot) - N \int_{\mathbb{R}^\mathsf{d}} \mathsf{g}(x - \Phi_t(y)) d\mu_0(y) \right) \\ &= \sum_{i=1}^N \int_{\mathbb{R}^\mathsf{d}} \nabla \mathsf{g}(x-y) \cdot \psi_t(\Phi_t(x_i)) \delta^{(\eta_i)}_{\Phi_t(x_i)}(y) - N \int_{\mathbb{R}^\mathsf{d}} \nabla \mathsf{g}(x-y) \cdot \psi_t(y) d\mu_t(y)\end{aligned}$$

hence

$$\begin{aligned} u^t(x) &= \int_0^t \int_{\mathbb{R}^\mathsf{d}} \nabla \mathsf{g}(x-y) \cdot \psi_s(y) d\left(\sum_{i=1}^N \delta^{(\eta_i)}_{\Phi_s(x_i)} - N\mu_s\right)(y)\, ds \\ (10.2.11)\qquad &\quad + \int_0^t \sum_{i=1}^N \int_{\mathbb{R}^\mathsf{d}} \nabla \mathsf{g}(x-y) \cdot (\psi_s(\Phi_s(x_i)) - \psi_s(y))\, \delta^{(\eta_i)}_{\Phi_s(x_i)}(y)\, ds.\end{aligned}$$

and

$$\nabla u^t(x) = -\frac{1}{\mathsf{c_d}} \int_0^t \int_{\mathbb{R}^\mathsf{d}} \nabla^{\otimes 2} \mathsf{g}(x-y) : \psi_s(y) \Delta h_N^s \, ds$$
$$+ \int_0^t \sum_{i=1}^N \int_{\mathbb{R}^\mathsf{d}} \nabla^{\otimes 2} \mathsf{g}(x-y) : (\psi_s(\Phi_s(x_i)) - \psi_s(y)) \, \delta_{\Phi_s(x_i)}^{(\eta_i)}(y) \, ds. \tag{10.2.12}$$

The function $\chi_x(y) := \nabla^{\otimes 2} \mathsf{g}(x - \Phi_s(y)) : \psi_s(y)$ is compactly supported in $\square_{\ell/2}$, moreover, it can be checked to satisfy

$$|\nabla \chi_x(y)| \le C \left(\frac{|\psi_s|_{L^\infty(\Omega)}}{\mathrm{dist}(x, \square_{\ell/2})^{\mathsf{s}+3}} + \frac{|\psi_s|_{C^1(\Omega)}}{\mathrm{dist}(x, Q_{\ell/2})^{\mathsf{s}+2}} \right). \tag{10.2.13}$$

hence for $x \in \Omega^c$,

$$\int_\Omega |\nabla \chi_x(y)|^2 dy \le (\|\psi_s\|_{L^\infty(\Omega)} + \ell |\psi_s|_{C^1(\Omega)})^2 \ell^{\mathsf{d}-2\mathsf{s}-6}. \tag{10.2.14}$$

Using this, Green's formula and the Cauchy-Schwarz inequality to control the first term in (10.2.12), and a rough bound for the second term, we are led to

$$|\nabla u^t(x)|^2$$
$$\le Ct \left(\int_0^t (\|\psi_s\|_{L^\infty} + \ell |\psi_s|_{C^1})^2 \ell^{\mathsf{d}-2\mathsf{s}-6} \int_\Omega |\nabla h_{N,\vec\eta}^s|^2 ds + \int_0^t |\psi_s|_{C^1}^2 \left(\sum_{i \in I_\Omega} \eta_i \ell^{-\mathsf{s}-2} \right)^2 ds \right).$$

Inserting that $\eta_i \le CN^{-1/\mathsf{d}}$, it follows that

$$\int_{\partial\Omega} |\nabla u^t|^2$$
$$\le C\ell^{\mathsf{d}-1} t \int_0^t (\|\psi_s\|_{L^\infty} + \ell |\psi_s|_{C^1})^2 \left(\ell^{\mathsf{d}-2\mathsf{s}-6} \int_\Omega |\nabla h_{N,\vec\eta}^s|^2 + N^{-\frac{2}{\mathsf{d}}} (\# I_\Omega)^2 \ell^{-2\mathsf{s}-6} \right) ds.$$

Combining with (10.2.10) and (10.2.9), and noting that $\int_{\Omega^c} |\nabla u^t|^2$ is an order $O(t^2)$ term, it follows that

$$\left| \frac{d}{dt}_{|t=0} \int_{\Omega^c} |\nabla h_{N,\vec\eta}^t|^2 \right|$$
$$\le C\ell^{\frac12} \|\nabla h_{N,\vec\eta}^0\|_{L^2(\Omega)} \ell^{\frac{\mathsf{d}-1}{2}} (\|\psi_0\|_{L^\infty} + \ell |\psi_0|_{C^1})$$
$$\times \left(\ell^{\mathsf{d}/2-\mathsf{s}-3} \|\nabla h_{N,\vec\eta}^0\|_{L^2(\Omega)} + N^{-\frac{1}{\mathsf{d}}} (\# I_\Omega) \ell^{-\mathsf{s}-3} \right).$$

On the other hand, using (4.1.18) and (4.1.27), we also have

$$\left| \frac{d}{dt}_{|t=0} \sum_{i \in I_\Omega} \int_{\mathbb{R}^\mathsf{d}} \mathsf{f}_{\eta_i}(x - \Phi_t(x_i)) d\mu_t(x) \right|$$
$$= \left| \sum_{i \in I_\Omega} \int_{\mathbb{R}^\mathsf{d}} \nabla \mathsf{f}_{\eta_i}(x - x_i) \cdot (\psi_0(x) - \psi_0(x_i)) d\mu_0(x) \right|$$
$$\le |\psi_0|_{C^1} \sum_{i \in I_\Omega} \int_{B(0,\eta_i)} |x|^{-\mathsf{s}} \le C \# I_\Omega |\psi_0|_{C^1} N^{-\frac{\mathsf{d}-\mathsf{s}}{\mathsf{d}}}.$$

Combining the above with (10.2.8) and (10.2.9), and by definition (6.1.2) and the commutator estimate in the form (6.1.38), after using Young's inequality, we are led to

$$\left|\frac{d}{dt}_{|t=0}\int_\Omega |\nabla h^t_{N,\vec\eta}|^2\right| \le C(\|\psi_0\|_{L^\infty}+\ell|\psi_0|_{C^1})$$
$$\times\left(\ell^{\mathsf{d}-\mathsf{s}-3}\int_\Omega |\nabla h^0_{N,\vec\eta}|^2+\ell^{-1}\int_{\square_{\ell/2}}|\nabla h^0_{N,\mathsf{r}}|^2+N^{-\frac{2}{\mathsf{d}}}(\#I_\Omega)^2\ell^{-\mathsf{s}-3}+\#I_\Omega\ell^{-1}N^{\frac{\mathsf{s}}{\mathsf{d}}}\right).$$

Arguing as in (6.1.39), we may bound

$$\int_{\square_{\ell/2}}|\nabla h^0_{N,\mathsf{r}}|^2\le 2\int_{\square_{\ell/2}}|\nabla h^0_{N,\vec\eta}|^2+C\sum_{i,\mathrm{dist}(x_i,\square_{\ell/2})\le\frac14\lambda}\mathsf{r}_i^{-\mathsf{s}},$$

and using (4.5.6) and $\mathsf{s}=\mathsf{d}-2$, we may absorb these terms into the others.

The same reasoning yields that the relation is true at any t, so we find

$$\left|\frac{d}{dt}\int_\Omega|\nabla h^t_{N,\vec\eta}|^2\right|$$
$$\le C(\ell^{-1}\|\psi_t\|_{L^\infty}+|\psi_t|_{C^1})$$
$$\times\left(\int_\Omega|\nabla h^t_{N,\vec\eta}|^2+\#I_\Omega N^{\frac{\mathsf{s}}{\mathsf{d}}}+N^{-\frac{2}{\mathsf{d}}}(\#I_\Omega)^2\ell^{-\mathsf{d}}\right).$$

Applying Gronwall's lemma, we deduce the result. □

10.2.2.2. *Main proof.* In view of (9.1.9) going further to obtain the CLT requires to precisely estimate $\mathbb{E}_{\mathbb{Q}_{N,\beta}(\mu_\theta)}(\mathsf{A}_1)$, which is what was called in [**LS18**] the *anisotropy term*. We do it by comparing with the second way of estimating $\log\mathsf{K}(\nu^t_\theta)-\log\mathsf{K}(\mu_\theta)$, by almost additivity estimates as in Chapter 8. The core of the proof is an improvement of (9.3.3), which relies on comparing that relation with the result (9.4.4) obtained by almost additivity.

For that we write that Corollary 9.10 provides an expansion of the form

(10.2.15)
$$\log\mathsf{K}_{N,\beta}(\mu_t)-\log\mathsf{K}_{N,\beta}(\mu_0)=N\beta\left(\mathcal{Z}(\beta,\mu_t)-\mathcal{Z}(\beta,\mu_0)\right)+O(\beta\chi(\beta)N\ell^{\mathsf{d}}\mathcal{R}_t)$$

where $\mathcal{R}_t$ is the rate that corresponds to the error rate in (9.4.1) for the measure μ_t, to which we add the rate for the measure μ_0. The form of $\mathcal{Z}$ is

$$\mathcal{Z}(\beta,\mu)=-\int_{\mathbb{R}^{\mathsf{d}}}\mu^{1+\frac{\mathsf{s}}{\mathsf{d}}}f_{\mathsf{d}}(\beta\mu^{\frac{\mathsf{s}}{\mathsf{d}}})-\frac{1}{2\mathsf{d}}\left(\int_{\mathbb{R}^{\mathsf{d}}}\mu\log\mu\right)\mathbf{1}_{\mathsf{s}=0}. \tag{10.2.16}$$

We will also denote by $\mathcal{B}(\beta,\mu_0,u)$ the derivative at $t=0$ of the function $\mathcal{Z}(\beta,\mu_t)$ when $\frac{d}{dt}|_{t=0}\mu_t=u$. For instance in view of (10.2.16), we compute that if $\mathsf{d}=2$,

$$\mathcal{B}(\beta,\mu_0,u)=-\frac14\int_{\mathbb{R}^{\mathsf{d}}}u\log\mu_0 \tag{10.2.17}$$

while if $\mathsf{d}=1$,

$$\mathcal{B}(\beta,\mu_0,u)=\int_{\mathbb{R}^{\mathsf{d}}}u\frac{f'_{\mathsf{d}}(\beta\mu_0^{-1})}{\mu_0^2}. \tag{10.2.18}$$

We also need to assume a quantitative differentiation property of the form

$$\mathcal{Z}(\beta,\mu_t)-\mathcal{Z}(\beta,\mu_0)=\mathcal{B}(\beta,\mu_0,u)+O(t^2\ell^{\mathsf{d}}\|u\|^2_{L^\infty}) \tag{10.2.19}$$

when $t\ell^{-2} \le \frac{1}{2c_d} m$ (where m bounds μ_0 from below), with O depending only on d and s. By a straightforward computation this condition holds for $\mathsf{d} = 2$. It holds in other dimensions if we know a bound for f_d''. This is in turn only known in one dimension, uniformly for $\beta > \beta_0$, thanks to the work of [**Kun74**] that proves that f_d is in fact analytic. In higher dimensions, f_d may not always be twice differentiable, values of β for which it fails to be correspond to phase-transitions.

Proposition 10.14 (Improved relative expansion and exponential moments of A_1). *Assume $\mathsf{s} = \mathsf{d} - 2$. Assume that μ_0 is a C^4 probability density over $\mathbb{R}^{\mathsf{d}}$, $\varphi \in C^6$ (or $\mu_0 \in C^1$ and $\varphi \in C^3$ if $\mathsf{d} = 1$) , $\square_\ell$ is a cube of sidelength $\ell \ge 4N^{-1/\mathsf{d}}$ such that $\operatorname{supp}\varphi \subset \square_{\ell/2} \subset \square_\ell$ with $\mu_0 \ge m > 0$ in $\square_\ell$. Assume also that there exists a constant M such that for $k \in [0,6]$, we have*

$$|\varphi|_{C^k} \le M\ell^{-k}. \tag{10.2.20}$$

Let then

$$\mu_t = \mu_0 + \frac{t}{\mathsf{c_d}}\Delta\varphi. \tag{10.2.21}$$

Assume that (10.2.19) holds with $u = \frac{1}{\mathsf{c_d}}\Delta\varphi$, and that the local laws (8.3.3) hold for $\mathbb{Q}_{N,\beta}(\mu_0)$ in $\square_\ell$. Let $\alpha = (\max_{|s|\le\ell^2}\mathcal{R}_s)^{\frac12}$. For every t such that $|t|\ell^{-2} \le \frac12\min(1,\alpha)$ and $|t| \le \frac{\mathsf{c_d}m}{2\|\Delta\varphi\|_{L^\infty}}$, we have

$$\log\frac{\mathsf{K}_{N,\beta}(\mu_t)}{\mathsf{K}_{N,\beta}(\mu_0)} = tN\beta\mathcal{B}\left(\beta,\mu_0,\frac{1}{\mathsf{c_d}}\Delta\varphi\right) + O\left(t^2\beta\chi(\beta)N\ell^{\mathsf{d}-4} + |t|\beta\chi(\beta)N\ell^{\mathsf{d}-2}\left(\max_{|s|\le\ell^2}\mathcal{R}_s\right)^{\frac12}\right) \tag{10.2.22}$$

and

$$\left|\log\mathbb{E}_{\mathbb{Q}_{N,\beta}(\mu_0)}\left(\exp\left(t\beta N^{-\frac{\mathsf{s}}{\mathsf{d}}}\mathsf{A}_1(X_N,\mu_0,\frac{1}{\mathsf{c_d}}\frac{\nabla\varphi}{\mu_0}) - t\beta N\mathcal{B}(\beta,\mu_0,\frac{1}{\mathsf{c_d}}\Delta\varphi)\right)\right)\right| \le C|t|\ell^{-2}\beta\chi(\beta)N\ell^{\mathsf{d}}\left(\max_{|s|\le\ell^2}\mathcal{R}_s\right)^{\frac12}. \tag{10.2.23}$$

with constants that depend on d, M, the bounds on μ_0, β_0 if $\mathsf{d} = 1$, but not N, ℓ, t.

The main point here is that compared to (10.2.15) we have gained a factor of t in the error terms, which will be very small when t is very small. Thus, we know the exponential moments of A_1 with good precision.

PROOF.

Step 1. Pushing the expansion to next order. Let us revisit the proof of Lemma 9.7. For $|s| \le |t|$, we define $\psi_s = -\frac{1}{\mathsf{c_d}}\frac{\nabla\varphi}{\mu_s}$ where $\mu_s = \mu_0 + \frac{s}{\mathsf{c_d}}\Delta\varphi$, and let Φ_s be given by (9.1.2). The assumptions $\mu_0 \ge m$ and $|t| \le \frac{\mathsf{c_d}m}{\|\Delta\varphi\|_{L^\infty}}$ ensure that μ_s remains a probability density.

By Lemma 9.6, we have $\mu_t = \Phi_t \# \mu_0$. Applying the change of variables (or transport) (9.1.14) with $\mu = \mu_0$ and $\phi_h = \Phi_t$, we find

(10.2.24)
$$\frac{\mathsf{K}_{N,\beta}(\mu_t)}{\mathsf{K}_{N,\beta}(\mu_0)} = \mathbb{E}_{\mathbb{Q}_{N,\beta}(\mu_0)}\left(\exp(-\beta N^{-\frac{\mathsf{s}}{\mathsf{d}}}(\mathsf{F}_N(\Phi_t(X_N), \Phi_t\#\mu_0) - \mathsf{F}_N(X_N, \mu_0)))\right).$$

Previously we have estimated this by the integral of A_1 (using (9.1.6)) and the commutator estimate, Theorem 6.1. In order to go beyond, we can get a more precise estimate by using Taylor's integral formula to second order, and writing in view of (9.1.6) and (9.1.7) that

$$\text{(10.2.25)}\quad \mathsf{F}_N(\Phi_t(X_N), \Phi_t\#\mu_0) - \mathsf{F}_N(X_N, \mu_0) \\ = t\mathsf{A}_1(X_N, \mu_0, \psi_0) + \int_0^t (t-s)\mathsf{A}_2(\Phi_s(X_N), \mu_s, \psi_s)\, ds.$$

Using the explicit form of ψ_s and the linearity of A_1 in its last variable, we obtain

$$\text{(10.2.26)}\quad \mathsf{F}_N(\Phi_t(X_N), \Phi_t\#\mu_0) - \mathsf{F}_N(X_N, \mu_0) \\ = -t\mathsf{A}_1\left(X_N, \mu_0, \frac{1}{\mathsf{c_d}}\frac{\nabla\varphi}{\mu_0}\right) + \int_0^t (t-s)\mathsf{A}_2\left(\Phi_s(X_N), \mu_s, \psi_s\right) ds.$$

Step 2. Inserting the second order commutator estimate. If $\mathsf{d} \geq 2$, the second order commutator estimate (6.1.40) bounds $\mathsf{A}_2(\Phi_s(X_N), \mu_s, \psi_s)$ by the energy, with factors depending on the norms of ψ_s and μ_s. If $\mathsf{d} = 1$ and $\mathsf{s} = -1$, A_2 is simply zero by definition (6.1.3), as remarked after Proposition 6.5, so no further regularity of φ is needed. To estimate these factors for $\mathsf{d} \geq 2$ we observe that for $k \leq 4$, we have

$$\text{(10.2.27)}\qquad |\psi_s|_{C^k} \leq C\ell^{-k-1}$$

where C depends on M, the lower bound m for μ_0, and the bounds on C^k norms ($k \leq 4$) on μ_0. Indeed, we first observe that by definition of μ_s, we have for $\sigma \leq 4$,

$$\text{(10.2.28)}\qquad |\mu_s|_{C^\sigma} \leq C(1 + |s||\varphi|_{C^{\sigma+2}}) \leq C(1 + |t|\ell^{-\sigma-2}) \leq C\ell^{-\sigma}$$

where we used that $|t|\ell^{-2} \leq 1$. Arguing as for (9.3.4), by definition of ψ_s we then easily deduce (10.2.27). With this and the assumption $\ell \geq 4N^{-1/\mathsf{d}}$ we find that the factor in the right-hand side of (6.1.40) applied to $v = \psi_s$ is bounded by $C\ell^{-4}$ and we thus obtain

$$\left|\int_0^t (t-s)\mathsf{A}_2\left(\Phi_s(X_N), \mu_s, \psi_s\right) ds\right| \leq Ct\ell^{-4}\int_0^t \Xi^{\square_\ell}(s) ds,$$

where Ξ is as in (10.2.6). To bound $\Xi^{\square_\ell}(s)$ we use Lemma 10.13 (or in the case of the whole space, simply Corollary 9.3), (10.2.27) and $|t|\ell^{-2} \leq C$, and conclude that

$$\left|\int_0^t (t-s)\mathsf{A}_2\left(\Phi_s(X_N), \mu_s, \psi_s\right) ds\right| \leq Ct^2\ell^{-4}\Xi^{\square_\ell}(0).$$

Inserting into (10.2.26), we have obtained

$$\mathsf{F}_N(\Phi_t(X_N), \Phi_t\#\mu_0) - \mathsf{F}_N(X_N, \mu_0) = -t\mathsf{A}_1\left(X_N, \mu_0, \frac{1}{\mathsf{c_d}}\frac{\nabla\varphi}{\mu_0}\right) + O(t^2\ell^{-4}\Xi^{\square_\ell}(0))$$

and inserting into (10.2.24) yields

$$\frac{\mathsf{K}_{N,\beta}(\mu_t)}{\mathsf{K}_{N,\beta}(\mu_0)}=\mathbb{E}_{\mathbb{Q}_{N,\beta}(\mu_0)}\left(\exp\left(\beta N^{-\frac{\mathsf{s}}{\mathsf{d}}}\left(t\mathsf{A}_1\left(X_N,\mu_0,\frac{1}{\mathsf{c_d}}\frac{\nabla\varphi}{\mu_0}\right)+O(t^2\ell^{-4}\Xi^{\square_\ell}(0))\right)\right)\right).$$

The exponential moments of $\Xi^{\square_\ell}(0)$ can be controlled thanks to (4.5.5) and a rescaling (based on (5.2.8)) of the local laws (8.3.3) that hold in $\square_\ell$. Thus, this can be rewritten

(10.2.29)

$$\frac{\mathsf{K}_{N,\beta}(\mu_t)}{\mathsf{K}_{N,\beta}(\mu_0)}=\mathbb{E}_{\mathbb{Q}_{N,\beta}(\mu_0)}\left(\exp\left(\beta N^{-\frac{\mathsf{s}}{\mathsf{d}}}t\mathsf{A}_1\left(X_N,\mu_0,\frac{1}{\mathsf{c_d}}\frac{\nabla\varphi}{\mu_0}\right)+O(t^2\beta\chi(\beta)N\ell^{\mathsf{d}-4})\right)\right).$$

Step 3. Comparison of the two formulas by Hölder trick and control of exponential moments of A_1.

The measures μ_t and μ_0 differ only in $\square_\ell$ and their difference integrates to 0, moreover since their densities are bounded below, up to changing ℓ by at most $O(N^{-1/\mathsf{d}})$, we may without loss of generality assume that $N\int_{\square_\ell}\mu_0=N\int_{\square_\ell}\mu_t$ is an integer, so we may apply Corollary 9.10 to find (10.2.15). Inserting (10.2.19), we obtain

(10.2.30)

$$\log\frac{\mathsf{K}_{N,\beta}(\mu_t)}{\mathsf{K}_{N,\beta}(\mu_0)}=t\beta N\mathcal{B}\left(\beta,\mu_0,\frac{1}{\mathsf{c_d}}\Delta\varphi\right)+O\left(\beta N\ell^{\mathsf{d}}t^2|\varphi|^2_{C^2}\right)+O(\beta\chi(\beta)N\ell^{\mathsf{d}}\mathcal{R}_t),$$

where O depends on m and also on β_0 such that $\beta>\beta_0$ if $\mathsf{d}=1$. We can now compare the two relations (10.2.29) and (10.2.30). The right-hand side of (10.2.30) consisting of constants, using (10.2.20) to absorb one error term into the other, the comparison implies that

$$\mathbb{E}_{\mathbb{Q}_{N,\beta}(\mu_0)}\left(\exp\left(tL+O\left(\beta\chi(\beta)N\ell^{\mathsf{d}}(\mathcal{R}_t+t^2\ell^{-4})\right)\right)\right)=1.$$

where we let

$$L:=\beta N^{-\frac{\mathsf{s}}{\mathsf{d}}}\mathsf{A}_1\left(X_N,\mu_0,\frac{1}{\mathsf{c_d}}\frac{\nabla\varphi}{\mu_0}\right)-\beta N\mathcal{B}\left(\beta,\mu_0,\frac{1}{\mathsf{c_d}}\Delta\varphi\right).\tag{10.2.31}$$

We wish to identify the terms of order t, in the regime where $t\ell^{-2}$ is small. To do so, let us see this relation in the form

$$\mathbb{E}_{\mathbb{Q}_{N,\beta}(\mu_0)}\left(e^{tL+\mathsf{Error}}\right)=1.$$

Using the Cauchy-Schwarz inequality, we may write

$$\begin{aligned}\mathbb{E}_{\mathbb{Q}_{N,\beta}(\mu_0)}\left(e^{\frac{1}{2}tL}\right)&=\mathbb{E}_{\mathbb{Q}_{N,\beta}(\mu_0)}\left(e^{\frac{1}{2}(tL+\mathsf{Error})-\frac{1}{2}\mathsf{Error}}\right)\\&\le\left(\mathbb{E}_{\mathbb{Q}_{N,\beta}(\mu_0)}\left(e^{tL+\mathsf{Error}}\right)\right)^{\frac{1}{2}}\left(\mathbb{E}_{\mathbb{Q}_{N,\beta}(\mu_0)}\left(e^{-\mathsf{Error}}\right)\right)^{\frac{1}{2}}\\&=\left(\mathbb{E}_{\mathbb{Q}_{N,\beta}(\mu_0)}\left(e^{-\mathsf{Error}}\right)\right)^{\frac{1}{2}}=e^{-\frac{1}{2}\mathsf{Error}}.\end{aligned}$$

We thus deduce that for all t such that $t\ell^{-2}$ is small enough, we have

$$\log\left(\mathbb{E}_{\mathbb{Q}_{N,\beta}(\mu_0)}\left(e^{\frac{1}{2}tL}\right)\right)\le O\left(\beta\chi(\beta)N\ell^{\mathsf{d}}(\mathcal{R}_t+t^2\ell^{-4})\right).\tag{10.2.32}$$

The idea is now to apply this relation to t as large as possible, then deduce information for smaller t by Hölder's inequality.

First, we apply (10.2.32) to $t = \pm\alpha\ell^2$ with α chosen such that

$$\alpha = \min\left(1, \left(\max_{|s|\le\ell^2} \mathcal{R}_s\right)^{\frac{1}{2}}\right).$$

Without loss of generality we may always assume α is small enough that $\alpha\ell^2 \le \frac{\mathsf{c_d} m}{2\|\Delta\varphi\|_{L^\infty}}$. This way, we find

$$\log \mathbb{E}_{\mathbb{Q}_{N,\beta}(\mu_0)}\left(e^{\pm\frac{1}{2}\alpha\ell^2 L}\right) \le O\left(\beta\chi(\beta)N\ell^{\mathsf{d}}\alpha^2\right).$$

Then by Hölder's inequality, we deduce that for $|t| \le \frac{1}{2}\alpha\ell^2$ we must have
(10.2.33)

$$\log \mathbb{E}_{\mathbb{Q}_{N,\beta}(\mu_0)}\left(e^{tL}\right) \le \frac{2|t|}{\alpha\ell^2}\log \mathbb{E}_{\mathbb{Q}_{N,\beta}(\mu_0)}\left(e^{\pm\frac{1}{2}\alpha\ell^2 L}\right) \le C|t|\ell^{-2}\beta\chi(\beta)N\ell^{\mathsf{d}}\left(\max_{|s|\le\ell^2}\mathcal{R}_s\right)^{\frac{1}{2}}.$$

In order to obtain a reverse inequality we may observe that by Hölder's inequality we have

$$1 \le \mathbb{E}(e^{tL})\mathbb{E}(e^{-tL})$$

thus applying (10.2.33) to $-t$, we obtain the converse inequality and reinserting (10.2.31) we can assert that for $|t| \le \frac{1}{2}\alpha\ell^2$, we have

(10.2.34)

$$\left|\log \mathbb{E}_{\mathbb{Q}_{N,\beta}(\mu_0)}\left(\exp\left(t\beta N^{-\frac{s}{\mathsf{d}}}\mathsf{A}_1(X_N,\mu_0,\frac{1}{\mathsf{c_d}}\frac{\nabla\varphi}{\mu_0}) - t\beta N\mathcal{B}\left(\beta,\mu_0,\frac{1}{\mathsf{c_d}}\Delta\varphi\right)\right)\right)\right| \le C|t|\ell^{-2}\beta\chi(\beta)N\ell^{\mathsf{d}}\left(\max_{|s|\le\ell^2}\mathcal{R}_s\right)^{\frac{1}{2}}.$$

Inserting this relation back into (10.2.29) and using Hölder's inequality again to separate terms, we obtain the expansion (10.2.22). □

Proof of the main theorems. We will run the proof for the one- and two-dimensional cases together as they only differ in localized places.

To obtain the main CLT, the proper choice of t in (10.1.4) is

$$t = \tau\ell^2\beta^{-\frac{1}{2}}(N^{\frac{1}{\mathsf{d}}}\ell)^{-1-\frac{\mathsf{d}}{2}} \tag{10.2.35}$$

for high temperatures (Theorems 10.7 and 10.11) and

$$t = \tau\ell^2\beta^{-1}(N^{\frac{1}{\mathsf{d}}}\ell)^{-1-\frac{\mathsf{d}}{2}} \tag{10.2.36}$$

for low temperatures (Theorems 10.9 and 10.12). We always assume that $|t| \le \frac{\mathsf{c_d} m}{2\|\Delta\varphi\|_{L^\infty}}$, so that ν_θ^t remains a probability density.

We note that by assumption on the form of φ, (10.2.20) holds. This is the only fact about φ that we will use, except for rephrasing the variance at the very end.

We may now return to (10.1.4) and insert (10.2.22) applied to $\mu_0 = \mu_\theta$ and $\mu_t = \nu_\theta^t$, (10.1.11) and (10.1.8). This yields

(10.2.37)
$$\begin{aligned}
&\log \mathbb{E}_{\mathbb{P}_{N,\beta}}\left(e^{-t\beta N^{1-\frac{s}{d}}\sum_{i=1}^N \varphi(x_i)}\right)\\
&= t\beta N^{2-\frac{s}{d}}\int_{\mathbb{R}^d}\varphi d\mu_\theta + \frac{\beta N^{2-\frac{s}{d}}t^2}{2\mathsf{c_d}}\left(\int_{\mathbb{R}^d}|\nabla\varphi|^2 - \frac{1}{\theta \mathsf{c_d}}\int_{\mathbb{R}^d}\frac{|\Delta\varphi|^2}{\mu_\theta}\right)\\
&+\frac{tN\beta}{\mathsf{c_d}}\left(-\frac14\left(\int_{\mathbb{R}^d}\Delta\varphi\log\mu_\theta\right)\mathbf{1}_{\mathsf{d}=2} + \left(\int_{\mathbb{R}^d}\Delta\varphi\frac{f_\mathsf{d}'(\beta\mu_\theta^{-1})}{\mu_\theta^2}\right)\mathbf{1}_{\mathsf{d}=1}\right)\\
&+\mathsf{Error}_1+\mathsf{Error}_2+\mathsf{Error}_3.
\end{aligned}$$

Here Error_1 is the error term in (10.1.8), with our assumption (10.2.20) it is $O(t^3\beta N^{2-\frac{s}{d}}\theta^{-1}\ell^{\mathsf{d}-6})$, Error_2 is the error term in (10.1.11), using (10.1.7) it is

$$\mathsf{Error}_2 = O\left(t\sqrt{\chi(\beta)}\beta N^{1+\frac{1}{d}}\ell^{\mathsf{d}-3}\theta^{-1} + t^2\theta^{-1}N\ell^{\mathsf{d}-6}\right).$$

Finally Error_3 is the error term in (10.2.22). We now insert $\theta = \beta N^{1-\frac{s}{d}}$, (10.2.35) and (9.4.2). We check that the choice of t of (10.2.35) satisfies $|t|\ell^{-2}\le \frac12\min(1,\alpha)$, for N large enough, under the assumption $\ell N^{1/\mathsf{d}}\rho_\beta^{-1}\to\infty$ (which is stronger than $\ell N^{\frac{1}{d}}\to\infty$). Indeed, in this regime, x, y in (9.4.2) tend to 0, and $\mathcal{R}\to 0$ at an algebraic rate, more precisely, we can bound

$$\max_{|s|\le\ell^2}\mathcal{R}_s \le C\left(\frac{N^{\frac{1}{d}}\ell}{\rho_\beta}\right)^{-\frac12}\left(\log\frac{N^{\frac{1}{d}}\ell}{\rho_\beta}\right)^{\frac{1}{d}}. \tag{10.2.38}$$

We then obtain that, in that limit,

$$\mathsf{Error}_1 = O\left(\tau^3\ell^{\mathsf{d}}N\beta^{-\frac32}(N^{\frac{1}{d}}\ell)^{-3-\frac{3\mathsf{d}}{2}}\right) = O(\tau^3(N^{\frac{1}{d}}\ell)^{\mathsf{d}-3-\frac{3\mathsf{d}}{2}}\beta^{-\frac32}).$$

We note that in view of (5.2.27) for both $\mathsf{d}=1$ and $\mathsf{d}=2$ (and $\mathsf{s}=\mathsf{d}-2$) the assumption $\ell N^{1/\mathsf{d}}\rho_\beta^{-1}\to\infty$ implies $N^{1/\mathsf{d}}\ell\beta^{3/4}\to\infty$ in dimension 1, and $N^{1/\mathsf{d}}\ell\beta^{1/2}$ in dimension 2, and

$$\mathsf{Error}_1 \le \begin{cases} O\left((\ell N^{1/\mathsf{d}}\rho_\beta^{-1})^{-\frac72}\right) & \text{if } \mathsf{d}=1\\ O\left((\ell N^{1/\mathsf{d}}\rho_\beta^{-1})^{-4}\right) & \text{if } \mathsf{d}=2.\end{cases} \tag{10.2.39}$$

For the second term, we use that $\chi(\beta)\beta^{-1}\le\rho_\beta^2$ and $\mathsf{s}=\mathsf{d}-2$ to obtain

$$\begin{aligned}
\mathsf{Error}_2 &= O\left(\sqrt{\chi(\beta)}N^{\frac{s}{d}+\frac{1}{d}}\ell^{\mathsf{d}-3}(\tau\ell^2\beta^{-\frac12}(N^{\frac1d}\ell)^{-1-\frac{\mathsf{d}}{2}}) + CN^{\frac{s}{d}}\ell^{\mathsf{d}-6}\beta^{-1}(\tau\ell^2\beta^{-\frac12}(N^{\frac1d}\ell)^{-1-\frac{\mathsf{d}}{2}})^2\right)\\
&= O\left(\tau\rho_\beta(N^{\frac1d}\ell)^{\frac{\mathsf{d}}{2}-2} + C\tau^2\left(\frac{N^{\frac1d}\ell}{\rho_\beta}\right)^{-4}\right) = o(1).
\end{aligned}$$

For the third error term, using (10.2.38) and using again $\chi(\beta)\beta^{-1}\le\rho_\beta^2$, we obtain

$$\begin{aligned}
\mathsf{Error}_3 &= O\left(\tau^2\beta^{-1}(N^{\frac1d}\ell)^{-2-\mathsf{d}}\beta\chi(\beta)N\ell^{\mathsf{d}} + \tau\beta^{-\frac12}(N^{\frac1d}\ell)^{-1-\frac{\mathsf{d}}{2}}\beta\chi(\beta)N\ell^{\mathsf{d}}\left(\max_{|s|\le\ell^2}\mathcal{R}_s\right)^{\frac12}\right)\\
&= O\left(\tau^2\beta\rho_\beta^2(N^{\frac1d}\ell)^{-2} + \tau\beta^{\frac12}\chi(\beta)(\rho_\beta)^{\frac{\mathsf{d}}{2}-1}\left(\frac{N^{\frac1d}\ell}{\rho_\beta}\right)^{\frac{\mathsf{d}}{2}-\frac54}\left(\log\frac{N^{\frac1d}\ell}{\rho_\beta}\right)^{\frac{1}{2\mathsf{d}}}\right).
\end{aligned}$$

When $\mathsf{d} \le 2$, this tends to 0 algebraically as soon as $\beta \le 1$, and if $\beta \ge 1$ (then $\rho_\beta = 1$) we use the assumption (10.2.1). This is where we change the definition of t to get convergence in the high β regime.

We thus see that all error terms tend to 0 at an algebraic rate when τ is fixed and $N^{\frac{1}{\mathsf{d}}}\ell \gg \rho_\beta$, if $\mathsf{d} \le 2$. The limitation to convergence in higher dimension is only due to the poor precision of $\mathcal{R}_s$.

Inserting the expression (10.2.35) into (10.2.37), we have thus found

$$\begin{aligned}
&\log \mathbb{E}_{\mathbb{P}_{N,\beta}} \left(\exp\left(-\tau\ell^2\beta^{-\frac{1}{2}}(N^{\frac{1}{\mathsf{d}}}\ell)^{-1-\frac{\mathsf{d}}{2}}\beta N^{1-\frac{s}{\mathsf{d}}}\,\mathrm{Fluct}_{\mu_\theta}(\varphi)\right)\right)\\
&= \frac{N^{1+\frac{2}{\mathsf{d}}}\tau^2\ell^4(N^{\frac{1}{\mathsf{d}}}\ell)^{-2-\mathsf{d}}}{2\mathsf{c}_{\mathsf{d}}}\left(\int_{\mathbb{R}^{\mathsf{d}}}|\nabla\varphi|^2 - \frac{1}{\theta\mathsf{c}_{\mathsf{d}}}\int_{\mathbb{R}^{\mathsf{d}}}\frac{|\Delta\varphi|^2}{\mu_\theta}\right)\\
&\quad + \frac{\tau\ell^2 N(N^{\frac{1}{\mathsf{d}}}\ell)^{-1-\frac{\mathsf{d}}{2}}\beta^{\frac{1}{2}}}{\mathsf{c}_{\mathsf{d}}}\left(-\frac{1}{4}\left(\int_{\mathbb{R}^{\mathsf{d}}}\Delta\varphi\log\mu_\theta\right)\mathbf{1}_{\mathsf{d}=2}\right.\\
&\qquad\qquad \left. + \left(\int_{\mathbb{R}^{\mathsf{d}}}\Delta\varphi\frac{f_{\mathsf{d}}'(\beta\mu_\theta^{-1})}{\mu_\theta^2}\right)\mathbf{1}_{\mathsf{d}=1}\right) + o(1).
\end{aligned}$$

Moreover, by assumption on φ, we have

$$\int_{\mathbb{R}^{\mathsf{d}}}|\nabla\varphi|^2 - \frac{1}{\theta\mathsf{c}_{\mathsf{d}}}\int_{\mathbb{R}^{\mathsf{d}}}\frac{|\Delta\varphi|^2}{\mu_\theta} = \ell^{\mathsf{d}-2}\int_{\mathbb{R}^{\mathsf{d}}}|\nabla\varphi_0|^2 + O\left(|\varphi_0|_{C^2}\frac{\ell^{\mathsf{d}-4}}{\theta}\right) \tag{10.2.40}$$

and since, as noted after (10.1.1) we have $\ell \gg \theta^{-1/2}$, the second term is $o(\ell^{\mathsf{d}-2})$. When $\mathsf{d} = 2$, we have thus obtained that

(10.2.41)
$$\log \mathbb{E}_{\mathbb{P}_{N,\beta}}\left(\exp\left(-\tau\beta^{\frac{1}{2}}\,\mathrm{Fluct}_{\mu_\theta}(\varphi)\right)\right) = \frac{\tau^2}{2\mathsf{c}_{\mathsf{d}}}\int_{\mathbb{R}^{\mathsf{d}}}|\nabla\varphi_0|^2 + \tau\frac{\beta^{\frac{1}{2}}}{4\mathsf{c}_{\mathsf{d}}}\int_{\mathbb{R}^{\mathsf{d}}}\Delta\varphi\log\mu_\theta + o(1).$$

This implies the result of Theorem 10.7 since the inverse Laplace transform of a Gaussian is a Gaussian.

In the case $\mathsf{d} = 1$, by change of variables, we check that

$$\frac{\tau\ell(N^{\frac{1}{\mathsf{d}}}\ell)^{-\frac{1}{2}}\beta^{\frac{1}{2}}}{\mathsf{c}_{\mathsf{d}}}\int_{\mathbb{R}}\Delta\varphi\frac{f_{\mathsf{d}}'(\beta\mu_\theta^{-1})}{\mu_\theta^2} = \frac{\tau(N^{\frac{1}{\mathsf{d}}}\ell)^{-\frac{1}{2}}\beta^{\frac{1}{2}}}{\mathsf{c}_{\mathsf{d}}}\int_{\mathbb{R}}\Delta\varphi_0(y)\frac{f_{\mathsf{d}}'(\beta\mu_\theta(\bar{x}+\ell y)^{-1})}{\mu_\theta(\bar{x}+\ell y)}dy,$$

and since $N^{1/\mathsf{d}}\ell \to +\infty$, this linear term tends to 0. We have then obtained for $\mathsf{d} = 1$ that

$$\begin{aligned}
&\log \mathbb{E}_{\mathbb{P}_{N,\beta}}\left(\exp\left(-\tau\beta^{\frac{1}{2}}(N^{\frac{1}{\mathsf{d}}}\ell)^{\frac{1}{2}}\,\mathrm{Fluct}_{\mu_\theta}(\varphi)\right)\right)\\
&= \frac{\tau^2\ell^{2-\mathsf{d}}}{2\mathsf{c}_{\mathsf{d}}}\left(\ell^{\mathsf{d}-2}\int_{\mathbb{R}^{\mathsf{d}}}|\nabla\varphi_0|^2 + o(\ell^{\mathsf{d}-2})\right) + o(1)\\
&= \frac{\tau^2}{2\mathsf{c}_{\mathsf{d}}}\int_{\mathbb{R}^{\mathsf{d}}}|\nabla\varphi_0|^2 + o(1),
\end{aligned} \tag{10.2.42}$$

and we conclude that Theorem 10.11 holds as well.

The proof for the low temperature case Theorems 10.9 and 10.12 is similar: we choose instead (10.2.36) which is equivalent to taking $\tau = s\beta^{-\frac{1}{2}}$ and obtain instead

$$\begin{aligned}\log \mathbb{E}_{\mathbb{P}_{N,\beta}}\left(\exp\left(-s(N^{\frac{1}{\mathsf{d}}}\ell)^{1-\frac{\mathsf{d}}{2}}\operatorname{Fluct}_{\mu_\theta}(\varphi)\right)\right) \\ = \frac{s^2\beta^{-1}}{2\mathsf{c_d}}\int_{\mathbb{R}^\mathsf{d}}|\nabla\varphi_0|^2 - \frac{s}{4\mathsf{c_d}}\left(\int_{\mathbb{R}^\mathsf{d}}\Delta\varphi\log\mu_\theta\right)\mathbf{1}_{\mathsf{d}=2} \\ + \frac{s(N^{\frac{1}{\mathsf{d}}}\ell)^{-\frac{1}{2}}}{\mathsf{c_d}}\left(\int_{\mathbb{R}}\Delta\varphi_0(y)\frac{f_\mathsf{d}'(\beta\mu_\theta(\bar{x}+\ell y)^{-1})}{\mu_\theta(\bar{x}+\ell y)}dy\right)\mathbf{1}_{\mathsf{d}=1} + o(1).\end{aligned}$$

□

Remark 10.15. The strategy of proof could work as well in higher dimensions, but encounters two obstacles. The first is that the proof requires to know a C^2 bound on f_d. This can be hypothetized away, as a sort of no phase-transition assumption, since phase transitions correspond to loss of regularity of the free energy and f_d is the thermodynamic limit, i.e., free energy per unit volume (the pressure). The second is that the error rate $\mathcal{R}$ needs to be quantitative and tend to 0 at a certain rate in order to compensate for the $(N^{\frac{1}{\mathsf{d}}}\ell)^{\mathsf{d}-2}$ factor that appears in the third error. Provided we know such a rate, and assume the regularity of f_d we thus can obtain by the same proof a CLT in higher dimension. The rate obtained in Proposition 9.9, which we do not believe to be optimal, is however not sufficient.

Remark 10.16. We note that the local laws are needed only to obtain the CLT at mesoscopic scales $\ell \ll 1$.

Remark 10.17 (The case of β-ensembles)**.** In the particular case of one-dimensional log gases $\mathsf{d}=1, \mathsf{s}=0$, we have

$$\mathsf{A}_1(X_N,\mu,\psi) = \iint_{\mathbb{R}\times\mathbb{R}\setminus\triangle}\frac{\psi(x)-\psi(y)}{x-y}d\left(\sum_{i=1}^N\delta_{x_i}-N\mu\right)(x)d\left(\sum_{i=1}^N\delta_{x_i}-N\mu\right)(y)$$

and the kernel involved in the integration $\frac{\psi(x)-\psi(y)}{x-y}$, is regular if ψ is regular enough, contrarily to all other cases with $\mathsf{s}\geq 0$ for which it is always singular. Moreover, one may rewrite the above as

(10.2.43)

$$\begin{aligned}\mathsf{A}_1(X_N,\mu,\psi) = \iint_{\mathbb{R}\times\mathbb{R}}\frac{\psi(x)-\psi(y)}{x-y}d\left(\sum_{i=1}^N\delta_{x_i}-N\mu\right)(x)d\left(\sum_{i=1}^N\delta_{x_i}-N\mu\right)(y) \\ -\sum_{i=1}^N\psi'(x_i)\end{aligned}$$

and the double integral on the right-hand side can be seen as a bilinear statistics with respect to the regular test function $\frac{\psi(x)-\psi(y)}{x-y}$. As we will see for instance in (10.1.4), fluctuations can themselves be reexpressed via ratio of partition functions, and fluctuations for regular functions are (in this case) of order 1 with respect to N. Thus, integrating (9.1.9), after finding the appropriate transport field ψ_t,[4] we

[4]This is equivalent to inverting a "master operator" as alluded to above.

can deduce that

$$\log \mathsf{K}_{N,\beta}(\mu_1) - \log \mathsf{K}_{N,\beta}(\mu_0) = \beta \int_0^1 \mathbb{E}_{\mathbb{Q}_{N,\beta}(\mu_t)} \left(\sum_{i=1}^N \psi_t'(x_i) \right) dt + O(1) \tag{10.2.44}$$
$$= \beta N \int_0^1 \int_{\mathbb{R}} \psi_t'(x) d\mu_t(x) dt + O(1),$$

which already gives the correct leading order to the free energy difference. In addition, since free energy differences encode fluctuations (again, via (10.1.4) or variants), the form of the leading order term allows to deduce the leading order (of order 1) of the expectation and variance of fluctuations of regular linear statistics. By a Fourier trick which turns double fluctuations into single fluctuation, one can then plug back this estimate for fluctuations into (10.2.43) to obtain the next order in the expansion of A_1, i.e., an expansion to $o(1)$, hence an expansion to $o(1)$ of (10.2.44), which in turns provides the order $1/N$ term in the expansion of fluctuations. This can be iterated to arbitrary order, providing expansions to all powers of $1/N$ of both free energy and fluctuations, provided ψ (hence in fact V) has enough derivatives. This is described in more detail in [**BLS18**, Appendix A] and is a reinterpretation of what is accomplished in the "topological recursion" of [**BG13**].

10.2.3. Approach via the usual equilibrium measure. Working with the usual equilibrium measure avoids having to analyze precisely the thermal equilibrium measure and thus having to use the results of Theorem 2.23, in particular the C^k bounds for the thermal equilibrium measure (2.5.32). The downside is that the results are less precise when β gets small, and do not allow to go down to the threshold scaling $\theta = \theta_0$, i.e., $\beta = N^{\frac{\mathsf{s}}{\mathsf{d}}-1}\theta_0$ for some fixed θ_0.

When working with the equilibrium measure, we use instead of $\mathsf{K}_{N,\beta}$ and $\mathbb{Q}_{N,\beta}$ the quantities $\tilde{\mathsf{K}}_{N,\beta}(\mu,\zeta)$ and $\tilde{\mathbb{Q}}_{N,\beta}(\mu,\zeta)$ in (5.1.5), (5.1.6). One may check that the analysis of Chapters 7 and 8, i.e., the screening, almost additivity and local laws, holds as well for fixed β when using $\mathbb{Q}_{N,\beta}(\mu,\zeta)$. It was first done this way in [**LS17**].

One can also argue that in view of (5.1.14) and (5.1.7), we have $\tilde{\mathbb{Q}}_{N,\beta}(\mu_V,\zeta) = \mathbb{Q}_{N,\beta}(\mu_\theta) = \mathbb{P}_{N,\beta}$, hence in view of Theorem 8.4, the local laws hold for $\tilde{\mathbb{Q}}_{N,\beta}(\mu_V,\zeta)$, under the sole assumption that $\mu_\theta \geq m > 0$, which is a much milder fact to check than (2.5.32)—a local uniform convergence of μ_θ to μ_V in Σ as $\theta \gg 1$ suffices.

Let us now continue explaining the approach under this assumption that local laws hold for $\mathbb{P}_{N,\beta}$ hence $\tilde{\mathbb{Q}}_{N,\beta}(\mu_V,\zeta)$. We assume for simplicity that the test function φ is supported in Σ. When working at mesoscales, we assume that $\operatorname{supp}\varphi \subset \square_{\ell/2} \subset \square_\ell \subset \hat{\Sigma}$ (as in (9.4.8)).

In the case where one works with μ_V, in view of (5.1.4) and with obvious notation we have, letting still $V_t = V + t\varphi$,

$$\mathbb{E}_{\mathbb{P}_{N,\beta}}\left(e^{-\beta t N^{1-\frac{\mathsf{s}}{\mathsf{d}}} \sum_{i=1}^N \varphi(x_i)} \right) = \frac{Z_{N,\beta}(V_t)}{Z_{N,\beta}(V)} \tag{10.2.45}$$
$$= \exp\left(-\beta N^{2-\frac{\mathsf{s}}{\mathsf{d}}} \left(\mathcal{E}^{V_t}(\mu_{V_t}) - \mathcal{E}^V(\mu_V)\right)\right) \frac{\tilde{\mathsf{K}}_{N,\beta}(\mu_{V_t},\zeta_{V_t})}{\tilde{\mathsf{K}}_{N,\beta}(\mu_V,\zeta_V)},$$

where $\mathcal{E}^V$ denotes the interaction energy functional (2.0.3) with the V-dependence highlighted.

Because φ is supported in Σ, by Theorem 2.8 and the unique characterization of the equilibrium measure, one may check that the perturbed equilibrium measure is simply equal to

$$\mu_{V_t} = \mu_V + \frac{t}{\mathsf{c_d}}\Delta\varphi,$$

and that $\zeta_{V_t} = \zeta_V$. This is the reason why the interior case is easier to treat than the case where the support of φ overlaps Σ^c, in which the support of the equilibrium measure changes (see [**SS18**]). Again we refer to [**LS18**] for the treatment of that more general situation.

We are in a situation where ζ is independent of t. Moreover, defining

$$\mu_0 = \mu_V, \quad \mu_s = \mu_0 + \frac{s}{\mathsf{c_d}}\Delta\varphi, \quad \psi_s = -\frac{\nabla\varphi}{\mathsf{c_d}\mu_s},$$

we have by Lemma 9.6 that

$$\mu_s = \Phi_s \# \mu_0,$$

with Φ_s as in (9.1.2).

Moreover, since for $|s| \le |t|$, we have $\mu_s = \mu_{V_s}$, the measure $\tilde{\mathbb{Q}}_{N,\beta}(\mu_s, \zeta)$ is equal to $\mathbb{P}_{N,\beta}$ for the potential V_s, and by the assumption made above, it also satisfies the local laws.

To take limits in (10.2.45), let us now examine the ratio of the reduced partition functions via the variant of the transport calculus in Lemma 9.4.

Lemma 10.18. *Let μ_s, ψ_s and φ be as above. We have*

$$\text{(10.2.46)} \quad \log \frac{\tilde{\mathsf{K}}_{N,\beta}(\mu_t,\zeta)}{\tilde{\mathsf{K}}_{N,\beta}(\mu_0,\zeta)} = -N\left(\int_{\mathbb{R}^\mathsf{d}} \mu_t \log \mu_t - \int_{\mathbb{R}^\mathsf{d}} \mu_0 \log \mu_0\right) + \int_0^t \mathbb{E}_{\tilde{\mathbb{Q}}_{N,\beta}(\mu_s,\zeta)}\left(-\beta N^{-\frac{s}{\mathsf{d}}}\mathsf{A}_1(X_N,\mu_s,\psi_s) + \mathrm{Fluct}_{\mu_s}(\operatorname{div}\psi_s)\right) ds$$

and

$$\text{(10.2.47)} \quad \log \frac{\tilde{\mathsf{K}}_{N,\beta}(\mu_t,\zeta)}{\tilde{\mathsf{K}}_{N,\beta}(\mu_0,\zeta)} = -N\left(\int_{\mathbb{R}^\mathsf{d}} \mu_t \log \mu_t - \int_{\mathbb{R}^\mathsf{d}} \mu_0 \log \mu_0\right)$$
$$+\mathbb{E}_{\tilde{\mathbb{Q}}_{N,\beta}(\mu_0,\zeta_0)}\left(\exp\left(-\beta N^{-\frac{s}{\mathsf{d}}}\left(\mathsf{A}_1(X_N,\mu_0,\psi_0) + \int_0^t (t-s)\mathsf{A}_2(\Phi_s(X_N),\mu_s,\psi_s)\right)\right.\right.$$
$$\left.\left.+\int_0^t \mathrm{Fluct}_{\mu_0}((\operatorname{div}\psi_s)\circ\Phi_s)\, ds\right)\right).$$

Proof. Since $\mu_s = \Phi_s\#\mu_0$, integrating (9.1.19) we obtain

$$\log \frac{\tilde{\mathsf{K}}_{N,\beta}(\mu_t,\zeta)}{\tilde{\mathsf{K}}_{N,\beta}(\mu_0,\zeta)} = \int_0^t \mathbb{E}_{\tilde{\mathbb{Q}}_{N,\beta}(\mu_s,\zeta)}\left(-\beta N^{-\frac{s}{\mathsf{d}}}\mathsf{A}_1(X_N,\mu_s,\psi_s) + \sum_{i=1}^N \operatorname{div}\psi_s(x_i)\right) ds.$$

We may rewrite

$$\sum_{i=1}^N \operatorname{div}\psi_s(x_i) = N\int_{\mathbb{R}^\mathsf{d}} \operatorname{div}\psi_s\, d\mu_s + \mathrm{Fluct}_{\mu_s}(\operatorname{div}\psi_s).$$

On the other hand, we observe that, using integration by parts,

$$\begin{aligned}(10.2.48)\quad \frac{d}{ds}\int_{\mathbb{R}^d}\mu_s\log\mu_s &= \int_{\mathbb{R}^d}\log\mu_s\frac{d}{ds}\mu_s = -\int_{\mathbb{R}^d}\mathrm{div}\,(\psi_s\mu_s)\log\mu_s = \int_{\mathbb{R}^d}\psi_s\mu_s\cdot\nabla\log\mu_s \\ &= \int_{\mathbb{R}^d}\psi_s\cdot\nabla\mu_s = -\int_{\mathbb{R}^d}\mathrm{div}\,\psi_s\,d\mu_s.\end{aligned}$$

Assembling these relations, we find that (10.2.46) holds.

Let us now write (9.1.20) with μ_0 in place of μ_t and Φ_t in place of ϕ_h, recalling that $\Phi_t\#\mu_0=\mu_t$, using that Φ_t is supported where ζ_t and ζ_0 vanish, and inserting (10.2.26), we find

$$\begin{aligned}&\frac{\tilde{\mathsf{K}}_{N,\beta}(\mu_t,\zeta)}{\tilde{\mathsf{K}}_{N,\beta}(\mu_0,\zeta)} \\ &= \mathbb{E}_{\tilde{\mathbb{Q}}_{N,\beta}(\mu_0,\zeta_0)}\Bigg(\exp\Big(-\beta N^{-\frac{s}{\mathsf{d}}}\Big(t\mathsf{A}_1(X_N,\mu_0,\psi_0)+\int_0^t(t-s)\mathsf{A}_2(\Phi_s(X_N),\mu_s,\psi_s)\,ds\Big) \\ &\qquad+\sum_{i=1}^N\log\det D\Phi_t(x_i)\Big)\Bigg).\end{aligned}$$

Using (9.1.12) we may also compute that $\frac{d}{ds}\log\det D\Phi_s=(\mathrm{div}\,\psi_s)\circ\Phi_s$, hence

$$\begin{aligned}&\sum_{i=1}^N\log\det D\Phi_t(x_i) \\ &\quad=\int_0^t\sum_{i=1}^N\mathrm{div}\,\psi_s(\Phi_s(x_i))ds \\ &\quad=\int_0^t\int_{\mathbb{R}^d}N\mathrm{div}\,\psi_s(\Phi_s)d\mu_0+\mathrm{Fluct}_{\mu_0}(\mathrm{div}\,\psi_s\circ\Phi_s)ds \\ &\quad=\int_0^t\int_{\mathbb{R}^d}N\mathrm{div}\,\psi_s\,d\mu_s+\mathrm{Fluct}_{\mu_0}(\mathrm{div}\,\psi_s\circ\Phi_s)ds \\ &\quad=-N\left(\int_{\mathbb{R}^d}\mu_t\log\mu_t-\int_{\mathbb{R}^d}\mu_0\log\mu_0\right)+\int_0^t\mathrm{Fluct}_{\mu_0}(\mathrm{div}\,\psi_s\circ\Phi_s)ds,\end{aligned}$$

where we used (10.2.48). We conclude that (10.2.47) holds. □

At this point, inserting this relation into (10.2.45) we are in a very similar situation as that of (10.1.4). First, we need to estimate $\mathcal{E}^{V_t}(\mu_t)-\mathcal{E}^V(\mu_V)$ and to bound $\mathrm{Fluct}_{\mu_V}(\mathrm{div}\,\psi_0)$. The former is done via the following explicit estimate, proved in Section 10.2.4.

Lemma 10.19. *Assume that φ is C^2 and supported in $\Sigma=\{\zeta=0\}$ (see notation in (2.1.12)), then*

$$\mathcal{E}^{V_t}(\mu_{V_t})-\mathcal{E}^V(\mu_V)=t\int_{\mathbb{R}^d}\varphi\,d\mu_V-\frac{t^2}{2\mathsf{c}_\mathsf{d}}\int_{\mathbb{R}^d}|\nabla\varphi|^2. \quad (10.2.49)$$

For the latter, we have the following, obtained by the rough fluctuations bounds of (4.3.1) combined with (4.2.16) (see Section 10.2.4 for the proof). Using that

(10.2.50)
$$|\psi_t|_{C^1} \le C\left((|\mu_V|_{C^1} + |t||\varphi|_{C^3})|\varphi|_{C^1} + |\varphi|_{C^2}\right)$$

(10.2.51)
$$|\psi_t|_{C^2} \le C\left((|\mu_V|_{C^2} + |t||\varphi|_{C^4})|\varphi|_{C^1} + (|\mu_V|_{C^1} + |t||\varphi|_{C^3})|\varphi|_{C^2} + |\varphi|_{C^3}\right),$$

we obtain the following,

Lemma 10.20. *Under the same assumptions,*

$$(10.2.52) \quad \left|\int_0^t \mathbb{E}_{\tilde{\mathbb{Q}}_{N,\beta}(\mu_s,\zeta)}(\mathrm{Fluct}_{\mu_s}(\operatorname{div}\psi_s))\right| \le C|t|\sqrt{\chi(\beta)}N^{\frac{1}{2}+\frac{\mathsf{s}}{2\mathsf{d}}}\ell^{\mathsf{d}}\big((|\mu_V|_{C^2} + |t||\varphi|_{C^4})|\varphi|_{C^1} + (|\mu_V|_{C^1} + |t||\varphi|_{C^3})|\varphi|_{C^2} + |\varphi|_{C^3}\big).$$

Here we need that ψ_s is C^2, which is implied by the conditions $\varphi \in C^3$, $\mu_V \in C^2$. In view of (2.1.17) the regularity on V needed is C^4.

Remark 10.21. We have not tried to optimize the regularity of ψ here, the estimates of (4.3.1) are in fact valid with less regularity on $\operatorname{div}\psi$.

To estimate the ratio in (10.2.45), we may first use (10.2.46). We directly estimate (as in (10.2.17)) that

$$(10.2.53) \quad \int_{\mathbb{R}^{\mathsf{d}}} \mu_{V_t}\log\mu_{V_t} - \int_{\mathbb{R}^{\mathsf{d}}} \mu_V \log\mu_V = \frac{t}{\mathsf{c}_{\mathsf{d}}}\int_{\mathbb{R}^{\mathsf{d}}} \Delta\varphi(\log\mu_V) + O(t^2|\varphi|_{C^2}\ell^{\mathsf{d}}).$$

One may then obtain a bound on $\int_0^t \mathbb{E}_{\tilde{\mathbb{Q}}_{N,\beta}(\mu_s,\zeta)}\left(-\beta N^{-\frac{\mathsf{s}}{\mathsf{d}}}\mathsf{A}_1(X_N,\mu_s,\psi_s)\right)$ as in (10.1.10), but with μ_θ replaced by μ_V, by simply inserting the commutator estimate of Theorem 6.1 combined with the local laws. The only regularity needed to apply the commutator estimate is that ψ_s be Lipschitz, which is implied by $\varphi \in C^2$ and $\mu_V \in C^1$.

Using (10.2.50) and (10.2.51), we then obtain the following control on fluctuations.

Theorem 10.22 (Control of fluctuations, Coulomb case, equilibrium measure). *Let* $\mathsf{d} \ge 1$. *Assume* (A1)–(A5). *Assume that* $\operatorname{supp}\varphi \subset \square_{\ell/2} \subset \square_\ell$, *with* ℓ *satisfying* (10.1.1), *and if* $\ell \to 0$, *assume also that* $\square_\ell$ *is a cube in which local laws hold. Assume* $\varphi \in C^3(Q_\ell)$ *and* $\mu_V \in C^2(Q_\ell)$. *Then, for any* t *such that* $|t||\varphi|_{C^2}$ *is small enough, we have*

$$(10.2.54) \quad \left|\log \mathbb{E}_{\mathbb{P}_{N,\beta}}\left(e^{\beta t N^{1-\frac{\mathsf{s}}{\mathsf{d}}}\mathrm{Fluct}_{\mu_V}(\varphi)}\right)\right|$$
$$\le Ct\sqrt{\chi(\beta)}N^{\frac{1}{2}+\frac{\mathsf{s}}{2\mathsf{d}}}\ell^{\mathsf{d}}\left((|\mu_V|_{C^2} + t|\varphi|_{C^4})|\varphi|_{C^1} + (|\mu_V|_{C^1} + |t||\varphi|_{C^3})|\varphi|_{C^2} + |\varphi|_{C^3}\right)$$
$$+ C|t|\beta\chi(\beta)N\ell^{\mathsf{d}}\left((|\mu_V|_{C^1} + |t||\varphi|_{C^3})|\varphi|_{C^1} + |\varphi|_{C^2}\right) + C(t^2|\varphi|_{C^2}N\ell^{\mathsf{d}} + t^2\beta N^{2-\frac{\mathsf{s}}{\mathsf{d}}}\ell^{\mathsf{d}}|\varphi|^2_{C^1}),$$

where the constants depend on $\mathsf{d}, \mathsf{s}, m, \|\mu_V\|_{L^\infty}$, *but not on* φ, β, N *or* t.

These estimates are as good as those of Theorem 10.4 when β is fixed, but not when $\beta \to 0$ as $N \to \infty$. One can then derive similar corollaries.

To prove the central limit theorems, we instead insert (10.2.47) into (10.2.45) and still use (10.2.49) and (10.2.52). We use again Proposition 6.5 to bound the

A_2 terms. The evaluation of $\mathbb{E}(\exp(-\beta N^{-\frac{s}{d}}\mathsf{A}_1(X_N,\mu_0,\psi_0)))$ we use should be done as in Proposition 10.14 except working with $\tilde{\mathbb{Q}}_{N,\beta}(\mu_0,\zeta)$ instead of $\mathbb{Q}_{N,\beta}(\mu_0)$. This way, choosing $t=\tau\beta^{-1}N^{-1}$, one can arrive, as in **[LS18]**, to a two-dimensional CLT in the form

$$\log \mathbb{E}_{\mathbb{P}_{N,\beta}}\left(e^{-\tau \operatorname{Fluct}_{\mu_V}(\varphi)}\right) \to -\tau \operatorname{mean}(\varphi)+\frac{\tau^2}{2}\operatorname{var}(\varphi)$$

with

$$\operatorname{mean}(\varphi)=\begin{cases}(\frac{1}{\beta}-\frac{1}{4})\frac{1}{\mathsf{c_d}}\int_{\mathbb{R}^\mathsf{d}}(\log\mu_V)\Delta\varphi_0, & \ell=1\\ 0 & \ell\to 0\end{cases}\qquad \operatorname{var}(\varphi)=\frac{1}{\beta\mathsf{c_d}}\int_{\mathbb{R}^\mathsf{d}}|\nabla\varphi_0|^2,$$

valid for fixed β, any $\ell\gg N^{-1/\mathsf{d}}$, and requiring only that $\mu_V\in C^2$ (hence $V\in C^4$ suffices) and $\varphi\in C^4$. One can check that this result is consistent with Theorem 10.4 in view the difference between μ_V and μ_θ given by (2.5.31).

10.2.4. Proof of auxiliary lemmas.

Proof of Lemma 10.1. With the quantities introduced, proceeding as in the proof of the splitting formula (5.1.1) except splitting with respect to the probability ν_θ^t instead of μ_θ and using (10.1.5), we find

(10.2.55)

$$\begin{aligned}\mathcal{H}_N^{V_t}(X_N)&=N^2\mathcal{E}^{V_t}(\nu_\theta^t)+N\int_{\mathbb{R}^\mathsf{d}}(\mathsf{g}*\nu_\theta^t+V_t)d\left(\sum_{i=1}^N\delta_{x_i}-N\nu_\theta^t\right)+\mathsf{F}_N(X_N,\nu_\theta^t)\\ &=N^2\mathcal{E}^{V_t}(\nu_\theta^t)+N\int_{\mathbb{R}^\mathsf{d}}(-\frac{1}{\theta}\log\nu_\theta^t+\varepsilon_t)d\left(\sum_{i=1}^N\delta_{x_i}-N\nu_\theta^t\right)+\mathsf{F}_N(X_N,\nu_\theta^t)\\ &=N^2\mathcal{E}_\theta^{V_t}(\nu_\theta^t)+\mathsf{F}_N(X_N,\nu_\theta^t)-\frac{N}{\theta}\sum_{i=1}^N\log\nu_\theta^t(x_i)+N\int_{\mathbb{R}^\mathsf{d}}\varepsilon_t\,d\left(\sum_{i=1}^N\delta_{x_i}-N\nu_\theta^t\right).\end{aligned}$$

Inserting into the definition of the Gibbs measure (1.1.5), we obtain (10.1.4). We now prove that ε_t, which measures how far ν_θ^t is from solving the same equation as $\mu_\theta^{V_t}$ (see (2.5.12)), is a small function as $\theta\gg 1$. Notice that ε_t is supported in $\operatorname{supp}\varphi$ and that

$$\mathsf{g}*(\nu_\theta^t-\mu_\theta)=-t\varphi. \tag{10.2.56}$$

Since $\mathsf{g}*\mu_\theta+V+\frac{1}{\theta}\log\mu_\theta=c_\theta$ by (5.1.8) and by definition (10.1.3), we deduce that

$$\varepsilon_t=\mathsf{g}*\nu_\theta^t+V+t\varphi+\frac{1}{\theta}\log\nu_\theta^t-c_\theta=\frac{1}{\theta}\log\frac{\nu_\theta^t}{\mu_\theta}=\frac{1}{\theta}\log\left(1+\frac{t}{\mathsf{c_d}}\frac{\Delta\varphi}{\mu_\theta}\right). \tag{10.2.57}$$

With direct computations and (2.5.32), it follows that (10.1.6) and (10.1.7) hold. □

PROOF OF LEMMA 10.2. We have

$$\begin{aligned}
&\mathcal{E}_\theta^{V_t}(\nu_\theta^t)-\mathcal{E}_\theta^V(\mu_\theta)\\
&=\Big(\frac12\iint_{\mathbb{R}^\mathsf{d}\times\mathbb{R}^\mathsf{d}}\mathsf{g}(x-y)d\nu_\theta^t(x)d\nu_\theta^t(y)\\
&\quad-\frac12\iint_{\mathbb{R}^\mathsf{d}\times\mathbb{R}^\mathsf{d}}\mathsf{g}(x-y)d\mu_\theta(x)d\mu_\theta(y)+\int_{\mathbb{R}^\mathsf{d}}V_td\nu_\theta^t-\int_{\mathbb{R}^\mathsf{d}}Vd\mu_\theta\Big)\\
&\quad+\frac1\theta\Big(\int_{\mathbb{R}^\mathsf{d}}\nu_\theta^t\log\nu_\theta^t-\int_{\mathbb{R}^\mathsf{d}}\mu_\theta\log\mu_\theta\Big)\\
&=\frac12\iint_{\mathbb{R}^\mathsf{d}\times\mathbb{R}^\mathsf{d}}\mathsf{g}(x-y)d(\nu_\theta^t-\mu_\theta)(x)d(\nu_\theta^t-\mu_\theta)(y)+\iint_{\mathbb{R}^\mathsf{d}\times\mathbb{R}^\mathsf{d}}\mathsf{g}(x-y)d(\nu_\theta^t-\mu_\theta)(x)d\mu_\theta(y)\\
&\quad+\int_{\mathbb{R}^\mathsf{d}}Vd(\nu_\theta^t-\mu_\theta)+t\int_{\mathbb{R}^\mathsf{d}}\varphi d\mu_\theta+t\int_{\mathbb{R}^\mathsf{d}}\varphi d(\nu_\theta^t-\mu_\theta)+\frac1\theta\Big(\int_{\mathbb{R}^\mathsf{d}}\nu_\theta^t\log\nu_\theta^t-\int_{\mathbb{R}^\mathsf{d}}\mu_\theta\log\mu_\theta\Big)\\
&=\frac12\iint_{\mathbb{R}^\mathsf{d}\times\mathbb{R}^\mathsf{d}}\mathsf{g}(x-y)d(\nu_\theta^t-\mu_\theta)(x)d(\nu_\theta^t-\mu_\theta)(y)+\int_{\mathbb{R}^\mathsf{d}}(\mathsf{g}*\mu_\theta+V+\frac1\theta\log\mu_\theta)d(\nu_\theta^t-\mu_\theta)\\
&\quad+t\int_{\mathbb{R}^\mathsf{d}}\varphi d\mu_\theta+t\int_{\mathbb{R}^\mathsf{d}}\varphi d(\nu_\theta^t-\mu_\theta)+\frac1\theta\int_{\mathbb{R}^\mathsf{d}}\nu_\theta^t(\log\nu_\theta^t-\log\mu_\theta).
\end{aligned}$$

The second term of the right-hand side vanishes by characterization of μ_θ in (2.5.12), and we are left with

$$\begin{aligned}
&\mathcal{E}_\theta^{V_t}(\nu_\theta^t)-\mathcal{E}_\theta^V(\mu_\theta)-t\int_{\mathbb{R}^\mathsf{d}}\varphi d\mu_\theta\\
&\quad=\frac{1}{2\mathsf{c_d}}\int_{\mathbb{R}^\mathsf{d}}|\nabla(\mathsf{g}*(\nu_\theta^t-\mu_\theta))|^2+t\int_{\mathbb{R}^\mathsf{d}}\varphi d(\nu_\theta^t-\mu_\theta)+\frac{1}{2\theta}\int_{\mathbb{R}^\mathsf{d}}\mu_\theta\left(\frac{\nu_\theta^t}{\mu_\theta}-1\right)^2\\
&\quad+O\left(\frac1\theta\int_{\mathbb{R}^\mathsf{d}}\left(\frac{\nu_\theta^t}{\mu_\theta}-1\right)^3\mu_\theta\right)
\end{aligned}$$

where we Taylor expanded the logarithm. We then use (10.1.3) to see that

$$|\nabla(\mathsf{g}*(\nu_\theta^t-\mu_\theta))|^2=t^2|\nabla\varphi|^2$$

and

$$\frac{\nu_\theta^t}{\mu_\theta}=1+t\frac{\Delta\varphi}{\mathsf{c_d}\mu_\theta}.$$

We thus find (10.1.8). Alternatively we can Taylor expand the log only to first order and get instead a bound by

$$Ct^2\left(\int_{\mathbb{R}^\mathsf{d}}|\nabla\varphi|^2+\frac1\theta\int_{\mathbb{R}^\mathsf{d}}\mu_\theta\left|\frac{\Delta\varphi}{\mu_\theta}\right|^2\right)$$

from which we deduce (10.1.9). □

PROOF OF LEMMA 10.3. By Theorem 8.4, local laws and concentration hold for $\mathbb{Q}_{N,\beta}(\nu_\theta^t)$ in $\hat\Sigma$ where ν_θ^t is bounded below provided $t|\varphi|_{C^2}$ is small enough. A rescaling of (8.4.6) yields that for any φ such that $\|\nabla\varphi\|_{L^\infty}\le N^{\frac1\mathsf{d}}$,

$$\left|\log\mathbb{E}_{\mathbb{Q}_{N,\beta}(\nu_\theta^t)}\left(\exp\frac{\beta}{CN\ell^\mathsf{d}}\left(\mathrm{Fluct}_{\nu_\theta^t}(\varphi)\right)^2\right)\right|\le C\beta\chi(\beta)N^{-\frac{\mathsf{s}}{\mathsf{d}}}\ell^\mathsf{d}\|\nabla\varphi\|_{L^\infty}^2$$

We may then apply this to $\varphi = \sqrt{C}\ell^{\frac{d}{2}} N^{\frac{1}{d}+\frac{1}{2}} \sqrt{\lambda}\varepsilon_t$. Thus, for any λ such that $\sqrt{\lambda C}\ell^{\frac{d}{2}} N^{\frac{1}{2}} |\varepsilon_t|_{C^1} \le 1$ (which ensures that $\|\nabla\varphi\|_{L^\infty} \le N^{1/d}$), using also Young's inequality to write

$$\theta \int \varepsilon_t \, d\left(\sum_{i=1}^N \delta_{x_i} - N\nu_\theta^t\right) \le \theta\lambda \left(\mathrm{Fluct}_{\nu_\theta^t}(\varepsilon_t)\right)^2 + \frac{\theta}{4\lambda},$$

we have

$$\log \mathbb{E}_{\mathbb{Q}_{N,\beta}(\nu_\theta^t)} \left(\exp\left(\theta \,\mathrm{Fluct}_{\nu_\theta^t}(\varepsilon_t)\right)\right) \le C\lambda\beta\chi(\beta) N^{1+\frac{2-s}{d}} \ell^{2d} |\varepsilon_t|_{C^1}^2 + \frac{\theta}{4\lambda}$$

and optimizing over $\lambda \le |\varepsilon_t|_{C^1}^{-2} (N\ell^d)^{-1}$ we find
(10.2.58)

$$\left|\log \mathbb{E}_{\mathbb{Q}_{N,\beta}(\nu_\theta^t)} \left(\exp\left(-\theta \,\mathrm{Fluct}_{\nu_\theta^t}(\varepsilon_t)\right)\right)\right| \le C\sqrt{\chi(\beta)}\beta N^{1+\frac{1}{d}} \ell^d |\varepsilon_t|_{C^1} + C\theta N \ell^d |\varepsilon_t|_{C^1}^2$$

hence the result (10.1.11).

We next turn to proving (10.1.12). This time we bound

$$\left|\mathrm{Fluct}_{\nu_\theta^t}(\varepsilon_t)\right| \le \|\varepsilon_t\|_{L^\infty} (\# I_{\square_\ell} + N\ell^d)$$

where $\# I_{\square_\ell}$ denotes the number of points in each configuration that fall in the set $\square_\ell$ containing the support of φ. We can in turn bound from above

$$\# I_{\square_\ell} \le N \int_{\square_\ell} d\nu_\theta^t + D(x, C\ell)$$

where $B(x, C\ell)$ is a ball that contains $\square_\ell$ and $D(x, C\ell) = \int_{B(x,C\ell)} \sum_{i=1}^N \delta_{x_i} - N d\mu$. Arguing as before, we write

$$\theta \|\varepsilon_t\|_{L^\infty} D(x, C\ell) \le \|\varepsilon_t\|_{L^\infty} \left(D^2(x, C\ell)\beta N^{\frac{2}{d}-1} \ell^{2-d} \lambda + \frac{\theta N \ell^{d-2}}{4\lambda}\right)$$

and thus using a rescaling of (8.4.2) and (3.2.5), we find,

$$\log \mathbb{E}_{\mathbb{Q}_{N,\beta}(\nu_\theta^t)} \left(\exp\left(\theta \|\varepsilon_t\|_{L^\infty} D(x, C\ell)\right)\right) \le C\|\varepsilon_t\|_{L^\infty} \lambda\beta\chi(\beta) N\ell^d + \frac{\beta \|\varepsilon_t\|_{L^\infty} N^{1+\frac{2}{d}} \ell^{d-2}}{4\lambda}.$$

Optimizing over $\lambda \le \|\varepsilon_t\|_{L^\infty}^{-1}$ we find

$$\begin{aligned}&\log \mathbb{E}_{\mathbb{Q}_{N,\beta}(\nu_\theta^t)} \left(\exp\left(\theta \|\varepsilon_t\|_{L^\infty} D(x, C\ell)\right)\right) \\ &\quad \le C\|\varepsilon_t\|_{L^\infty} \sqrt{\chi(\beta)}\beta N^{1+\frac{1}{d}} \ell^{d-1} + C\|\varepsilon_t\|_{L^\infty}^2 \beta N^{1+\frac{2}{d}} \ell^{d-2}.\end{aligned}$$

After observing that $\sqrt{\chi(\beta)} N^{-\frac{1}{d}} \ell^{-1} \le 1$ by (10.1.1) and (8.0.2). It follows that
(10.2.59)

$$\left|\log \mathbb{E}_{\mathbb{Q}_{N,\beta}(\nu_\theta^t)} \left(\exp\left(-\theta \,\mathrm{Fluct}_{\nu_\theta^t}(\varepsilon_t)\right)\right)\right| \le C\|\varepsilon_t\|_{L^\infty} \theta N \ell^d + C\|\varepsilon_t\|_{L^\infty}^2 \theta N \ell^{d-2},$$

hence the result. □

PROOF OF LEMMA 10.19. By definition (2.0.3), we compute that

$$\begin{aligned}\mathcal{E}^{V_t}(\mu_{V_t}) - \mathcal{E}^V(\mu_V) &= \frac{1}{2} \iint_{\mathbb{R}^d \times \mathbb{R}^d} \mathsf{g}(x-y) d\mu_{V_t}(x) d\mu_{V_t}(y) \\ &\quad - \frac{1}{2} \iint \mathsf{g}(x-y) d\mu(x) d\mu(y) + \int_{\mathbb{R}^d} V d(\mu_t - \mu) + t \int_{\mathbb{R}^d} \varphi d\mu_t.\end{aligned}$$

Inserting that $\mu_{V_t} = \mu_V + \frac{t}{\mathsf{c_d}}\Delta\varphi$ and that $\mathsf{g} * \Delta\varphi = -\mathsf{c_d}\varphi$, we obtain

$$\begin{aligned}\mathcal{E}^{V_t}(\mu_{V_t}) - \mathcal{E}^V(\mu_V) &= \frac{t}{\mathsf{c_d}}\iint_{\mathbb{R}^\mathsf{d}\times\mathbb{R}^\mathsf{d}} \mathsf{g}(x-y)d\mu_V(x)d\Delta\varphi(y)\\ &+\frac{t}{\mathsf{c_d}}\int_{\mathbb{R}^\mathsf{d}} V\Delta\varphi + t\int_{\mathbb{R}^\mathsf{d}}\varphi d\mu_V\\ &+\frac{1}{2}\frac{t^2}{\mathsf{c_d^2}}\iint_{\mathbb{R}^\mathsf{d}\times\mathbb{R}^\mathsf{d}} \mathsf{g}(x-y)\Delta\varphi(x)\Delta\varphi(y) + \frac{t^2}{\mathsf{c_d}}\int_{\mathbb{R}^\mathsf{d}}\varphi\Delta\varphi\\ &= \frac{t}{\mathsf{c_d}}\int_{\mathbb{R}^\mathsf{d}}(h^{\mu_V}+V)\Delta\varphi - \frac{t^2}{2\mathsf{c_d}}\int_{\mathbb{R}^\mathsf{d}}\varphi\Delta\varphi + \frac{t^2}{\mathsf{c_d}}\int_{\mathbb{R}^\mathsf{d}}\varphi\Delta\varphi + t\int_{\mathbb{R}^\mathsf{d}}\varphi d\mu_V\\ &= \frac{t}{\mathsf{c_d}}\int_{\mathbb{R}^\mathsf{d}}(\zeta + c)\Delta\varphi + \frac{t^2}{2\mathsf{c_d}}\int_{\mathbb{R}^\mathsf{d}}|\nabla\varphi|^2 + t\int_{\mathbb{R}^\mathsf{d}}\varphi d\mu_V.\end{aligned}$$

Using that φ is supported in the set $\{\zeta = 0\}$ and φ is compactly supported, we conclude that (10.2.49) holds. $\square$

PROOF OF LEMMA 10.20. Since local laws and concentration hold for $\mathbb{Q}_{N,\beta}(\mu_s, \zeta)$ in $\hat{\Sigma}$ where μ_s is bounded below, a rescaling of (8.4.6) yields, after application of Jensen's inequality, that for any φ,

$$\left|\mathbb{E}_{\mathbb{Q}_{N,\beta}(\mu_s,\zeta)}\left((\mathrm{Fluct}_{\mu_s}(\varphi))^2\right)\right| \le C\chi(\beta)N\ell^\mathsf{d}N^{\frac{\mathsf{s}}{\mathsf{d}}}\ell^\mathsf{d}\|\nabla\varphi\|_{L^\infty}^2$$

We may then apply this to $\varphi = \mathrm{div}\,\psi_s$. Thus, for any λ, since we have

$$|\,\mathrm{Fluct}_{\mu_s}(\mathrm{div}\,\psi_s)| \le \lambda\left(\mathrm{Fluct}_{\mu_s}(\mathrm{div}\,\psi_s)\right)^2 + \frac{1}{4\lambda},$$

we also have

$$\left|\mathbb{E}_{\mathbb{Q}_{N,\beta}(\mu_s,\zeta)}\left(\mathrm{Fluct}_{\mu_s}(\mathrm{div}\,\psi_s)\right)\right| \le C\lambda\chi(\beta)N^{1+\frac{\mathsf{s}}{\mathsf{d}}}\ell^{2\mathsf{d}}|\mathrm{div}\,\psi_s|_{C^1}^2 + \frac{1}{4\lambda}$$

and optimizing over λ we find

$$\left|\mathbb{E}_{\mathbb{Q}_{N,\beta}(\mu_s,\zeta)}\left(\mathrm{Fluct}_{\mu_s}(\mathrm{div}\,\psi_s)\right)\right| \le C\sqrt{\chi(\beta)}N^{\frac{1}{2}+\frac{\mathsf{s}}{2\mathsf{d}}}\ell^\mathsf{d}|\mathrm{div}\,\psi_s|_{C^1}$$

hence the result. $\square$

10.3. Nonsmooth test functions

Understanding fluctuations for less smooth test functions, in particular for indicator functions of sets, which correspond to fluctuations of number of points or charge fluctuations, is generally harder. Such fluctuations are expected to be larger than those for regular test functions. It is generally not known what the threshold of regularity to have, say, boundedness of fluctuations, is. However, in the one-dimensional logarithmic case this is proven: the variance of of the number of points in an interval is logarithmic in the size of the interval [**NV21**], see also the circular Riesz case in [**Boub**].

It was conjectured for a long time in the statistical mechanics literature [**Mar88**, **MY80**, **Leb83**, **LWL00**, **JLM93**, **Tor18**] that the two-component Coulomb gas is "hyperuniform" (in the terminology of [**Tor18**] for all $\beta > 0$, i.e., the charge fluctuations in a ball has much smaller variance than that of a Poisson point process, i.e., than the area (after zooming). This was proven in [**Leb**] in the following.

Theorem 10.23 (Hyperuniformity of the two-dimensional Coulomb gas [**Leb**]). *Let* $\mathsf{d} = 2, \mathsf{s} = 0$, *and* $\beta > 0$. *Let* $\{x_N\}_N$ *be a sequence of points in* Σ *with* $\mathrm{dist}(x_N, \partial\Sigma) \geq c > 0$ *and* $R = R_N$ *such that* $R_N \gg N^{-1/2}$ *as* $N \to \infty$. *Then, under* $\mathbb{P}_{N,\beta}$ *the variance of the number of points in* $B(x_N, R)$ *is* $o((N^{1/2}R)^2)$.

In the determinantal case $\beta = 2$ this was already known at the level of the Ginibre point process [**Shi06, OS08, FL22**] and at the level of the Coulomb gas [**Cha24, ACCL24**], with the stronger bound $O(N^{1/2}R)$ corresponding to the perimeter, which is conjectured to be the variance order for all $\beta > 0$. In fact the more precise

$$\mathrm{Var}(\#\{X_N \cap \Omega\}) = CN^{1/2}|\partial\Omega|$$

is conjectured, see [**MY80**]. The proof of Theorem 10.23 follows the overall road map introduced in [**NSV08**] in the context of zeroes of Gaussian Analytic Functions. Its implementation relies on all the tools presented so far (electric formulation, transport, screening) but also new ideas such as isotropic averaging and the analysis of "subsystems".

In dimension 2, precise conjectures, the so-called Jancovici-Lebowitz-Manificat laws, are also given for the deviations of charge discrepancies in balls in [**JLM93**]. They assert that, asymptotically, and in blown-up scale,

$$\log \mathbb{P}_{N,\beta}\left(D(x,R) \geq R^\alpha\right) \simeq -R^{\varphi(\alpha)} + o(1) \quad \text{as } R \to \infty$$

where $D(x,R)$ is the discrepancy in $B(x,R)$ and

$$\varphi(\alpha) = \begin{cases} 2\alpha - 1 & \text{if } \frac{1}{2} < \alpha \leq 1, \\ 3\alpha - 2 & \text{if } 1 \leq \alpha \leq 2, \\ 2\alpha & \text{if } \alpha \geq 2. \end{cases}$$

We saw in Section 5.3.2 that part of these conjectures were proved in [**Tho24**]. It was also proven very recently in [**NY24**] that they hold for the hierarchical Coulomb gas studied in [**Cha19, GS20**] and alluded to in Section 10.1.3.

Part 4

Microscopic behavior and local limits

CHAPTER 11

The jellium renormalized energy

In this last part, we return to the general Riesz case (2.0.1).

We work in blow-up coordinates and examine how we can derive an infinite volume version of the next order energy F, called "renormalized jellium energy." This function was first introduced in [**SS12**] in the context of Ginzburg-Landau vortices, which corresponds to the two-dimensional Coulomb case. The definition and the name were inspired by the work of [**BBH94**] on Ginzburg-Landau vortices. In [**RS15**], a definition based on smearing the charges was provided for the general Coulomb case, and in [**PS17**] it was generalized to the Riesz case $\mathsf{d}-2 \leq \mathsf{s} < \mathsf{d}$ based on the truncation procedure. We now present a new and simpler definition, relying on the idea of letting the truncation depend on the point as in Chapter 4. This definition avoids the need to introduce an extra parameter η and take a double limit as in [**RS15, PS17**].

11.1. Motivation

Let us return to the electric formulation for F, via the electric potential. In the blown-up setting, we have from (5.2.15) and (5.2.16)

$$-\mathrm{div}\,(|y|^\gamma \nabla h_N) = \mathsf{c}_{\mathsf{d},\mathsf{s}} \left(\sum_{i=1}^N \delta_{x_i'} - \mu' \delta_{\mathbb{R}^\mathsf{d}}\right) \quad \text{in } \mathbb{R}^{\mathsf{d}+\mathsf{k}} \tag{11.1.1}$$

where μ' is the blown-up reference probability measure (typically either μ_V or μ_θ) around some origin x_0, $X_N' = N^{1/\mathsf{d}}(X_N - x_0)$, $\mu'(x') = \mu(x' N^{-1/\mathsf{d}} + x_0)$. Let us assume that $\mathsf{F}(X_N', \mu') \leq CN$, which is true for typical configurations by (5.2.59). Then for most blow-up centers x_0, the number of points becomes infinite and they fill up the whole space, with their local density remaining bounded. In such a situation $\sum_{i=1}^N \delta_{x_i}$ converges to a distribution $\mathcal{C} \in \mathcal{X}(\mathbb{R}^\mathsf{d})$, where for A a Borel set of $\mathbb{R}^\mathsf{d}$ we denote by $\mathcal{X}(A)$ the set of (possibly infinite) locally finite point configurations in A or equivalently the set of nonnegative, purely atomic Radon measures on A giving an integer mass to singletons (see [**DVJ88**]). We mostly use $\mathcal{C}$ for denoting a point configuration in $\mathcal{X}(\mathbb{R}^\mathsf{d})$ and we will write $\mathcal{C}$ for $\sum_{p\in\mathcal{C}} \delta_p$ and also $\mathcal{C}(U)$ for $\int_U \sum_{p\in\mathcal{C}} \delta_p$. Note that points could appear with multiplicity. We say that $\mathcal{C}$ is simple if no point of the configuration has multiplicity > 1.

Taking $N \to \infty$ in (11.1.1) and assuming that the density μ is continuous, we expect to obtain a relation of the form

$$-\mathrm{div}\,(|y|^\gamma \nabla H) = \mathsf{c}_{\mathsf{d},\mathsf{s}} \left(\sum_{p\in\mathcal{C}} \delta_p - m \delta_{\mathbb{R}^\mathsf{d}}\right) \quad \text{in } \mathbb{R}^{\mathsf{d}+\mathsf{k}} \tag{11.1.2}$$

with $m = \mu(x_0)$, for some limit infinite point configuration $\mathcal{C}$. This corresponds to what is called in physics a *jellium*, that is an infinite configuration of discrete positive charges in a uniform neutralizing bath, with uniform density m. This is also called uniform electron gas and used as a toy model for solid matter seen as a gas of electrons in a uniform background charge representing the atomic nuclei density.

The goal of this chapter is to define a Coulomb/Riesz jellium energy for such an infinite system. It is a priori not clear how to do this because of the infinite size of the system and because of the lack of local charge neutrality of the system. Computing the sum of pair interactions between the charges in the jellium leads to a priori divergent sums. The definition we present instead replaces it with (renormalized variants of) the extensive quantity $\int |y|^\gamma |\nabla H|^2$ (see (4.1.6) and the comments following it).

The name *renormalized* energy reflects the fact that the integral of $|y|^\gamma |\nabla H|^2$ is infinite, and is computed in a renormalized way by first applying a truncation and then removing the appropriate divergent part $\mathsf{c}_{\mathsf{d},\mathsf{s}}\mathsf{g}(\eta)$, as in Chapter 4.

11.2. Definitions and first properties

As in Chapter 7 we consider more generally the electric field $E = \nabla H$ and, forgetting that E comes from a gradient, we wish to define the energy associated to it.

The reason for considering the electric field as the variable is that it will be the object on which we have compactness.

Definition 11.1. For any $m \geq 0$, we say that an electric field E is compatible with $(\mathcal{C}, m)$ if it satisfies an equation of the form

$$-\mathrm{div}\,(|y|^\gamma E) = \mathsf{c}_{\mathsf{d},\mathsf{s}}\,(\mathcal{C} - m\delta_{\mathbb{R}^\mathsf{d}}) \quad \text{in } \mathbb{R}^{\mathsf{d}+\mathsf{k}} \tag{11.2.1}$$

where $\mathcal{C} \in \mathcal{X}(\mathbb{R}^\mathsf{d})$. We denote by $\mathrm{Elec}(\mathcal{C}, m)$ the set of such vector fields, by Elec_m the union over all configurations for fixed m, and $\mathrm{Elec} = \cup_{m>0}\mathrm{Elec}_m$.

The knowledge of E and m suffices to recover the configuration, via

$$\sum_{p\in\mathcal{C}} \delta_p = -\frac{1}{\mathsf{c}_{\mathsf{d},\mathsf{s}}}\mathrm{div}\,(|y|^\gamma E) + m\delta_{\mathbb{R}^\mathsf{d}} := \mathrm{Conf}_m(E), \tag{11.2.2}$$

thus for any $E \in \mathrm{Elec}_m$, there exists a unique underlying configuration $\mathcal{C}$ such that E is compatible with $(\mathcal{C}, m)$. We denote it $\mathrm{Conf}_m(E)$.

Note that if $E \in \mathrm{Elec}(\mathcal{C}, m)$ then so does $E + X$ whenever $|y|^\gamma X$ is divergence-free. In other words, by formulating in terms of electric fields, we have not retained the information that E was a gradient. Also, when $\gamma = 0$ for instance, we can add to E any gradient of a harmonic function and still satisfy (11.2.1). In other words, when taking $N \to \infty$, we have lost the "boundary conditions at infinity" or far field.

We now follow the same steps as in Chapter 4 and Chapter 7 to define an energy in this infinite volume setting. The difficulty is that the energy will be computed over growing cubes, and care needs to be taken near the boundary of the cubes: in order to provide an energy that is bounded below, the definition needs to avoid favoring accumulation of points near the boundary of the cubes.

Given a configuration $\mathcal{C}$ of points in $\mathbb{R}^{\mathsf{d}}$ and a density $m > 0$, we first define
(11.2.3)
$$\mathsf{r}_p = \begin{cases} \frac{1}{4}\min_{q\in\mathcal{C},q\neq p}\left(|q-p|, m^{-1/\mathsf{d}}\right) & \text{if } p \text{ is a simple point of } \mathcal{C} \text{ or } \mathsf{s}<0 \\ 0 & \text{if } p \text{ appears with multiplicity and } \mathsf{s}\geq 0, \end{cases}$$

the (truncated) minimal distance to nearest neighbors in the infinite configuration. Given a closed cube $\square_R$, analogously to (7.1.25) we define the minimal distance relative to the cube by
(11.2.4)
$$\tilde{\mathsf{r}}_p = \begin{cases} \mathsf{r}_p & \text{if } p\in\square_R,\ \mathrm{dist}(p,\partial\square_R)\geq 2m^{-1/\mathsf{d}} \\ \frac{m^{-1/\mathsf{d}}}{4} & \text{if } \mathrm{dist}(p,\partial\square_R)\leq m^{-1/\mathsf{d}} \text{ or } p\notin\square_R \\ t\mathsf{r}_p+(1-t)\frac{m^{-1/\mathsf{d}}}{4} & \text{if } p\in\square_R,\ \mathrm{dist}(p,\partial\square_R)=(1+t)m^{-1/\mathsf{d}}, t\in[0,1]. \end{cases}$$

The reason for this definition is that, as in Chapter 7, it makes the energy superadditive over cells. The definition also makes $\tilde{\mathsf{r}}$ a continuous function of the point configuration.

Let $\mathcal{C}$ be a point configuration, for $m \geq 0$ let E be in $\mathrm{Elec}(\mathcal{C}, m)$. For any $\vec{\eta}$, family of positive numbers indexed by $\mathcal{C}$, we define the truncation of E with parameters $\vec{\eta}$ as

$$E_{\vec{\eta}}(x) := E(x) - \sum_{p\in\mathcal{C}} \nabla \mathsf{f}_{\eta_p}(x-p), \tag{11.2.5}$$

where f_η is as in (4.1.18). In particular we define E_{r} to be $E(x) - \sum_{p\in\mathcal{C}} \nabla\mathsf{f}_{\mathsf{r}_p}(x-p)$. Note that here there are contributions from points that lie outside $\square_R$. We observe that

$$-\mathrm{div}\left(|y|^\gamma E_{\mathsf{r}}\right) = \mathsf{c}_{\mathsf{d}}\left(\sum_{p\in\mathcal{C}} \delta_p^{(\mathsf{r}_p)} - m\delta_{\mathbb{R}^{\mathsf{d}}}\right) \quad \text{in } \mathbb{R}^{\mathsf{d}+\mathsf{k}}. \tag{11.2.6}$$

This procedure is exactly the same, at the level of the electric fields, as the truncation procedure described in Chapter 4.

11.2.1. Monotonicity property and lower bound. For $E \in \mathrm{Elec}(\mathcal{C}, m)$ and $\square_R$ the *closed* cube centered at 0 and of sidelength R, we let

(11.2.7)
$$\mathcal{F}^{\square_R}(E,m) := \begin{cases} \frac{1}{2\mathsf{c}_{\mathsf{d},\mathsf{s}}}\int_{\square_R\times\mathbb{R}^{\mathsf{k}}} |y|^\gamma|E_{\tilde{\mathsf{r}}}|^2 - \frac{1}{2}\sum_{p\in\mathcal{C}\cap\square_R} \mathsf{g}(\tilde{\mathsf{r}}_p) - m\sum_{p\in\mathcal{C}\cap\square_R}\int_{\mathbb{R}^{\mathsf{d}}} \mathsf{f}_{\tilde{\mathsf{r}}_p}(x-p)\,dx \\ \qquad \text{if all } \mathsf{r}_p>0 \text{ for } p\in\square_R \text{ or if } \mathsf{s}<0 \\ +\infty \qquad \text{otherwise.} \end{cases}$$

Let us emphasize that the contribution of points that lie on the boundary is counted in $\sum_{p\in\mathcal{C}\cap\square_R} \mathsf{g}(\tilde{\mathsf{r}}_p)$. This is meant to make $\mathcal{F}^{\square_R}$ lower semicontinuous.

Let us next present a rewriting of Lemma 7.8 in this context.

Lemma 11.2 (Monotonicity). *Let $\square_R$ be a closed cube of size R and $\tilde{\mathsf{r}}$ be the minimal distance of the configuration $\mathcal{C}$ relative to $\square_R$ as in* (11.2.4). *For any $\vec{\eta}$*

such that $\eta_p \geq \tilde{\mathsf{r}}_p$ for all p, with equality unless $B(p,\eta_p) \subset \square_R$, we have

$$\frac{1}{2} \sum_{\substack{p \neq q \in \mathcal{C} \cap \square_R \\ \mathrm{dist}(p, \partial \square_R) \geq \eta_p}} (\mathsf{g}(p-q) - \mathsf{g}(\eta_p))_+ + \frac{1}{2\mathsf{c}_{\mathsf{d},\mathsf{s}}} \int_{\square_R \times \mathbb{R}^{\mathsf{k}}} |y|^\gamma |E_{\vec{\eta}}|^2 - \frac{1}{2} \sum_{p \in \mathcal{C} \cap \square_R} \mathsf{g}(\eta_p) - m \sum_{p \in \mathcal{C} \cap \square_R} \int_{\mathbb{R}^{\mathsf{d}}} \mathsf{f}_{\eta_p}(x-p)dx \leq \mathcal{F}^{\square_R}(E,m) \tag{11.2.8}$$

with equality if $\mathcal{C}$ is simple and all the $B(p,\eta_p)$'s are disjoint.

Proposition 11.3. *For any simple configuration $\mathcal{C}$, it holds that*

$$\sum_{p \in \mathcal{C} \cap \square_R} \mathsf{g}(\tilde{\mathsf{r}}_p) \leq C_1 \left(\mathcal{F}^{\square_R}(E,m) + \left(C_0 m^{\frac{\mathsf{s}}{\mathsf{d}}} + (\frac{1}{\mathsf{d}} \log m) \mathbf{1}_{\mathsf{s}=0} \right) \mathcal{C}(\square_R) \right) \tag{11.2.9}$$

and

$$\int_{\square_R \times \mathbb{R}^{\mathsf{k}}} |y|^\gamma |E_{\tilde{\mathsf{r}}}|^2 \leq C \left(\mathcal{F}^{\square_R}(E,m) + \left(C_0 m^{\frac{\mathsf{s}}{\mathsf{d}}} + (\frac{1}{2\mathsf{d}} \log m) \mathbf{1}_{\mathsf{s}=0} \right) \mathcal{C}(\square_R) \right) \tag{11.2.10}$$

for some $C_1 > 0$ equal to 2 in the case $\mathsf{s} = 0$ and some $C_0 > 0$ depending only on d and s.

We note in particular that

$$\mathcal{F}^{\square_R}(E,m) + \left((\frac{1}{2\mathsf{d}} \log m) \mathbf{1}_{\mathsf{s}=0} + C_0 m^{\frac{\mathsf{s}}{\mathsf{d}}} \right) \mathcal{C}(\square_R) \geq 0. \tag{11.2.11}$$

Proof. It follows the proof of Proposition 4.28. Let us choose $\eta_p = \frac{1}{4} m^{-1/\mathsf{d}}$ for all p in (11.2.8), and observe that for each p such that $\mathrm{dist}(p, \partial \square_R) \geq \frac{1}{4} m^{-1/\mathsf{d}}$, by definition (11.2.4) there exists $q \neq p$ such that

$$(\mathsf{g}(p-q) - \mathsf{g}(\frac{1}{4} m^{-1/\mathsf{d}}))_+ \geq (\mathsf{g}(4\mathsf{r}_p) - \mathsf{g}(\frac{1}{4} m^{-1/\mathsf{d}}))_+.$$

Using that $\int |\mathsf{f}_{\tilde{\mathsf{r}}}| \leq C_{\mathsf{s},\mathsf{d}} \tilde{\mathsf{r}}_p^{\mathsf{d}-\mathsf{s}}$ by (4.2.3) and $\tilde{\mathsf{r}}_p \leq m^{-1/\mathsf{d}}$, we may thus write that

$$\frac{1}{2} \sum_{p \in \mathcal{C} \cap \square_R, \mathrm{dist}(p, \partial \square_R) \geq \frac{1}{4} m^{-1/\mathsf{d}}} (\mathsf{g}(4\mathsf{r}_p) - \mathsf{g}(\frac{1}{4} m^{-1/\mathsf{d}}))_+ \leq \mathcal{F}^{\square_R}(E,m) - \frac{1}{2\mathsf{c}_{\mathsf{d},\mathsf{s}}} \int_{\square_R \times \mathbb{R}^{\mathsf{k}}} |y|^\gamma |E_{\vec{\eta}}|^2 + \frac{1}{2} \sum_{p \in \mathcal{C} \cap \square_R} \mathsf{g}(\frac{1}{4} m^{-1/\mathsf{d}}) + C m^{\frac{\mathsf{s}}{\mathsf{d}}} \mathcal{C}(\square_R). \tag{11.2.12}$$

After rearranging terms, and adding back $\mathsf{g}(\tilde{\mathsf{r}}_p)$ for the p's such that $\mathrm{dist}(p, \partial \square_R) \leq \frac{1}{4} m^{-1/\mathsf{d}}$, we obtain (11.2.9). Inserting into the definition of $\mathcal{F}^{\square_R}(E,m)$, we deduce (11.2.10). □

11.2.2. The jellium energy. In the sequel, $\square_R$ still denotes the *closed* cube $[-R/2, R/2]^{\mathsf{d}}$.

Definition 11.4 (Jellium renormalized energy for electric fields)**.** Let C_0 be the constant of (11.2.9) and (11.2.10). Let $E \in \mathrm{Elec}_m$, i.e., satisfying (11.2.1). The Coulomb/Riesz renormalized energy of E with background m is defined by

$$\mathcal{W}(E, m) := \limsup_{R\to\infty} \frac{1}{R^{\mathsf{d}}} \left(\mathcal{F}^{\square_R}(E, m) + \left((\frac{1}{2\mathsf{d}} \log m)\mathbf{1}_{\mathsf{s}=0} + C_0 m^{\frac{\mathsf{s}}{\mathsf{d}}}\right) (\mathcal{C}(\square_R) - mR^{\mathsf{d}})\right) \tag{11.2.13}$$

where $\mathcal{F}^{\square_R}$ is as in (11.2.7).

Note that the terms added to $\mathcal{F}^{\square_R}(E, m)$ serve to control the number of points and use (11.2.11) but will eventually bring no contribution to the quantity as soon as we know that $\mathcal{C}(\square_R) \sim mR^{\mathsf{d}}$, which we will show in Lemma 11.9 below.

Remark 11.5. We note that this renormalized energy controls both the electric field and the number of points. Indeed, in view of (11.2.9) and (11.2.10), there exists a constant $C > 0$ such that

$$\limsup_{R\to\infty} \frac{1}{R^{\mathsf{d}}} \left(\int_{\square_R \times \mathbb{R}^{\mathsf{k}}} |y|^\gamma |E_{\tilde{\mathsf{r}}}|^2 + \mathcal{C}(\square_R)\right) < C\left(\mathcal{W}(E, m) + m\left(C_0 m^{\frac{\mathsf{s}}{\mathsf{d}}} + (\frac{1}{2\mathsf{d}} \log m)\mathbf{1}_{\mathsf{s}=0}\right)\right). \tag{11.2.14}$$

We also note that $\mathcal{W}$ is insensitive to a compact perturbation of E.

Definition 11.6 (Jellium renormalized energy for infinite point configurations)**.** Let $\mathcal{C}$ be a point configuration. We define the renormalized energy of $\mathcal{C}$ with background $m \geq 0$ as

$$\mathbb{W}(\mathcal{C}, m) := \inf\{\mathcal{W}(E, m),\ E \in \mathrm{Elec}(\mathcal{C}, m),\ E \text{ is a gradient}\} \tag{11.2.15}$$

with the convention $\inf(\varnothing) = +\infty$.

In view of (11.2.11) and the definition, we have

Remark 11.7. $\mathcal{W}(\cdot, m)$ and $\mathbb{W}(\cdot, m)$ are bounded below by a constant depending only on d, s and m.

Lemma 11.8. *Let* $\mathcal{C}$ *be fixed. Two elements of* $\mathrm{Elec}(\mathcal{C}, m)$ *with finite energy which are gradients differ by a constant vector field. In particular the* inf *in* (11.2.15) *is a* min.

PROOF. The proof is borrowed from [**LS17**, Lemma 2.3] and relies on elliptic regularity theory. If E_1, E_2 are two gradient vector fields in $\mathrm{Elec}(\mathcal{C}, m)$ then $E_1 - E_2 = \nabla u$ with u solving

$$-\mathrm{div}\left(|y|^\gamma \nabla u\right) = 0 \qquad \text{in } \mathbb{R}^{\mathsf{d}+\mathsf{k}}. \tag{11.2.16}$$

Moreover, the finiteness of $\mathcal{W}(E_1, m)$ and $\mathcal{W}(E_2, m)$ directly imply that

$$\limsup_{R\to\infty} \frac{1}{R^{\mathsf{d}}} \int_{\square_R \times \mathbb{R}^{\mathsf{k}}} |y|^\gamma |\nabla u|^2 < \infty. \tag{11.2.17}$$

The equation (11.2.16) is divergence form equation with weight $|y|^\gamma$ that belongs to the so-called A_2-Muckenhoupt class, which makes it amenable to elliptic regularity theory. The result of [**FKS82**, Theorem 2.3.13] gives that there exists $\lambda > 0$ such that for any $X \in \mathbb{R}^{\mathsf{d}+\mathsf{k}}$

$$\operatorname{osc}(\nabla_{\mathbb{R}^\mathsf{d}} u, B(X,r)) \le C \left(\frac{1}{\int_{B(X,R)} |y|^\gamma} \int_{B(X,R)} |y|^\gamma |\nabla_{\mathbb{R}^\mathsf{d}} u|^2 \right)^{\frac{1}{2}} \left(\frac{r}{R}\right)^\lambda$$

where $\operatorname{osc}(u, B(X,r)) = \max_{B(X,r)} u - \min_{B(X,r)} u$. In the Coulomb case for which $\mathsf{k} = 0$ and $\gamma = 0$, this is just standard regularity estimates for harmonic functions. Inserting (11.2.17) and letting $R \to +\infty$, we deduce that $\operatorname{osc}(\nabla_{\mathbb{R}^\mathsf{d}} u, B(X,r)) = 0$ which means that $\nabla_{\mathbb{R}^\mathsf{d}} u$ is constant on every compact set of $\mathbb{R}^{\mathsf{d}+\mathsf{k}}$. In the case $\mathsf{k} = 0$ this proves that u is affine, and that E_1 and E_2 differ by a constant vector. In the case $\mathsf{k} = 1$ this implies that u is an affine function of the $\mathbb{R}^\mathsf{d}$ variables for each y. Writing $u(x,y) = a(y) \cdot x + b(y)$ and inserting back into (11.2.16), combining with the fact that $\int_\mathbb{R} |y|^\gamma |\partial_y u|^2 dy$ is convergent, finally yields that u is constant, hence the result. □

Let us now record some natural scaling relations. If $E \in \mathrm{Elec}_m$, we define $\sigma_m E$ by

$$\sigma_m E := m^{-\frac{\mathsf{s}+1}{\mathsf{d}}} E\left(\frac{\cdot}{m^{1/\mathsf{d}}}\right). \tag{11.2.18}$$

We have $\sigma_m E \in \mathrm{Elec}_1$ and

$$\begin{cases} \mathcal{W}(E,m) = m^{1+\frac{\mathsf{s}}{\mathsf{d}}} \mathcal{W}(\sigma_m E, 1) & \text{if } \mathsf{s} \neq 0 \\ \mathcal{W}(E,m) = m\mathcal{W}(\sigma_m E, 1) - \frac{1}{2\mathsf{d}} m \log m & \text{if } \mathsf{s} = 0 \end{cases} \tag{11.2.19}$$

and in the same way

$$\begin{cases} \mathbb{W}(\mathcal{C},m) = m^{1+\frac{\mathsf{s}}{\mathsf{d}}} \mathbb{W}(\sigma_m \mathcal{C}, 1) & \text{if } \mathsf{s} \neq 0 \\ \mathbb{W}(\mathcal{C},m) = m\mathbb{W}(\sigma_m \mathcal{C}, 1) - \frac{1}{2\mathsf{d}} m \log m & \text{if } \mathsf{s} = 0. \end{cases} \tag{11.2.20}$$

11.2.3. Discrepancy and neutrality. The next property expresses that if the energy $\mathcal{W}(E,m)$ is finite then we control well the discrepancy in large boxes of the underlying system $(\mathcal{C}, m)$, implying that the system is on average neutral and the density of points on large boxes converges to m.

Lemma 11.9 (Asymptotic neutrality for configurations with finite energy). *Assume $\mathcal{W}(E,m) < \infty$ and let $\mathcal{C}$ be associated via (11.2.1). There exists $\varepsilon > 0$ depending only on d and s, and $C > 0$ depending only on d, s and m, such that*

$$|\mathcal{C}(\square_R) - mR^\mathsf{d}|^2 \le CR^{2\mathsf{d}-\varepsilon}(1 + \mathcal{W}(E,m)). \tag{11.2.21}$$

In particular,

$$\lim_{R\to+\infty} \frac{\mathcal{C}(\square_R)}{R^\mathsf{d}} = m. \tag{11.2.22}$$

Proof. By scaling, it suffices to prove it for $m = 1$. We argue exactly as in Lemma 4.25 (with μ instead of $N\mu$). Recalling that $\gamma = \mathsf{s} + 2 - \mathsf{k} - \mathsf{d}$ from (2.2.5), we have $\gamma < 1$, so we may find $\varepsilon > 0$ such that $\gamma + 2\varepsilon < 1$. Taking $\delta = R^{\gamma+\varepsilon}$ in the proof of Lemma 4.25, we obtain that, letting $D(\square_R) = \mathcal{C}(\square_R) - |\square_R|$,

$$|D(\square_R)|^2 \le CR^{2(\mathsf{d}-1+\gamma+\varepsilon)} + \frac{R^{\mathsf{d}+\gamma}}{R^{\gamma+\varepsilon}} \int_{(\square_{R+1}\setminus\square_{R-1})\times\mathbb{R}^\mathsf{k}} |y|^\gamma |E_{\tilde{\mathsf{r}}}|^2. \tag{11.2.23}$$

In view of (11.2.14) we have

$$\int_{\square_{R+1}\times\mathbb{R}^{\mathsf{k}}} |y|^\gamma |E_{\tilde{\mathsf{r}}}|^2 \le CR^{\mathsf{d}}(\mathcal{W}(E,1)+1).$$

Inserting this and $\gamma+\varepsilon<1-\varepsilon$ into (11.2.23), we obtain

$$|D(\square_R)|^2 \le CR^{2\mathsf{d}-\varepsilon} + CR^{2\mathsf{d}-\varepsilon}\mathcal{W}(E,1).$$

Dividing by R^{d} and letting $R\to\infty$, we get (11.2.22) as well. □

In view of the definition (11.2.13), we thus have

Corollary 11.10. *If $\mathcal{W}(E,m)<\infty$ then*

$$\mathcal{W}(E,m) = \limsup_{R\to\infty} \frac{1}{R^{\mathsf{d}}} \mathcal{F}^{\square_R}(E,m). \tag{11.2.24}$$

The work [**GS21**] presents criteria for $\mathbb{W}$ to be finite in the case $\mathsf{s}=0$, $\mathsf{d}=2$. For instance, discrepancies in balls of radius R that grow less than $R^{1-\varepsilon}$ suffice. Recently, [**HL24**] pushed the question further by examining, still in two dimensions, the link between hyperuniformity, finite $\mathbb{W}$ and Wasserstein distance to the Lebesgue measure, for stationary point processes. In particular they show that finite $\mathbb{W}$ is equivalent to a certain quantitative hyperuniformity property.

11.3. The case of periodic configurations

For periodic configurations, $\mathbb{W}$ can be computed and expressed as a sum of pairwise periodized Coulomb or Riesz interactions between the points. By periodic configuration, we mean a configuration on the fundamental cell of a torus, repeated periodically, which can be viewed as a configuration of N points on a torus (cf. Figure 11.3.1).

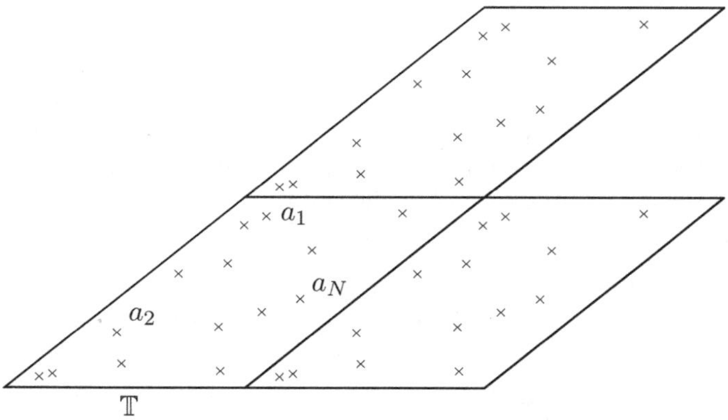

FIGURE 11.3.1. Periodic configurations

Proposition 11.11 (Explicit form of the energy in the periodic case)**.** *Let $a_1,\dots,a_N$ be N points in a torus $\mathbb{T}$ of volume N in $\mathbb{R}^{\mathsf{d}}$. Let C denote the infinite configuration on $\mathbb{R}^{\mathsf{d}}$ obtained by reproducing periodically $a_1,\dots,a_N$. If $\mathsf{s}\ge 0$ and there is* *a multiple point, then for any* $E \in \mathrm{Elec}(\mathcal{C}, 1)$, *we have* $\mathcal{W}(E) = +\infty$. *Otherwise, letting* H *be solution to*

$$-\mathrm{div}\,(|y|^\gamma \nabla H) = \mathsf{c}_{\mathsf{d},\mathsf{s}}\Big(\sum_{i=1}^N \delta_{a_i} - \delta_{\mathbb{R}^\mathsf{d}}\Big) \quad \text{in } \mathbb{T}\times\mathbb{R}^\mathsf{k}, \qquad \int_{\mathbb{T}} H = 0, \tag{11.3.1}$$

then seen as a periodic function in $\mathbb{R}^\mathsf{d}\times\mathbb{R}^\mathsf{k}$, *any periodic* $E \in \mathrm{Elec}(\mathcal{C}, 1)$ *satisfies*

$$\mathcal{W}(E, 1) \ge \mathcal{W}(\nabla H, 1) = \frac{\mathsf{c}_{\mathsf{d},\mathsf{s}}}{2N}\sum_{i\neq j} G(a_i - a_j) + \frac{\mathsf{c}_{\mathsf{d},\mathsf{s}}}{2}\lim_{x\to 0}\Big(G - \frac{\mathsf{g}}{\mathsf{c}_{\mathsf{d},\mathsf{s}}}\Big) \tag{11.3.2}$$

where G, *the fractional Green function of the torus, solves*

$$(-\Delta)^{\frac{\mathsf{d}-\mathsf{s}}{2}} G = \delta_0 - \frac{1}{|\mathbb{T}|} \quad \text{over } \mathbb{T}, \qquad \int_{\mathbb{T}} G = 0. \tag{11.3.3}$$

Note the similarity with "doubly periodic" Coulomb gases such as studied in [**For06**].

Remark 11.12.

(1) The limit appearing in this definition is called the Madelung constant of the torus.

(2) In the case $\mathsf{d} = 1$ and $\mathsf{s} = 0$, an explicit formula for G is available (see for instance [**BS13**]):

$$G(x) = -\frac{1}{2\pi}\log\left|2\sin\frac{\pi x}{N}\right|.$$

For other values of s in $\mathsf{d} = 1$, we have (see [**PS17**])

$$G(x) = 2\frac{N^{2\alpha-1}}{(2\pi)^{2\alpha}\Gamma(2\alpha)}\int_0^\infty \frac{t^{2\alpha-1}(e^t\cos(\frac{2\pi}{N}x) - 1)}{1 - 2e^t\cos(\frac{2\pi}{N}x) + e^{2t}}dt.$$

(3) One can expand G in Fourier series, then Eisenstein series appear, in particular there is a direct expression of $\mathbb{W}$ for a lattice configuration ($N = 1$) in terms of the Eisenstein series of the lattice, see the proof of Theorem 11.18 below.

One can also use these formulae valid in the periodic situation and extend them to nonperiodic configurations in order to provide alternative definitions of a Coulomb/Riesz jellium energy, and compare them with the definition in $\mathbb{W}$. This, and the comparison, is done in the context of random point processes [**Leb16**]. In [**BS13**] this was used to compute $\mathbb{W}$ and its variants for some explicit point processes and use it as a measure of their rigidity.

Proof. It suffices to consider the case of simple points. The first inequality in (11.3.2) follows from the projection lemma, Lemma 7.1, adapted to the periodic setting: set $X = E - \nabla H$ and observe that X satisfies $\mathrm{div}\,(|y|^\gamma X) = 0$. Then write

$$\int_{\mathbb{T}\times\mathbb{R}^\mathsf{k}} |y|^\gamma |E_\mathsf{r}|^2 = \int_{\mathbb{T}\times\mathbb{R}^\mathsf{k}} |y|^\gamma |\nabla H_\mathsf{r} + X|^2 = \int_{\mathbb{T}\times\mathbb{R}^\mathsf{k}} |y|^\gamma |\nabla H_\mathsf{r}|^2 + |y|^\gamma |X|^2 + 2|y|^\gamma \nabla H_\mathsf{r}\cdot X$$

and the last term in the right-hand side vanishes after integration by parts since $\mathrm{div}\,(|y|^\gamma X) = 0$. This proves that $\mathcal{W}(E, 1) \ge \mathcal{W}(\nabla H, 1)$.

Let us now turn to the proof of the equality in (11.3.2). Let $\eta < \min_{i=1}^N \mathsf{r}_i$ and $\vec{\eta}$ the vector in $\mathbb{R}^n$ of entries all equal to η. Let H be the solution of (11.3.1). As in Sections 2.2.1 and 4.1.3, the function G solving (11.3.3) also satisfies

$$-\mathrm{div}\,(|y|^\gamma \nabla G) = \mathsf{c}_{\mathsf{d},\mathsf{s}}\left(\delta_0 - \frac{1}{N}\delta_{\mathbb{R}^\mathsf{d}}\right) \quad \text{in } \mathbb{T}\times\mathbb{R}^\mathsf{k}. \tag{11.3.4}$$

We then deduce that

$$H(x) = \mathsf{c_{d,s}} \sum_{i=1}^{N} G(x - a_i)$$

with G the Green function defined in the proposition, and thus

$$H_{\vec{\eta}}(x) = \mathsf{c_{d,s}} \sum_{i=1}^{N} G(x - a_i) - \sum_{i=1}^{N} \mathsf{f}_{\eta_i}(x - a_i).$$

Also $G = \frac{1}{\mathsf{c_{d,s}}}\mathsf{g} + \phi$ with ϕ a continuous function. Since the $B(a_i, \mathsf{r}_i)$ are disjoint, by definition (11.2.13) and the equality case in (11.2.8), we have

(11.3.5)

$$\mathcal{W}(\nabla H, 1) = \frac{1}{|\mathbb{T}|}\left(\frac{1}{2\mathsf{c_{d,s}}}\int_{\mathbb{T}\times\mathbb{R}^{\mathsf{k}}} |y|^\gamma |\nabla H_{\mathsf{r}}|^2 - \frac{1}{2}\sum_{i=1}^{N}\mathsf{g}(\mathsf{r}_i) - \sum_{i=1}^{N}\int \mathsf{f}_{\mathsf{r}_i}(x - a_i)dx\right)$$
$$= \frac{1}{|\mathbb{T}|}\left(\frac{1}{2\mathsf{c_{d,s}}}\int_{\mathbb{T}\times\mathbb{R}^{\mathsf{k}}} |y|^\gamma |\nabla H_{\vec{\eta}}|^2 - \frac{1}{2}\sum_{i=1}^{N}\mathsf{g}(\eta) - \sum_{i=1}^{N}\int \mathsf{f}_{\eta}(x - a_i)dx\right).$$

Inserting the above and using Green's formula, we have on the other hand

$$(11.3.6)\quad \int_{\mathbb{T}\times\mathbb{R}^{\mathsf{k}}} |y|^\gamma |\nabla H_{\vec{\eta}}|^2 = -\int_{\mathbb{T}\times\mathbb{R}^{\mathsf{k}}} H_{\vec{\eta}} \operatorname{div}(|y|^\gamma \nabla H_{\vec{\eta}})$$
$$= \mathsf{c_{d,s}} \int_{\mathbb{T}\times\mathbb{R}^{\mathsf{k}}} \left(\mathsf{c_{d,s}}\sum_{i=1}^{N} G(x - a_i) - \sum_{i=1}^{N}\mathsf{f}_\eta(x - a_i)\right)\left(\sum_{j=1}^{N}\delta_{a_j}^{(\eta)} - \delta_{\mathbb{R}^{\mathsf{d}}}\right)(x)$$
$$= \mathsf{c_{d,s}} N(\mathsf{g}(\eta) + \mathsf{c_{d,s}}\phi(0)) + \mathsf{c_{d,s}^2}\sum_{i\neq j} G(a_i - a_j) + \mathsf{c_{d,s}}\int_{B(0,\eta)} \mathsf{f}_\eta + o_\eta(1)$$

where we used the disjointness of the balls, that f_η vanishes on $\partial B(0, \eta)$ where $\delta_0^{(\eta)}$ is supported, and that $\int_{\mathbb{T}} G = 0$. Letting $\eta \to 0$, in view of (4.1.27), we obtain (11.3.2). □

11.4. Existence of minimizers

In Corollary 12.4 in the next chapter, we will prove that the minima of $\mathcal{W}(\cdot, 1)$ and $\mathbb{W}(\cdot, 1)$ are achieved and are equal. We now show the following, which relies on the screening procedure.

Proposition 11.13. *For any $m > 0$, any s, d with $\mathsf{s} \in [\mathsf{d} - 2, \mathsf{d})$, $\inf \mathcal{W}(\cdot, m)$ is finite. Moreover, F being as in (7.1.22), we have*

$$(11.4.1)\qquad \lim_{R\to\infty}\left(\frac{1}{R^{\mathsf{d}}}\inf \mathsf{F}(\cdot, m, \square_R)\right) \le \min \mathcal{W}(\cdot, m)$$

and

$$(11.4.2)\qquad \min \mathcal{W}(\cdot, m) = \lim_{R\to\infty}\ \min_{E \text{ is } (R\mathbb{Z})^{\mathsf{d}}\text{-periodic}} \mathcal{W}(E, m),$$

with the limits taken along sequences of R's such that $R^{\mathsf{d}} m$ are integers.

PROOF. By scaling we reduce to the case $m = 1$. We note that the periodic case of Proposition 11.11 implies that $\inf \mathcal{W}(\cdot, 1) < +\infty$.

Let $(E_k)_{k\in\mathbb{N}}$ be a sequence such that

$$\mathcal{W}(E_k, 1) \le \inf \mathcal{W}(\cdot, 1) + \frac{1}{k}$$

and let $(\mathcal{C}_k)_k$ be the associated configurations. In view of (11.2.13) and Lemma 11.9 there must thus exist $C > 0$ such that for all $R > 1$

$$|\mathcal{C}_k(\square_R)| \le CR^{\mathsf{d}} \tag{11.4.3}$$

and

$$\mathcal{F}^{\square_R}(E_k, 1) \le \left(\inf \mathcal{W}(\cdot, 1) + \frac{1}{k} + o_R(1)\right) R^{\mathsf{d}}. \tag{11.4.4}$$

In the Coulomb case, we may now argue exactly as in the proof of Theorem 8.2 and apply the screening result of Proposition 7.19 in any cube $\square_R$ of quantized volume. It provides us with a screened configuration X_R such that

$$\mathsf{F}(X_R, 1, \square_R) \le \mathcal{F}^{\square_R}(E_k, 1) + o(R^{\mathsf{d}}) \le \left(\inf \mathcal{W}(\cdot, 1) + \frac{1}{k} + o_R(1)\right) R^{\mathsf{d}} \tag{11.4.5}$$

where F is defined in (7.1.22) in terms of the solution to the Neumann potential (7.1.1) which we denote by $u_{k,R}$. This proves (11.4.1).

The configuration X_R can then be periodized into an infinite periodic configuration $\mathcal{C}_{k,R}$ on $\mathbb{R}^{\mathsf{d}}$, and $\nabla u_{k,R}$ periodized into a periodic vector field $E_{k,R} \in \mathrm{Elec}(\mathcal{C}_{k,R}, 1)$ over $\mathbb{R}^{\mathsf{d}}$. By periodicity and definition (11.2.13) we have

$$\mathcal{W}(E_{k,R}, 1) = \mathsf{F}(X_R, 1, \square_R).$$

In view of (11.4.5), letting $k \to \infty$ and $R \to \infty$, we deduce the inequality

$$\lim_{R\to\infty} \min_{E \text{ is } (R\mathbb{Z})^{\mathsf{d}}\text{-periodic}} \mathcal{W}(E, 1) \le \inf \mathcal{W}(\cdot, 1)$$

while the converse inequality is trivial.

In the Riesz cases, we may use instead the screening result of [**PS17**] or [**PS**]. ☐

Remark 11.14. In the definition (11.2.13) we have used cubes, where other shapes could be used (for instance balls, ellipsoids...). One may show, using screening and the method of the following chapter, that while the value of $\mathcal{W}(E, m)$ may depend on the shape used, the minimum $\min \mathcal{W}$ does not, as long as the shapes remain nondegenerate.

11.5. Minimization of $\mathbb{W}$ and the crystallization conjecture

We have seen that the minima of $\mathbb{W}$ can be achieved as limits of the minima over periodic configurations (with respect to larger and larger tori). On the other hand, Proposition 11.11 provides a more explicit expression for periodic configurations. In the one-dimensional case only, we know how to use this expression (11.3.2) to identify the minimum over periodic configurations: a convexity argument, for which we refer to [**SS15a**, Prop. 2.3] and [**Leb15**], shows that the minimum is achieved when the points are equally spaced, in other words for the lattice or crystalline distribution $\mathbb{Z}$. There is no uniqueness of minimizers since as we have seen, $\mathbb{W}$ is unchanged under a compact perturbation of the configuration, however a uniqueness result can be proven when viewing $\mathbb{W}$ as a function of stationary point processes, cf. [**Leb15, EHL21**].

In higher dimension, determining the value of $\min \mathbb{W}$ is a much more delicate question. The solution is known only in dimension 8 and 24, thanks to the resolution of the Cohn-Kumar conjecture by Cohn-Kumar-Miller-Radchenko-Viazovska [**CKM$^+$22**].

To describe this conjecture, let us define a point configuration $\mathcal{C}$ to be a nonempty, discrete, closed subset of Euclidean space $\mathbb{R}^\mathsf{d}$. For $p : \mathbb{R}_+ \to \mathbb{R}$ any function, let the (lower) p-energy of $\mathcal{C}$ be

$$E_p(\mathcal{C}) := \liminf_{R\to\infty} \frac{1}{|\mathcal{C} \cap B_R|} \sum_{x,y\in\mathcal{C}\cap B_R, x\neq y} p(|x-y|) \tag{11.5.1}$$

where B_R is the ball of center 0 and radius R in $\mathbb{R}^\mathsf{d}$. (Note this E_p has nothing to do with the electric field E encountered previously). Finding the minimum of E_p belongs to the wider class of (in general difficult) crystallization problems, see [**BL15**] for a review. Such questions are fundamental in order to understand the crystalline structure of matter. They also arise in the arrangement of Fekete points [**SK97**]. One should immediately stress that there are very few positive results in that direction in the literature, in fact it is very rare to have a proof that the solution to any minimization problem is periodic. Some exceptions include the two-dimensional sphere packing problem, for which Radin [**Rad81**] showed that the minimizer is the triangular lattice, and an extension of this by Theil [**The06**] for a class of very short range Lennard-Jones potentials. The techniques used there do not apply to Coulomb interactions, which are much longer range. Another positive result from [**BPT14**] shows the optimality of the triangular lattice for the question of minimizing the (normalized) $\|\sum_p \delta_p - 1\|_{\mathrm{Lip}^*}$ over point configurations.

Let us now describe the Cohn-Kumar conjecture.

Definition 11.15. We say that p is a completely monotone function of the squared distance when $p(r) = g(r^2)$ with g a smooth completely monotone function on $\mathbb{R}_+$, i.e., satisfying $(-1)^k g^{(k)}(r) \geq 0$ for all $r \geq 0$ for every integer $k \geq 0$.

This includes for instance Gaussians.

Let Λ_0 denote the triangular lattice A_2 in dimension 2, the E_8 lattice in dimension 8 and the Leech lattice in dimension 24, dilated so that their fundamental cell has volume 1. We do not give here the precise definitions of the E_8 and Leech lattices, but suffice it to say that these are Bravais lattices, which means that they are of the form $\sum_{i=1}^{\mathsf{d}} u_i\mathbb{Z}$ for some vectors $u_i \in \mathbb{R}^\mathsf{d}$, and that the triangular lattice in dimension 2 is the one spanned by two vectors of same norm forming an angle $\pi/3$. The triangular lattice (also often called the hexagonal lattice) is what is called the Abrikosov lattice in the context of superconductivity, cf. Chapter 1.

Conjecture 11.16 (Cohn-Kumar [**CK07**]). *In dimension* $\mathsf{d} = 2, 8$, *resp.* 24, *the lattice* Λ_0 *is universally minimizing in the sense that it minimizes* E_p *among all possible point configurations of density* 1 *for all* p*'s that are completely monotone functions of the squared distance.*

The conjecture is not true without the complete monotonicity assumption, as shown for instance in [**BP19**]. It was proven in dimension 8 and 24 in [**CKM$^+$22**] but remains open in dimension 2. Very recently, [**Leb25**] showed the remarkable fact that the triangular lattice is a local minimizer of E_p in dimension 2 (as long as p is smooth).

In [**PS20**] it was shown that the Cohn-Kumar conjecture implies the result on the minimum of $\mathbb{W}$.

Theorem 11.17 ([**PS20**]). *If the Cohn-Kumar conjecture holds, then Λ_0 achieves the minimum of $\mathbb{W}$, for $\mathsf{d} = 2, 8, 24$.*

This implication relies on the reduction to the periodic case for large tori of Proposition 11.13 and the following representation formula: for Λ a given lattice of covolume 1 and n an integer

$$G_{n\Lambda}(x) = \frac{1}{\Gamma(\frac{\mathsf{d}-\mathsf{s}}{2})} \int_0^\infty \left(\sum_{v \in n\Lambda} \Psi_t(x-v) - \frac{1}{n} \right) t^{\frac{\mathsf{d}-\mathsf{s}}{2}-1}\, dt \tag{11.5.2}$$

where $\Psi_t(x)$ is the standard heat kernel $(4\pi t)^{-\frac{\mathsf{d}}{2}} e^{-|x|^2/(4t)}$, and $G_{n\Lambda}$ is the periodic fractional Green function on the torus of volume n. This way, $G_{n\Lambda}$ is written as a superposition of heat kernels, which are completely monotonic functions of the square distance, to which the Cohn-Kumar conjecture applies.

In other dimensions, there does not always exist a universally minimizing lattice, i.e., minimizers are expected to depend on the interaction. This is for instance the case for Riesz interactions in three dimensions, see [**BL15**] and [**Lew22**]. Identifying the minimizer remains an open question, even though it would suffice in principle to be able to minimize in the class of periodic configurations with larger and larger period, using the formula (11.3.2). Common wisdom is that in sufficiently large dimension, minimizers should not be lattices. Indeed, counterexamples made with superpositions of two different lattices can be built, see [**CS99**].

In [**RNS15**], we showed the equivalence between several ways of phrasing the minimization of $\mathbb{W}$ in dimension 2 over a finite size box: minimization with prescribed boundary trace and minimization among periodic configurations. In all cases, we were able to prove, in the spirit of [**ACO09**], that the energy density and the points were uniformly distributed at any scale $\gg 1$, in good agreement with (but of course much weaker than!) the conjecture of periodicity of the minimizers. This was generalized to higher dimensional Coulomb cases in [**PRN18**]. The Riesz case is harder and a conditional result is given in the same paper but can now be obtained unconditionally via the methods of [**PS**], see comments therein.

One question that is answered is that of the minimization over the restricted class of pure lattice configurations, in dimension $\mathsf{d} = 2$ only, this means over vector fields which are gradient of functions that are periodic with respect to a lattice $\mathbb{Z}\vec{u} + \mathbb{Z}\vec{v}$ with $\det(\vec{u}, \vec{v}) = 1$, corresponding to configurations of points that can be identified with $\mathbb{Z}\vec{u} + \mathbb{Z}\vec{v}$.

Theorem 11.18 (The triangular lattice is the minimizer over lattices in 2D). *The minimum of $\mathcal{W}$ over this class of vector fields is achieved uniquely by the one corresponding to the triangular (Abrikosov) lattice A_2.*

When restricted to lattices, $\mathbb{W}$ corresponds to a "height" of the associated flat torus in Arakelov geometry. With that point of view, the result was already known since [**OPS88**]. The same result was also obtained in [**CO07**] for a similar energy.

We next give a sketch of the proof of Theorem 11.18 from [**SS12**], which is not very difficult thanks to the fact that it reduces (as [**OPS88**] does) to the same question for a certain modular function, which was solved by number theorists in the 50's and 60's.

PROOF OF THEOREM 11.18. Proposition 11.11, more specifically (11.3.2), provides an explicit formula for the renormalized energy of such periodic configurations. Denoting by H_Λ the periodic solution associated with (11.3.1), and expressing G as a Fourier series, we find that

$$\mathcal{W}(\nabla H_\Lambda, 1) = \lim_{x\to 0}\left(\sum_{\vec{k}\in\Lambda^*\setminus\{0\}} \frac{e^{2i\pi\vec{k}\cdot\vec{x}}}{4\pi^2|\vec{k}|^2} + 2\pi\log x\right). \tag{11.5.3}$$

By using either the first Kronecker limit formula (cf. [**Lan87**]) or a direct computation, one shows that in fact

$$\mathcal{W}(\nabla H_\Lambda, 1) = C_1 + C_2 \lim_{x\to 0, x>0}\left(\sum_{\vec{k}\in\Lambda^*\setminus\{0\}} \frac{1}{|\vec{k}|^{2+x}} - \int_{\mathbb{R}^2}\frac{dy}{1+|y|^{2+x}}\right), \tag{11.5.4}$$

where C_1 and $C_2 > 0$ are constants. The series $\sum_{\vec{k}\in\Lambda^*\setminus\{0\}} \frac{1}{|\vec{k}|^{2+x}}$ that appears is now the Epstein Zeta function of the dual lattice Λ^*. The first Kronecker limit formula allows to pass from one modular function, the Eisenstein series, to another, the Epstein Zeta function. Note that both formulas (11.5.3) and (11.5.4), when $x \to 0$, correspond to two different ways of regularizing the divergent series $\sum_{p\in\Lambda^*\setminus\{0\}} \frac{1}{|p|^2}$, and they are in fact explicitly related.

The question of minimizing $\mathcal{W}$ among lattices is then reduced to minimizing the Epstein Zeta function

$$\Lambda \mapsto \zeta_\Lambda(x) := \sum_{\vec{k}\in\Lambda\setminus\{0\}} \frac{1}{|\vec{k}|^{2+x}}$$

as $x \to 0$. But results from [**Cas59, Ran53, Enn64a, Enn64b, Dia64, Mon88**] assert that

$$\zeta_\Lambda(x) \geq \zeta_{A_2}(x),\ \forall x > 0 \tag{11.5.5}$$

and the equality holds if and only if $\Lambda = A_2$ (the triangular lattice). Because that lattice is self-dual, it follows that it is the unique minimizer. □

As we have seen, the Cohn-Kumar conjecture implies that this triangular lattice does achieve the global minimum of $\mathbb{W}$. This was also conjectured in [**SS12**] on the basis that the triangular lattice is observed in superconductors (the Abrikosov lattice in physics language) combined with the fact that $\mathbb{W}$ can be derived as the limiting minimization problem of the Ginzburg-Landau functional. It was also proven in [**BS18**] that this conjecture is equivalent to a conjecture derived by analytic continuation by Brauchart-Hardin-Saff [**BHS14**] on the next order term in the asymptotic expansion of the minimal logarithmic energy on the sphere (an important problem in approximation theory, also related to Smale's 7th problem for the twenty-first century), which is obtained by formal analytic continuation, hence by very different arguments. Moreover, the triangular lattice for the two-dimensional Coulomb gas appears in [**AJ81**]. All this reinforces the plausibility of this conjecture in dimension 2.

In dimension $\mathsf{d} \geq 3$ the computation of the renormalized energy restricted to the class of lattices holds but the meaning of (11.5.4) is not clear. The minimization of the Epstein Zeta function over lattices is then an open number theoretic question (except in dimensions 8 and 24). In dimension 3, both the FCC (face centered cubic) and BCC (body centered cubic) lattices (cf. Figure 11.5.2) are expected to play the role of the triangular lattice. It is only conjectured that FCC is a local minimizer in [**SS06**], and so by duality BCC can be expected to minimize $\mathbb{W}$ in the Coulomb case. As mentioned above, the Riesz case is different, as described in [**BL15**] BCC is expected to be the global minimizer when $\mathsf{s} < 3/2$, while it is expected to be FCC when $\mathsf{s} > 3/2$.

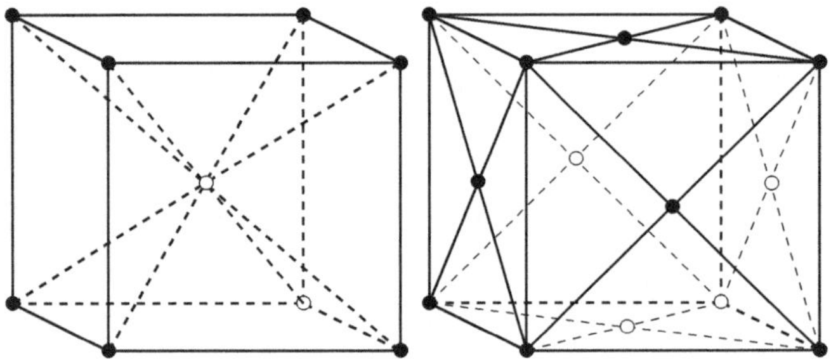

FIGURE 11.5.2. BCC and FCC lattices

CHAPTER 12

Microscopic characterization of minimizers via the jellium renormalized energy

The main goal of this chapter is to obtain a next order asymptotic lower bound for the energy $\mathcal{H}_N$. While a lower bound was already derived in Corollary 5.5, we look here for a more precise lower bound by the jellium renormalized energy $\mathbb{W}$ of the previous chapter, as a way toward showing that $\mathbb{W}$ is the infinite volume limit of F. The general lower bound valid for all configurations, in the spirit of Γ-convergence, will be expressed in terms of local limits, which are point processes living at the microscale. This lower bound can then be complemented by upper bounds thanks to the screening procedure, thus providing a next order expansion and local limit description of energy minimizers. The main result is roughly that, after blow-up of a minimizer, the limit point configuration obtained minimizes $\mathbb{W}$. This recovers results first obtained in [**SS15b, SS15a, RS15, PS17**]. This connects directly to the crystallization questions of the previous chapter. In particular, in dimensions 1, 8, 24 at least, minimizers should exhibit lattice patterns, and it is also expected to happen in dimension 2.

In the case with temperature, the upper and lower bounds we obtain can be inserted into the Gibbs measure. This will be done in the next chapter, and, when combined with an analysis of entropic effects, it allows to derive a full large deviations principle in terms of local point processes.

12.1. Tagged empirical field

Given $X_N = (x_1, \ldots, x_N)$ in $(\mathbb{R}^{\mathsf{d}})^N$, we recall that we define X'_N as the finite configuration rescaled by a factor $N^{1/\mathsf{d}}$, $X'_N = N^{1/\mathsf{d}}(x_1, \ldots, x_N)$.

One observable we wish to describe is the *tagged empirical field* defined as

$$\bar{P}_N[X_N] := \frac{1}{|\Sigma|} \int_\Sigma \delta_{(x,\, \theta_{N^{1/\mathsf{d}}x} \cdot X'_N)} dx, \tag{12.1.1}$$

where θ_x denotes the translation by $-x$. It is a probability measure on $\Sigma \times \mathcal{X}$, where Σ is the support of the equilibrium measure and $\mathcal{X}$ is the space of locally finite point configurations as in the previous chapter. For any $x \in \Sigma$, the term $\theta_{N^{1/\mathsf{d}}x} \cdot X'_N$ is an element of $\mathcal{X}$ which represents the N-tuple of particles X_N centered at x and seen at microscopic scale (or, equivalently, seen at microscopic scale and then centered at $N^{1/\mathsf{d}}x$). In particular any information about this point configuration in a given ball (around the origin) translates to an information about X'_N around x. We may thus think of $\theta_{N^{1/\mathsf{d}}x} \cdot X'_N$ as encoding the behavior of X'_N around "the tag" x. This terminology is inspired from classical works on random point processes [**Geo93, GZ93**]; see Figure 12.1.1. Limits of the tagged point processes $\bar{P}_N[X_N]$ are naturally tagged point processes $\bar{P} \in \mathcal{P}(\Sigma \times \mathcal{X})$, whose first marginal is the normalized Lebesgue measure on Σ. As a consequence we may consider the

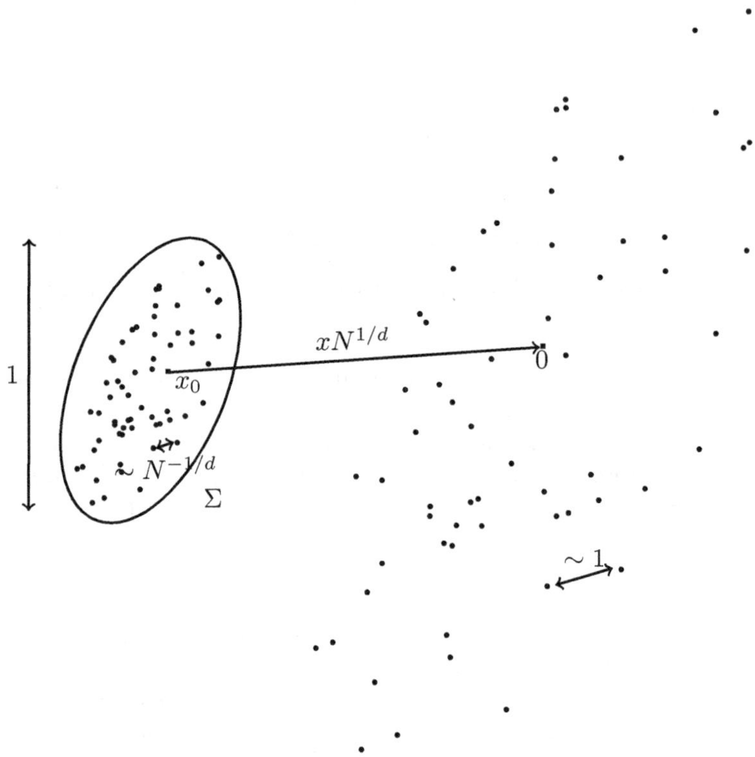

FIGURE 12.1.1. An arbitrary blown-up configuration

disintegration measures[1] $\{\bar{P}^x\}_{x\in\Sigma}$ of $\bar{P}$. For any $x \in \Sigma$, $\bar{P}^x$ is a probability measure on $\mathcal{X}$ and we have, for any $\Phi \in C^0(\Sigma \times \mathcal{X})$

$$\mathbb{E}_{\bar{P}}[\Phi] = \frac{1}{|\Sigma|} \int_\Sigma \mathbb{E}_{\bar{P}^x}[\Phi(x, \cdot)]dx.$$

Finally, the tagged point processes obtained at the limit will naturally be *stationary*, i.e., translation-invariant in their second entry. We denote by $\mathcal{P}_s(\mathcal{X})$ the set of translation-invariant (or stationary) point processes. We also call stationary a tagged point process such that the disintegration measure $\bar{P}^x$ is stationary for (Lebesgue-)a.e. $x \in \Sigma$ and we denote by $\mathcal{P}_s(\Sigma \times \mathcal{X})$ the set of stationary tagged point processes.

Finally, for any density μ on Σ, we denote by $\mathcal{P}_{s,\mu}(\Sigma \times \mathcal{X})$ the set of stationary tagged point processes such that for almost every $x \in \Sigma$, $\bar{P}^x$ has intensity $\mu(x)$.

We can also and will consider local versions of the empirical field. Given x_0, we define the "local empirical field" averaged in a cube of microscopic scale size R around x_0 by

$$P_N^{x_0,R}[X_N] := \fint_{\square_R(N^{1/d}x_0)} \delta_{\theta_x \cdot X'_N|_{\square_R(N^{1/d}x_0)}} dx \tag{12.1.2}$$

[1] We refer, e.g., to [**AGS08**, Section 5.3] for a definition.

where $|_{\square_R(N^{1/\mathsf{d}}x_0)}$ denotes the restriction of the configuration to $\square_R(N^{1/\mathsf{d}}x_0)$, the cube of size R centered at $N^{1/\mathsf{d}}x_0$. In other words we look at a spatial average at scale R of the (deterministic) point process formed by the configuration.

To give a proper meaning to the convergence, we must specify the topology used. First we endow $\mathcal{X}$ with the topology induced by the topology of weak convergence of Radon measures, which makes it a metrizable Polish space. This is a convergence against compactly supported continuous functions, which only allows to test local (i.e., compactly supported) properties. We endow $\mathcal{P}(\mathcal{X})$ and $\mathcal{P}(\Sigma\times\mathcal{X})$ with the usual topology of weak convergence of probability measures (for the Borel σ-algebra on $\mathcal{X}$). This defines a notion of convergence corresponding to the weak convergence of probability distributions on $\mathcal{X}$. Another natural topology on $\mathcal{P}(\mathcal{X})$ is convergence of the finite distributions [**DVJ88**, Section 11.1]—sometimes also called convergence with respect to vague topology for the counting measure of the point process. These topologies coincide as stated in [**DVJ88**, Theorem 11.1.VII]. We metrize it with an appropriate distance. For more detail see [**LS17**].

Remark 12.1. Strictly speaking, elements of $\mathcal{X}$ are *point processes* and elements of $\mathcal{P}(\mathcal{X})$ are *laws of point processes.* However, in these notes, elements of $\mathcal{X}$ are called *point configurations* (as above), a point process is defined as an element of $\mathcal{P}(\mathcal{X})$, and a tagged point process is a probability measure on $\Sigma\times\mathcal{X}$.

12.2. Energy lower bound in terms of the empirical field

The first main result we wish to prove is that the renormalized energy $\mathbb{W}$ of the previous chapter naturally arises as a Γ-liminf (in the terminology of Chapter 3) of the next order energy F of (5.2.9), and can be seen as an infinite volume limit of F. In order to see this concretely, it suffices to use Fubini's theorem to rewrite $\mathsf{F}(X_N',\mu')$ (where μ' is again the blown-up of μ defined by $\mu'(x)=\mu(xN^{-1/\mathsf{d}})$) as

$$\mathsf{F}(X_N',\mu')\simeq\frac{1}{R^{\mathsf{d}}}\int_{\mathbb{R}^{\mathsf{d}}}\mathsf{F}^{\square_R(z)}(X_N',\mu')dz$$

where R is any large scale (details will be given in the proof below) and $\mathsf{F}^{\square_R}$ is as in (7.1.23). The idea is then to let $N\to\infty$ so that X_N' converges to an infinite point configuration, while μ' converges to a uniform neutralizing background m, and then to let $R\to\infty$ so as to formally transform $\lim_{R\to\infty}\frac{1}{R^{\mathsf{d}}}\mathsf{F}^{\square_R(z)}(X_N',\mu')$ into $\mathbb{W}(\mathcal{C},m)$ by definition (11.2.13) and (11.2.15). The way to properly take this double limit and obtain this way a lower bound is to use the empirical field process defined above, show its tightness, and then deduce the result by Fatou's lemma.

In [**Ser15**, Chap. 5] we describe in more detail how this can be seen as a special case of an abstract method for proving lower bounds on energies which depend on two scales for which Γ-convergence is known at the small scale.

The first main result we will prove is the following. For simplicity, we present the result in the setting of the whole space, but it would work without change when working with the Neumann energy in a set U.

Throughout we will use the notation

$$\overline{\mathbb{W}}(\bar{P},\mu):=|\Sigma|\int_{\Sigma}\left(\int\mathbb{W}(\mathcal{C},\mu(x))d\bar{P}^x(\mathcal{C})\right)dx \tag{12.2.1}$$

for $\bar{P}$ a tagged empirical point process in $\mathcal{P}_s(\Sigma\times\mathcal{X})$, where $\bar{P}^x$ is the disintegration of $\bar{P}$.

Proposition 12.2.

Lower bound by the jellium renormalized energy. *Let* μ *be a bounded probability density on* $\mathbb{R}^{\mathsf{d}}$ *with compact support* Σ *with* $\partial\Sigma \in C^1$ *and* μ *continuous in the interior of* Σ. *Assume* $\{\mu_N\}_N$ *is a uniformly bounded sequence of probability measures on* $\mathbb{R}^{\mathsf{d}}$ *converging to* μ *locally uniformly in* Σ *as* $N \to \infty$, *and let* μ'_N *be the blown-up densities* $\mu'_N(x) = \mu_N(xN^{-\frac{1}{\mathsf{d}}})$. *Let* $\{X_N\}_N$ *be a sequence of* N*-tuples of points in* $\mathbb{R}^{\mathsf{d}}$, *and let* $X'_N = N^{1/\mathsf{d}} X_N$ *be the blown-up points.*

Global version. *Assume that* $N^{-1}\mathsf{F}(X'_N, \mu'_N)$ *is bounded independently of* N, *where* F *is as in* (5.2.9). *Then, up to extraction, the sequence* $\{\bar{P}_N[X_N]\}_N$ *as in* (12.1.1) *converges to some* $\bar{P}$ *in* $\mathcal{P}_{s,\mu}(\Sigma \times \mathcal{X})$, *and we have*

$$\liminf_{N\to\infty} \frac{1}{N}\mathsf{F}(X'_N, \mu'_N) \geq \overline{\mathbb{W}}(\bar{P}, \mu). \tag{12.2.2}$$

Local version. *Let* $\{x_0^N\}_N$ *be a sequence of points converging to* x_0, *a point in the interior of* Σ. *There exists a constant* $M > 0$ *depending only on* d, s *and* $\|\mu_N\|_{L^\infty}$ *such that, if* $1 \ll L \ll N^{1/\mathsf{d}}$ *as* $N \to \infty$ *and if*

$$\frac{1}{L^{\mathsf{d}}}\mathsf{F}^{\square_L(N^{1/\mathsf{d}}x_0^N)}(X'_N, \mu'_N) + M\#\{X'_N \cap \square_L(N^{1/\mathsf{d}}x_0^N)\}$$

is bounded independently of N, *then, up to extraction, the sequence* $\{P_N^{x_0^N, L}[X_N]\}_N$ *as in* (12.1.2) *converges to some* $P \in \mathcal{P}_s(\mathcal{X})$, *and we have*

$$\liminf_{N\to\infty} \frac{1}{L^{\mathsf{d}}}\mathsf{F}^{\square_L(N^{1/\mathsf{d}}x_0^N)}(X'_N, \mu'_N) \geq \int \mathbb{W}(\mathcal{C}, \mu(x_0))dP(\mathcal{C}). \tag{12.2.3}$$

Assuming that $\min \mathbb{W}(\cdot, m)$ is achieved, which we will prove below, the quantity in the right-hand side of (12.2.2) can be further bounded below by

$$\overline{\mathbb{W}}(\bar{P}, \mu) \geq \int_\Sigma \min \mathbb{W}(\cdot, \mu(x))dx. \tag{12.2.4}$$

But this expression can be further transformed by the scaling relation (11.2.20) which yields
(12.2.5)

$$\int_\Sigma \min \mathbb{W}(\cdot, \mu(x))dx = \min \mathbb{W}(\cdot, 1)\int_\Sigma \mu(x)^{1+\frac{\mathsf{s}}{\mathsf{d}}}\,dx - \mathbf{1}_{\mathsf{s}=0}\left(\frac{1}{2\mathsf{d}}\int_\Sigma \mu(x)\log\mu(x)\,dx\right).$$

12.2.1. Tagged electric field process and compactness. For the proof, we will use tagged electric field processes, which encode more information than just the configuration, since a configuration can be deduced from its electric field and not the converse.

We define an *electric field process* as an element of $\mathcal{P}(\mathrm{Elec})$ (definition of Elec in 11.1), usually denoted by P^{e} (with e like electric) . We say that P^{e} is *stationary* when it is invariant under the (push-forward by) translations $\theta_x \cdot E := E(\cdot - x)$ for any $x \in \mathbb{R}^{\mathsf{d}} \subset \mathbb{R}^{\mathsf{d}} \times \{0\}^k$. We say that P^{e} is compatible with (P, m), where P is a point process, provided P^{e} is concentrated on Elec_m and the push-forward of P^{e} by the map Conf_m coincides with P (see the notation in Definition 11.1).

Finally, we define a *tagged electric field process* as an element of $\mathcal{P}(\Sigma \times \mathrm{Elec})$, usually denoted by $\bar{P}^{\mathrm{e}}$, whose first marginal is the normalized Lebesgue measure on Σ. We say that $\bar{P}^{\mathrm{e}}$ is *stationary* if for a.e. $x \in \Sigma$, the disintegration measure $\bar{P}^{\mathrm{e},x}$ is stationary (in the previous sense). From the knowledge of a tagged electric field process $\bar{P}^{\mathrm{e}}$ we naturally deduce that of a tagged point process $\bar{P}$.

For each configuration X_N in $\mathbb{R}^{\mathsf{d}}$, we now precisely define the tagged electric field process $P_N^{\mathrm{e}}[X_N]$ (then we will drop the X_N) by

$$P_N^{\mathrm{e}}[X_N](x, E) := \frac{1}{|\Sigma|} \int_{\Sigma} \delta_{(x, \theta_{N^{1/\mathsf{d}} x} \cdot E_N)}\, dx \tag{12.2.6}$$

where $E_N = \nabla u_N$ as in (7.1.1) and θ_y is the action of translation by y.

We will need the following tightness criterion for tagged electric field processes.

Lemma 12.3 (Compactness for sets with bounded electric energy and number of points). *Let $\{E^{(n)}\}_n$ be a sequence of vector fields in $\mathbb{R}^{\mathsf{d}+\mathsf{k}}$, and let $\{\mathcal{C}^{(n)}\}_n$ be a sequence of point configurations in $\mathbb{R}^{\mathsf{d}}$ such that*

$$-\mathrm{div}\,(|y|^\gamma E^{(n)}) = \mathsf{c}_{\mathsf{d},\mathsf{s}} \left(\mathcal{C}^{(n)} - \mu^{(n)} \delta_{\mathbb{R}^{\mathsf{d}}}\right) \quad \text{in } \mathbb{R}^{\mathsf{d}+\mathsf{k}}, \tag{12.2.7}$$

for some $\mu^{(n)}$ converging pointwise to a constant $m \geq 0$. Assume that there exists a constant C_R such that

$$\sup_n \int_{\square_R \times \mathbb{R}^{\mathsf{k}}} |y|^\gamma |E_{\tilde{\mathsf{r}}}^{(n)}|^2 + |\mathcal{C}^{(n)}(\square_R)| < C_R, \tag{12.2.8}$$

where $\tilde{\mathsf{r}}$ is computed as in (11.2.4) with respect to $\square_R$, the closed centered cube of sidelength R in $\mathbb{R}^{\mathsf{d}}$.

Then there exists a vector field E in $\mathbb{R}^{\mathsf{d}+\mathsf{k}}$, which is a gradient if the $E^{(n)}$'s are gradients, and $\mathcal{C}$ a configuration in $\mathbb{R}^{\mathsf{d}}$, satisfying

$$-\mathrm{div}\,(|y|^\gamma E) = \mathsf{c}_{\mathsf{d},\mathsf{s}} \left(\mathcal{C} - m \delta_{\mathbb{R}^{\mathsf{d}}}\right) \quad \text{in } \mathbb{R}^{\mathsf{d}+\mathsf{k}}, \tag{12.2.9}$$

and such that, after extraction, for any compact set $K \subset \mathbb{R}^{\mathsf{d}}$, $E^{(n)}$ converges weakly to E in $L^p_{\mathrm{loc}}(K \times \mathbb{R}^{\mathsf{k}})$, for $p < \frac{\mathsf{d}+\mathsf{k}}{\mathsf{s}+1}$, $\mathcal{C}^{(n)}$ converges to $\mathcal{C}$ in $\mathcal{X}(K)$ and for every $R > 0$,

(12.2.10)
$$\liminf_{n \to \infty} \left(\frac{1}{2\mathsf{c}_{\mathsf{d},\mathsf{s}}} \int_{\square_R \times \mathbb{R}^{\mathsf{k}}} |y|^\gamma |E_{\tilde{\mathsf{r}}}^{(n)}|^2 - \frac{1}{2} \sum_{p \in \mathcal{C}^{(n)} \cap \square_R} \mathsf{g}(\tilde{\mathsf{r}}_p) - m \sum_{p \in \mathcal{C}^{(n)} \cap \square_R} \int_{\mathbb{R}^{\mathsf{d}}} \mathsf{f}_{\tilde{\mathsf{r}}_p}(x - p) \mu^{(n)}(x)\, dx \right) \geq \mathcal{F}^{\square_R}(E, m)$$

with $\mathcal{F}^{\square_R}$ as defined in (11.2.7).

Moreover, if $\{P^{(n)}\}_n$ is a sequence of probability measures on $L^p_{\mathrm{loc}}(\mathbb{R}^{\mathsf{d}+\mathsf{k}})$, $p < \frac{\mathsf{d}+\mathsf{k}}{\mathsf{s}+1}$, satisfying

$$\forall R > 1, \quad \int \left(\int_{\square_R \times \mathbb{R}^{\mathsf{k}}} |y|^\gamma |E_{\tilde{\mathsf{r}}}|^2 + \mathcal{C}(\square_R) \right) dP^{(n)}(E) < C_R \tag{12.2.11}$$

with C_R independent of n, then the sequence $\{P^{(n)}\}_n$ is tight in $L^p_{\mathrm{loc}}(\mathbb{R}^{\mathsf{d}+\mathsf{k}})$ and any weak limit point P satisfies that P-a.e. $E \in \mathrm{Elec}_m$ and is a gradient.

Proof. From (12.2.8), after diagonal extraction we have that for every $k \in \mathbb{N}$, $\mathcal{C}^{(n)}$ converges in $\mathcal{X}(\square_k)$ to some $\mathcal{C}$, and $E_{\tilde{\mathsf{r}}}^{(n)}$ converges weakly in $L^2_{|y|^\gamma}(\square_k \times \mathbb{R}^{\mathsf{k}})$ to some vector field X such that, for every $R > 0$,

$$\liminf_{n \to \infty} \int_{\square_R \times \mathbb{R}^{\mathsf{k}}} |y|^\gamma |E_{\tilde{\mathsf{r}}}^{(n)}|^2 \geq \int_{\square_R \times \mathbb{R}^{\mathsf{k}}} |y|^\gamma |X|^2. \tag{12.2.12}$$

On the other hand, the bound on $\int |y|^\gamma |E_{\tilde{\mathsf{r}}}^{(n)}|^2$ and the argument of Proposition 4.23 provide a bound on E in $L^p(\square_k \times \mathbb{R}^\mathsf{k})$ for $1 \le p < \frac{\mathsf{d}+\mathsf{k}}{\mathsf{s}+1}$, hence we may find a weak limit point E in $\cap_{k\in\mathbb{N}} L^p(\square_k \times \mathbb{R}^\mathsf{k})$ which satisfies (12.2.9) by taking the limit as $n \to \infty$ in (12.2.7) in the distributional sense. In addition, if $E^{(n)} = \nabla h^{(n)}$ for some $h^{(n)}$, by taking weak limits we also have that E is a gradient. Since by definition (11.2.5) we have

$$E_{\tilde{\mathsf{r}}}^{(n)} = E^{(n)} - \sum_{p\in\mathcal{C}^{(n)}} \nabla \mathsf{f}_{\tilde{\mathsf{r}}_p}(\cdot - p),$$

with $\tilde{\mathsf{r}}$ computed with respect to $\square_R$, taking the weak $n \to \infty$ limit and using the continuity of $\tilde{\mathsf{r}}_p$ with respect to the configuration, we must have

$$X = E - \sum_{p\in\mathcal{C}} \nabla \mathsf{f}_{\tilde{\mathsf{r}}_p}(\cdot - p),$$

and we recognize the right-hand side as equal to $E_{\tilde{\mathsf{r}}}$ (relative to $\square_R$). From (12.2.12) we thus have for every $R > 0$,

$$\liminf_{n\to\infty} \int_{\square_R\times\mathbb{R}^\mathsf{k}} |y|^\gamma |E_{\tilde{\mathsf{r}}}^{(n)}|^2 \ge \int_{\square_R\times\mathbb{R}^\mathsf{k}} |y|^\gamma |E_{\tilde{\mathsf{r}}}|^2. \tag{12.2.13}$$

The upper bound of the left-hand side implies that the right-hand side is finite, which we claim implies in the situation $\mathsf{s} \ge 0$ that r_p cannot be 0 for any point p in $\mathcal{C}$. Indeed, if $\mathsf{r}_p = 0$ this means by definition (11.2.3) that the point p comes with multiplicity $q \ge 2$, and that $E_\mathsf{r} = E$ in a neighborhood of p (because then $\mathsf{f}_{\mathsf{r}_p} = 0$). We would thus have a vector field defined in a ball $B(p,r)$ for some $r > 0$ satisfying

$$-\mathrm{div}\,(|y|^\gamma E) = \mathsf{c}_{\mathsf{d},\mathsf{s}}\,(q\delta_p - m\delta_{\mathbb{R}^\mathsf{d}}) \quad \text{in } B(p,r)\times\mathbb{R}^\mathsf{k}$$

and such that $\int_{B(p,r)} |y|^\gamma |E|^2 < \infty$. We may reason as in (4.4.17)–(4.4.18) to assert that if r is small enough, by Cauchy-Schwarz and Green's theorem we have

$$\begin{aligned}\int_{B(p,r)} |y|^\gamma |E|^2 &\ge \int_0^r \int_{\partial B(p,t)} |y|^\gamma |E\cdot\nu|^2 \\ &\ge \int_0^r \left(\int_{\partial B(p,t)} |y|^\gamma E\cdot\nu\right)^2 \frac{1}{\int_{\partial B(p,t)} |y|^\gamma} dt \\ &\ge \mathsf{c}_{\mathsf{d},\mathsf{s}} \int_0^r \frac{1}{t^{\mathsf{d}-1+\mathsf{k}+\gamma}} dt = \mathsf{c}_{\mathsf{d},\mathsf{s}} \int_0^r t^{-\mathsf{s}-1} dt\end{aligned} \tag{12.2.14}$$

by (2.2.5), The right-hand side is always infinite if $\mathsf{s} \ge 0$, a contradiction.

We now know that if $\mathsf{s} \ge 0$, $\tilde{\mathsf{r}}_p > 0$ for the points of the limiting configuration in $\square_R$, thus the definition of $\mathcal{F}^{\square_R}(E,m)$ only involves the first case of (11.2.7). To prove (12.2.10), in view of (12.2.13), it thus remains to check that in the case $\mathsf{s} \ge 0$ (otherwise these terms do not appear) $-\frac{1}{2}\sum_{p\in\square_R} \mathsf{g}(\tilde{\mathsf{r}}_p)$ and $-m\sum_{p\in\square_R} \int \mathsf{f}_{\tilde{\mathsf{r}}_p}(x-p)$ are lower semicontinuous with respect to the convergence of the configuration, which they are by continuity of the definition of $\tilde{\mathsf{r}}$ and closedness of the cubes.

We now turn to the statement about tightness. For any $\varepsilon > 0$ and integer k, defining

$$\mathcal{K}_{\varepsilon,k} = \left\{(x,E), \int_{\square_k\times\mathbb{R}^\mathsf{k}} |y|^\gamma |E_{\tilde{\mathsf{r}}}|^2 + \mathcal{C}(\square_k) < \frac{2^k C_k}{\varepsilon}\right\},$$

we have that $P^{(n)}(\mathcal{K}^c_{\varepsilon,k}) < \varepsilon 2^{-k}$ hence

$$P^{(n)}\left((\cap_k \mathcal{K}_{\varepsilon,k})^c\right) = P^{(n)}\left(\cup_k \mathcal{K}^c_{\varepsilon,k}\right) < \varepsilon.$$

But $\cap_k \mathcal{K}_{\varepsilon,k}$ is a compact set in L^p_{loc} by the above results. Thus we conclude that the sequence is tight, and it satisfies the stated properties in view of the above. □

12.2.2. Proof of Proposition 12.2.

Step 1. Fubini rewriting. For each configuration X_N, let $P^{\mathsf{e}}_N[X_N]$ be as in (12.2.6). The idea of the Fubini rewriting can be explained by the following calculation, simplifying F into $\int_{\mathbb{R}^{\mathsf{d}}} |\nabla u_N|^2$ where $E_N = \nabla u_N$: by definition (12.2.6) it holds that for any $R > 0$, $\square_R$ being the closed centered cube of sidelength R,

$$\int \left(\fint_{\square_R} |E|^2 \right) dP^{\mathsf{e}}_N[X_N](x, E) = \frac{1}{|\Sigma|} \int_\Sigma \left(\fint_{\square_R} |\nabla u_N(N^{1/\mathsf{d}} x + \cdot)|^2 \right) dx.$$

Then, using the change of variables $z = N^{1/\mathsf{d}} x + y$ and Fubini's theorem, we may write

$$\begin{aligned}
\int \left(\fint_{\square_R} |E|^2 \right) dP^{\mathsf{e}}_N[X_N](x, E) &= \frac{1}{R^{\mathsf{d}}|\Sigma|} \int_\Sigma \left(\int \mathbf{1}_{\square_R}(y) |\nabla u_N(N^{1/\mathsf{d}} x + y)|^2 dy \right) dx \\
&\le \frac{1}{R^{\mathsf{d}}|\Sigma|} \int \left(\int_{\mathbb{R}^{\mathsf{d}}} |\nabla u_N(N^{1/\mathsf{d}} x + y)|^2 dx \right) \mathbf{1}_{\square_R}(y) dy \\
&= \frac{1}{N|\Sigma|} \int_{\mathbb{R}^{\mathsf{d}}} |\nabla u_N|^2
\end{aligned} \tag{12.2.15}$$

and the right-hand side is our simplified version of $\frac{1}{N|\Sigma|} \mathsf{F}(X_N, \mu'_N)$. Thus to bound from below $\frac{1}{N}\mathsf{F}(X_N, \mu'_N)$ it suffices to take the limit of $P^{\mathsf{e}}_N[X_N]$ and use Fatou's lemma.

Let us now implement this precisely. The main point is that we need to work with positive and coercive quantities, in order to obtain tightness and use Fatou's lemma. To do so, we add to the energy a large constant times the number of points.

For any $z \in \square_R(0)$, using the superadditivity property (4.5.3) (equivalent to (7.1.37) in the general Riesz case) we find that

$$\mathsf{F}(X'_N, \mu'_N) \ge \sum_{q \in (R\mathbb{Z})^{\mathsf{d}}} \mathsf{F}^{\square_R(z+q)}(X'_N, \mu'_N) \tag{12.2.16}$$

where by definition (4.5.2)

$$\begin{aligned}
&\mathsf{F}^{\square_R(z)}(X'_N, \mu'_N) \\
&= \frac{1}{2\mathsf{c}_{\mathsf{d},\mathsf{s}}} \int_{\square_R(z) \times \mathbb{R}^k} |y|^\gamma |\nabla u_{N,\tilde{\mathsf{r}}}|^2 - \frac{1}{2} \sum_{x'_i \in \square_R(z)} \mathsf{g}(\tilde{\mathsf{r}}_i) - \sum_{x'_i \in \square_R(z)} \int_{\mathbb{R}^{\mathsf{d}}} \mathsf{f}_{\tilde{\mathsf{r}}_i}(x'_i - y) d\mu'_N(y)
\end{aligned} \tag{12.2.17}$$

with $\tilde{\mathsf{r}}$ computed relative to $\square_R(z)$ as in (7.1.25) (with $U = \mathbb{R}^{\mathsf{d}}$). Thus, after averaging over $z \in \square_R(0)$,

$$\begin{aligned}
\mathsf{F}(X'_N, \mu'_N) &\ge \frac{1}{R^{\mathsf{d}}} \int_{\square_R(0)} \sum_{q \in (R\mathbb{Z})^{\mathsf{d}}} \mathsf{F}^{\square_R(z+q)}(X'_N, \mu'_N)\, dz \\
&= \frac{1}{R^{\mathsf{d}}} \int_{\mathbb{R}^{\mathsf{d}}} \mathsf{F}^{\square_R(z)}(X'_N, \mu'_N)\, dz,
\end{aligned} \tag{12.2.18}$$

and arguing in the same way, for any $M > 0$,
(12.2.19)

$$\mathsf{F}(X'_N, \mu'_N) + MN \geq \frac{1}{R^{\mathsf{d}}} \int_{\mathbb{R}^{\mathsf{d}}} \mathsf{F}^{\square_R(z)}(X'_N, \mu'_N)\, dz + \frac{M}{R^{\mathsf{d}}} \int_{\mathbb{R}^{\mathsf{d}}} \#\{X'_N \cap \square_R(z)\}\, dz.$$

By the local uniform convergence assumption on $\{\mu'_N\}_N$ and the continuity assumption on μ, we may write that for $x'_i \in \square_R(z)$, for R fixed and as $N \to \infty$,

$$\int \mathsf{f}_{\tilde{\mathsf{r}}_i}(x'_i - y) d\mu'_N(y) = \left(\mu(zN^{-1/\mathsf{d}}) + o_N(1)\right) \int \mathsf{f}_{\tilde{\mathsf{r}}_i}(x'_i - y) dy.$$

Using that $\int |\mathsf{f}_{\tilde{\mathsf{r}}_i}(x'_i - y) dy| \leq C \tilde{\mathsf{r}}_i^{\mathsf{d}-\mathsf{s}} \leq C$ by (4.1.27), we may thus write

$$\mathsf{F}^{\square_R(z)}(X'_N, \mu'_N) = \frac{1}{2\mathsf{c}_{\mathsf{d},\mathsf{s}}} \int_{\square_R(z) \times \mathbb{R}^{\mathsf{k}}} |y|^\gamma |\nabla u_{N,\tilde{\mathsf{r}}}|^2 - \frac{1}{2} \sum_{x'_i \in \square_R(z)} \mathsf{g}(\tilde{\mathsf{r}}_i) - \mu(zN^{-1/\mathsf{d}}) \sum_{x'_i \in \square_R(z)} \int \mathsf{f}_{\tilde{\mathsf{r}}_i}(x'_i - y) dy + o_N(1) \mathcal{C}(\square_R(z)). \tag{12.2.20}$$

Let us now define for $x \in \Sigma$, $E \in L^p_{\mathrm{loc}}(\mathbb{R}^{\mathsf{d}+\mathsf{k}})$, $p < \frac{\mathsf{d}+\mathsf{k}}{\mathsf{s}+1}$,

(12.2.21)

$$f_{N,R}(x, E) := \chi_N(x, E) \frac{1}{R^{\mathsf{d}}} \left(\frac{1}{2\mathsf{c}_{\mathsf{d},\mathsf{s}}} \int_{\square_R(0) \times \mathbb{R}^{\mathsf{k}}} |y|^\gamma |E_{\tilde{\mathsf{r}}}|^2 - \frac{1}{2} \sum_{p \in \mathcal{C} \cap \square_R(0)} \mathsf{g}(\tilde{\mathsf{r}}_p) - \mu(x) \sum_{p \in \mathcal{C} \cap \square_R(0)} \int \mathsf{f}_{\tilde{\mathsf{r}}_p}(p - y) dy + M\mathcal{C}(\square_R) \right)$$

and

$$f_R(x, E) := \chi(x, E) \frac{1}{R^{\mathsf{d}}} \left(\frac{1}{2\mathsf{c}_{\mathsf{d},\mathsf{s}}} \int_{\square_R(0) \times \mathbb{R}^{\mathsf{k}}} |y|^\gamma |E_{\tilde{\mathsf{r}}}|^2 - \frac{1}{2} \sum_{p \in \mathcal{C} \cap \square_R(0)} \mathsf{g}(\tilde{\mathsf{r}}_p) - \mu(x) \sum_{p \in \mathcal{C} \cap \square_R(0)} \int \mathsf{f}_{\tilde{\mathsf{r}}_p}(p - y) dy + M\mathcal{C}(\square_R) \right) \tag{12.2.22}$$

where χ_N and χ are defined as

$$\chi_N(x, E) := \begin{cases} 1 & \text{if } E = \nabla u_N(N^{1/\mathsf{d}} x + \cdot) \\ +\infty & \text{otherwise} \end{cases}$$

and

$$\chi(x, E) := \begin{cases} 1 & \text{if } E \text{ is a gradient and s.t. } -\mathrm{div}\,(|y|^\gamma E) = \mathsf{c}_{\mathsf{d},\mathsf{s}}(\sum_{p \in \mathcal{C}} \delta_p - \mu(x) \delta_{\mathbb{R}^{\mathsf{d}}}) \\ & \text{with only simple points in } \square_R \text{ if } \mathsf{s} \geq 0 \\ +\infty & \text{otherwise} \end{cases}$$

and $\tilde{\mathsf{r}}$ is as in (11.2.4).

Choosing $M > 1 + C_0 + C \sup_N \|\mu'_N\|_{L^\infty}$ with C depending only on d, s and C_0 the constant of (11.2.10), in view of that inequality, there exists a constant $C > 0$ such that

$$C f_{N,R}(x, E) \geq \frac{1}{R^{\mathsf{d}}} \left(\int_{\square_R \times \mathbb{R}^{\mathsf{k}}} |y|^\gamma |E_{\tilde{\mathsf{r}}}|^2 + \mathcal{C}(\square_R) \right) \geq 0. \tag{12.2.23}$$

In view of (12.2.6), after using Fubini's theorem as in (12.2.15), we may rewrite (12.2.19) and (12.2.20) as

$$\mathsf{F}(X'_N,\mu'_N)+MN\geq N|\Sigma|(1+o_N(1))\int f_{N,R}(x,E)\,dP^{\mathrm{e}}_N(x,E). \tag{12.2.24}$$

Step 2. Tightness and lower bound. The upper bound assumed on $\mathsf{F}(X'_N,\mu'_N)$ and (12.2.24) imply that $\int f_{N,R}(x,E)\,dP^{\mathrm{e}}_N(x,E)$ is bounded independently of N. In view of (12.2.23), it follows that

$$\int\left(\int_{\square_R\times\mathbb{R}^{\mathsf{k}}}|y|^\gamma|E_{\tilde{r}}|^2+\mathcal{C}(\square_R)\right)dP^{\mathrm{e}}_N(x,E)<CR^{\mathsf{d}}$$

where C is a constant independent of N. By the second part of Lemma 12.3, we deduce that, up to extraction, $P^{\mathrm{e}}_N\to\bar{P}^{\mathrm{e}}$ weakly for some tagged electric field process $\bar{P}^{\mathrm{e}}$.

In addition the results of Lemma 12.3 and the definition of χ_N and χ ensure that if $x_N\to x$ and $E_N\to E$ (weakly in L^p_{loc}), we have

$$\liminf_{N\to\infty}f_{N,R}(x_N,E_N)\geq f_R(x,E).$$

Combining this with the weak convergence of P^{e}_N to $\bar{P}^{\mathrm{e}}$ we deduce in view of (12.2.23) that for each $R\geq 1$, we have

$$\liminf_{N\to\infty}\int f_{N,R}(x,E)dP^{\mathrm{e}}_N(x,E)\geq\int f_R(x,E)d\bar{P}^{\mathrm{e}}(x,E),$$

in other words, returning to (12.2.24),

$$\liminf_{N\to\infty}\frac{1}{N}\mathsf{F}(X'_N,\mu'_N)+M\geq|\Sigma|\int f_R(x,E)\,d\bar{P}^{\mathrm{e}}(x,E). \tag{12.2.25}$$

Since $f_R(x,E)$ is defined to be $+\infty$ for $E\notin\mathrm{Elec}_{\mu(x)}$, we also deduce that $\bar{P}^{\mathrm{e}}$-a.e. (x,E), we have $E\in\mathrm{Elec}_{\mu(x)}$ and E is a gradient. In the same way, if $\mathsf{s}\geq 0$, for $\bar{P}^{\mathrm{e}}$-a.e. (x,E) the points of $\mathrm{Conf}_{\mu(x)}(E)$ (see Definition 11.1) are simple.

Step 3. Stationarity and conclusion. To finish, let us justify that $\bar{P}^{\mathrm{e}}$ is stationary. Observe that for every x, thanks to the assumed regularity of $\partial\Sigma$, we have

$$\lim_{\varepsilon\to 0}\frac{|(\Sigma+\varepsilon x)\triangle\Sigma|}{|\Sigma|}=0 \tag{12.2.26}$$

where here $\triangle$ denotes the symmetric difference of sets. Consider a test function $\Phi\in C^0(\Sigma\times L^p_{\mathrm{loc}}(\mathbb{R}^{\mathsf{d}+\mathsf{k}}))$ and $y\in\mathbb{R}^{\mathsf{d}}$. Since $\bar{P}^{\mathrm{e}}=\lim_{N\to\infty}P^{\mathrm{e}}_N$, we may thus write

$$\begin{aligned}\int\Phi(x,\theta_yE)\,d\bar{P}^{\mathrm{e}}&=\lim_{N\to\infty}\frac{1}{|\Sigma|}\int_\Sigma\Phi(x,\theta_{N^{1/\mathsf{d}}x+y}E_N)\,dx\\&=\lim_{N\to\infty}\frac{1}{|\Sigma|}\int_\Sigma\Phi(x,\theta_{N^{1/\mathsf{d}}(x+yN^{-1/\mathsf{d}})}E_N)\,dx\\&=\lim_{N\to\infty}\int\Phi(x,E)\,dP^{\mathrm{e}}_N=\lim_{N\to\infty}\int\Phi(x,E)\,d\bar{P}^{\mathrm{e}},\end{aligned}$$

hence $\bar{P}^{\mathrm{e}}$ is stationary. Pushing forward by the map

$$E\mapsto\left(-\frac{1}{\mathsf{c}_{\mathsf{d},\mathsf{s}}}\mathrm{div}\,(|y|^\gamma E)+\mu'_N\delta_{\mathbb{R}^{\mathsf{d}}}\right)$$

naturally yields the convergence of $\bar{P}_N[X_N]$, defined in (12.1.1) to some $\bar{P} \in \mathcal{P}_s(\Sigma \times \mathcal{X})$.

By stationarity of $\bar{P}^e$, we now also have

$$\int f_R(x,E)\, d\bar{P}^e(x,E) = \int \left(\lim_{R\to\infty} f_R(x,E)\right) d\bar{P}^e(x,E).$$

Inserting into (12.2.25), we obtain

$$\liminf_{N\to\infty} \frac{1}{N}\mathsf{F}(X_N', \mu_N') + M \geq |\Sigma| \int \left(\lim_{R\to\infty} f_R(x,E)\right) d\bar{P}^e(x,E). \tag{12.2.27}$$

We may then rewrite f_R as $\chi\left(\frac{1}{R^{\mathsf{d}}}\mathcal{F}^{\square_R}(E,\mu(x)) + \frac{M}{R^{\mathsf{d}}}\mathcal{C}(\square_R)\right)$ in the notation of (11.2.7). Thus by definition (11.2.13) and Lemma 11.9,

$$\lim_{R\to\infty} f_R(x,E) = \mathcal{W}(E,\mu(x)) + M\mu(x), \quad \bar{P}^e(x,E) - a.e., \tag{12.2.28}$$

hence we have obtained

$$\liminf_{N\to\infty} \frac{1}{N}\mathsf{F}(X_N', \mu_N') + M \geq |\Sigma| \int (\mathcal{W}(E,\mu(x)) + M\mu(x))\, d\bar{P}^e(x,E). \tag{12.2.29}$$

Since the first marginal of $\bar{P}^e$ is the normalized Lebesgue measure on Σ and since $\int \mu(x)dx = 1$, it follows that

$$\liminf_{N\to\infty} \frac{1}{N}\mathsf{F}(X_N', \mu_N') \geq |\Sigma| \int \mathcal{W}(E,\mu(x)) d\bar{P}^e(x,E). \tag{12.2.30}$$

Using the definition (11.2.15) we get the further lower bound

$$\liminf_{N\to\infty} \frac{1}{N}\mathsf{F}(X_N', \mu_N') \geq |\Sigma| \int \mathbb{W}(\mathcal{C},\mu(x)) d\bar{P}^e(x,E) \tag{12.2.31}$$

and projecting this onto the tagged point processes $\bar{P}$, with (12.2.1) we obtain the result (12.2.2).

Step 4. Localized version. We next turn to the localized version of the result. Given the cube $\square_L(N^{1/\mathsf{d}}x_0^N)$ with $1 \ll L \ll N^{1/\mathsf{d}}$, let now

$$P_N^{x_0^N,L,e}(E) := \fint_{\square_L(N^{1/\mathsf{d}}x_0^N)} \delta_{\theta_x \cdot E_N}\, dx.$$

Instead of (12.2.19) we write that for any $R < L$, we have

$$\begin{aligned}\mathsf{F}^{\square_L(N^{1/\mathsf{d}}x_0^N)}(X_N', \mu_N') &+ M\#I_{\square_L(N^{1/\mathsf{d}}x_0^N)} \\ &\geq \frac{1}{R^{\mathsf{d}}}\int_{z\in\square_{L-R}(N^{1/\mathsf{d}}x_0^N)} \left(\mathsf{F}^{\square_R(z)}(X_N', \mu_N') + M\#I_{\square_R(z)}\right) dz.\end{aligned}$$

With (12.2.22), we may rewrite this as

$$\begin{aligned}\mathsf{F}^{\square_L(N^{1/\mathsf{d}}x_0^N)}(X_N', \mu_N') &+ M\#I_{\square_L(N^{1/\mathsf{d}}x_0^N)} \\ &\geq (L-R)^{\mathsf{d}}(1+o_N(1)) \int f_R(x_0^N, E) dP_N^{x_0^N,L-R}(E).\end{aligned}$$

Since we assume that the left-hand side is bounded by CL^{d}, if $L \gg R$ the same proof as in Step 2 implies that $\{P_N(E)^{x_0^N,L-R,e}\}_N$ is tight, hence has a subsequential limit P^e which is proven as in Step 3 to be stationary, and

$$\liminf_{N\to\infty} \frac{1}{L^{\mathsf{d}}}\left(\mathsf{F}^{\square_L(N^{1/\mathsf{d}}x_0^N)}(X_N', \mu_N') + M\#I_{\square_L(N^{1/\mathsf{d}}x_0^N)}\right) \geq \int f_R(x_0, E) dP^e.$$

Letting $N \to \infty$, then $R \to \infty$ with $R \ll L$, and using (12.2.28), we obtain

$$\liminf_{N\to\infty} \frac{1}{L^{\mathsf{d}}} \left(\mathsf{F}^{\square_L(N^{1/\mathsf{d}}x_0^N)}(X'_N, \mu'_N) + M\#I_{\square_L(N^{1/\mathsf{d}}x_0^N)} \right) \geq \int \mathcal{W}(E, \mu(x_0))dP^{\mathrm{e}}(E) + M\mu(x_0).$$

Moreover, we note that when $R \ll L$, $P_N(E)^{x_0^N, L-R, \mathrm{e}}$ and $P_N(E)^{x_0^N, L, \mathrm{e}}$ have the same limit, P^{e}. Projecting onto point processes we obtain that, P being the (subsequential) limit of $P_N^{x_0^N, L}[X_N]$,

$$\liminf_{N\to\infty} \frac{1}{L^{\mathsf{d}}} \left(\mathsf{F}^{\square_L(N^{1/\mathsf{d}}x_0^N)}(X'_N, \mu'_N) + M\#I_{\square_L(N^{1/\mathsf{d}}x_0^N)} \right) \geq \int \mathbb{W}(\mathcal{C}, \mu(x_0))dP(\mathcal{C}) + M\mu(x_0).$$

As above, in view of Lemma 11.9, we also have $L^{-\mathsf{d}}\#I_{\square_L(N^{1/\mathsf{d}}x_0^N)} \to \mu(x_0)$ as $N \to \infty$ if the left-hand side is finite. We can thus finally conclude that (12.2.3) holds.

12.3. Lower semicontinuity and existence of minimizers for $\mathbb{W}$

With the same proof idea we can now finally give the proof of existence of minimizers of the renormalized energy.

Corollary 12.4 (Existence of minimizers for $\mathcal{W}$ and $\mathbb{W}$). *The functions* $\mathcal{W}(\cdot, 1)$ *and* $\mathbb{W}(\cdot, 1)$ *admit a minimizer and* $\min_{\text{gradients}} \mathcal{W}(\cdot, 1) = \min \mathbb{W}(\cdot, 1)$.

PROOF. The proof is analogous to the previous one: returning to the definitions (11.2.13) and (11.2.15), let E^n be a sequence of gradients in $L^p_{\mathrm{loc}}(\mathbb{R}^{\mathsf{d}+\mathsf{k}})$, $p < \frac{\mathsf{d}+\mathsf{k}}{\mathsf{s}+1}$, such that $\mathcal{W}(E^n, 1) \to \inf \mathbb{W}(\cdot, 1)$, let $\mathcal{C}^n$ be the associated configuration, and let $R_n \to \infty$ be such that

$$\lim_{n\to\infty} \frac{1}{(R_n)^{\mathsf{d}}} \left(\int_{\square_{R_n}} \frac{1}{2\mathsf{c}_{\mathsf{d}}} |E^n_{\tilde{\mathsf{r}}}|^2 - \sum_{p\in\square_{R_n}} \mathsf{g}(\tilde{\mathsf{r}}_p) - \sum_{p\in\square_{R_n}} \int \mathsf{f}_{\tilde{\mathsf{r}}_p}(x-p)dx \right) + C_0 \left(\frac{\mathcal{C}^n(\square_{R_n})}{(R_n)^{\mathsf{d}}} - 1 \right) = \inf \mathbb{W}(\cdot, 1).$$

Defining P_n to be

$$\fint_{\square_{R_n}} \delta_{\theta_x \cdot E^n}\, dx,$$

we obtain, exactly as in the proof above (noting that here it was important to make the definition such that the quantity $\mathcal{F}^{\square_R}$ is superadditive)

$$\inf_{\text{gradients}} \mathcal{W}(\cdot, 1) = \inf \mathbb{W}(\cdot, 1) = \lim_{n\to\infty} \mathcal{W}(E^n, 1) \geq \int \mathcal{W}(E, 1)\, dP^{\mathrm{e}}(E) \geq \int \mathbb{W}(\mathcal{C}, 1) dP(\mathcal{C}),$$

where P^{e} is a stationary limit point (up to extraction) of P_n, and P is its push-forward by Conf_m or $E \mapsto -\mathsf{c}_{\mathsf{d},\mathsf{s}}^{-1}\mathrm{div}\,(|y|^\gamma E) + \delta_{\mathbb{R}^{\mathsf{d}}}$. Moreover, P_n being supported on gradient vector-fields, P also is. Thus minimizers must exist, and P^{e} must be concentrated on minimizers of $\mathcal{W}(\cdot,1)$ over gradients and P on minimizers of $\mathbb{W}(\cdot,1)$. We can prove in the same way the existence of minimizers of $\mathcal{W}(\cdot,1)$ without restriction to gradients. □

Using again the same proof we obtain the following lemma which will be important for the proof of the large deviations principle in the next chapter.

Lemma 12.5. *The maps*

$$P \mapsto \int \mathbb{W}(\mathcal{C},m)dP(\mathcal{C}), \qquad P \mapsto \overline{\mathbb{W}}(\bar{P},\mu),$$

are lower semicontinuous on the space $\mathcal{P}_s(\mathcal{X})$, respectively $\mathcal{P}_s(\Sigma\times\mathcal{X})$. In addition, their sublevel sets are compact on these spaces. Thus, they are good rate functions in the sense of Definition 3.9.

Before proving the lemma, we need the following.

Lemma 12.6 (Lifting stationary point processes to electric processes)**.** *Let $m \geq 0$ and $P \in \mathcal{P}_s(\mathcal{X})$ such that $\int \mathbb{W}(\mathcal{C},m)dP(\mathcal{C}) < \infty$. There exists P^{e} a stationary probability measure on* Elec_m*, concentrated on gradients*[2] *such that the push-forward of P^{e} by* Conf_m *equals P and such that*

$$\int \mathcal{W}(E,m)dP^{\mathrm{e}}(E) = \int \mathbb{W}(\mathcal{C},m)dP(\mathcal{C}).$$

PROOF. Let P be as in the assumption. For P-a.e. $\mathcal{C}$, the energy $\mathbb{W}(\mathcal{C},m)$ is finite and according to Lemma 11.8 we may find a gradient electric field $E \in \mathrm{Elec}(\mathcal{C},m)$ such that $\mathcal{W}(E,m) = \mathbb{W}(\mathcal{C},m)$. Let P^{e} be the push-forward of P by this map $\mathcal{C} \mapsto E$. It may happen that P^{e} is not stationary. In that case we consider a stationarizing sequence, namely a sequence of averages of translations of P^{e} over large hypercubes. Each element of this sequence is still compatible with P (because P is stationary) and has the correct energy. Any limit point of that sequence is stationary, has the correct energy, and is still compatible with P. □

PROOF OF LEMMA 12.5. We only prove the first result, the second one easily follows from the first using Fatou's lemma and the fact that $\mathbb{W}$ is bounded below (Remark 11.7). Assume that $P_n \to P$ in $\mathcal{P}_s(\mathcal{X})$. We may assume without loss of generality that $\sup_n \int \mathbb{W}(\mathcal{C},m)dP_n(\mathcal{C}) < C$. By Lemma 12.6, we may lift P_n into P_n^{e} which is stationary and such that

$$\int \mathbb{W}(\mathcal{C},m)dP_n(\mathcal{C}) = \int \mathcal{W}(E,m)dP_n^{\mathrm{e}}(E) = \int \limsup_{R\to\infty} \frac{\mathcal{F}^{\square_R}(E,m)}{R^{\mathsf{d}}} dP_n^{\mathrm{e}}(E)$$

in view of the formula (11.2.24). Since P_n^{e} is stationary, we may also write that for all $R > 0$,

$$\int \limsup_{R\to\infty} \frac{\mathcal{F}^{\square_R}(E,m)}{R^{\mathsf{d}}} dP_n^{\mathrm{e}}(E) = \frac{1}{R^{\mathsf{d}}} \int \mathcal{F}^{\square_R}(E,m)dP_n^{\mathrm{e}}(E).$$

Since we assumed $\int \mathbb{W}(\mathcal{C},m)dP_n(\mathcal{C}) < C$, in view of (11.2.14), we have

$$\frac{1}{R^{\mathsf{d}}} \int \left(\int_{\square_R\times\mathbb{R}^{\mathsf{k}}} |y|^\gamma |E_{\mathsf{r}}|^2 + \mathcal{C}(\square_R) \right) dP_n^{\mathrm{e}}(E) < C'.$$

---

[2]We recall that a probability measure P is concentrated on S if for every S', $P(S') = 0$ if $S \cap S' = \varnothing$.

By Lemma 12.3, we then deduce that $\{P_n^e\}_n$ is tight, and we may assume, up to extraction that $\{P_n^e\}_n$ converges to a limit P^e, which is stationary and concentrated on gradients, hence we obtain the compactness of the sublevel sets. By compatibility, we must have that the push-forward of P^e by Conf_m equals P. We may then conclude with the lower semicontinuity of $\mathcal{F}^{\square_R}$ (application of (12.2.10) with $\mu^{(n)} = m$) and using the stationarity of P, that

$$\liminf_{n\to\infty} \int \mathbb{W}(\mathcal{C}, m) dP_n(\mathcal{C}) = \liminf_{n\to\infty} \frac{1}{R^{\mathsf{d}}} \int \mathcal{F}^{\square_R}(E, m) dP_n^e(E)$$
$$\geq \frac{1}{R^{\mathsf{d}}} \int \mathcal{F}^{\square_R}(E, m) dP^e(E) = \int \mathcal{W}(E, m) dP^e(E).$$

Since $\int \mathcal{W}(E, m) dP^e(E) \geq \int \mathbb{W}(\mathcal{C}, m) dP(\mathcal{C})$ by definition of $\mathbb{W}$, the proof of lower semicontinuity is complete. □

12.4. Next order asymptotics for the minimal energy

We may now complete Theorem 8.2 and conclude the next order asymptotics of $\mathcal{H}_N$ at the level of minimizers. First we have

Corollary 12.7. *Assume* (2.0.1). *With the notation of Theorem* 8.2, *we have*

$$\mathsf{f_d}(\infty) = \min \mathbb{W}(\cdot, 1) = \lim_{R\to\infty} \frac{\mathsf{E}_\infty(1, \square_R)}{R^{\mathsf{d}}} \tag{12.4.1}$$

where E_∞ *is defined in* (7.1.57).

We present here the proof for $\mathsf{s} = \mathsf{d}-2$, which relies on the screening procedure presented in these notes, and remark that (8.2.1) provided a rate $1/R$ of convergence of the limit in (12.4.1). In the case $\mathsf{s} \in (\mathsf{d}-2, \mathsf{d})$, it suffices to use the Riesz screening procedure of [**PS17**] instead since we do not need the quantitative convergence provided by (8.2.1).

Proof. One inequality can be obtained by comparing (8.2.6) to (12.2.3) (for instance in the case $\mu_N = \mu = 1$ in Σ). The converse inequality is (11.4.1) combined with Corollary 12.4 and (8.2.1). □

We next turn to completing the analysis of minimizers of $\mathcal{H}_N$ or of $\mathsf{F}(X_N, \mu)$, which in view of the splitting formula (5.1.1) is the same question. The result (12.2.2) of Proposition 12.2 combined with (12.2.4) and (12.2.5) provides a lower bound. This lower bound is sharp because a matching upper bound can be obtained thanks to the screening procedure: this consists in partitioning the support of μ into quantized hyperrectangles Q_i of large microscopic size, and pasting in each of them screened minimizers of $\mathbb{W}(\cdot, \mu_i)$ (as done for showing (11.4.1)) where μ_i is the average value of μ in Q_i. One has to separately treat the boundary layer, which cannot be exactly tiled by hyperrectangles, via a rougher bound. In the general Riesz case, it can be done in the same way, using the Riesz screening, see [**PS17**]. An additional difficulty arising in the Riesz case is that the screening procedure requires a lower bound on the density, here on μ_V, but in Riesz cases, as seen in Remark 2.15 typical equilibrium measures vanish like $\mathrm{dist}(x, \partial\Sigma)^{1-\frac{\mathsf{d}-\mathsf{s}}{2}}$ as one approaches the boundary of their support. To deal with this, a boundary layer must be removed and treated separately with rougher estimates. We refer to [**PS17**, Section 7] for details.

The fact that the upper and lower bounds match implies that for minimizers there must be equality in (12.2.2) and also in (12.2.3). This allows to identify the limits of the empirical fields $\bar{P}$ or P as minimizers of the corresponding version of $\mathbb{W}$. We now recount all this in the following.

Theorem 12.8 (The case of minimizers). *Assume* $\mathsf{s} \in [\mathsf{d}-2, \mathsf{d})$. *Assume* V *satisfies* (A1)–(A3) *so that the equilibrium measure* μ_V *exists and is compactly supported in* Σ. *Assume also that* μ_V *is Hölder continuous in* Σ *and that* $\partial\Sigma$ *is* C^1. *In the Coulomb case* $\mathsf{s} = \mathsf{d}-2$, *also assume that* $\mu_V(x) \geq m > 0$ *for all* $x \in \Sigma$. *We have*

$$(12.4.2)\quad \min \mathcal{H}_N = N^2\mathcal{E}(\mu_V) - \frac{1}{2\mathsf{d}}(N\log N)\mathbf{1}_{\mathsf{s}=0} + N^{1+\frac{\mathsf{s}}{\mathsf{d}}}\int_\Sigma \mu_V(x)^{1+\frac{\mathsf{s}}{\mathsf{d}}}\min \mathbb{W}(\cdot,1) - \mathbf{1}_{\mathsf{s}=0}\left(\frac{N}{2\mathsf{d}}\int_\Sigma \mu_V(x)\log\mu_V(x)\right) + o(N^{1+\frac{\mathsf{s}}{\mathsf{d}}}).$$

Moreover, if X_N *is a minimizer of* $\mathcal{H}_N$, *then up to extraction,* $\bar{P}_N[X_N]$ *as defined in* (12.1.1) *converges to some* $\bar{P} \in \mathcal{P}_{s,\mu_V}(\Sigma \times \mathcal{X})$ *which is such that for almost every* $x \in \Sigma$, *the disintegration* $\bar{P}^x$ *minimizes* $P \mapsto \int \mathbb{W}(\mathcal{C}, \mu_V(x))dP(\mathcal{C})$.

We note that in the Coulomb case $\mathsf{s} = \mathsf{d}-2$ we can obtain a more quantitative estimate of the $o(N^{1+\frac{\mathsf{s}}{\mathsf{d}}})$, as a power of N, see [**AS21**].

Theorem 12.9 (The case of minimizers, local result). *Assume* $\mathsf{s} = \mathsf{d}-2$ *and the assumptions of the above theorem. If* X_N *is a minimizer of* $\mathcal{H}_N$, *then for* $R \gg 1$ *as* $N \to \infty$, *for any* $x_0^N \to x_0 \in \Sigma$ *such that* $\operatorname{dist}(x_0^N, \partial\Sigma) \geq CN^{-\frac{2}{\mathsf{d}(\mathsf{d}+2)}}$, *we have that, up to extraction,* $P_N^{x_0^N,R}[X_N] \to P$ *as* $N \to \infty$ *with* P*-a.e.* $\mathcal{C}$ *minimizing* $\mathbb{W}(\mathcal{C}, \mu_V(x_0))$. *Moreover,*

$$(12.4.3)\qquad \lim_{N\to\infty} \frac{1}{R^{\mathsf{d}}}\mathsf{F}^{\square_R(N^{1/\mathsf{d}}x_0^N)}(X_N', \mu_V') = \min \mathbb{W}(\cdot, \mu_V(x_0)).$$

Proof of Theorem 12.8 in the Coulomb case. We note that (12.4.2) was already obtained as (9.4.5) in Theorem 9.12 assuming that μ_V is Lipschitz on its support. The Lipschitz assumption there allowed to obtain a precise rate, but could be relaxed to just Hölder continuous.

Next, combining (5.1.1) and (5.2.8) we have

$$(12.4.4)\quad \mathcal{H}_N(X_N) = N^2\mathcal{E}(\mu_V) - \left(\frac{N}{2\mathsf{d}}\log N\right)\mathbf{1}_{\mathsf{s}=0} + N\sum_{i=1}^N \zeta(x_i) + N^{\frac{\mathsf{s}}{\mathsf{d}}}\mathsf{F}(X_N', \mu_V'),$$

and inserting the result of Proposition 12.2 and using that $\zeta \geq 0$, we are led to

$$(12.4.5)\quad \mathcal{H}_N(X_N) \geq N^2\mathcal{E}(\mu_V) - \left(\frac{N}{2\mathsf{d}}\log N\right)\mathbf{1}_{\mathsf{s}=0} + N^{1+\frac{\mathsf{s}}{\mathsf{d}}}\overline{\mathbb{W}}(\bar{P}, \mu_V) + o(N^{1+\frac{\mathsf{s}}{\mathsf{d}}})$$

where $\bar{P} \in \mathcal{P}_s(\Sigma \times \mathcal{X})$ is the limit (up to extraction) of $\bar{P}_N[X_N]$. Combining this with (12.2.4) and (12.2.5) provides the lower bound of (12.4.2). Comparing with the upper bound of (12.4.2) means that there must be equality in (12.4.5). It follows that if X_N minimizes $\mathcal{H}_N$, the limit points $\bar{P}$ of the empirical field $\bar{P}_N[X_N]$ obtained above must minimize $\overline{\mathbb{W}}(\bar{P}, \mu_V)$. In particular this implies that for a.e. $x \in \Sigma$, $\bar{P}^x$ minimizes $P \mapsto \int \mathbb{W}(\mathcal{C}, \mu_V(x))dP(\mathcal{C})$. □

Proof of Theorem 12.9. By (8.2.2) we already know that the number of points in $\square_R(N^{1/\mathsf{d}}x_0^N)$ is controlled by CR^{d}. This yields the subsequential convergence of $P_N^{x_0^N,R}$ to some P, with (12.2.3). Moreover, (12.4.3) is a consequence of (8.2.6) in Theorem 8.2 combined with (9.4.3) and Corollary 12.7. Thus, there must be equality in (12.2.3), hence P-a.e. $\mathcal{C}$ minimizes $\mathbb{W}(\cdot, \mu_V(x_0))$. $\square$

We have obtained that the local limit processes, after some averaging at a large microscale, minimize $\mathbb{W}$. In view of the discussion of the previous chapter, and modulo a form of uniqueness of minimizers, we expect these local limits to be lattices in dimensions 1, 2, 8 and 24.

CHAPTER 13

LDP for empirical fields

We now turn to the case with temperature and prove as a counterpart of Theorems 12.8 and 12.9 a global and local large deviations principle for the push-forward of the Gibbs measure by the empirical field map, i.e., at the level of (12.1.1) or (12.1.2), in terms of local point processes. This was first obtained in [**LS17**], but here the presentation is simplified by the new definition of $\mathbb{W}$ which avoids the use of two extra parameters in [**LS17**]. We also prove the local version originally found in [**AS21**]. This will provide a variational interpretation for the free energy per unit volume (or pressure) $f_{\mathsf{d}}(\beta)$ introduced in Theorem 8.14 and a variational characterization of the sine-β and Ginibre point processes.

While the energetic aspects are similar to the case of minimizers (lower bound via Proposition 12.2 and upper bound by screening), we now have to deal with the entropic effects. For that we need an analogue of the entropy to use in the way Sanov's theorem was used in Chapter 3. The adapted notion of entropy is the specific relative entropy, and the first step will be to show how it appears as a large deviations rate function for the reference measure.

13.1. Specific relative entropy

13.1.1. Definitions. Let us start by defining the analogue of the entropy at the level of point processes, which is the *specific relative entropy* with respect to the Poisson point process. It can be found in early papers on empirical fields, for instance [**FO88**], and we refer to the books [**RAS09, FV18**] for more detail.

We recall that the Poisson point process with intensity m is the point process characterized by the fact that for any bounded Borel set B in $\mathbb{R}^{\mathsf{d}}$, we have

$$\mathbb{P}\left(N(B)=n\right)=\frac{(m|B|)^n}{n!}e^{-m|B|}$$

where $N(B)$ denotes the number of points in B. The expectation of the number of points in B can then be computed to be $m|B|$, and one also observes that the number of points in two disjoint sets are independent, thus the points "do not interact".

For any $m \geq 0$, we denote by Π^m the (law of the) Poisson point process of intensity m in $\mathbb{R}^{\mathsf{d}}$, it is an element of $\mathcal{P}_s(\mathcal{X})$. Let P be in $\mathcal{P}_s(\mathcal{X})$. We define the specific relative entropy of P with respect to Π^1 as

$$\mathsf{ent}[P|\Pi^1] := \lim_{R\to\infty} \frac{1}{R^{\mathsf{d}}}\mathrm{ent}[P_{\square_R}|\Pi^1_{\square_R}], \tag{13.1.1}$$

where $P_{\square_R}, \Pi^1_{\square_R}$ denote the restriction of the respective processes to the hypercube $\square_R$. Here, we recall that as in (3.2.4), $\mathrm{ent}[\cdot|\cdot]$ denotes the *usual* relative entropy of two probability measures defined on the same probability space, namely

$$\mathrm{ent}[\mu|\nu] := \int \log \frac{d\mu}{d\nu}\, d\mu$$

if μ is absolutely continuous with respect to ν, and $+\infty$ otherwise.

Lemma 13.1. *The following properties are known:*

(1) *If P is stationary, the limit in* (13.1.1) *exists.*
(2) *The map $P \mapsto \mathsf{ent}[P|\Pi^1]$ is affine and lower semicontinuous on $\mathcal{P}_s(\mathcal{X})$.*
(3) *The sublevel sets of $\mathsf{ent}[\cdot|\Pi^1]$ are compact in $\mathcal{P}_s(\mathcal{X})$ (it is a good rate function).*
(4) *We have $\mathsf{ent}[P|\Pi^1] \geq 0$ with equality if and only if $P = \Pi^1$.*
(5) *We have the following scaling results: σ_m being as in* (11.2.18),

$$\mathsf{ent}[P|\Pi^1] = m\,\mathsf{ent}[(\sigma_m P)|\Pi^1] + m\log m + 1 - m, \tag{13.1.2}$$

and

$$\mathsf{ent}[P|\Pi^1] = \mathsf{ent}[P|\Pi^m] + m\log m + 1 - m. \tag{13.1.3}$$

Proof. We refer to [**RAS09**, Chapter 6] or [**FV18**] for a proof. The first point follows from sub-additivity, the third and fourth ones from usual properties of the relative entropy. The fact that $\mathsf{ent}[\cdot|\Pi^1]$ is an *affine* map, whereas the classical relative entropy is strictly convex, is due to the infinite-volume limit taken in (13.1.1). The scaling result is in [**LS17**, Lemma 4.4]. □

Next, if $\bar{P}$ is in $\mathcal{P}_s(\Sigma \times \mathcal{X})$ as defined in the previous chapter, given a density $\mu(x)$ over Σ, we define the tagged specific relative entropy as

$$\overline{\mathsf{ent}}[\bar{P}|\Pi^\mu] := \int_\Sigma \mathsf{ent}[\bar{P}^x|\Pi^{\mu(x)}]dx. \tag{13.1.4}$$

13.1.2. LDP for empirical fields without interaction. The main result we will use is a large deviation principle (recall the terminology from Section 3.2.1) for the tagged empirical field (12.1.1), when the points are distributed according to a reference measure on $(\mathbb{R}^{\mathsf{d}})^N$ with no interaction. This is a microscopic or "type III" (in the LDP jargon) analogue of Sanov's theorem for the empirical measures.

Proposition 13.2 (Large Deviations for the reference measure). *Let $\{\mu_N\}_N$ be a sequence of probability densities on $\mathbb{R}^{\mathsf{d}}$ converging locally uniformly to μ in Σ, where $\partial\Sigma \in C^1$, μ is continuous and bounded below by a positive constant in Σ. Then the push-forward of $\mu_N^{\otimes N}$ by the map $X_N \mapsto \bar{P}_N[X_N]$ of* (12.1.1) *satisfies a LDP at speed N with good rate function $\overline{\mathsf{ent}}[\bar{P}|\Pi^\mu]$. In particular, for any $\bar{P} \in \mathcal{P}_{s,\mu}(\Sigma \times \mathcal{X})$, we have*

$$\limsup_{\varepsilon\to 0}\limsup_{N\to\infty} \frac{1}{N}\log(\mu_N)^{\otimes N}\left(X_N, \bar{P}_N[X_N] \in B(\bar{P},\varepsilon)\right) \leq -\int_\Sigma \mathsf{ent}[\bar{P}^x|\Pi^{\mu(x)}]dx \tag{13.1.5}$$

and for any $\bar{P} \in \mathcal{P}_s(\Sigma \times \mathcal{X})$ we have

$$\liminf_{\varepsilon\to 0}\liminf_{N\to\infty} \frac{1}{N}\log(\mu_N)^{\otimes N}\left(X_N, \bar{P}_N[X_N] \in B(\bar{P},\varepsilon)\right) \geq -\int_\Sigma \mathsf{ent}[\bar{P}^x|\Pi^{\mu(x)}]dx \tag{13.1.6}$$

where the balls are taken for a distance metrizing the weak topology on $\mathcal{P}(\Sigma \times \mathcal{X})$.

Early LDPs for empirical fields can be found in [**Var88, F88**], the specific relative entropy is formalized in [**FO88**] (for the noninteracting discrete case), [**Geo93, Oll88**] (for the interacting discrete case) and [**GZ93**] (for the interacting continuous case). In the light of these results, Proposition 13.2 is not surprising, but there are some technical differences. In our case, the reference measure $\mu_N^{\otimes N}$ is not the restriction of a Poisson point process to a hypercube and is not uniform. Moreover we want to study large deviations for tagged point processes (our *tags* are not the same as the *marks* in [**GZ93**]) which requires an additional argument. These adaptations are largely drawing on ingredients from [**LS17**]. The main starting point is the following result from [**GZ93**].

Proposition 13.3 (LDP for Poisson's empirical field, [**GZ93**]). *Let $\{\Lambda_N\}_N$ be a sequence of cubes increasing to $\mathbb{R}^{\mathsf{d}}$ and let R_N be the push-forward of Π^1 by the map*

$$\mathcal{C} \mapsto \frac{1}{|\Lambda_N|}\int_{\Lambda_N} \delta_{\theta_x\cdot\mathcal{C}}dx$$

where θ_x denotes the translation by x. Then $\{R_N\}_N$ satisfies a LDP at speed $|\Lambda_N|$ with rate function $\mathsf{ent}[\cdot|\Pi^1]$.

In [**LS17**] it is adapted to the case of tagged point processes and the case of more general shapes than cubes. We recall that the N-point Bernoulli process in Λ is the law of N points chosen uniformly and independently in Λ. In particular the following is proven in [**LS17**, Lemma 7.8].

Lemma 13.4 (LDP for Bernoulli point processes). *Let Λ be a compact set of $\mathbb{R}^{\mathsf{d}}$ with C^1 boundary and nonempty interior and let S_N be the push-forward of the N-point Bernoulli process in $N^{\frac{1}{\mathsf{d}}}\Lambda$ by the map*

$$\mathcal{C} \mapsto \frac{1}{N|\Lambda|}\int_{N^{\frac{1}{\mathsf{d}}}\Lambda} \delta_{(N^{-1/\mathsf{d}}x,\theta_x\cdot\mathcal{C})}dx.$$

Then $\{S_N\}_N$ satisfies a LDP at speed N with good rate function

$$\bar{P} \mapsto \int_\Lambda \mathsf{ent}[\bar{P}^x|\Pi^{|\Lambda|^{-1}}]dx.$$

In [**LS17**] the rate function is written as $\int_\Lambda \mathsf{ent}[\bar{P}^x|\Pi^1]dx + \log|\Lambda| - |\Lambda| + 1$. In view of (13.1.3) and viewing it as a process of intensity $m = |\Lambda|^{-1}$ we can also rewrite it exactly as

$$\int_\Lambda \mathsf{ent}[\bar{P}^x|\Pi^1]dx + \log|\Lambda| - |\Lambda| + 1 = \int_\Lambda \mathsf{ent}[\bar{P}^x|\Pi^m]dx.$$

We will use the following slightly more general version.

Lemma 13.5. *Let Λ be a compact set of $\mathbb{R}^{\mathsf{d}}$ with C^1 boundary and nonempty interior. Also assume that $\{n\}_N$ is a sequence of integers such that*

$$\lim_{N\to\infty}\frac{n}{N|\Lambda|} = m.$$

Let S_N be the push-forward of the n-point Bernoulli process in $N^{\frac{1}{\mathsf{d}}}\Lambda$ by the map

$$\mathcal{C} \mapsto \frac{1}{N|\Lambda|}\int_{N^{\frac{1}{\mathsf{d}}}\Lambda} \delta_{(N^{-1/\mathsf{d}}x,\theta_x\cdot\mathcal{C})}dx.$$

Then $\{S_N\}_N$ satisfies a LDP at speed N with good rate function

$$\bar{P} \mapsto \int_\Lambda \mathsf{ent}[\bar{P}^x|\Pi^m]dx.$$

Let R_N be the push-forward of the n-point Bernoulli process in $N^{\frac{1}{\mathsf{d}}}\Lambda$ by the map

$$\mathcal{C} \mapsto \frac{1}{N|\Lambda|}\int_{N^{\frac{1}{\mathsf{d}}}\Lambda} \delta_{\theta_x\cdot\mathcal{C}}dx.$$

Then $\{R_N\}_N$ satisfies a LDP at speed $N|\Lambda|$ with good rate function $\mathsf{ent}[\cdot|\Pi^m]$.

Before we get to the proof of Proposition 13.2, let us state a lemma that allows to reduce to the uniformly convergent situation.

Lemma 13.6. *Assume that $\Sigma_\eta \subset \Sigma$ is such that $|\Sigma\backslash\Sigma_\eta| \to 0$ as $\eta \to 0$. Then letting*

$$\bar{P}_{N,\eta}[X_N] := \frac{1}{|\Sigma_\eta|}\int_{\Sigma_\eta} \delta_{(x,\theta_{N^{1/\mathsf{d}}x}\cdot X'_N)}dx,$$

we have that

$$\mathrm{dist}_{\mathcal{P}_s(\Sigma\times\mathcal{X})}(\bar{P}_{N,\eta}[X_N], \bar{P}_N[X_N]) \to 0, \quad \text{as } \eta \to 0.$$

Let $\mathfrak{P}_{N,\eta}$ be the push-forward of some probability measure $\mathbb{P}_N$ on $(\mathbb{R}^{\mathsf{d}})^N$ by $X_N \mapsto \bar{P}_{N,\eta}[X_N]$, respectively, $\mathfrak{P}_N$ be the push-forward of $\mathbb{P}_N$ by $X_N \mapsto \bar{P}_N[X_N]$. If $\mathfrak{P}_{N,\eta}$ satisfies a LDP at speed a_N with good rate function I_η, $I_\eta \to I$ pointwise as $\eta \to 0$, and $\mathfrak{P}_N$ is exponentially tight at speed a_N, then $\mathfrak{P}_N$ satisfies a LDP with good rate function I.

PROOF. The first statement follows from the definition of $\bar{P}_N[X_N]$ (12.1.1) and the fact that the topology on $\mathcal{P}_s(\Sigma \times \mathcal{X})$ only allows to test against bounded functions. The second statement follows from the definition of LDPs after reducing to statements over balls by exponential tightness, see Corollary 3.13. □

PROOF OF PROPOSITION 13.2. First, using Lemma 13.6, we can work in a subset $\Sigma_\eta \subset \Sigma$ such that $\partial\Sigma_\eta$ is piecewise C^1, $|\Sigma\backslash\Sigma_\eta| \to 0$ as $\eta \to 0$ and $\mu_N \to \mu$ uniformly in Σ_η, as $N \to \infty$. We thus reduce to a situation where $\mu_N \to \mu$ uniformly, and will from now on drop the η and assume that we are in the uniform convergent situation in Σ.

Step 1. Reduction to the piecewise constant case. Let $\eta > 0$ (different from the η just above). We may first partition Σ into cells Q_i with piecewise C^1 boundary, $i = 1, \dots, p$, of diameter $\le \eta$ and aspect ratios bounded above and below, requiring for instance that each Q_i is included in a ball of radius η and contains a ball of radius $\frac{1}{2}\eta$. We then let

$$\mu_{N,\eta} = \sum_{i=1}^p \mathbf{1}_{Q_i} m_{N,i}, \qquad m_{N,i} := \fint_{Q_i} \mu_N,$$

and

$$\mu_\eta = \sum_{i=1}^p \mathbf{1}_{Q_i} m_i, \qquad m_i = \fint_{Q_i} \mu,$$

i.e., $\mu_{N,\eta}$ is a piecewise constant approximation of μ_N, with "mesh size" η, and the same for μ_η. By the assumed uniform convergence of μ_N to μ, we have $\mu_{N,\eta} \to \mu_\eta$ uniformly. Since $\mu_\eta \to \mu$ as $\eta \to 0$, and since μ is bounded below in Σ, given $\delta > 0$

we may choose $\eta > 0$ small enough so that the Radon-Nikodym derivative of $\mu_{N,\eta}$ with respect to μ_N satisfies, for N large enough,

$$1 - \delta < \left| \frac{\mu_{N,\eta}}{\mu_N} \right| < 1 + \delta. \tag{13.1.7}$$

The uniform convergence also implies that

$$\lim_{N \to \infty} m_{N,i} = m_i. \tag{13.1.8}$$

We also let $\bar{\text{n}}_i$ be an integer equal to $N \int_{Q_i} \mu_N$ up to an error ≤ 1. We may choose them so that $\sum_{i=1}^p \bar{\text{n}}_i = N$.

In view of (13.1.7), given an event A we may write that

$$\log \left((1-\delta)^N (\mu_{N,\eta})^{\otimes N}(A) \right) \leq \log \mu_N^{\otimes N}(A) \leq \log \left((1+\delta)^N (\mu_{N,\eta})^{\otimes N}(A) \right)$$

thus

$$(1-\delta) + \frac{1}{N} \log \mu_{N,\eta}^{\otimes N}(A) \leq \frac{1}{N} \log \mu_N^{\otimes N}(A) \leq (1+\delta) + \frac{1}{N} \log \mu_{N,\eta}^{\otimes N}(A), \tag{13.1.9}$$

hence since in order to prove (13.1.5) and (13.1.6) we need to evaluate limits as $N \to \infty$ of $\frac{1}{N} \log \mu_N^{\otimes N}(A)$, it suffices to do so with μ_N replaced by $\mu_{N,\eta}$, then let $\delta \to 0$.

Step 2. Lower bound. Let $\bar{P} \in \mathcal{P}_s(\Sigma \times \mathcal{X})$. Let $X_{\bar{\text{n}}_i}$ be configurations of $\bar{\text{n}}_i$ points in Q_i and let $\bar{P}_{N,i}[X_{\bar{\text{n}}_i}]$ be the associated tagged empirical process in $\mathcal{P}_s(Q_i \times \mathcal{X})$ as in (12.1.1) with Σ replaced by Q_i. Let $\bar{P}_i$ be the restriction of $\bar{P}$ to Q_i, i.e., an element of $\mathcal{P}_s(Q_i \times \mathcal{X})$ obtained by restricting the first variable to Q_i. We note that $\frac{1}{(N \int_{Q_i} \mu_N)^{\bar{\text{n}}_i}} (N \mu_{N,\eta}|_{Q_i})^{\otimes \bar{\text{n}}_i}$ can be identified with an $\bar{\text{n}}_i$-point Bernoulli process in Q_i, and apply Lemma 13.5 (after zooming by $N^{1/\mathsf{d}}$) with $n = \bar{\text{n}}_i \to \infty$ to obtain, in view of (13.1.8), that

$$\begin{aligned} \liminf_{\varepsilon \to 0} \liminf_{N \to \infty} \frac{1}{N} \log \frac{(N \mu_{N,\eta}|_{Q_i})^{\otimes \bar{\text{n}}_i}}{(N \int_{Q_i} \mu_N)^{\bar{\text{n}}_i}} \left(X_{\bar{\text{n}}_i}, \bar{P}_{N,i}[X_{\bar{\text{n}}_i}] \in B(\bar{P}_i, \varepsilon) \right) \\ \geq - \int_{Q_i} \mathsf{ent}[(\bar{P}_i)^x | \Pi^{m_i}] dx. \end{aligned} \tag{13.1.10}$$

We can glue together the $X_{\bar{\text{n}}_i}$ to obtain a configuration X_N in $\cup_{i=1}^p Q_i$. Since $\bar{P}_{N,i}[X_{\bar{\text{n}}_i}] \in B(\bar{P}_i, \varepsilon)$ for every i and p depends only on η, we find that $\bar{P}_N[X_N] \in B(\bar{P}, \varphi(\varepsilon))$ where $\varphi(\varepsilon)$ is a function that tends to 0 as $\varepsilon \to 0$, which depends on η but not on N. We will use that notation throughout the proof, with a φ that may change.

Using that $\mu_{N,\eta} = \sum_{i=1}^p \mu_{N,\eta}|_{Q_i}$ and expanding $(\sum_i \mu_{N,\eta}|_{Q_i})^{\otimes N}$, retaining only the terms that correspond to $\bar{\text{n}}_i$ points for each $i \in [1,p]$, we find

$$(\mu_{N,\eta})^{\otimes N} \geq \frac{N!}{\prod_{i=1}^p \bar{\text{n}}_i!} (\mu_{N,\eta}|_{Q_i})^{\otimes \bar{\text{n}}_i} = \frac{N!}{\prod_{i=1}^p \bar{\text{n}}_i!} \frac{(N \mu_{N,\eta}|_{Q_i})^{\otimes \bar{\text{n}}_i}}{N^N}.$$

We may thus write that

$$\begin{aligned}(13.1.11)\qquad &\frac{1}{N}\log(\mu_{N,\eta})^{\otimes N}\left(X_N, \bar P_N[X_N]\in B(\bar P,\varphi(\varepsilon))\right)\\ &\ge \sum_{i=1}^p \frac{1}{N}\log\frac{(N\mu_{N,\eta}|_{Q_i})^{\otimes \bar{\mathrm{n}}_i}}{(N\int_{Q_i}\mu_N)^{\bar{\mathrm{n}}_i}}\left(X_{\bar{\mathrm{n}}_i}, \bar P_{N,i}[X_{\bar{\mathrm{n}}_i}]\in B(\bar P|_{Q_i},\varepsilon)\right)\\ &+\frac{1}{N}\log\left(\frac{N!}{N^N}\prod_{i=1}^p\frac{(N\int_{Q_i}\mu_N)^{\bar{\mathrm{n}}_i}}{\bar{\mathrm{n}}_i!}\right).\end{aligned}$$

Using Stirling's formula, $|N\int_{Q_i}\mu_N-\bar{\mathrm{n}}_i|\le 1$ and $\sum_i\bar{\mathrm{n}}_i=N$ and (13.1.10), we deduce that

$$\begin{aligned}(13.1.12)\qquad &\liminf_{\varepsilon\to0}\liminf_{N\to\infty}\frac{1}{N}\log(\mu_{N,\eta})^{\otimes N}\left(X_N,\bar P_N[X_N]\in B(\bar P,\varphi(\varepsilon))\right)\\ &\qquad\qquad\ge -\sum_{i=1}^p\int_{Q_i}\mathsf{ent}[(\bar P_i)^x|\Pi^{m_i}]dx.\end{aligned}$$

We next use (13.1.9), the continuity of $\int_\Sigma \mathsf{ent}[\cdot|\Pi^m]dx$ with respect to m, which is implied by (13.1.2), and let $\eta\to0$ to obtain
(13.1.13)

$$\liminf_{\varepsilon\to0}\liminf_{N\to\infty}\frac{1}{N}\log(\mu_N)^{\otimes N}\left(X_N,\bar P_N[X_N]\in B(\bar P,C\varepsilon)\right)\ge-\int_\Sigma\mathsf{ent}[\bar P^x|\Pi^{\mu(x)}]dx,$$

as desired.

Step 3. Upper bound. Let $\bar P\in\mathcal{P}_{s,\mu}(\Sigma\times\mathcal{X})$, see the definition in Section 12.1. We know that for Lebesgue-a.e. $x\in\Sigma$, $\bar P^x$ has intensity $\mu(x)$, hence we have

$$(13.1.14)\qquad \int\#(\mathcal{C}\cap\square_1)d\bar P^x(\mathcal{C})=\mu(x).$$

Let X_N be a configuration of N points with $\bar P_N[X_N]\in B(\bar P,\varepsilon)$. We let $\bar P_{N,i}$ be as above the tagged empirical fields of the restriction of the configuration in Q_i. When X_N is drawn from $(\mu_{N,\eta})^{\otimes N}$, these are independent. Moreover, since $\bar P_N[X_N]\in B(\bar P,\varepsilon)$, then $\bar P_{N,i}\in B(\bar P_i,\varphi(\varepsilon))$. Indeed we remark that

$$\fint_\Lambda \delta_{(x,\theta_{N^{1/\mathsf{d}}x}\cdot X_N'|_\Lambda)}dx \quad\text{and}\quad \fint_\Lambda\delta_{(x,\theta_{N^{1/\mathsf{d}}x}\cdot X_N')}dx,$$

i.e., the empirical fields with restricted or unrestricted configurations are close in the local topology when $\partial\Lambda$ is piecewise C^1.

We now wish to evaluate the number of points $n_i(X_N)$ a configuration with similar tagged empirical field has in the set Q_i by using (13.1.14) and the closeness of $\bar P_{N,i}$ to $\bar P_i$.

The idea is to use the fact that by definition of $\bar P_{N,i}$,

$$\begin{aligned}(13.1.15)\qquad &\int\#(\mathcal{C}\cap\square_1)d\bar P_{N,i}=\frac{1}{|Q_i|}\int_{Q_i}\#(\mathcal{C}\cap\square_1)\delta_{(x,\theta_{N^{1/\mathsf{d}}x}\cdot X_N')}(\mathcal{C})dx\\ &=\frac{1}{|Q_i|}\int_{Q_i}\#\{\theta_{N^{1/\mathsf{d}}x}\cdot X_N'\cap\square_1\}dx=\frac{1}{|Q_i|}\int\#\{X_N\cap(x+\square_{N^{-1/\mathsf{d}}})\}dx\end{aligned}$$

and that by the fact that ∂Q_i is piecewise C^1 and Fubini's theorem and $|\square_{N^{-1/\mathsf{d}}}|=N^{-1}$, the right-hand side is equal (up to a boundary error which is $o_N(1)$) to

$|Q_i|^{-1}N^{-1}$ times the number of points of X_N in Q_i. We then wish to use the closeness of $\bar{P}_{N,i}$ to $\bar{P}_i$ to deduce, by comparing (13.1.15) and (13.1.14) that

$$\frac{n_i(X_N)}{N|Q_i|} = \int_{Q_i} \mu(x)dx + o_N(1) + o_\varepsilon(1). \tag{13.1.16}$$

This is however not quite correct, because the ε-closeness we know is in the local topology and we can only test against bounded and continuous local functions, and $\#\{\mathcal{C} \cap \square_1\}$ is neither bounded nor continuous for the local topology. To remedy this, we can argue as in the proof of Lemma 7.8 in [**LS17**] by first approximating this function by continuous ones (integrating against a smooth cutoff instead of the indicator of $\square_1$) and then truncating the function at level S, and checking that the errors due to the truncation become negligible as $S \to \infty$.

We now deduce from (13.1.16), the definition of $\bar{\mathrm{n}}_i$ and the convergence of μ_N to μ that

$$|n_i(X_N) - \bar{\mathrm{n}}_i| \le N(o_N(1) + o_\varepsilon(1)) := N\rho_{N,\varepsilon}. \tag{13.1.17}$$

Again, $\frac{(N\mu_{N,\eta}|_{Q_i})^{\otimes n_i}}{(N\int_{Q_i}\mu_N)^{n_i}}$ can be identified with an n_i-Bernoulli point process in Q_i, hence Lemma 13.5 gives this time

$$\begin{aligned} \frac{1}{N}\log \frac{(N\mu_{N,\eta}|_{Q_i})^{\otimes n_i}}{(N\int_{Q_i}\mu_N)^{n_i}} &\left(X_{n_i}, \bar{P}_{N,i}[X_{n_i}] \in B(\bar{P}|_i, \varepsilon)\right) \\ &\le -\int_{Q_i} \mathsf{ent}[(\bar{P}_i)^x|\Pi^{m_i}]dx + o_N(1) + o_\varepsilon(1). \end{aligned} \tag{13.1.18}$$

In view of (13.1.17) we have

$$\mu_{N,\eta}^{\otimes N}(\bar{P}_N[X_N] \in B(\bar{P},\varepsilon)) = \sum_{\substack{n_i, \sum_i n_i = N \\ |n_i - \bar{\mathrm{n}}_i| \le N\rho_{N,\varepsilon}}} \mu_{N,\eta}^{\otimes N}(B(\bar{P},\varepsilon)\mathbf{1}_{\cap_{i=1}^p \{\#\{X_N \cap Q_i\} = n_i\}}) \tag{13.1.19}$$

but by independence and the above, we have

$$\begin{aligned} &\mu_{N,\eta}^{\otimes N}(B(\bar{P},\varepsilon)\mathbf{1}_{\cap_{i=1}^p \{\#\{X_N \cap Q_i\} = n_i\}}) \\ &\le \frac{N!}{\prod_{i=1}^p n_i!}\prod_{i=1}^p (\mu_{N,\eta}|_{Q_i})^{\otimes n_i}(\bar{P}_{N,i}[X_{n_i}] \in B(\bar{P}_i, \varphi(\varepsilon))). \end{aligned}$$

Inserting (13.1.18), then reinserting into (13.1.19), we obtain

$$\begin{aligned} &\frac{1}{N}\log \mu_{N,\eta}^{\otimes N}(\bar{P}_N[X_N] \in B(\bar{P},\varepsilon)) \\ &\le \max_{\substack{n_i, \sum_i n_i = N \\ |n_i - \bar{\mathrm{n}}_i| \le N\rho_{N,\varepsilon}}} \left(-\sum_{i=1}^p \int_{Q_i} \mathsf{ent}[(\bar{P}_i)^x|\Pi^{m_i}]dx + \frac{1}{N}\log\frac{N!\prod_{i=1}^p \bar{\mathrm{n}}_i^{n_i}}{N^N \prod_{i\in I} n_i!}\right) + o_N(1) + o_\varepsilon(1) \end{aligned}$$

hence the desired result after using Stirling's formula, letting $N \to \infty$, $\varepsilon \to 0$ then $\eta \to 0$.

We note that the second statement (about non tagged processes) can be deduced from the first one (about tagged processes) by applying the forgetful map $\phi : \mathcal{P}(\Sigma \times \mathcal{X}) \to \mathcal{P}(\mathcal{X})$ obtained by pushing forward by $(x, \mathcal{C}) \mapsto \mathcal{C}$, and using that the specific relative entropy is affine. $\square$

13.2. LDP for empirical fields

13.2.1. Statements and consequences on limit point processes. Let us introduce the rate function of [**LS17**], it is defined over the set of stationary point processes of intensity m (equipped with the topology of weak convergence) by

$$\mathcal{I}_\beta^m(P) := \beta \int \mathbb{W}(\mathcal{C}, m) dP(\mathcal{C}) + \mathsf{ent}[P|\Pi^m], \tag{13.2.1}$$

where Π^m is the (law of the) Poisson process of intensity m over $\mathbb{R}^{\mathsf{d}}$, and ent is the specific relative entropy. In view of Lemmas 12.5 and 13.1, it is a good rate function.

To minimize (13.2.1) there is a competition (depending on β) between the energy term $\mathbb{W}$ which prefers ordered configurations (remember that $\mathbb{W}$-minimizing configurations are expected to be crystalline in low enough dimensions, as seen in Chapter 11) and the relative entropy term which favors disorder hence configurations that are more Poissonnian. The choice of temperature scaling that we made in (1.1.5) is precisely the one for which these two competing effects are of comparable strength for fixed β. The limiting regime where θ of (3.2.5) is constant is treated in [**PG**] in a similar way.

Theorem 13.7 (Global large deviations principle). *Assume* (2.0.1). *Let* $\{\mu_N\}_N$ *be a sequence of probability densities on* $\mathbb{R}^{\mathsf{d}}$ *satisfying* (4.1.1) *and converging locally uniformly to* μ *in a compact set* Σ *such that* $\partial\Sigma \in C^1$, μ *is continuous and bounded below by a positive constant in* Σ. *Assume that* β *is such that* $\mathsf{K}_{N,\beta}(\mu_N)$ *is a convergent integral for each* N *large enough.*

Let $\mathfrak{P}_{N,\beta}$ *be the push-forward of* $\mathbb{Q}_{N,\beta}(\mu_N)$ *(defined in* (5.1.13)*) by* $X_N \mapsto \bar{P}_N[X_N]$ *as defined in* (12.1.1). *Then, assuming* $\theta = \beta N^{1-\frac{\mathsf{s}}{\mathsf{d}}} \to +\infty$ *as* $N \to \infty$, *we have the following.*

- *If* β *is independent of* N, *the sequence* $\{\mathfrak{P}_{N,\beta}\}_N$ *satisfies a LDP at speed* N *with good rate function*

$$\int_\Sigma \mathcal{I}_\beta^{\mu(x)}(\bar{P}^x)\, dx - \min \int_\Sigma \mathcal{I}_\beta^{\mu(x)}(\cdot)\, dx.$$

Moreover,

$$\lim_{N\to\infty} \frac{1}{N} \log \mathsf{K}_{N,\beta}(\mu_N) - \left(\frac{\beta}{2\mathsf{d}} \log N\right) \mathbf{1}_{\mathsf{s}=0} = - \min_{\mathcal{P}_{s,\mu}(\Sigma\times\mathcal{X})} \int_\Sigma \mathcal{I}_\beta^{\mu(x)}(\bar{P}^x) dx. \tag{13.2.2}$$

- *If* $\beta \to 0$ *as* $N \to \infty$, *then* $\{\mathfrak{P}_{N,\beta}\}_N$ *satisfies a LDP at speed* N *with good rate function*

$$\int_\Sigma \mathsf{ent}[\bar{P}^x|\Pi^{\mu(x)}]\, dx.$$

- *If* $\beta \to \infty$ *as* $N \to \infty$, *then* $\{\mathfrak{P}_{N,\beta}\}_N$ *satisfies a LDP at speed* βN *with good rate function*

$$\overline{\mathbb{W}}(\bar{P}, \mu) - \min_{\mathcal{P}_{s,\mu}(\Sigma\times\mathcal{X})} \overline{\mathbb{W}}(\bar{P}, \mu).$$

Moreover,

$$\lim_{N\to\infty} \frac{1}{\beta N} \log \mathsf{K}_{N,\beta}(\mu_N) - \left(\frac{1}{2\mathsf{d}} \log N\right) \mathbf{1}_{\mathsf{s}=0} = - \min_{\mathcal{P}_{s,\mu}(\Sigma\times\mathcal{X})} \overline{\mathbb{W}}(\bar{P}, \mu). \tag{13.2.3}$$

We next state the analogous local result in the Coulomb case (valid for point processes averaged at any scale larger than microscopic). The same result can be obtained in the Riesz case (2.0.1) by combining the method of proof with the results of [**PS**].

Theorem 13.8 (Local large deviations principle in the Coulomb case). *Let* $\mathsf{s} = \mathsf{d}-2$. *Assume* μ *is continuous and bounded below by a positive constant in* Σ. *Let* R, β *be such that* $N^{\frac{1}{\mathsf{d}}} \gg R \gg \rho_\beta$ *as* $N \to \infty$ *and* $x_0^N \to x_0 \in \Sigma$ *satisfies* $x_0^N \in \Sigma$ *and* $\mathrm{dist}(x_0^N, \partial\Sigma) \geq \bar{d}_0$ *as in* (9.3.1). *Assume that* $\{\mu_N\}_N$ *be a sequence of probability densities on* $\mathbb{R}^\mathsf{d}$ *satisfying* (4.1.1), *bounded below by a positive constant in* Σ, *and such that* $\mathsf{K}_{N,\beta}(\mu_N)$ *is a convergent integral for each* N *large enough. Assume that* μ_N *converges uniformly to* μ *in* $\square_R(N^{1/\mathsf{d}}x_0^N)$. *Let* μ'_N *denote the blown-up density* $\mu'_N(x) = \mu_N(x_0^N + N^{1/\mathsf{d}}x)$.

Let $\mathfrak{P}_{N,\beta}^{x_0,R}$ *be the push-forward of* $\mathbb{Q}_{N,\beta}(\mu_N)$ *by* $X_N \mapsto P_N^{x_0^N,R}[X_N]$ *defined in* (12.1.2).

Then we have the following.

- *If* β *is independent of* N, *the sequence* $\{\mathfrak{P}_{N,\beta}^{x_0,R}\}_N$ *satisfies a LDP at speed* R^d *with good rate function* $\mathcal{I}_\beta^{\mu(x_0)} - \min \mathcal{I}_\beta^{\mu(x_0)}$. *Moreover, if* $\int_{\square_R} \mu'_N$ *is an integer, we have*

$$\lim_{N\to\infty} \frac{1}{R^\mathsf{d}} \log \mathsf{K}_\beta(\mu'_N, \square_R) = - \min_{P \in \mathcal{P}_s(\mathcal{X})} \mathcal{I}_\beta^{\mu(x_0)}. \tag{13.2.4}$$

- *If* $\beta \to 0$ *as* $N \to \infty$, *then* $\{\mathfrak{P}_{N,\beta}^{x_0,R}\}_N$ *satisfies a LDP at speed* R^d *with good rate function* $\mathsf{ent}[P|\Pi^{\mu(x_0)}]$. *Moreover, if* $\int_{\square_R} \mu'_N$ *is an integer, we have*

$$\lim_{N\to\infty} \frac{1}{R^\mathsf{d}} \log \mathsf{K}_\beta(\mu'_N, \square_R) = 0. \tag{13.2.5}$$

- *If* $\beta \to \infty$ *as* $N \to \infty$, *then* $\{\mathfrak{P}_{N,\beta}^{x_0,R}\}_N$ *satisfies a LDP at speed* βR^d *with good rate function* $\mathbb{W}(\cdot, \mu(x_0)) - \min \mathbb{W}(\cdot, \mu(x_0))$. *Moreover, if* $\int_{\square_R} \mu'_N$ *is an integer, we have*

$$\lim_{N\to\infty} \frac{1}{\beta R^\mathsf{d}} \log \mathsf{K}_\beta(\mu'_N, \square_R) = - \min_{P \in \mathcal{P}_s(\mathcal{X})} \int \mathbb{W}(\mathcal{C}, \mu(x_0)) dP(\mathcal{C}). \tag{13.2.6}$$

In view of (5.2.14), applying to $\mu_N = \mu_\theta$ and $\mu = \mu_V$, with $\Sigma = \mathrm{supp}\, \mu_V$, and using Theorem 2.23 for the local uniform convergence of μ_N to μ, we obtain the following.

Corollary 13.9. *Assume the hypotheses of Theorem* 2.23. *Assume that the equilibrium measure* μ_V *is compactly supported in* Σ *and Hölder continuous in* Σ, *where* $\partial\Sigma \in C^1$. *Then* $\mathfrak{P}_{N,\beta}$, *the push-forward of* $\mathbb{P}_{N,\beta}$ *of* (1.1.5) *by* $X_N \mapsto \bar{P}_N[X_N]$, *respectively* $X_N \mapsto P_N^{x_0,R}[X_N]$, *satisfies the results of the corresponding theorem above.*

The theorem will be proven in the Coulomb case, but the statement is also true in the Riesz case $\mathsf{s} \in (\mathsf{d}-2, \mathsf{d})$. Note that a similar LDP for empirical fields is also proven in the hypersingular case $\mathsf{s} > \mathsf{d}$ in [**HLSS18**]. That regime is quite different in the sense that there is no equilibrium measure and no splitting formula, and the macroscopic density optimizer is determined implicitly via a local density approximation that involves an optimization over the microscopic distribution of points—macroscopic and microscopic arrangement of points play at the same order in the energy. On the other hand, the short range nature of the interaction allows to easily get almost additivity of the energy without the need for the screening procedure.

The results above allow to assert that local point processes obtained for instance in [**Tho24**] must minimize $\mathcal{I}_\beta$. As a corollary, we also obtain that the known point processes, Ginibre for $\mathsf{d}=2, \mathsf{s}=0$ and $\beta=2$, sine-β for $\mathsf{d}=1, \mathsf{s}=0$, and Riesz-$\beta$ obtained in [**Boua**] in the case $\mathsf{d}=1, \mathsf{s}\in(0,1)$ must minimize $\mathcal{I}_\beta$.

Corollary 13.10 (Variational characterization of the classical point processes). *The point processes sine-β, Ginibre, and Riesz-β minimize the corresponding $\mathcal{I}_\beta^1$ among stationary point processes of intensity* 1.

Applying to $\mu_N=\mu=1$ in Σ, comparing (13.2.2) and (13.2.3) with (9.4.6) and using (11.2.20), we find the analogue of (12.4.1) in the case with temperature, relating $f_{\mathsf{d}}(\beta)$ of Theorem 8.14 with $\inf \mathcal{I}_\beta$. This way the formulae (13.2.4)–(13.2.6) are the same as (9.4.1) except without the Lipschitz assumption on μ.

Corollary 13.11 (Variational interpretation of the pressure). *For any $\beta\in[0,+\infty)$ we have*

$$\beta f_{\mathsf{d}}(\beta)=\min_{P\in\mathcal{P}_s(\mathcal{X})}\mathcal{I}_\beta^1(P), \tag{13.2.7}$$

and for any $m\ge 0$,

$$\min_{P\in\mathcal{P}_s(\mathcal{X})}\mathcal{I}_\beta^m(P)=\beta m^{1+\frac{\mathsf{s}}{\mathsf{d}}} f_{\mathsf{d}}(\beta m^{\frac{\mathsf{s}}{\mathsf{d}}})+\left(1-\frac{\beta}{2\mathsf{d}}\mathbf{1}_{\mathsf{s}=0}\right) m\log m. \tag{13.2.8}$$

If $\mathsf{s}\neq 0$ an effective temperature $\beta m^{\frac{\mathsf{s}}{\mathsf{d}}}$ depending on the density of points appears here (as well as every time the density dependence is kept explicit).

We recall that the understanding of the function $f_{\mathsf{d}}(\beta)$ and its smoothness is one of the keys to understanding phase transitions. Note that [**KK16**] proposes explicit expressions for it. We have provided here a variational interpretation for it, which offers another potential angle of analysis for it. Unfortunately, we do not know if and when minimizers of $\mathcal{I}_\beta^1$ are unique (nonuniqueness would be another manifestation of a phase transition), as $\mathcal{I}_\beta^1$ is not in general convex. A notable exception is the one-dimensional logarithmic case, for which uniqueness is proven in [**EHL21**] by a displacement convexity argument combined with screening.

13.2.2. Proof outline. We will only prove Theorem 13.7 in the Coulomb case (and then we also assume $\mu_V\ge m>0$ on Σ for simplicity), the result is also true in the Riesz case and requires using the Riesz screening procedure of [**PS17**]. We refer the interested reader to [**LS17**]. Let us now give an idea of the steps of the proof, which are parallel to that of the minimizers case, Theorem 12.8.

From (5.1.13) and (5.2.8), we may rewrite $\mathbb{Q}_{N,\beta}$ in blown-up scale as

$$\mathbb{Q}_{N,\beta}(\mu_N)=\frac{1}{N^N\mathsf{K}_\beta(\mu_N')}\exp\left(-\beta\mathsf{F}(X_N',\mu_N')\right)d(\mu_N')^{\otimes N}(X_N'). \tag{13.2.9}$$

As usual, proving an LDP requires to prove exponential tightness (which in our case is easy thanks to the a priori bound on the number of points), prove a large deviations upper bound and a large deviations lower bound for balls, more precisely for

$$\mathbb{Q}_{N,\beta}(\mu_N)\{X_N,\bar{P}_N[X_N]\in B(\bar{P},\varepsilon)\}.$$

The upper bound is quite straightforward with the results we already have in hand. Indeed, inserting (12.2.2) into (13.2.9) and combining it with the lower semicontinuity of Lemma 12.5 we find that

$$\mathbb{Q}_{N,\beta}(\mu_N)\{X_N, \bar{P}_N[X_N] \in B(\bar{P},\varepsilon)\} \\ \le \frac{1}{\mathsf{K}_\beta(\mu'_N)} \exp\left(-N\left(\beta\overline{\mathbb{W}}(\bar{P},\mu) + o_{N,\varepsilon}(1)\right)\right) \int_{X_N, \bar{P}_N[X_N]\in B(\bar{P},\varepsilon)} d(\mu_N)^{\otimes N}(X_N),$$

and we then obtain directly in view of (13.1.5) the following upper bound

$$\log \mathbb{Q}_{N,\beta}(\mu_N)\{X_N, \bar{P}_N[X_N] \in B(\bar{P},\varepsilon)\} \\ \le -\log \mathsf{K}_\beta(\mu'_N) - N\beta\overline{\mathbb{W}}(\bar{P},\mu) - N\int_\Sigma \mathsf{ent}[\bar{P}^x|\Pi^{\mu(x)}]dx + No_{N,\varepsilon}(1). \tag{13.2.10}$$

We note here that we are able to apply (13.1.5) thanks to the fact that we may restrict our attention to $\bar{P} \in \mathcal{P}_{s,\mu}(\Sigma \times \mathcal{X})$ in view of Proposition 12.2. This is the needed upper bound—the $\log \mathsf{K}_\beta(\mu'_N)$ term is shown later to be the minimum of the rate function. The competition between the energy and entropy terms is straightforward.

The lower bound is much more delicate. It requires producing enough configurations whose empirical field are in $B(\bar{P},\varepsilon)$ and whose energy is bounded above by $\overline{\mathbb{W}}(\bar{P},\mu)$ up to $o(N)$. To do so, the starting point is to draw configurations at random from the law $\mu_N^{\otimes N}$. By Proposition 13.2, the probability that the empirical field (12.1.1) of such configurations drawn from $\mu_N^{\otimes N}$ resembles $\bar{P}$ is like $\exp\left(-N\int_\Sigma \mathsf{ent}[\bar{P}^x|\Pi^{\mu(x)}]dx\right)$, which is the desired volume estimate. But the control from above of the energy of these random configurations in terms of $\overline{\mathbb{W}}(\bar{P},\mu)$ is delicate for two reasons: $\mathbb{W}$ is defined as a limit over increasing cubes, and the background density varies, which requires to localize the estimates. To do so, exactly as in the upper bound for Theorem 12.8, we partition the domain into large microscale hyperrectangles Q_i in which $\mu'_N(Q_i)$ is integer, and draw configurations in each. We then screen the configurations in each Q_i, which allows to bound from above the energy F by the sum of the Neumann energies over Q_i thanks to (7.1.28). The only data that we have is that $\bar{P}_N[X_N]$ is close to $\bar{P}$, and we need to use this to control the energy by $\overline{\mathbb{W}}(\bar{P},\mu)$, this requires an upper semicontinuity property.

The screening procedure modifies the original random configuration in a boundary layer near the boundary of each Q_i. Since that boundary has small volume, the empirical field of the configuration does not get modified much, so it remains close to $\bar{P}$. On the other hand, modifying the configuration also modifies the volume estimate, but the relative error, which is handled by Proposition 8.5, is in the end shown to be small. A delicate task is to handle the screenability condition since not all configurations are screenable.

13.2.3. Main proofs.

Proof of Theorem 13.7 in the Coulomb case. In the proof we will abbreviate $\overline{\mathbb{W}}(\bar{P},\mu)$ into $\overline{\mathbb{W}}(\bar{P})$. As outlined just above we may combine (13.2.9), (12.2.2), Lemma 12.5 and (13.1.5) to obtain (13.2.10). This concludes the upper bound and we now turn to the lower bound.

Lower bound.

Step 1: Setup and partitioning. First, we note that in view of Lemma 13.6 we may reduce to the situation where $\mu_N \to \mu$ uniformly in Σ. Indeed, exponential tightness of $\mathbb{Q}_{N,\beta}(\mu_N)$ is an easy consequence of the fact that the total number of points in Σ is bounded by N.

Let $\bar{P} \in \mathcal{P}_s(\Sigma \times \mathcal{X})$ be such that $\overline{\mathbb{W}}(\bar{P}) < \infty$. We may lift $\bar{P}$ into a stationary tagged gradient electric process $\bar{P}^{\mathsf{e}}$ as in Lemma 12.6. By stationarity of $\bar{P}$, lifting and definition (11.2.13), we have for any $R > 0$,

(13.2.11)
$$\overline{\mathbb{W}}(\bar{P}) = \int_\Sigma \int \mathcal{W}(E, \mu(x)) d\bar{P}^{\mathsf{e},x}(E) dx = \int_\Sigma \int \frac{1}{R^{\mathsf{d}}} \mathcal{F}^{\square_R}(E, \mu(x)) d\bar{P}^{\mathsf{e},x}(E) dx.$$

Indeed, Lemma 11.9 and the stationarity imply that we can remove the $\frac{\mathcal{C}(\square_R)}{R^{\mathsf{d}}} - \mu(x)$ term present in the definition of $\mathcal{W}$.

Given $R > 1$ independent of N, let us apply Lemma 5.13 and partition $\Sigma' := N^{1/\mathsf{d}}\Sigma$ so that

$$\Sigma' = \cup_{i \in I} Q_i \cup \omega$$

where Q_i are hyperrectangles of size $\in [R, R + CR^{1-\mathsf{d}}]$ included in Σ' such that $\int_{Q_i} \mu'_N := \bar{\mathrm{n}}_i$, an integer, and ω is a remaining boundary layer such that $|\omega| \le o(N)$. We let

$$\Sigma'_{\mathrm{int}} = \cup_{i \in I} Q_i.$$

We would like to replace the Q_i's by translations of a fixed square. For that we may find for each i a square $\square_i$ of center x_i and sidelength exactly R and included in Q_i. By Lemma 5.13 the difference between R and the sidelength of Q_i is bounded by $O(R^{1-\mathsf{d}})$. This way $|Q_i \backslash \square_i| = o(R^{\mathsf{d}})$. Also we can pick $\square_i$ in such a way that for any $\tilde{\ell} \ge CR^{1-\mathsf{d}}$ (with the same C appearing in the sidelength of Q_i) we have

$$\{x \in Q_i, \mathrm{dist}(x, \partial Q_i) \ge \tilde{\ell}\} \subset \square_i. \tag{13.2.12}$$

For a configuration, we define its discrete average empirical field relative to the partition Q_i by

$$\bar{D}_N[X_N] := \frac{1}{\# I} \sum_{i \in I} \delta_{(N^{-1/\mathsf{d}} x_i, \theta_{x_i} \cdot X'_N|_{Q_i})} \tag{13.2.13}$$

where we recall the x_i's are the centers of the $\square_i$'s.

Step 2. Comparing discrete and continuous averages. Since we will need to work with the discrete averages (13.2.13) it is important to be able to show that they are close to the continuous averages.

The idea is that if one knows that a discrete average over large hypercubes is very close to some point process P, then the continuous average over much smaller hypercubes is also close to P since it can be re-written using the discrete average up to a small error. More precisely for any fixed $\delta > 0$ establishing that a point process is in $B(P, \delta)$ can be done by testing against local functions in Loc_k (the space of functions of $\mathcal{X}$ which are only functions of $\mathcal{C} \cap \square_k$) for some k large enough. For R, N large enough, an overwhelming majority of all translates of $\square_k$ by a point in Σ' is included in one of the hypercubes Q_i (this follows from the definitions and the tiling).

For any such local function $f \in \mathrm{Loc}_k$ we have

$$\frac{1}{|\Sigma'|}\int_{\Sigma'} f(\theta_x \cdot \mathcal{C}) \approx \frac{1}{\#I}\sum_{i\in I}\frac{1}{R^{\mathsf{d}}}\int_{Q_i} f(\theta_x \cdot \mathcal{C})dx, \tag{13.2.14}$$

which allows us to pass from the assumption that the discrete average (in the right-hand side of (13.2.14)) of a configuration is close to P to the fact that the continuous average (in the left-hand side of (13.2.14)) is close to P. These considerations are easily adapted to the situation of tagged point processes. We conclude that if $\bar{D}_N[X_N] \in B(\bar{P}, \varepsilon)$ then $\bar{P}_N[X_N] \in B(\bar{P}, \varphi(\varepsilon))$ for some function $\varphi(\varepsilon)$ tending to 0 as $\varepsilon \to 0$, and vice-versa.

Step 3. Good controls for configurations near $\bar{P}$.

Substep 3.1. Point and energy control. We will consider point configurations such that $\bar{D}_N[X_N]$ and $\bar{P}_N[X_N]$ are close to $\bar{P}$, more precisely are in some ball $B(\bar{P}, \varepsilon)$ for the weak local topology.

For that we may use Lemma 13.6 to replace $\bar{P}$ by its restriction to $\Sigma_{\mathrm{int}} = N^{-1/\mathsf{d}}\Sigma'_{\mathrm{int}}$, modification that we assume has been made from now on.

We first derive consequences of the fact that $\bar{D}_N[X_N] \in B(\bar{P}, \varepsilon)$. We will use the following variant of (13.1.6) for discrete averages, which follows from (13.1.6) in view of the conclusion of Step 2,

$$\liminf_{\varepsilon\to 0}\liminf_{N\to\infty}\frac{1}{N}\log(\mu_N)^{\otimes N}\left(X_N, \bar{D}_N[X_N] \in B(\bar{P}, \varepsilon)\right) \geq -\int_{\Sigma} \mathsf{ent}[\bar{P}^x|\Pi^{\mu(x)}]dx. \tag{13.2.15}$$

Secondly, the relation (11.2.21) implies that

$$\int |\mathcal{C}(\square_R) - \mu(x)R^{\mathsf{d}}|^2 d\bar{P}(\mathcal{C}) \leq CR^{2\mathsf{d}-2\kappa}(1 + \overline{\mathbb{W}}(\bar{P})), \tag{13.2.16}$$

for some $\kappa > 0$ depending only on d. We may extend these relations to probabilities close to $\bar{P}$. If $\bar{D}_N[X_N] \in B(\bar{P}, \varepsilon)$, since we may test against bounded continuous local functions, approximating the indicator function of $\square_R$ by continuous functions, we in particular deduce

$$\int |\mathcal{C}(\square_R) - \mu(x)R^{\mathsf{d}}|^2 d\bar{D}_N[X_N](x, \mathcal{C}) \leq CR^{2\mathsf{d}-2\kappa}(1 + \overline{\mathbb{W}}(\bar{P})), \tag{13.2.17}$$

which means by definition (13.2.13) that

$$\frac{1}{\#I}\sum_{i\in I}\left|\#\{X'_N \cap \square_i\} - \mu'(x_i)R^{\mathsf{d}}\right|^2 \leq CR^{2\mathsf{d}-2\kappa}(1 + \overline{\mathbb{W}}(\bar{P})). \tag{13.2.18}$$

By uniform convergence of μ_N to μ and continuity of μ, we also have

$$|\mu'(x_i)R^{\mathsf{d}} - \bar{\mathrm{n}}_i| = \left|\mu'(x_i)R^{\mathsf{d}} - \int_{Q_i}\mu'_N\right| \leq o_N(1)R^{\mathsf{d}},$$

and thus

$$\frac{1}{\#I}\sum_{i\in I}\left|\#\{X'_N \cap \square_i\} - \bar{\mathrm{n}}_i\right|^2 \leq CR^{2\mathsf{d}-2\kappa}(1 + \overline{\mathbb{W}}(\bar{P})) + o_N(1)R^{2\mathsf{d}}. \tag{13.2.19}$$

Thirdly, the relation (13.2.11) ensures that for any $R > 0$ we have

$$\int \frac{1}{R^{\mathsf{d}}}\mathcal{F}^{\square_R}(E, \mu(x))d\bar{P}^{\mathsf{e}}(x, E) \leq \overline{\mathbb{W}}(\bar{P}), \tag{13.2.20}$$

in fact we have equality. We now need upper semicontinuity (for the local topology) and boundedness to be able to replace $\bar{P}^{\mathsf{e}}$ by $\bar{D}_N[X_N]$ in the inequality (13.2.20).

Let $M \geq \overline{\mathbb{W}}(\bar{P}) + 1$. First, we may truncate at level M and write that

$$\int \left(\frac{1}{R^{\mathsf{d}}} \mathcal{F}^{\square_R}(E, \mu(x)) \wedge M \right) d\bar{P}^{\mathsf{e}}(x, E) \leq \overline{\mathbb{W}}(\bar{P}). \tag{13.2.21}$$

Next, we note that by definition (11.2.7), and generalizing the definition in the obvious way to open cubes, either $\mathcal{F}^{\square_R}(E, m) = +\infty$ or

$$\mathcal{F}^{\square_R}(E, m) = \mathcal{F}^{\mathring{\square}_R}(E, m) + r(\mathcal{C}, m) \tag{13.2.22}$$

where

$$r(\mathcal{C}, m) = -\frac{1}{2} \sum_{p \in \partial \square_R} \mathsf{g}(\tilde{\mathsf{r}}_p) - m \sum_{p \in \partial \square_R} \int_{\mathbb{R}^{\mathsf{d}}} \mathsf{f}_{\tilde{\mathsf{r}}_p}(x - p),$$

in particular $\mathcal{F}^{\square_R}(E, m) \leq \mathcal{F}^{\mathring{\square}_R}(E, m)$. Here we have isolated in r the contribution of points that fall exactly on the boundary and which prevent the function $\mathcal{F}^{\square_R}$ from being upper semicontinuous. But it is negligible: by the boundedness of $\tilde{\mathsf{r}}_p$ for $p \in \partial \square_R$, r is bounded by the number of points on the boundary, and in view of Lemma 11.9 we must have that $\int r(\mathcal{C}, \mu(x)) d\bar{P}(\mathcal{C}) = O(R^{\mathsf{d}-\kappa})$. Thus (13.2.21) can be changed into

$$\int \left(\frac{1}{R^{\mathsf{d}}} \mathcal{F}^{\mathring{\square}_R}(E, \mu(x)) \wedge M \right) d\bar{P}^{\mathsf{e}}(x, E) \leq \overline{\mathbb{W}}(\bar{P}) + O(R^{-\kappa}). \tag{13.2.23}$$

In addition, $\mathcal{F}^{\mathring{\square}_R}(E, m)$ coincides with the definitions of Chapter 7 (7.1.22) for gradient vector-fields, while $\bar{P}^{\mathsf{e}}$ is concentrated on gradient vector-fields (by definition and finiteness of $\overline{\mathbb{W}}(\bar{P})$).

Substep 3.2. Preparing for screening. With the goal of applying the screening procedure of Definition 7.15 and Proposition 7.19, we introduce parameters $\max(\beta^{-\frac{1}{\mathsf{d}-\mathsf{s}}} \mathbf{1}_{\mathsf{s} \leq 0}, 1) < \ell \leq \tilde{\ell} \leq R$ such that

$$\ell^{\mathsf{d}+1} \geq \frac{2\mathsf{c}_{\mathsf{d}} C M R^{\mathsf{d}}}{\tilde{\ell}} \tag{13.2.24}$$

for the constant C of (7.2.7). More specifically we let $\tilde{\ell} = R^{1-\kappa}$ and $\ell = R^{1-2\kappa}$ with κ small, where R may depend on β. We may in particular choose κ small enough (depending on d) such that (13.2.24) holds. In addition, making κ smaller if necessary, we can assume it is the same κ as in (13.2.19).

We will apply the screening with $\Omega = Q_i$ and $\Omega' = \square_i$,[1] for that we need $\{x \in Q_i, \mathrm{dist}(x, \partial Q_i) \geq \tilde{\ell}\} \subset \square_i$ which in view of (13.2.12) is guaranteed as long as $\tilde{\ell} \geq C R^{1-\mathsf{d}}$ which can be reduced to $\tilde{\ell} \geq C$ and with our choice, to R being large enough.

Let us assume that $\mathcal{F}^{\mathring{\square}_R}(E, \mu(x)) \leq M R^{\mathsf{d}}$. Since we may assume that E is a gradient (see (11.2.15)), then $E = \nabla w$ for some w which is automatically inner screenable in $\square_R$ according to Definition 7.15, in particular (7.2.7) is satisfied in

[1] The need to deal with Q_i and $\square_i$ being slightly different is the reason why the screening was including the possibility for the sets Ω and Ω' to be different.

view of the condition (13.2.24). We deduce from the definition (7.2.9) and (13.2.23) that

$$\int \left(\frac{1}{R^{\mathsf{d}}} \mathsf{G}^{\mathrm{inn}}_{\mathbb{R}^{\mathsf{d}}}(\mathcal{C}, \mu(x), \mathring{\square}_R) \wedge M \right) d\bar{P}(x, \mathcal{C}) \leq \overline{\mathbb{W}}(\bar{P}) + O(R^{-\kappa}). \tag{13.2.25}$$

In the rest of the proof we will drop the inn superscript and $\mathbb{R}^{\mathsf{d}}$ subscript and just write G instead of $\mathsf{G}^{\mathrm{inn}}_{\mathbb{R}^{\mathsf{d}}}$.

Substep 3.3. Upper semicontinuity of G. We now argue that for any m, $\mathsf{G}(\cdot, m, \mathring{\square}_R)$ is upper semicontinuous within the class of configurations such that $\mathsf{G}(\mathcal{C}, m, \mathring{\square}_R) \leq MR^{\mathsf{d}}$ with M as above. Let $\{\mathcal{C}^k\}_k$ be a sequence of configurations in $\mathbb{R}^{\mathsf{d}}$ converging to $\mathcal{C}$ for the local topology. We may assume that $\mathcal{C}$ has only simple points in $\mathring{\square}_R$ otherwise $\mathsf{G}(\mathcal{C}, m, \mathring{\square}_R) = +\infty$ and the desired result is true. Let G be the Dirichlet Green's function of $\square_R$, which is given by

$$G(x, y) = \mathsf{g}(x - y) - \mathsf{g}(x - y^*) \tag{13.2.26}$$

with y^* being a reflection of y through the boundary of $\square_R$.

Let w achieve the min in the definition (7.2.9) relative to $\mathring{\square}_R$, and let us now change notation and denote by $p_1^k, \ldots, p_n^k$ the points of $\mathcal{C}^k \cap \mathring{\square}_R$ (for k large enough, we can assume the number of points is constant independent of k). After extraction of a subsequence, we have $p_i^k \to p_i \in \square_R$ as $k \to \infty$. We then wish to build a competitor for the definition of (7.2.9) of $\mathsf{G}(\mathcal{C}^k, m, \mathring{\square}_R)$. Let

$$w^k(x) = w_{\hat{\mathsf{r}}}(x) + \int_{\square_R} G(x, y) d\left(\sum_{i=1}^n \delta_{p_i^k}^{(\hat{\mathsf{r}}_i^k)} - \delta_{p_i}^{(\hat{\mathsf{r}}_i)} \right)(y) + \sum_{i=1}^n \mathsf{f}_{\hat{\mathsf{r}}_i^k}(\cdot - p_i^k).$$

We may check that $-\Delta(w^k - w) = \mathsf{c_d}\left(\sum_{i=1}^n \delta_{p_i^k} - \delta_{p_i}\right)$, hence w^k is compatible with the points p_i^k and admissible in the definition of $\mathsf{G}(\mathcal{C}^k, m, \mathring{\square}_R)$ provided it is screenable. Moreover, by construction, $w_{\hat{\mathsf{r}}}^k = w_{\hat{\mathsf{r}}}$ on $\partial \square_R$, hence integrating by parts, we find that

$$\begin{aligned}
&\frac{1}{2\mathsf{c_d}} \int_{\square_R} |\nabla w_{\hat{\mathsf{r}}}^k|^2 - \frac{1}{2\mathsf{c_d}} \int_{\square_R} |\nabla w_{\hat{\mathsf{r}}}|^2 \\
&\quad = \frac{1}{2} \int_{\square_R} (w_{\hat{\mathsf{r}}}^k - w_{\hat{\mathsf{r}}}) \left(\sum_{i=1}^n \delta_{p_i^k}^{(\hat{\mathsf{r}}_i^k)} + \delta_{p_i}^{(\hat{\mathsf{r}}_i)} + 2\sum_j \delta_{x_j}^{(\eta_j)} - 2m \right) \\
&\quad = \frac{1}{2} \int_{\square_R \times \square_R} G(x, y) \\
&\qquad \times d\left(\sum_{i=1}^n \delta_{p_i^k}^{(\hat{\mathsf{r}}_i^k)} - \delta_{p_i}^{(\hat{\mathsf{r}}_i)} \right)(y) d\left(\sum_{i=1}^n \delta_{p_i^k}^{(\hat{\mathsf{r}}_i^k)} + \delta_{p_i}^{(\hat{\mathsf{r}}_i)} + 2\sum_j \delta_{x_j}^{(\eta_j)} - 2m \right)(x).
\end{aligned}$$

for some points $x_j \notin \mathring{\square}_R$. Expanding and using the symmetry of G, we find that

$$\begin{aligned}&\frac{1}{2\mathsf{c_d}}\int_{\square_R}|\nabla w_{\hat{\mathsf{r}}}^k|^2-\frac{1}{2\mathsf{c_d}}\int_{\square_R}|\nabla w_{\hat{\mathsf{r}}}|^2\\&=\frac{1}{2}\sum_{i=1}^n\int G(x,y)d\delta_{p_i^k}^{(\hat{\mathsf{r}}_i^k)}(x)d\delta_{p_i^k}^{(\hat{\mathsf{r}}_i^k)}(y)-\int G(x,y)d\delta_{p_i}^{(\hat{\mathsf{r}}_i)}(x)d\delta_{p_i}^{(\hat{\mathsf{r}}_i)}(y)\\&\quad+\frac{1}{2}\sum_{i\neq l}\int_{\square_R\times\square_R}G(x,y)d(\delta_{p_i^k}^{(\hat{\mathsf{r}}_i^k)}-\delta_{p_i}^{(\hat{\mathsf{r}}_i)})(y)d(\delta_{p_l^k}^{(\hat{\mathsf{r}}_l^k)}+\delta_{p_l}^{(\hat{\mathsf{r}}_l)})(x)\\&\quad+\int_{\square_R\times\square_R}G(x,y)d\left(\sum_{i=1}^n\delta_{p_i^k}^{(\hat{\mathsf{r}}_i^k)}-\delta_{p_i}^{(\hat{\mathsf{r}}_i)}\right)(y)d\Big(\sum_j\delta_{x_j}^{(\eta_j)}-m\Big)(x).\end{aligned}$$

Since $p_i^k\to p_i$ and the points $p_i\in\mathring{\square}_R$ are all distinct, G being continuous away from the diagonal and the radii $\hat{\mathsf{r}}$ being continuous functions of the point locations, these terms all converge to 0 as long as $p_i,p_l\in\mathring{\square}_R$. For terms involving p_i or p_l that belong to $\partial\square_R$ we use (13.2.26) to write that

$$\begin{aligned}\int G(x,y)d\delta_{p_i^k}^{(\hat{\mathsf{r}}_i^k)}(y)&=\int \mathsf{g}(x-y)d\delta_{p_i^k}^{(\hat{\mathsf{r}}_i^k)}(y)-\int\mathsf{g}(x-y^*)d\delta_{p_i^k}^{(\hat{\mathsf{r}}_i^k)}(y)\\&=\mathsf{g}_{\hat{\mathsf{r}}_i^k}(x-p_i^k)-\mathsf{g}_{\hat{\mathsf{r}}_i^k}(x-(p_i^k)^*)\end{aligned}$$

which then tends to 0 uniformly as $k\to\infty$ if $p_i\in\partial\square_R$. We can thus conclude that

$$\lim_{k\to\infty}\int_{\square_R}|\nabla w_{\hat{\mathsf{r}}}^k|^2=\int_{\square_R}|\nabla w_{\hat{\mathsf{r}}}|^2 \tag{13.2.27}$$

and

$$\begin{aligned}&\limsup_{k\to\infty}\frac{1}{2\mathsf{c_d}}\int_{\square_R}|\nabla w_{\hat{\mathsf{r}}}^k|^2-\frac12\sum_{i=1}^n\mathsf{g}(\hat{\mathsf{r}}_i^k)-m\sum_{i=1}^n\int\mathsf{f}_{\hat{\mathsf{r}}_i^k}(x-p_i^k)dx\\&\quad\le\frac{1}{2\mathsf{c_d}}\int_{\square_R}|\nabla w_{\hat{\mathsf{r}}}|^2-\frac12\sum_{p_i\in\mathring{\square}_R}\mathsf{g}(\hat{\mathsf{r}}_i)-m\sum_{p_i\in\mathring{\square}_R}\int\mathsf{f}_{\hat{\mathsf{r}}_i}(x-p_i)dx\le MR^{\mathsf{d}},\end{aligned}$$

since at worst, when $p_i^k\to\partial\square_R$ the points are not counted in the limit. It follows that w^k is screenable thanks to (13.2.24), hence w^k is admissible in the definition of the minimum in (7.2.9) for $\mathsf{G}(\mathcal{C}^k,m,\mathring{\square}_R)$ and thus, using that w achieves the min in the definition of $\mathsf{G}(\mathcal{C},m,\mathring{\square}_R)$, we conclude that

$$\limsup_{k\to\infty}\mathsf{G}(\mathcal{C}^k,m,\mathring{\square}_R)\le\mathsf{G}(\mathcal{C},m,\mathring{\square}_R) \tag{13.2.28}$$

which is the claimed upper semicontinuity.

From this, we deduce that $\frac{1}{R^{\mathsf{d}}}\mathsf{G}(\cdot,\mu(x),\mathring{\square}_R)\wedge M$ is upper semicontinuous. Finally, we conclude from (13.2.25) and this upper semicontinuity that if $\bar{D}_N[X_N]\in B(\bar{P},\varepsilon)$, we have

$$\int\left(\frac{1}{R^{\mathsf{d}}}\mathsf{G}(\mathcal{C},\mu(x),\square_R)\wedge M\right)d\bar{D}_N[X_N](x,\mathcal{C})\le\overline{\mathbb{W}}(\bar{P})+O(R^{-\kappa})+o_\varepsilon(1) \tag{13.2.29}$$

or in other words, by definition (13.2.13),

$$\frac{1}{\#I}\sum_{i\in I}\frac{1}{R^{\mathsf{d}}}\mathsf{G}(X'_N|_{\square_i},\mu'(x_i),\square_i)\wedge M\le \overline{\mathbb{W}}(\bar P)+O(R^{-\kappa})+o_\varepsilon(1). \tag{13.2.30}$$

Step 4. Rectifying the background. We next wish to change

$$\mathsf{G}(X'_N|_{\square_i},\mu'(x_i),\square_i)$$

into $\mathsf{G}(X'_N|_{\square_i},\mu'_N,\square_i)$. For that it suffices to add to w achieving the min in the definition (7.2.9) of G, the function u solving

$$\begin{cases}-\Delta u=\mathsf{c}_{\mathsf{d},\mathsf{s}}(\mu'_N-\mu'(x_i)) & \text{in } \square_i\\ u=0 & \text{on } \partial\square_i.\end{cases}$$

By elliptic estimates (arguing by reflection through the boundary), the uniform convergence of μ_N to μ and the continuity of μ, we have that

$$\|\nabla u\|_{L^\infty(\square_i)}\le C_R\|\mu'_N-\mu'(x_i)\|_{L^\infty(\square_i)}\le C_R o_N(1). \tag{13.2.31}$$

By definition of G we thus find that

$$\begin{aligned}&|\mathsf{G}(X'_N|_{\square_i},\mu'_N,\square_i)-\mathsf{G}(X'_N|_{\square_i},\mu'(x_i),\square_i)|\\ &\qquad\le C\|\nabla u\|^2_{L^2(\square_i)}+C\|\nabla u\|_{L^\infty(\square_i)}\|\nabla w_{\hat{\mathsf{r}}}\|_{L^1(\square_i)}.\end{aligned} \tag{13.2.32}$$

Since we are using $\mathsf{G}_U^{\mathrm{inn}}$ in the case $U=\mathbb{R}^{\mathsf{d}}$, the radii $\hat{\mathsf{r}}$ coincide with those of (7.1.25). The relation (7.1.35) can be checked to be valid for G as well as for F, i.e., we have the control

$$\int_{\square_i}|\nabla w_{\hat{\mathsf{r}}}|^2\le C\left(\mathsf{G}(X'_N,\mu'(x_i),\square_i)+C_0\#\{X'_N\cap\square_i\}\right).$$

Combining with (13.2.30), (13.2.31), (13.2.32) and (13.2.19), we thus deduce that

$$\sum_{i\in I}|\mathsf{G}(X'_N|_{\square_i},\mu'_N,\square_i)-\mathsf{G}(X'_N|_{\square_i},\mu'(x_i),\square_i)|\le o(N). \tag{13.2.33}$$

In view of (13.2.30) and since, in view of the tiling procedure, $\#I$ is of order $\frac{N}{R^{\mathsf{d}}}$, we have found that

$$\frac{1}{\#I}\sum_{i\in I}\frac{1}{R^{\mathsf{d}}}\mathsf{G}(X'_N|_{\square_i},\mu'_N,\square_i)\wedge M\le \overline{\mathbb{W}}(\bar P)+O(R^{-\kappa})+o_\varepsilon(1)+o_N(1). \tag{13.2.34}$$

Step 5. Screening. We are now in a position to apply the screening procedure to such configurations drawn near $\bar P$. For any X_N such that $\bar D_N[X_N]\in B(\bar P,\varepsilon)$, we have all the properties described in the previous steps, and we proceed as follows. We let I_1 be the subset of I such that $\mathsf{G}(X'_N|_{\square_i},\mu'(x_i),\square_i)\le MR^{\mathsf{d}}$, in particular $X'_N|_{\square_i}$ is screenable, and let $I_2=I\backslash I_1$. Moreover, (13.2.34) implies that $\#I_2/\#I\le\frac{\overline{\mathbb{W}}(\bar P)}{M}$.

If $i\in I_1$, we apply Proposition 7.19 with μ'_N as the reference measure, $\Omega=Q_i$ and $\Omega'=\Omega''=\square_i$, and the $\ell,\tilde\ell$ chosen above. This provides a configuration $Y_{\bar{\mathrm{n}}_i}$ in Q_i, coinciding with the restriction of $X'_N|_{\square_i}$ in some set $\mathcal{O}_i$ pasted with a configuration $Z_{\bar{\mathrm{n}}_i-n_{\mathcal{O}_i}}$ in $\mathcal{N}_i=Q_i\backslash\{x,\mathrm{dist}(x,\mathcal{O}_i)\le\eta\}$, and a positive measure $\tilde\mu_i$

in $\mathcal{N}_i$, such that

$$\begin{aligned}&\mathsf{F}(Y_{\bar{\mathrm{n}}_i},\mu'_N,Q_i)\\&\quad\le \mathsf{G}(X_N|_{\square_i},\mu'_N,\square_i)\\&\quad\quad+C\left(\frac{\ell M R^{\mathsf{d}}}{\tilde\ell}+R^{\mathsf{d}-1}\tilde\ell+\mathsf{F}(Z_{\bar{\mathrm{n}}_i-n_{\mathcal{O}_i}},\tilde\mu_i,\mathcal{N}_i)+|\bar{\mathrm{n}}_i-n_i|+\sum_{k,j\in J_i}\mathsf{g}(x_k-z_j)\right),\end{aligned}$$

where n_i is $\#\{X'_N\cap\square_i\}$. We emphasize here that the sets $\mathcal{O}_i$ and $\mathcal{N}_i$ depend on $X'_N|_{\square_i}$.

If $i\in I_2$, we let $\mathcal{O}_i=\varnothing$, $\mathcal{N}_i=Q_i$ and $\tilde\mu_i=(\mu'_N)|_{Q_i}$, i.e., we delete the configuration in Q_i and replace it with a generic configuration $Y^i_{\bar{\mathrm{n}}_i}=Z_{\bar{\mathrm{n}}_i}$. Finally, we do the same for the configuration in $\mathbb{R}^{\mathsf{d}}\backslash\Sigma'_{\mathrm{int}}$ and replace it with a generic configuration of $\bar{\mathrm{n}}_0:=N-\sum_{i\in I}\bar{\mathrm{n}}_i$ points. We simplify notation by writing $Q_0=\mathbb{R}^{\mathsf{d}}\backslash\Sigma'_{\mathrm{int}}$.

Pasting together the configurations obtained over all the Q_i's gives a configuration Y_N of N points in $\mathbb{R}^{\mathsf{d}}$. The set of all configurations obtained this way when X_N varies in $A:=\{X_N,\bar D_N[X_N]\in B(\bar P,\varepsilon)\}$ is denoted A'.

The important fact that we can check is that since $\#I_2=O(\frac{1}{M}\#I)$ and the configurations are unchanged in each $\mathcal{O}_i$ subset of Q_i, whose area is $|Q_i|-o(R^{\mathsf{d}})$, the total configuration is modified in only a vanishing fraction of the volume as $M\to\infty$. In view of this and of the result of Step 2, the set A' consists of configurations whose empirical field $\bar P_N[Y_N]\in B(\bar P,\varphi(\varepsilon))$ for N large enough and M large enough, where φ is some function tending to 0 as $\varepsilon\to 0$.

Step 6. Integrating and volume estimates. We next integrate the energy inequality over the possible choices of $Z_{\bar{\mathrm{n}}_i-n_{\mathcal{O}_i}}$ in each Q_i, with respect to the measure $(\mu'_N)|_{Q_i}$, and then over the choices of the initial $X_N\in A$ as above. This leads us to a calculation entirely similar to the proof of Proposition 8.5, in particular using the result of Lemma 8.6, we may write that

(13.2.35)

$$\begin{aligned}&\int_{A'}\exp\left(-\beta\mathsf{F}(Y_N,\mu'_N,\mathbb{R}^{\mathsf{d}})\right)d(\mu'_N)^{\otimes N}(Y_N)\\&\ge\int_{X_N\in A}\frac{\prod_{i\in I}n_i!}{N!}\frac{N!}{\prod_{i\in I}\bar{\mathrm{n}}_i!}\\&\times\exp\left(-\beta\sum_{i\in I_1}\left(\mathsf{G}(X'_N|_{\square_i},\mu'_N,\square_i)+C(\frac{MR^{\mathsf{d}}\ell}{\tilde\ell}+\chi(\beta)R^{\mathsf{d}-1}\tilde\ell+|\bar{\mathrm{n}}_i-n_i|)\right)\right)\\&\times\prod_{i\in I_1}\frac{\bar{\mathrm{n}}_i!(n-k_i)!}{n_i!(\bar{\mathrm{n}}_i-k_i)!}\frac{(\bar{\mathrm{n}}_i-n_{\mathcal{O}_i})^{\bar{\mathrm{n}}_i-k_i}}{(\alpha'_i)^{n_i-k_i}}e^{\alpha'_i-\alpha_i-C(\frac{MR^{\mathsf{d}}}{\ell\tilde\ell}+\ell^{-1}R^{\mathsf{d}-1}+\log R)}d(\mu'_N)^{\otimes n_i}(X'_N|_{Q_i})\\&\times\prod_{i\in I_2}\frac{1}{\bar{\mathrm{n}}_i!}\int_{Q_i^{\bar{\mathrm{n}}_i}}\exp\left(-\beta\mathsf{F}(Y_{\bar{\mathrm{n}}_i},\mu'_N,Q_i)\right)d(\mu'_N)^{\otimes\bar{\mathrm{n}}_i}(Y_{\bar{\mathrm{n}}_i})\\&\times\frac{1}{\bar{\mathrm{n}}_0!}\int_{Q_0^{\bar{\mathrm{n}}_0}}\exp\left(-\beta\mathsf{F}(Y_{\bar{\mathrm{n}}_0},\mu'_N,Q_0)\right)d(\mu'_N)^{\otimes\bar{\mathrm{n}}_0}(Y_{\bar{\mathrm{n}}_0}),\end{aligned}$$

where k_i is an integer $\le \bar{n}_i$, $\alpha_i = \bar{n}_i - k_i$ and α_i' satisfy (with the choices $\ell = R^{1-2\kappa}$ and $\tilde{\ell} = R^{1-\kappa}$)
(13.2.36)

$$\left|\frac{\alpha_i'}{\alpha_i} - 1\right| \le C\left(\frac{1}{\tilde{\ell}} + \frac{MR}{\tilde{\ell}^2}\right) \le CMR^{-\kappa}, \quad \frac{1}{C}R^{\mathsf{d}-\kappa} \le \frac{1}{C}\tilde{\ell}R^{\mathsf{d}-1} \le \alpha_i \le C\tilde{\ell}R^{\mathsf{d}-1} \le CR^{\mathsf{d}-\kappa}$$

and $\alpha_i + n_i - \bar{n}_i \ge 0$.

In view of the a priori bounds (7.1.55), we can bound the last integrals, corresponding to $i \in I_2$, by

$$\prod_{i \in I_2} \bar{n}_i^{\bar{n}_i} \frac{1}{\bar{n}_i!} \mathsf{K}_\beta(Q_i, \mu_N') \le \exp\left(\sum_{i \in I_2} (\bar{n}_i + C\beta\chi(\beta)R^{\mathsf{d}})\right) \le \exp\left(C(1 + \beta\chi(\beta))\frac{N}{M}\right),$$

where we used that in view of the tiling procedure, we have $\#I \le C\frac{N}{R^{\mathsf{d}}}$. To control the last integral on the last line of (13.2.35), we can use again (7.1.55) to obtain

$$\frac{1}{\bar{n}_0!}\int_{Q_0^{\bar{n}_0}} \exp\left(-\beta \mathsf{F}(Y_{\bar{n}_0}, \mu_N', Q_0)\right) d(\mu_N')^{\otimes \bar{n}_0}(Y_{\bar{n}_0}) \le \exp\left(C\beta\chi(\beta)o(N)\right).$$

Using (13.2.19) to control $\sum_i |\bar{n}_i - n_i|$ and (13.2.36), we arrive at

(13.2.37)
$$\begin{aligned}
&\int_{A'} \exp\left(-\beta \mathsf{F}(Y_N, \mu_N', \mathbb{R}^{\mathsf{d}})\right) d(\mu_N')^{\otimes N}(Y_N) \\
&\ge \exp\left(-C\left(\beta\chi(\beta)(NR^{-\kappa} + o(N)) + MNR^{-\kappa} + (1+\beta\chi(\beta))\frac{N}{M} + \frac{N}{R^{\mathsf{d}}}\log R\right)\right) \\
&\qquad \times \int_{X_N \in A} \prod_{i \in I_2, i=0} \frac{n_i!}{\bar{n}_i!} \prod_{i \in I_1} \frac{(n_i - k_i)!}{(\bar{n}_i - k_i)!} \frac{(\bar{n}_i - k_i)^{\bar{n}_i - k_i}}{(\alpha_i')^{n_i - k_i}} \\
&\qquad\qquad \exp\left(-\beta \sum_{i \in I_1} \mathsf{G}(X_N'|_{\square_i}, \mu_N'(x_i), \square_i)\right) d(\mu_N')^{\otimes N}(X_N').
\end{aligned}$$

First, for the contribution of the indices in I_2 and $i = 0$, we may write that $\log \frac{n_i!}{\bar{n}_i!} \le C|\bar{n}_i - n_i| \log R$ and use (13.2.19) to control this by $No_R(1)$.

Second, for the contribution of the indices in I_1, by Stirling's formula, $\alpha_i = \bar{n}_i - k_i$ and (13.2.36), we have

$$\begin{aligned}
&\log \prod_{i \in I_1} \frac{(n_i - k_i)!}{(\bar{n}_i - k_i)!} \frac{(\bar{n}_i - k_i)^{\bar{n}_i - k_i}}{(\alpha_i')^{n_i - k_i}} \\
&= \sum_{i \in I_1} (\bar{n}_i - n_i) + (n_i - k_i)\log(n - k_i) \\
&\qquad - (n_i - k_i)\log \alpha_i' + \frac{1}{2}\log(n_i - k_i) - \frac{1}{2}\log(\bar{n}_i - k_i) + O(1) \\
&= \sum_{i \in I_1} (\bar{n}_i - n_i) + (\alpha_i + n_i - \bar{n}_i + \frac{1}{2})\log(1 + \frac{n_i - \bar{n}_i}{\alpha_i}) + (\alpha_i + n_i - \bar{n}_i)\log\frac{\alpha_i}{\alpha_i'} + O(1).
\end{aligned}$$

We next use (13.2.36) and the convexity of the function $x \log x$, then (13.2.19) and $\#I \le C\frac{N}{R^{\mathsf{d}}}$, to obtain

$$\begin{aligned}
&\log \prod_{i\in I_1} \frac{(n_i-k_i)!}{(\bar{\mathrm{n}}_i-k_i)!}\frac{(\bar{\mathrm{n}}_i-k_i)^{\bar{\mathrm{n}}_i-k_i}}{(\alpha_i')^{n_i-k_i}}\\
&\ge -C\Big(\sum_{i\in I_1}|\bar{\mathrm{n}}_i-n_i| + R^{\mathsf{d}-\kappa}\sum_{i\in I_1}(1+\frac{n_i-\bar{\mathrm{n}}_i}{\alpha_i})\log\left(\frac{1}{\#I_1}\sum_{i\in I_1}(1+\frac{n_i-\bar{\mathrm{n}}_i}{\alpha_i})\right)\\
&\qquad + \sum_{i\in I_1}(R^{\mathsf{d}-\kappa}+|n_i-\bar{\mathrm{n}}_i|)MR^{-\kappa}\Big)\\
&\ge -CN(1+M)R^{-\kappa} - o(N) - \frac{\sum_{i\in I_1}|\bar{\mathrm{n}}_i-n_i|^2}{R^{\mathsf{d}-\kappa}} \ge -CNMR^{-\kappa}-o(N)
\end{aligned}$$

for some C that depends on $\bar{P}$. Inserting into (13.2.37), and recalling that for configurations in A', the continuous empirical field is close to $\bar{P}$, we obtain that for N large enough, after absorbing some terms,

$$\begin{aligned}
(13.2.38)\quad &\int_{Y_N,\bar{P}_N[Y_N]\in B(\bar{P},\varphi(\varepsilon))} \exp\left(-\beta\mathsf{F}(Y_N,\mu_N',\mathbb{R}^{\mathsf{d}})\right) d(\mu_N')^{\otimes N}(Y_N)\\
&\ge \exp\left(-C\left(\beta\chi(\beta)NR^{-\kappa}+MNR^{-\kappa}+(1+\beta\chi(\beta))\frac{N}{M}\right)\right)\\
&\quad\times\int_{X_N\in A}\exp\left(-\beta\sum_{i\in I_1}\mathsf{G}(X_N'|_{\square_i},\mu_N'(x_i),\square_i)\right) d(\mu_N')^{\otimes N}(X_N').
\end{aligned}$$

Combining with (13.2.34) and using that $\#IR^{\mathsf{d}} = N(1+o(1))$ because the Q_i's are almost hypercubes, we are led to

$$\begin{aligned}
(13.2.39)\quad &\int_{Y_N,\bar{P}_N[Y_N]\in B(\bar{P},\varphi(\varepsilon))} \exp\left(-\beta\mathsf{F}(Y_N,\mu_N',\mathbb{R}^{\mathsf{d}})\right) d(\mu_N')^{\otimes N}(Y_N)\\
&\ge \exp\Big(-\beta N(\overline{\mathbb{W}}(\bar{P})+o_\varepsilon(1)) - C\Big((\beta\chi(\beta)+M)NR^{-\kappa}+(\beta\chi(\beta)+1)\frac{N}{M}\Big)\Big)\\
&\qquad\times\int_A d(\mu_N')^{\otimes N}(X_N').
\end{aligned}$$

Step 7. Conclusion by large deviations estimate. Recalling that $A = \{X_N, \bar{D}_N[X_N]\in B(\bar{P},\varepsilon)\}$ and combining (13.2.15) with (13.2.39), we obtain

$$\begin{aligned}
&\frac{1}{N}\log\left(\frac{1}{N^N}\int_{Y_N,\bar{P}_N[Y_N]\in B(\bar{P},\varphi(\varepsilon))}\exp\left(-\beta\mathsf{F}(Y_N,\mu_N',\mathbb{R}^{\mathsf{d}})\right) d(\mu_N')^{\otimes N}(Y_N)\right)\\
&\ge -\beta\overline{\mathbb{W}}(\bar{P}) - \int_\Sigma \mathsf{ent}[\bar{P}^x|\Pi^{\mu(x)}]\,dx + o_\varepsilon(1)\\
&\quad - C\left((\beta\chi(\beta)+M)R^{-\kappa}+(\beta\chi(\beta)+1)\frac{1}{M}\right),
\end{aligned}$$

thus in view of (13.2.9), we have obtained

$$\begin{aligned}(13.2.40)\quad \frac{1}{N}\log \mathbb{Q}_{N,\beta}(\mu_N)\left\{Y_N, \bar{P}_N[Y_N]\in B(\bar{P},\varphi(\varepsilon))\right\} \\ \geq -\frac{1}{N}\log \mathsf{K}_\beta(\mu'_N) - \beta\overline{\mathbb{W}}(\bar{P}) - \int_\Sigma \mathsf{ent}[\bar{P}^x|\Pi^{\mu(x)}]\,dx \\ + o_\varepsilon(1) - C\left((\beta\chi(\beta)+M)R^{-\kappa} + (\beta\chi(\beta)+1)\frac{1}{M}\right).\end{aligned}$$

Conclusion. Combining this lower bound with the upper bound (13.2.10), letting $N\to\infty$, then $\varepsilon\to 0$, then $R\to\infty$, and $M\to\infty$, we then obtain, if β is independent of N,

$$\begin{aligned}&-\beta\overline{\mathbb{W}}(\bar{P}) - \int_\Sigma \mathsf{ent}[\bar{P}^x|\Pi^{\mu(x)}]\,dx \\ &\leq \lim_{\varepsilon\to 0}\liminf_{N\to\infty}\left(\frac{1}{N}\log \mathbb{Q}_{N,\beta}(\mu_N)\left\{X_N, \bar{P}_N[X_N]\in B(\bar{P},\varepsilon)\right\} + \frac{1}{N}\log \mathsf{K}_\beta(\mu'_N)\right) \\ &\leq \lim_{\varepsilon\to 0}\limsup_{N\to\infty}\left(\frac{1}{N}\log \mathbb{Q}_{N,\beta}(\mu_N)\left\{X_N, \bar{P}_N[X_N]\in B(\bar{P},\varepsilon)\right\} + \frac{1}{N}\log \mathsf{K}_\beta(\mu'_N)\right) \\ &\qquad\qquad\leq -\beta\overline{\mathbb{W}}(\bar{P}) - \int_\Sigma \mathsf{ent}[\bar{P}^x|\Pi^{\mu(x)}]\,dx.\end{aligned}$$

If $\beta\to 0$ as $N\to\infty$ we need to take R to be β-dependent in such a way that $R^{1-\kappa}\geq \beta^{-\frac{1}{\mathsf{d}-\mathsf{s}}}\mathbf{1}_{\mathsf{s}\leq 0}$ (see the assumption in Proposition 8.5). Since $\beta\chi(\beta)$ tends to 0 faster than $\beta^{1/2}|\log\beta|$ by (5.2.27), choosing $R=\beta^{-1}$ for instance ensures all error terms tend to 0 as soon as $\kappa<\frac{1}{2}$. We then obtain, after letting $N\to\infty$ then $\varepsilon\to 0$ and $M\to\infty$,

$$\begin{aligned}&-\int_\Sigma \mathsf{ent}[\bar{P}^x|\Pi^{\mu(x)}]\,dx \\ &\leq \lim_{\varepsilon\to 0}\liminf_{N\to\infty}\left(\frac{1}{N}\log \mathbb{Q}_{N,\beta}(\mu_N)\left\{X_N, \bar{P}_N[X_N]\in B(\bar{P},\varepsilon)\right\} + \frac{1}{N}\log \mathsf{K}_\beta(\mu'_N)\right) \\ &\leq \lim_{\varepsilon\to 0}\limsup_{N\to\infty}\left(\frac{1}{N}\log \mathbb{Q}_{N,\beta}(\mu_N)\left\{X_N, \bar{P}_N[X_N]\in B(\bar{P},\varepsilon)\right\} + \frac{1}{N}\log \mathsf{K}_\beta(\mu'_N)\right) \\ &\leq -\int_\Sigma \mathsf{ent}[\bar{P}^x|\Pi^{\mu(x)}]\,dx.\end{aligned}$$

If $\beta\to\infty$, we obtain instead

$$\begin{aligned}-\overline{\mathbb{W}}(\bar{P}) &\leq \lim_{\varepsilon\to 0}\liminf_{N\to\infty}\left(\frac{1}{N\beta}\log \mathbb{Q}_{N,\beta}(\mu_N)\{X_N, \bar{P}_N[X_N]\in B(\bar{P},\varepsilon)\}\right. \\ &\qquad\qquad\left. + \frac{1}{N\beta}\log \mathsf{K}_\beta(\mu'_N)\right) \\ &\leq \lim_{\varepsilon\to 0}\limsup_{N\to\infty}\left(\frac{1}{N\beta}\log \mathbb{Q}_{N,\beta}(\mu_N)\{X_N, \bar{P}_N[X_N]\in B(\bar{P},\varepsilon)\}\right. \\ &\qquad\qquad\left. + \frac{1}{N\beta}\log \mathsf{K}_\beta(\mu'_N)\right) \leq -\overline{\mathbb{W}}(\bar{P}).\end{aligned}$$

Exponential tightness of $\mathbb{Q}_{N,\beta}(\mu_N)$ is an easy consequence of the fact that the total number of points in Σ is bounded by N. As explained in Corollary 3.13 and using the result of Step 2, it allows to upgrade these results into results about arbitrary sets $A \in \mathcal{P}_s(\Sigma \times \mathcal{X})$:

$$\begin{aligned}
-\inf_{\overset{\circ}{A}} &\left(\beta\overline{\mathbb{W}}(\bar{P}) + \int_\Sigma \mathsf{ent}[\bar{P}^x|\Pi^{\mu(x)}]\,dx\right)\\
&\le \liminf_{N\to\infty}\left(\frac{1}{N}\log \mathbb{Q}_{N,\beta}(\mu_N)\left\{X_N, \bar{P}_N[X_N]\in A\right\} + \frac{1}{N}\log \mathsf{K}_\beta(\mu_N')\right)\\
&\le \limsup_{N\to\infty}\left(\frac{1}{N}\log \mathbb{Q}_{N,\beta}(\mu_N)\left\{X_N, \bar{P}_N[X_N]\in A\right\} + \frac{1}{N}\log \mathsf{K}_\beta(\mu_N')\right)\\
&\qquad\qquad \le -\inf_{\overline{A}\cap\mathcal{P}_{s,\mu}(\Sigma\times\mathcal{X})}\left(\beta\overline{\mathbb{W}}(\bar{P}) + \int_\Sigma \mathsf{ent}[\bar{P}^x|\Pi^{\mu(x)}]\,dx\right)
\end{aligned}$$

and respectively the same for the other regimes. Applying to A equals to the whole space, and using (5.2.12) we obtain (13.2.2) in the regime of fixed β, or

$$\lim_{N\to\infty}\frac{1}{N}\log \mathsf{K}_{N,\beta}(\mu_N) + \left(\frac{1}{2\mathsf{d}}\log N\right)\mathbf{1}_{\mathsf{s}=0} = 0$$

in the regime $\beta \to 0$ (which we already knew), resp. (13.2.3) if $\beta\to\infty$.

Inserting into the above relations, we have obtained the full LDP results. Note that the "goodness" of the rate functions is a consequence of Corollary 12.5 and Lemma 13.1. □

PROOF OF THEOREM 13.8. Let us consider P a stationary probability measure on infinite point configurations with intensity $\mu(x_0)$, and $B(P,\varepsilon)$ a ball for some distance that metrizes the weak topology. We focus on proving upper and lower bounds on $\log \mathfrak{P}_{N,\beta}^{x_0,R}(B(P,\varepsilon))$. For simplicity, let us denote $\square_R$ for $\square_R(N^{1/\mathsf{d}}x_0^N)$.

Step 1: Reducing to good number of points and good energy. Since R is large enough, we may include $\square_R$ in a hyperrectangle Q_R such that $\bar{\mathrm{n}} = \mu(Q_R)$ is an integer and $|Q_R| - |\square_R| = O(R^{\mathsf{d}-1}) = o(R^{\mathsf{d}})$.

Let us denote by n the number of points a configuration has in Q_R. Since we assume $R \gg \rho_\beta \ge C\max\left(\beta^{-\frac12}\chi(\beta)^{\frac12}, 1\right)$, for κ small enough we have from (5.2.27) that $R^{2-3\kappa} \ge \chi(\beta)$ in all dimensions, hence in view of the local laws in the form (8.4.2) and (8.4.3) we may write that for some $\kappa > 0$

$$\mathbb{Q}_{N,\beta}(\mu_N)\left\{|n-\bar{\mathrm{n}}| \ge R^{\mathsf{d}-\kappa}\right\} \le \exp\left(-C\beta R^{\mathsf{d}+\kappa}\right) \tag{13.2.41}$$

and

$$\mathbb{Q}_{N,\beta}(\mu_N)\left\{\sup_x \int_{\square_{R^{1+\kappa/\mathsf{d}}}} |\nabla u_{\tilde{\mathsf{r}}}|^2 \ge C\chi(\beta)R^{\mathsf{d}+\kappa}\right\} \le \exp\left(-\chi(\beta)\beta R^{\mathsf{d}+\kappa}\right) \tag{13.2.42}$$

for some C large enough independent of R and β. Hence we may restrict the study to the event

$$\mathcal{B} = \left\{|n-\bar{\mathrm{n}}| \le R^{\mathsf{d}-\kappa},\ \sup_x \int_{\square_{R^{1+\kappa/\mathsf{d}}}} |\nabla u_{\tilde{\mathsf{r}}}|^2 \le \chi(\beta)R^{\mathsf{d}+\kappa}\right\}, \tag{13.2.43}$$

since the complement has a probability which is negligible in the speed we are interested in. In particular, such configurations are screenable.

Step 2: Upper bound. We recall that $P_N^{x_0^N,R}$ is defined in (12.1.2). Using (7.1.37) and Lemma 7.18, for $\mathcal{B}$ as in (13.2.43), we have

$$\begin{aligned}
&\mathfrak{P}_{N,\beta}^{x_0,R}(B(P,\varepsilon)\cap P_N^{x_0^N,R}[\mathcal{B}]) \\
&= \frac{1}{N^N \mathsf{K}_\beta(\mu'_N)} \int_{\{P_N^{x_0^N,R}[X_N]\in B(P,\varepsilon)\}\cap\mathcal{B}} \exp\left(-\beta \mathsf{F}(X'_N,\mu'_N,\mathbb{R}^{\mathsf{d}})\right) d(\mu'_N)^{\otimes N}(X_N) \\
&\le \frac{1}{N^N \mathsf{K}_\beta(\mu'_N)} \int_{\{P_N^{x_0^N,R}[X_N]\in B(P,\varepsilon)\}\cap\mathcal{B}} \\
&\quad \times \exp\left(-\beta \mathsf{G}^{\mathrm{inn}}_{\mathbb{R}^{\mathsf{d}}}(X'_N|_{Q_R},\mu'_N,Q_R) - \beta \mathsf{G}^{\mathrm{out}}_{\mathbb{R}^{\mathsf{d}}}(X'_N|_{Q_R^c},\mu'_N,Q_R^c)\right) d(\mu'_N)^{\otimes N}(X_N).
\end{aligned}$$

Splitting up the events as in the proof of Theorem 8.4 with n being the number of points of the configuration which belong to Q_R, and using that $P_N^{x_0^N,R}[X_N]$ depends only on the configuration in $\square_R$ hence in Q_R, we may then write

(13.2.44)

$$\begin{aligned}
&\mathfrak{P}_{N,\beta}^{x_0,R}(B(P,\varepsilon)\cap P_N^{x_0^N,R}[\mathcal{B}]) \\
&\le \frac{1}{N^N \mathsf{K}_\beta(\mu'_N)} \sum_{n=\bar{\mathrm{n}}-R^{\mathsf{d}-\kappa}}^{\bar{\mathrm{n}}+R^{\mathsf{d}-\kappa}} \frac{N!}{n!(N-n)!} \\
&\quad \times \int_{\mathcal{B}_n\cap(Q_R^c)^{N-n}} \exp\left(-\beta \mathsf{G}^{\mathrm{out}}_{\mathbb{R}^{\mathsf{d}}}(\cdot,\mu'_N,Q_R^c)\right) d(\mu'_N)^{\otimes(N-n)} \\
&\quad \times \int_{(Q_R)^n\cap\{P_N^{x_0^N,R}[N^{-1/\mathsf{d}}X_n]\in B(P,\varepsilon)\}} \exp\left(-\beta \mathsf{G}^{\mathrm{inn}}_{\mathbb{R}^{\mathsf{d}}}(\cdot,\mu'_N,Q_R)\right) d(\mu'_N)^{\otimes n},
\end{aligned}$$

where $\mathcal{B}_n$ is $\mathcal{B}$ intersected with the event that X'_N has n points in Q_R. Then, (8.3.29) applied with L such that $R \gg L \gg \rho_\beta$ and combined with Remark 8.7 yields

$$\begin{aligned}
&\int_{\mathcal{B}_n\cap(Q_R^c)^{N-n}} \exp\left(-\beta \mathsf{G}^{\mathrm{out}}_{\mathbb{R}^{\mathsf{d}}}(\cdot,\mu'_N,Q_R^c)\right) d(\mu'_N)^{\otimes(N-n)} \\
&\qquad \le \frac{(N-n)!(N-\bar{\mathrm{n}})^{N-\bar{\mathrm{n}}}}{(N-\bar{\mathrm{n}})!} \mathsf{K}_\beta(\mu'_N,Q_R^c)\exp\left(C(\beta\chi(\beta)+1)o(R^{\mathsf{d}})\right),
\end{aligned}$$

with C independent of β.

We next apply Lemma 13.5 in Q_R with $m=\mu(x_0)$ to obtain that

$$\begin{aligned}
(13.2.45)\quad &\frac{1}{|Q_R|}\log\left(\frac{1}{\bar{\mathrm{n}}^n}(\mu'_N)^{\otimes n}\{X_n\in(Q_R)^n, P_N^{x_0^N,R}[N^{-1/\mathsf{d}}X_n]\in B(P,\varepsilon)\}\right) \\
&\qquad = -\mathsf{ent}[P|\Pi^{\mu(x_0)}] + o_\varepsilon(1) + o_N(1).
\end{aligned}$$

Technically, we used that the restriction of $(\mu'_N/\int_{Q_R}\mu'_N)^{\otimes n}$ to Q_R is asymptotic to an n-point Bernoulli process, the fact that $\bar{\mathrm{n}}=\mu(Q_R)$ and $|n-\bar{\mathrm{n}}|=o(R^{\mathsf{d}})$. We may rewrite this as

$$\begin{aligned}
(13.2.46)\quad &\log(\mu'_N)^{\otimes n}\{X_n\in(Q_R)^n, P_N^{x_0^N,R}[N^{-1/\mathsf{d}}X_n]\in B(P,\varepsilon)\} \\
&\qquad = -R^{\mathsf{d}}\mathsf{ent}[P|\Pi^{\mu(x_0)}] + n\log\bar{\mathrm{n}} + R^{\mathsf{d}}(o_\varepsilon(1)+o_N(1)).
\end{aligned}$$

Moreover, the same proof as that of Proposition 12.2 (applied to $\mathsf{G}^{\mathrm{inn}}_{\mathbb{R}^{\mathsf{d}}}$ instead of F) yields that

$$\liminf_{N\to\infty} \frac{1}{|Q_R|} \mathsf{G}^{\mathrm{inn}}_{\mathbb{R}^{\mathsf{d}}}(X_n, \mu'_N, Q_R) \geq \int \mathbb{W}(\mathcal{C}, \mu(x_0)) dP'(\mathcal{C})$$

where P' is the limit, up to extraction, of $P_N^{x_0^N, R}[X_n]$. In addition, from the lower semicontinuity of Lemma 12.5, we deduce with the fact that $|Q_R| = R^{\mathsf{d}} + o(R^{\mathsf{d}})$, that if $P_N^{x_0^N, R}[X_n] \in B(P, \varepsilon)$ then

$$\liminf_{N\to\infty} \frac{1}{R^{\mathsf{d}}} \mathsf{G}^{\mathrm{inn}}_{\mathbb{R}^{\mathsf{d}}}(X_n, \mu'_N, Q_R) \geq \int \mathbb{W}(\mathcal{C}, \mu(x_0)) dP(\mathcal{C}) - o_\varepsilon(1).$$

Denoting $\mathbb{W}(P, m)$ for $\int \mathbb{W}(\mathcal{C}, m) dP(\mathcal{C})$, combining this with (13.2.46) and inserting both into (13.2.44) leads to

$$\begin{aligned} &\mathfrak{P}_{N,\beta}^{x_0,R}(B(P,\varepsilon) \cap P_N^{x_0,R}(\mathcal{B})) \\ &\leq \exp\Big(-R^{\mathsf{d}}\Big(\beta\mathbb{W}(P,\mu(x_0)) + \mathsf{ent}[P|\Pi^{\mu(x_0)}] + (1+\beta)o_{\varepsilon,N}(1) + C\beta\chi(\beta)R^{-\kappa}\Big)\Big) \\ &\quad \times \sum_{n=\bar{\mathrm{n}}-R^{\mathsf{d}-\kappa}}^{\bar{\mathrm{n}}+R^{\mathsf{d}-\kappa}} \frac{1}{N^N \mathsf{K}_\beta(\mu'_N, \mathbb{R}^{\mathsf{d}})} \frac{N!}{(N-\bar{\mathrm{n}})!} \frac{(N-\bar{\mathrm{n}})^{N-\bar{\mathrm{n}}}\bar{\mathrm{n}}^n}{n!} \mathsf{K}_\beta(\mu'_N, Q_R^c). \end{aligned} \tag{13.2.47}$$

On the other hand, using (7.1.17) we have

$$\mathsf{K}_\beta(\mu'_N) \geq \frac{N!N^{-N}}{\mathrm{n}!(N-\mathrm{n})!\bar{\mathrm{n}}^{-\bar{\mathrm{n}}}(N-\bar{\mathrm{n}})^{-(N-\bar{\mathrm{n}})}} \mathsf{K}_\beta(Q_R, \mu'_N) \mathsf{K}_\beta(\mu'_N, Q_R^c),$$

and inserting this into (13.2.47), we find

$$\begin{aligned} &\mathfrak{P}_{N,\beta}^{x_0,R}(B(P,\varepsilon) \cap P_N^{x_0^N,R}(\mathcal{B})) \\ &\leq \sum_{n=\bar{\mathrm{n}}-R^{\mathsf{d}-\kappa}}^{\bar{\mathrm{n}}+R^{\mathsf{d}-\kappa}} \exp\Big(-R^{\mathsf{d}}\Big(\beta\mathbb{W}(P,\mu(x_0)) + \mathsf{ent}[P|\Pi^{\mu(x_0)}] \\ &\qquad\qquad + (1+\beta)o_{\varepsilon,N}(1)\Big)\Big) \frac{\bar{\mathrm{n}}!}{n!} \bar{\mathrm{n}}^{n-\bar{\mathrm{n}}} \frac{1}{\mathsf{K}_\beta(\mu'_N, Q_R)}. \end{aligned}$$

We note that $\log \bar{\mathrm{n}}^{n-\bar{\mathrm{n}}} = O(R^{\mathsf{d}-\kappa}\log R) = o(R^{\mathsf{d}})$ and similarly $\log \frac{\bar{\mathrm{n}}!}{n!} = o(R^{\mathsf{d}})$ for $|n - \bar{\mathrm{n}}| \leq R^{\mathsf{d}-\kappa}$, hence bounding the sum by the number of terms which is $O(R^{\mathsf{d}-\kappa})$ and reabsorbing this factor into the errors, we find the upper bound

$$\begin{aligned} &\log \mathfrak{P}_{N,\beta}^{x_0,R}(B(P,\varepsilon) \cap P_N^{x_0^N,R}[\mathcal{B}]) \\ &\leq -R^{\mathsf{d}}\left(\beta\mathbb{W}(P,m) + \mathsf{ent}[P|\Pi^m] + (1+\beta)o_{\varepsilon,N}(1)\right) - \log \mathsf{K}_\beta(\mu'_N, Q_R) \end{aligned} \tag{13.2.48}$$

Step 3: Lower bound. We claim that given any P such that $\mathbb{W}(P, \mu(x_0)) + \mathsf{ent}[P|\Pi^{\mu(x_0)}]$ is finite, we can construct a family A of configurations $X_{\bar{\mathrm{n}}}$ of $\bar{\mathrm{n}}$ points in Q_R such that $P_N^{x_0^N,R}[X_{\bar{\mathrm{n}}}] \in B(P, \varepsilon)$,

$$\mathsf{F}(X_{\bar{\mathrm{n}}}, \mu'_N, Q_R) \leq R^{\mathsf{d}}\mathbb{W}(P, \mu(x_0)) + o(R^{\mathsf{d}}) \tag{13.2.49}$$

uniformly in A, and

$$\log(\mu'_N)^{\otimes \bar{\mathrm{n}}}(A) = -R^{\mathsf{d}}\mathsf{ent}[P|\Pi^{\mu(x_0)}] + \bar{\mathrm{n}}\log\bar{\mathrm{n}} + R^{\mathsf{d}}(o_\varepsilon(1) + o_N(1)). \tag{13.2.50}$$

This follows the same steps as the lower bound in the proof of Theorem 13.7, i.e., it is done by sampling configurations whose local empirical field $P_N^{x_0^N,R}$ is close to

P and screening them in Q_R which still keeps $P_N^{x_0^N,R}$ close to P. It is however an easier setting since we do not have to partition into rectangles.

We may thus write with the help of (7.1.28)

$$\begin{aligned}
&\mathfrak{P}_{N,\beta}^{x_0^N,R}(B(P,\varepsilon))\\
&= \frac{1}{N^N \mathsf{K}_\beta(\mu_N')} \int_{P_N^{x_0^N,R}(X_N)\in B(P,\varepsilon)} \exp\left(-\beta \mathsf{F}(X_N,\mu_N',\mathbb{R}^{\mathsf{d}})\right) d(\mu_N')^{\otimes N}(X_N)\\
&\geq \frac{1}{N^N \mathsf{K}_\beta(\mu_N')} \frac{N!}{\bar{\mathrm{n}}!(N-\bar{\mathrm{n}})!} \int_{(Q_R^c)^{N-\bar{\mathrm{n}}}} \exp\left(-\beta \mathsf{F}(\cdot,\mu_N',Q_R^c)\right) d(\mu_N')^{\otimes (N-\bar{\mathrm{n}})}(X_N)\\
&\quad \times \int_A \exp\left(-\beta \mathsf{F}(X_{\bar{\mathrm{n}}},\mu_N',Q_R)\right) d(\mu_N')^{\otimes \bar{\mathrm{n}}}(X_{\bar{\mathrm{n}}})\\
&= \frac{\mathsf{K}_\beta(\mu_N',Q_R^c)}{N^N \mathsf{K}_\beta(\mu_N')} \frac{N!}{\bar{\mathrm{n}}!(N-\bar{\mathrm{n}})!(N-\bar{\mathrm{n}})^{-(N-\bar{\mathrm{n}})}} \int_A \exp\left(-\beta \mathsf{F}(X_{\bar{\mathrm{n}}},\mu_N',Q_R)\right) d(\mu_N')^{\otimes \bar{\mathrm{n}}}(X_{\bar{\mathrm{n}}}).
\end{aligned}$$

But in view of (8.4.9) we have

$$\log \mathsf{K}_\beta(\mu_N') = \log \mathsf{K}_\beta(\mu_N',Q_R) + \log \mathsf{K}_\beta(\mu_N',Q_R^c) + o((1+\beta\chi(\beta))R^{\mathsf{d}})$$

so in view of (13.2.49) and (13.2.50) and using Stirling's formula, we conclude that

$$\begin{aligned}
&\log \mathfrak{P}_{N,\beta}^{x_0,R}(B(P,\varepsilon)) \\
&\geq -\log \mathsf{K}_\beta(\mu_N',Q_R) - R^{\mathsf{d}}\left(\beta \mathbb{W}(P,\mu(x_0)) + \mathsf{ent}[P|\Pi^{\mu(x_0)}] + (1+\beta)o_{\varepsilon,N}(1)\right).
\end{aligned} \tag{13.2.51}$$

Step 4: Conclusion. Exponential tightness at speed R^{d} follows from the fact that the number of points is essentially bounded by CR^{d} by Theorem 8.4. Then by Corollary 3.13, we may upgrade the conclusions of the previous steps to a strong LDP result: for any Borel set E, it holds that, as $N\to\infty$,

(13.2.52)

$$\begin{aligned}
&\log \mathfrak{P}_{N,\beta}^{x_0,R}(E)\\
&\leq -R^{\mathsf{d}} \inf_{P\in \bar{E}} \left(\beta \mathbb{W}(P,\mu(x_0)) + \mathsf{ent}[P|\Pi^{\mu(x_0)}]\right) - \log \mathsf{K}_\beta(\mu_N',Q_R) + (1+\beta)o(R^{\mathsf{d}})
\end{aligned}$$

and

(13.2.53)

$$\begin{aligned}
&\log \mathfrak{P}_{N,\beta}^{x_0,R}(E)\\
&\geq -R^{\mathsf{d}} \inf_{P\in \mathring{E}} \left(\beta \mathbb{W}(P,\mu(x_0)) + \mathsf{ent}[P|\Pi^{\mu(x_0)}]\right) - \log \mathsf{K}_\beta(\mu_N',Q_R) + (1+\beta)o(R^{\mathsf{d}}).
\end{aligned}$$

Applying this relation to E equal the whole space, we find

$$\log \mathsf{K}_\beta(\mu_N',Q_R) = -R^{\mathsf{d}} \inf_{P\in \mathcal{P}_s(\mathcal{X})} \left(\beta \mathbb{W}(P,\mu(x_0)) + \mathsf{ent}[P|\Pi^{\mu(x_0)}]\right) + (1+\beta)o(R^{\mathsf{d}}).$$

Reinserting into (13.2.52) and (13.2.53), the stated LDP result follows if β is fixed. The generalization to $\beta\to 0$ or $\beta\to\infty$ is straightforward from (13.2.48) and (13.2.51). This concludes the proof of Theorem 13.8. □

APPENDIX A

Proof of the screening result

The goal of this appendix is to prove the screening result of Proposition 7.19. We follow here closely [**AS21**, Appendix C]. This consists in adapting and optimizing the procedure from [**SS15b, RS15, PS17**], in particular [**PS17**] simplified to the Coulomb case.

Let us first describe things informally, for the inner screening. Let w solve (7.2.4) and let $E = \nabla w$ be the associated electric field, which satisfies a relation of the form

$$\begin{cases} -\mathrm{div}\, E = \mathsf{c_d}\left(\sum_{i=1}^n \delta_{x_i} - \mu\right) & \text{in } \Omega'' \\ E \cdot \nu = 0 & \text{on } \partial U \cap \Omega''. \end{cases} \tag{A.0.1}$$

A truncated version of E is defined as in (7.1.26) by

$$E_{\hat{\mathsf{r}}} = E - \sum_{i=1}^n \nabla \mathsf{f}_{\hat{\mathsf{r}}_i}(x - x_i) \tag{A.0.2}$$

where $\hat{\mathsf{r}}_i$ is as in (7.1.24). The precise choice of truncation is unimportant, we may use $\mathsf{r}, \tilde{\mathsf{r}}$ or $\hat{\mathsf{r}}$, since in the boundary layer where the definitions differ, the configuration will be deleted.

Given a configuration X_n in Ω', together with its electric field E, assume that we have a good control on its energy near the boundary of a hyperrectangle Q_t with t close to R. The goal of the screening is to modify the configuration X_n and the electric field E only outside of Q_{t-1} and extend them into a screened configuration $Y_{\bar{\mathrm{n}}}$ and a screened electric field E^{scr} in $\Omega = Q_R \cap U$ in such a way that

$$\begin{cases} -\mathrm{div}\, E^{\mathrm{scr}} = \mathsf{c_d}\left(\sum_{p \in Y_{\bar{\mathrm{n}}}} \delta_p - \mu\right) & \text{in } \Omega \\ E^{\mathrm{scr}} \cdot \nu = 0 & \text{on } \partial U \cap \Omega \end{cases}$$

This implies in particular that the screened system is neutral, i.e., the number of points of $Y_{\bar{\mathrm{n}}}$ must be equal to $\bar{\mathrm{n}} = \mu(\Omega)$, an integer. We note that in the Neumann case where Ω can intersect ∂U, the desired boundary condition is already satisfied for the original field on ∂U, so there is no need to modify it near ∂U.

The screened electric field E^{scr} may not be a gradient, however thanks to Lemma 7.5 its energy provides an upper bound for computing $\mathsf{F}(Y_{\bar{\mathrm{n}}}, \Omega)$. The goal of the construction is to show that we can build E^{scr} and $Y_{\bar{\mathrm{n}}}$ without adding too much energy to that of the original configuration, which will allow to bound $\mathsf{F}(Y_{\bar{\mathrm{n}}}, \Omega)$ in terms of $\mathsf{G}_U^{\mathrm{inn}}(X_n, \Omega)$. In order to accomplish this, we will split the region to be filled into cells where we solve appropriate elliptic problems and estimate the energies by elliptic regularity estimates. In order to "absorb" and screen the effect of the possibly rough data on ∂Q_t, we need a certain distance ℓ, which has to be large enough in terms of the energy of E. This leads to the "screenability condition" bound on ℓ, as previously mentioned.

A.0.1. Finding a good boundary. We focus on the outer screening proof, the proof of the inner case is analogous (for details of what to do near the corners, one may refer to [**RNS15**]).

Assume then that $\Omega = Q_R \cap U$. Since U is assumed to be a disjoint union of parallel hyperrectangles, Ω is itself a hyperrectangle.

We are given $\tilde{\ell} \geq \ell \geq C$, a configuration X_n in Ω', $E = \nabla w$ with the notation $E_{\hat{\mathsf{r}}}$ defined in (A.0.2). We recall there are two variants of the construction depending on which term ensures the screenability condition is met.

- In the first case, where screenability is achieved by (7.2.7), by a mean value argument we can find $\Gamma = \partial Q_t$ for some $t \in [R - 2\tilde{\ell} + 2, R - \tilde{\ell} - 2]$ such that

$$\int_{\Gamma \cap U} |E_{\hat{\mathsf{r}}}|^2 \leq \frac{S(X_n, w)}{\tilde{\ell}} \tag{A.0.3}$$

  and

$$\int_{(Q_{t+2} \backslash Q_{t-2}) \cap U} |E_{\hat{\mathsf{r}}}|^2 \leq C \frac{S(X_n, w)}{\tilde{\ell}}. \tag{A.0.4}$$

- In the second case where the screenability is ensured by (7.2.8), using a mean-value argument we can find $t \in [R - 2\tilde{\ell}, R - \tilde{\ell} - \ell]$ such that

$$\int_{(Q_{t+\ell} \backslash Q_t) \cap U} |E_{\hat{\mathsf{r}}}|^2 \leq C \frac{S(X_n, w)\ell}{\tilde{\ell}} \tag{A.0.5}$$

  and then, by a covering argument and a mean-value argument in the strip $Q_{t+\ell} \backslash Q_t$, we can find a piecewise affine boundary Γ, included in $Q_{t+\ell} \backslash Q_t$ for some $t \in [R - 2\tilde{\ell} + \ell + 1, R - \tilde{\ell} - 1]$, with faces parallel to those of Q_R, of sidelengths bounded above and below by constants times ℓ, such that

$$\int_{\Gamma \cap U} |E_{\hat{\mathsf{r}}}|^2 \leq C \frac{S(X_n, w)}{\tilde{\ell}}, \qquad \sup_x \int_{\Gamma \cap U \cap \square_\ell(x)} |E_{\hat{\mathsf{r}}}|^2 \leq C S'(X_n, w) \tag{A.0.6}$$

  and

$$\int_{\Gamma_1} |E_{\hat{\mathsf{r}}}|^2 \leq C \frac{S(X_n, w)}{\tilde{\ell}}, \tag{A.0.7}$$

  where Γ_1 denotes the 1-neighborhood of Γ.

In both cases, we let $M = C\frac{S(X_n, w)}{\tilde{\ell}}$, and in the second case we let $M_\ell = CS'(X_n, w)$, for the largest C appearing in the right-hand side. We also can make t a measurable function of X_n by requiring it to be the largest t in the admissible interval such that the inequalities hold (and same for the parameters that define the facets).

We note that as soon as $\tilde{\ell}$ is large enough, we only consider regions at distance ≥ 1 from $\partial\Omega \backslash \partial U$, so there is no difference between $\hat{\mathsf{r}}$ and $\tilde{\mathsf{r}}$ there.

We denote by $\mathcal{O}$ (like "old") the part of Ω delimited by Γ and ∂U, and by $\mathcal{N}$ (like "new") the set $\Omega \backslash \mathcal{O}$. By construction, we have $\mathcal{O} \subset \Omega''$. We keep X_n and E unchanged in $\mathcal{O}$ and discard the points of X_n in $\mathcal{O}^c$ to replace them by new ones. The good boundary Γ may intersect some $B(x_i, \hat{\mathsf{r}}_i)$ balls centered at points of X_n. These balls will need to be "completed", i.e., the contributions of $\delta_{x_i}^{(\hat{\mathsf{r}}_i)} \mathbf{1}_{\mathcal{N}}$ retained.

A.0.2. Preliminary lemmas. We start with a series of preliminary results which will be the building blocks for the construction of E^{scr}.

Lemma A.1 (Correcting fluxes on rectangles)**.** *Let H be a hyperrectangle of $\mathbb{R}^{\mathsf{d}}$ with sidelengths in $[\ell, C\ell]$ with C depending only on d. Let $g \in L^2(\partial H)$. Then there exists a constant C depending only on d such that the mean zero solution of*

$$\begin{cases} -\Delta h = \frac{1}{|H|}\int_{\partial H} g & \text{in } H \\ \frac{\partial h}{\partial \nu} = g & \text{on } \partial H \end{cases} \tag{A.0.8}$$

satisfies the estimate

$$\int_H |\nabla h|^2 \le C\ell \int_{\partial H} |g|^2. \tag{A.0.9}$$

Proof. We test the equation (A.0.8) by h, and using Green's theorem and the fact that h integrates to 0 in H, we obtain that

$$\int_H |\nabla h|^2 = \int_{\partial H} gh$$

where the last integral can be understood in the duality pairing of $H^{-1/2}(\partial H)$ and $H^{1/2}(\partial H)$.

Since $g \in L^2$, by trace theorem and scaling we obtain

$$\int_H |\nabla h|^2 \le C\ell^{\frac{1}{2}} \left(\int_H |\nabla h|^2\right)^{\frac{1}{2}} \left(\int_{\partial H} |g|^2\right)^{\frac{1}{2}},$$

and the result follows. □

We note that the result holds as well in a set that is not a hyperrectangle but is an ℓ-rescaling of a nondegenerate fixed polygon, as we need in the second case.

The next lemma serves to complete the smeared charges which were cut into two pieces by the choice of the good boundary.

Lemma A.2 (Completing charges near the boundary)**.** *Let H be a hyperrectangle of $\mathbb{R}^{\mathsf{d}}$ with sidelengths in $[\ell, C\ell]$ with C depending only on d. Let F be a face of H. Let $\{x_i\}_{i\in I}$ be points contained in an $1/4$-neighborhood of F. Let c be a constant such that*

$$c|F| = \mathsf{c_d} \int_H \sum_{i\in I} \delta_{x_i}^{(\hat{\mathsf{r}}_i)}. \tag{A.0.10}$$

The mean-zero solution to

$$\begin{cases} -\Delta h = \mathsf{c_d} \sum_{i\in I} \delta_{x_i}^{(\hat{\mathsf{r}}_i)} & \text{in } H\,, \\ \frac{\partial h}{\partial \nu} = 0 & \text{on } \partial H \setminus F\,, \\ \frac{\partial h}{\partial \nu} = c & \text{on } F \end{cases} \tag{A.0.11}$$

satisfies

$$\int_H |\nabla h|^2 \le C\left((\#I)^2 \ell^{2-\mathsf{d}} + \sum_{i\ne j} \mathsf{g}(x_i - x_j) + \sum_{i\in I} \mathsf{g}(\hat{\mathsf{r}}_i)\right) \tag{A.0.12}$$

where C depends only on d.

The result holds as well in a set that is not a hyperrectangle but is an ℓ-rescaling of a nondegenerate fixed polygon, as we need in the second case.

PROOF. Integrating (A.0.11) over H, we find that $c|F| \le \mathsf{c_d}\#I$ hence $c \le C(\#I)\ell^{1-\mathsf{d}}$. We may then split h into $h = u + v$ where

$$\begin{cases} -\Delta u = \mathsf{c_d}\sum_i \delta_{x_i}^{(\hat{\mathsf{r}}_i)} - c\frac{|F|}{|H|} & \text{in } H \\ \frac{\partial u}{\partial \nu} = 0 & \text{on } \partial H, \end{cases}$$

and

$$\begin{cases} -\Delta v = c\frac{|F|}{|H|} & \text{in } H \\ \frac{\partial v}{\partial \nu} = 0 & \text{on } \partial H \setminus F \\ \frac{\partial v}{\partial \nu} = c & \text{on } F. \end{cases}$$

For the v part, Lemma A.1 yields that $\int_H |\nabla v|^2 \le Cc^2\ell^{\mathsf{d}} \le C(\#I)^2\ell^{2-\mathsf{d}}$. For the u part, we observe that

$$u = \mathsf{c_d}\sum_{i\in I}\int_H G_H(x,y)d\delta_{x_i}^{(\hat{\mathsf{r}}_i)}(y)$$

where $G_H(x,y)$ is the Neumann Green function of the hyperrectangle with average 0. From [**AS21**, Proposition A.1], or by reflection through the boundary, we have

$$G_H(x,y) \le C\mathsf{g}(x-y)$$

hence, using that G_H has mean zero, we deduce that

$$\begin{aligned}\int_H |\nabla u|^2 &= \mathsf{c_d^2}\sum_{i\in I}\iint G_H(x,y)d\delta_{x_i}^{(\hat{\mathsf{r}}_i)}(x)d\left(\sum_{j\in I}\delta_{x_j}^{(\hat{\mathsf{r}}_j)} - c\frac{|F|}{|H|}\right)(y)\\ &= \mathsf{c_d^2}\sum_{i,j\in I}\iint G_H(x,y)d\delta_{x_i}^{(\hat{\mathsf{r}}_i)}(x)d\delta_{x_j}^{(\hat{\mathsf{r}}_j)}\\ &\le C\sum_{i\ne j}\iint \mathsf{g}(x-y)d\delta_{x_i}^{(\hat{\mathsf{r}}_i)}(x)d\delta_{x_j}^{(\hat{\mathsf{r}}_j)}(y) + C\sum_{i\in I}\mathsf{g}(\hat{\mathsf{r}}_i)\end{aligned}$$

and the result follows from Newton's theorem and disjointness of the balls. □

A.0.3. Main proof.

We let I_∂ be the indices corresponding to the points of X_n whose smeared charges touch Γ, i.e.,

(A.0.13) $$I_\partial = \{i \in [1,n] : \ B(x_i, \hat{\mathsf{r}}_i) \cap \Gamma \ne \varnothing\}$$

and define

$$n_{\mathcal{O}} = \#I_\partial + \#(\{i, x_i \in \mathcal{O}\}\backslash I_\partial).$$

The goal of the construction is to place an additional $\bar{\mathrm{n}} - n_{\mathcal{O}}$ points in $(Q_R \cap U)\backslash\mathcal{O}$, where $\bar{\mathrm{n}} = \mu(Q_R \cap U)$.

By construction of Γ, we may partition $(Q_R \cap U)\backslash\mathcal{O}$ into hyperrectangles H_k with sidelengths $\in [\ell/C, C\ell]$ for some positive constant $C > 0$. We build the H_k's in such a way that, letting m_k be the constant such that

(A.0.14) $$m_k|H_k| = \frac{1}{\mathsf{c_d}}\left(\int_{\Gamma\cap\partial H_k} E_{\hat{\mathsf{r}}}\cdot\nu - n_k\right),$$

with ν denoting the outer unit normal to $\mathcal{O}$ and

$$n_k := \mathsf{c_d} \int_{H_k} \sum_{i \in I_\partial} d\delta_{x_i}^{(\hat{\mathsf{r}}_i)},$$

we have $\int_{H_k} (\mu + m_k) \in \mathbb{N}$. This is possible if $|m_k| < \frac{1}{2}m$ (recall $\mu \geq m$) and can be done by constructing successive strips as in Lemma 5.13, as soon as $\ell > C$ for some $C > 0$ depending only on d and m.

We will give below a condition to ensure that $|m_k| < \frac{1}{2}m$. Now define

$$\tilde{\mu} = \mu + \sum_k \mathbf{1}_{H_k} m_k. \tag{A.0.15}$$

It is a nonnegative density supported in $\mathcal{N}$. Since

$$n_{\mathcal{O}} = -\frac{1}{\mathsf{c_d}} \int_\Gamma E_{\hat{\mathsf{r}}} \cdot \nu + \frac{1}{\mathsf{c_d}} \sum_k n_k + \int_{\mathcal{O}} d\mu$$

and $\bar{\mathrm{n}} = \mu(\Omega)$, in view of (A.0.14) we may check that

$$\int_{\mathcal{N}} \tilde{\mu} = \bar{\mathrm{n}} - n_{\mathcal{O}}. \tag{A.0.16}$$

Step 1: Defining E^{scr}.
We define E^{scr} by adding to E a sum $E_1 + E_2 + E_3$, some of these terms being zero except for H_k that has some boundary in common with Γ, then denoted F_k.

The first vector field contains the contribution of the completion of the smeared charges belonging to I_∂. We let

$$E_1 := \sum_k \mathbf{1}_{H_k} \nabla h_{1,k}$$

where $h_{1,k}$ is the solution of

$$\begin{cases} -\Delta h_{1,k} = \mathsf{c_d} \sum_{i \in I_\partial} \delta_{x_i}^{(\hat{\mathsf{r}}_i)} & \text{in } H_k, \\ \frac{\partial h_{1,k}}{\partial \nu} = 0 & \text{on } \partial H_k \setminus \Gamma\ , \\ \frac{\partial h_{1,k}}{\partial \nu} = \frac{-n_k}{|F_k|} & \text{on } F_k, \end{cases} \tag{A.0.17}$$

We note that the definition of n_k makes this equation solvable.

The second vector field is defined to be $E_2 = \sum_k \mathbf{1}_{H_k} \nabla h_{2,k}$ with

$$\begin{cases} -\Delta h_{2,k} = \mathsf{c_d} m_k \mathbf{1}_{H_k} & \text{in } H_k\ , \\ \frac{\partial h_{2,k}}{\partial \nu} = g_k & \text{on } \partial H_k, \end{cases}$$

where we let $g_k = 0$ if H_k has no face in common with Γ and otherwise

$$g_k = -E_{\hat{\mathsf{r}}} \cdot \nu + \frac{n_k}{|F_k|} \tag{A.0.18}$$

with $E_{\hat{\mathsf{r}}} \cdot \vec{\nu}$ taken with respect to the outer normal to $\mathcal{O}$. We note that this is solvable in view of (A.0.14).

The third vector field consists in the potential generated by a sampled configuration $Z_{\bar{\mathrm{n}}-n_{\mathcal{O}}}$ in $\mathcal{N}$: we let $E_3 = (\nabla h_3)\mathbf{1}_{\mathcal{N}}$ where h_3 solves

$$\begin{cases} -\Delta h_3 = \mathsf{c_d}\left(\sum_{j=1}^{\bar{\mathrm{n}}-n_{\mathcal{O}}} \delta_{z_j} - \tilde{\mu}\right) & \text{in } \mathcal{N} \\ \frac{\partial h_3}{\partial \nu} = 0 & \text{on } \partial\mathcal{N}. \end{cases} \tag{A.0.19}$$

This equation is solvable since (A.0.16) holds. We then define

$$E^{\mathrm{scr}} = (E_1 + E_2 + E_3)\mathbf{1}_{\mathcal{N}} + E_{\hat{\mathsf{r}}}\mathbf{1}_{\mathcal{O}} + \sum_{i \in I_\partial} \nabla \mathsf{f}_{\hat{\mathsf{r}}_i}(x - x_i) \tag{A.0.20}$$

and $Y_{\bar{\mathrm{n}}} = \{x_i \in X_n, B(x_i, \hat{\mathsf{r}}_i) \cap \mathcal{O} \neq \varnothing\} \cup \{Z_{\bar{\mathrm{n}}-n_{\mathcal{O}}}\}$.

We then let $\bar{\mathsf{r}}_i$ be the minimal distances as in (7.1.21) of $Y_{\bar{\mathrm{n}}}$. Note that for the points near Γ, these may not correspond to the previous minimal distances for the configuration X_n or $Z_{\bar{\mathrm{n}}-n_{\mathcal{O}}}$, which is why we use a different notation.

We note that the normal components are always constructed to be continuous across interfaces, so that no divergence is created there, and so, since $\mathcal{O} \subset \Omega''$ where w satisfies (7.2.4), E^{scr} thus defined satisfies

$$\begin{cases} -\operatorname{div} E^{\mathrm{scr}} = \mathsf{c_d}(\sum_{p \in Y_{\bar{\mathrm{n}}}} \delta_p - \mu) & \text{in } \Omega \\ E^{\mathrm{scr}} \cdot \nu = 0 & \text{on } \partial\Omega. \end{cases} \tag{A.0.21}$$

Step 2: Controlling m_k. First we control the n_k. Observe that

$$n_k \le n_k' := \mathsf{c_d}\#\{i, B(x_i, \hat{\mathsf{r}}_i) \cap (\partial H_k \cap \Gamma) \neq \varnothing\}.$$

Applying the discrepancy estimates (4.4.7)–(4.4.8) (in rescaled version) in balls B_α of radius of order 1 near Γ that form a covering of Γ (there are then $O(\ell^{\mathsf{d}-1})$ such balls), we find

$$n_{k,\alpha} \le C\left(\|\mu\|_{L^\infty} + \sqrt{\int_{H_k \cap B_\alpha} |E_{\bar{\mathsf{r}}}|^2}\right)$$

where $n_{k,\alpha}$ is the number of points in each B_α. Summing over α and using Cauchy-Schwarz, using a covering with finite overlap, in view of (A.0.4) or (A.0.7) we can then bound

$$\begin{cases} n_k \le n_k' \le \sum_\alpha n_{k,\alpha} \le C\left(\|\mu\|_{L^\infty}\ell^{\mathsf{d}-1} + \ell^{\frac{\mathsf{d}-1}{2}}\sqrt{\int_{H_k \cap \Gamma_1} |E_{\hat{\mathsf{r}}}|^2}\right) \\ \qquad\qquad\qquad\qquad \le C\left(\|\mu\|_{L^\infty}\ell^{\mathsf{d}-1} + \ell^{\frac{\mathsf{d}-1}{2}}\sqrt{\min(M, M_\ell)}\right) \\ \sum_k n_k \le C(\|\mu\|_{L^\infty} R^{\mathsf{d}-1} + R^{\frac{\mathsf{d}-1}{2}}\sqrt{M}). \end{cases} \tag{A.0.22}$$

We note that we can show in the same way that

$$\#I_\partial \le C(\|\mu\|_{L^\infty} R^{\mathsf{d}-1} + R^{\frac{\mathsf{d}-1}{2}}\sqrt{M}). \tag{A.0.23}$$

To control m_k we write that in view of (A.0.14),

$$|m_k| \le C\left(\ell^{-\mathsf{d}}\int_{\Gamma \cap \partial H_k} |E_{\hat{\mathsf{r}}}| + |n_k|\ell^{-\mathsf{d}}\right). \tag{A.0.24}$$

Using the Cauchy-Schwarz inequality and (A.0.3) or (A.0.6), we bound

$$\int_{\Gamma \cap \partial H_k} |E_{\hat{\mathsf{r}}}| \le \ell^{\frac{\mathsf{d}-1}{2}}\sqrt{\min(M, M_\ell)}, \qquad \sum_k \int_{\Gamma \cap \partial H_k} |E_{\hat{\mathsf{r}}}| \le R^{\frac{\mathsf{d}-1}{2}} M^{\frac{1}{2}}. \tag{A.0.25}$$

Combining with (A.0.22), we conclude that

$$|m_k| \le C\left(\ell^{-\mathsf{d}}\ell^{\frac{\mathsf{d}-1}{2}}\sqrt{\min(M, M_\ell)} + \|\mu\|_{L^\infty}\ell^{-1}\right). \tag{A.0.26}$$

Using Young's inequality, the condition $|m_k| < \frac{1}{2}m$ is thus implied by

$$\frac{C}{m}\|\mu\|_{L^\infty} < \frac{\ell}{4}, \quad \text{and } C\ell^{-\frac{\mathsf{d}+1}{2}}\sqrt{\min(M, M_\ell)} < \frac{1}{4}m.$$

This is guaranteed by the screenability condition (7.2.7), resp. (7.2.8) (provided ℓ is large enough).

Moreover, starting from (A.0.15) and (A.0.14), and using (A.0.25), (A.0.22), we can also bound

$$\begin{aligned}\left|\int_{\mathcal{N}} \mu - \tilde{\mu}\right| &= \frac{1}{\mathsf{c_d}}\left|\sum_k \left(\int_{\Gamma\cap\partial H_k} E_{\hat{\mathsf{r}}}\cdot\nu - n_k\right)\right| \\ &\le C(R^{\frac{\mathsf{d}-1}{2}}M^{\frac{1}{2}} + R^{\mathsf{d}-1}) \le CR^{\mathsf{d}-1} + C\frac{S(X_n)}{\tilde{\ell}},\end{aligned} \tag{A.0.27}$$

using Young's inequality, (A.0.3) or (A.0.6), thus completing the proof of (7.2.14). In the same way, we have

$$\int_{\mathcal{N}} (\mu - \tilde{\mu})^2 = \sum_k m_k^2 |H_k|,$$

while, returning to (A.0.24) and (A.0.22) and using Cauchy-Schwarz, we also have that

$$\begin{aligned} m_k^2 &\le C\ell^{-2\mathsf{d}}\int_{\Gamma\cap\partial H_k} |E_{\hat{\mathsf{r}}}|^2\ell^{\mathsf{d}-1} + Cn_k^2\ell^{-2\mathsf{d}} \\ &\le C\ell^{-\mathsf{d}-1}\left(\int_{\Gamma\cap\partial H_k}|E_{\hat{\mathsf{r}}}|^2 + \int_{\Gamma_1\cap H_k}|E_{\hat{\mathsf{r}}}|^2\right) + C\ell^{-2}\end{aligned}$$

and thus using (A.0.6) and (A.0.7), we obtain

$$\int_{\mathcal{N}} (\mu - \tilde{\mu})^2 \le C\frac{S(X_n)}{\tilde{\ell}\ell} + CR^{\mathsf{d}-1}\ell^{-1}.$$

Also, by construction

$$\|\mu - \tilde{\mu}\|_{L^\infty(\mathcal{N})} \le \max_k |m_k| \le \frac{1}{2}m$$

hence (7.2.15) is proven.

Step 3: Estimating the energy of E^{scr}. To estimate the energy of E^{scr} we need to evaluate $\int_\Omega |E_{\hat{\mathsf{r}}}^{\mathrm{scr}}|^2$. First, for E_1 we use Lemma A.2 and combine it with (7.1.36) applied with $\alpha_i = \frac{1}{4}$ to bound $\sum_{p\neq q, |p-q|\le\frac{1}{8}} \mathsf{g}(p-q)$ by the energy in a slightly larger set and bound $\sum_{|p-q|\ge\frac{1}{8}} \mathsf{g}(p-q)$ by Cn_k^2, thus we are led to

$$\begin{aligned}\int_{\mathcal{N}} |(E_1)_{\hat{\mathsf{r}}}|^2 &\le C\left(\ell^{2-\mathsf{d}}\sum_k (n_k')^2 + CM\right) \\ &\le C\ell^{2-\mathsf{d}}(R^{\mathsf{d}-1}\ell^{\mathsf{d}-1} + \ell^{\mathsf{d}-1}M) + CM \le C(\ell M + R^{\mathsf{d}-1}\ell),\end{aligned}$$

where we have used (A.0.4) and (A.0.22).

For E_2 we use Lemma A.1 to get

$$\int_{H_k} |E_2|^2 \le C\ell \left(\int_{\partial H_k \cap \Gamma} |E_{\hat{\mathsf{r}}}|^2 + \frac{n_k^2}{|F_k|} \right).$$

Summing over k and using (A.0.3) and (A.0.22), we obtain

$$\sum_k \int_{H_k} |E_2|^2 \le C \left(\ell M + \ell^{2-\mathsf{d}} \sum_k n_k^2 \right) \le C \left(\ell M + R^{\mathsf{d}-1}\ell + \ell^{\mathsf{d}+2} \frac{R^{\mathsf{d}-1}}{\ell^{\mathsf{d}-1}} \right).$$

For E_3 we use that, by definition of F and using (4.1.27),

$$\int_{\mathcal{N}} |\nabla h_{3,\hat{\mathsf{r}}}|^2 \le 2\mathsf{c}_{\mathsf{d}} \mathsf{F}(Z_{\bar{\mathsf{n}}-n_{\mathcal{O}}}, \tilde{\mu}, \mathcal{N}) + \mathsf{c}_{\mathsf{d}} \sum_{j=1}^{\bar{\mathsf{n}}-n_{\mathcal{O}}} \mathsf{g}(\hat{\mathsf{r}}_j) + C(\bar{\mathsf{n}} - n_{\mathcal{O}}) \tag{A.0.28}$$

with the $\hat{\mathsf{r}}_j$ defined relative to $\mathcal{N}$. In view of (7.1.33), the inequality still holds when $\hat{\mathsf{r}}_j$ are replaced by larger balls.

Combining the prior estimates, by definition (A.0.20) and Young's inequality, we deduce that

$$\int_{\Omega} |E_{\hat{\mathsf{r}}}^{\mathrm{scr}}|^2 \le \int_{\mathcal{O}} |\nabla w_{\hat{\mathsf{r}}}|^2 + C(\ell M + R^{\mathsf{d}-1}\ell) + C\mathsf{F}(Z_{\bar{\mathsf{n}}-n_{\mathcal{O}}}, \tilde{\mu}, \mathcal{N}) + C \sum_{j=1}^{\bar{\mathsf{n}}-n_{\mathcal{O}}} \mathsf{g}(\hat{\mathsf{r}}_j) + C(\bar{\mathsf{n}} - n_{\mathcal{O}}). \tag{A.0.29}$$

To estimate $\mathsf{F}(Y_{\bar{\mathsf{n}}}, \mu, \Omega)$ we use Lemma 7.5, the definition of F and (7.1.33), which tells us that to go from the r_i and $\hat{\mathsf{r}}_j$ which lead to possibly intersecting balls, to $\bar{\mathsf{r}}$ the minimal distances of $Y_{\bar{\mathsf{n}}}$, we just need to add the new interactions $\sum_{(i,j)\in J} \mathsf{g}(x_i - z_j)$. This yields

$$\begin{aligned}\mathsf{F}(Y_{\bar{\mathsf{n}}}, \mu, \Omega) \le \frac{1}{2\mathsf{c}_{\mathsf{d}}} \int_{\mathcal{O}} |\nabla w_{\hat{\mathsf{r}}}|^2 - \frac{1}{2} \sum_{i=1}^{\bar{\mathsf{n}}} \mathsf{g}(\hat{\mathsf{r}}_i) - \sum_{i=1}^{\bar{\mathsf{n}}} \int_{\Omega} \mathsf{f}_{\hat{\mathsf{r}}_i}(y - y_i) d\mu(y) + C \sum_{(i,j)\in J} \mathsf{g}(x_i - z_j) \\ + C(\ell M + R^{\mathsf{d}-1}\ell) + C \left(\mathsf{F}(Z_{\bar{\mathsf{n}}-n_{\mathcal{O}}}, \tilde{\mu}, \mathcal{N}) + \sum_{j=1}^{\bar{\mathsf{n}}-n_{\mathcal{O}}} \mathsf{g}(\hat{\mathsf{r}}_j) + (\bar{\mathsf{n}} - n_{\mathcal{O}}) \right).\end{aligned}$$

We may use Remark 7.9 and (7.1.39) to absorb $\sum_{j=1}^{\bar{\mathsf{n}}-n_{\mathcal{O}}} \mathsf{g}(\hat{\mathsf{r}}_j)$ into $\mathsf{F}(Z_{\bar{\mathsf{n}}-n_{\mathcal{O}}}, \tilde{\mu}, \mathcal{N}) + (\bar{\mathsf{n}} - n_{\mathcal{O}})$. It follows that

$$\begin{aligned}\mathsf{F}(Y_{\bar{\mathsf{n}}}, \mu, \Omega) - \left(\frac{1}{2\mathsf{c}_{\mathsf{d}}} \int_{\Omega'} |\nabla w_{\hat{\mathsf{r}}}|^2 - \frac{1}{2} \sum_{i=1}^{n} \mathsf{g}(\hat{\mathsf{r}}_i) - \sum_{i=1}^{n} \int_{\Omega'} \mathsf{f}_{\hat{\mathsf{r}}_i}(x - x_i) d\mu(x) \right) \\ \le -\frac{1}{2\mathsf{c}_{\mathsf{d}}} \int_{\Omega' \setminus \mathcal{O}} |\nabla w_{\hat{\mathsf{r}}}|^2 + \frac{1}{2} \sum_{\{i\in\{1,\dots,n\} \,:\, x_i \notin \mathcal{O}\}} \mathsf{g}(\hat{\mathsf{r}}_i) \\ + C \sum_{(i,j)\in J} \mathsf{g}(x_i - z_j) + C(\ell M + R^{\mathsf{d}-1}\ell) \\ + C\left(\mathsf{F}(Z_{\bar{\mathsf{n}}-n_{\mathcal{O}}}, \tilde{\mu}, \mathcal{N}) + (n - n_{\mathcal{O}}) + (\bar{\mathsf{n}} - n_{\mathcal{O}})\right).\end{aligned} \tag{A.0.30}$$

On the other hand, since $\mathcal{O}$ contains $Q_{t-2}\cap\Omega$, we have in the first screening situation

$$\begin{aligned}(\text{A.0.31})\quad &\frac{1}{2\mathsf{c_d}}\left(-\int_{\Omega'\setminus\mathcal{O}}|\nabla w_{\hat{\mathsf{r}}}|^2+\mathsf{c_d}\sum_{\{i\in\{1,\dots,n\}\,:\,x_i\notin\mathcal{O}\}}\mathsf{g}(\hat{\mathsf{r}}_i)\right)\\ &\le\frac{1}{2\mathsf{c_d}}\int_{(Q_{t+2}\setminus Q_{t-2})\cap U}|\nabla w_{\hat{\mathsf{r}}}|^2\\ &\quad+\frac{1}{2\mathsf{c_d}}\left(\mathsf{c_d}\sum_{\{i\in\{1,\dots,n\}\,:\,x_i\notin\mathcal{O}\}}\mathsf{g}(\hat{\mathsf{r}}_i)-\int_{\Omega'\setminus Q_{t-2}}|\nabla w_{\hat{\mathsf{r}}}|^2\right)\\ &\le\frac{M}{2\mathsf{c_d}}+C(n-n_{\mathcal{O}}),\end{aligned}$$

where we bounded the second term in the right-hand side by using (7.1.33) to change $\hat{\mathsf{r}}$ into $\frac{1}{4}$ and then bounded $\sum\mathsf{g}(\frac{1}{4})$ for $x_i\notin\mathcal{O}$ by the number of points not in $\mathcal{O}$. In the second situation, we replace $Q_{t+2}\setminus Q_{t-2}$ by $Q_{t+\ell}\setminus Q_t$ and use (A.0.5) instead.

Inserting (A.0.31) into (A.0.30) and using (A.0.4), we find in all cases that

$$\begin{aligned}&\mathsf{F}(Y_{\bar{\mathrm{n}}},\mu,\Omega)-\left(\frac{1}{2\mathsf{c_d}}\int_{\Omega'}|\nabla w_{\hat{\mathsf{r}}}|^2-\frac{1}{2}\sum_{i=1}^n\mathsf{g}(\hat{\mathsf{r}}_i)-\sum_{i=1}^n\int_{\Omega'}\mathsf{f}_{\hat{\mathsf{r}}_i}(x-x_i)d\mu(x)\right)\\ &\le C\ell\frac{S(X_n)}{\tilde{\ell}}+CR^{\mathsf{d}-1}\ell+C\mathsf{F}(Z_{\bar{\mathrm{n}}-n_{\mathcal{O}}},\tilde{\mu},\mathcal{N})+C\sum_{(i,j)\in J}\mathsf{g}(x_i-z_j)+C(|n-\bar{\mathrm{n}}|+|\bar{\mathrm{n}}-n_{\mathcal{O}}|).\end{aligned}$$

Using (7.2.14) and $\mu(\mathcal{N})\le C\tilde{\ell}R^{\mathsf{d}-1}$ allows us to bound the last term on the right side, and then we get (7.2.16).

Bibliography

[ABG12] R. Allez, J.-P. Bouchaud, and A. Guionnet, *Invariant beta ensembles and the Gauss-Wigner crossover*, Phys. Rev. Lett., **109**, Aug 2012, 094102.

[Abr57] A. Abrikosov, *On the magnetic properties of superconductors of the second type*, Soviet Phys. JETP, **5** (1857), 1174–1182.

[ABZ23] F. Augeri, R. Butez, and O. Zeitouni, *A CLT for the characteristic polynomial of random Jacobi matrices, and the* GβE, Probab. Theory Related Fields **186** (2023), no. 1-2, 1–89, DOI 10.1007/s00440-023-01194-9. MR4586217

[AC22] Y. Ameur and J. Cronvall, *Szegő type asymptotics for the reproducing kernel in spaces of full-plane weighted polynomials*, Comm. Math. Phys. **398** (2023), no. 3, 1291–1348, DOI 10.1007/s00220-022-04539-y. MR4561803

[ACC23] Y. Ameur, C. Charlier, and J. Cronvall, *Free energy and fluctuations in the random normal matrix model with spectral gaps*, `arXiv:2312.13904`, 2023.

[ACC24] Y. Ameur, C. Charlier, and J. Cronvall, *Random normal matrices: eigenvalue correlations near a hard wall*, J. Stat. Phys. **191** (2024), no. 8, Paper No. 98, 49, DOI 10.1007/s10955-024-03314-8. MR4783708

[ACC25] Y. Ameur, C. Charlier, and J. Cronvall, *The two-dimensional Coulomb gas: fluctuations through a spectral gap*, Arch. Ration. Mech. Anal. **249** (2025), no. 6, Paper No. 63, 62, DOI 10.1007/s00205-025-02133-9. MR4967138

[ACCL24] Y. Ameur, C. Charlier, J. Cronvall, and J. Lenells, *Disk counting statistics near hard edges of random normal matrices: the multicomponent regime*, Adv. Math. **441** (2024), Paper No. 109549, 55, DOI 10.1016/j.aim.2024.109549. MR4708154

[ACO09] G. Alberti, R. Choksi, and F. Otto, *Uniform energy distribution for an isoperimetric problem with long-range interactions*, J. Amer. Math. Soc. **22** (2009), no. 2, 569–605, DOI 10.1090/S0894-0347-08-00622-X. MR2476783

[AD14] R. Allez and L. Dumaz, *From sine kernel to Poisson statistics*, Electron. J. Probab. **19** (2014), no. 114, 25, DOI 10.1214/EJP.v19-3742. MR3296530

[Adh18] K. Adhikari, *Hole probabilities for β-ensembles and determinantal point processes in the complex plane*, Electron. J. Probab. **23** (2018), Paper No. 48, 21, DOI 10.1214/18-ejp176. MR3814242

[ADK+19] S. Agarwal, A. Dhar, M. Kulkarni, A. Kundu, S. N. Majumdar, D. Mukamel, and G. Schehr, *Harmonically confined particles with long-range repulsive interactions*, Phys. Rev. Lett. **123** (2019), no. 10, 100603, 6, DOI 10.1103/PhysRevLett.123.100603. MR4008477

[AGS08] L. Ambrosio, N. Gigli, and G. Savaré, *Gradient flows in metric spaces and in the space of probability measures*, 2nd ed., Lectures in Mathematics ETH Zürich, Birkhäuser Verlag, Basel, 2008. MR2401600

[AGZ10] G. W. Anderson, A. Guionnet, and O. Zeitouni, *An introduction to random matrices*, Cambridge Studies in Advanced Mathematics, vol. 118, Cambridge University Press, Cambridge, 2010. MR2760897

[AH96] D. R. Adams and L. I. Hedberg, *Function spaces and potential theory*, Grundlehren der mathematischen Wissenschaften [Fundamental Principles of Mathematical Sciences], vol. 314, Springer-Verlag, Berlin, 1996, DOI 10.1007/978-3-662-03282-4. MR1411441

[AHM11] Y. Ameur, H. Hedenmalm, and N. Makarov, *Fluctuations of eigenvalues of random normal matrices*, Duke Math. J. **159** (2011), no. 1, 31–81, DOI 10.1215/00127094-1384782. MR2817648

[AHM15] Y. Ameur, H. Hedenmalm, and N. Makarov, *Random normal matrices and Ward identities*, Ann. Probab. **43** (2015), no. 3, 1157–1201, DOI 10.1214/13-AOP885. MR3342661

[AJ81] A. Alastuey and B. Jancovici, *On the classical two-dimensional one-component Coulomb plasma*, J. Physique **42** (1981), no. 1, 1–12, DOI 10.1051/jphys:019810042010100. MR604143

[Ala86] A. Alastuey, *Propriétés d'équilibre du plasma classique à une composante en trois et deux dimensions*, Annales de Physique, **11** (1986), no. 6, 653–738.

[AM25] A. Alastuey and P. Martin, *Statistical mechanics of Coulomb systems*, EPFL Press, Lausanne, Switzerland, 2025.

[AM81] M. Aizenman and P. A. Martin, *Structure of Gibbs states of one-dimensional Coulomb systems*, Comm. Math. Phys. **78** (1980/81), no. 1, 99–116. MR597033

[Ame18] Y. Ameur, *Repulsion in low temperature β-ensembles*, Comm. Math. Phys. **359** (2018), no. 3, 1079–1089, DOI 10.1007/s00220-017-3027-2. MR3784540

[Ame21] Y. Ameur, *A localization theorem for the planar Coulomb gas in an external field*, Electron. J. Probab. **26** (2021), Paper No. 46, 21, DOI 10.1214/21-ejp613. MR4244340

[AMR22] Y. Ameur, F. Marceca, and J. L. Romero, *Gaussian beta ensembles: The perfect freezing transition and its characterization in terms of Beurling–Landau densities*, Ann. Inst. Henri Poincaré Probab. Stat. **62** (2026), no. 1, DOI 10.1214/24-aihp1524. MR5036541

[AOC12] Y. Ameur and J. Ortega-Cerdà, *Beurling-Landau densities of weighted Fekete sets and correlation kernel estimates*, J. Funct. Anal. **263** (2012), no. 7, 1825–1861, DOI 10.1016/j.jfa.2012.06.011. MR2956927

[AR23] Y. Ameur and J. L. Romero, *The planar low temperature Coulomb gas: separation and equidistribution*, Rev. Mat. Iberoam. **39** (2023), no. 2, 611–648, DOI 10.4171/rmi/1340. MR4575375

[AS16] S. N. Armstrong and C. K. Smart, *Quantitative stochastic homogenization of convex integral functionals* (English, with English and French summaries), Ann. Sci. Éc. Norm. Supér. (4) **49** (2016), no. 2, 423–481, DOI 10.24033/asens.2287. MR3481355

[AS21] S. Armstrong and S. Serfaty, *Local laws and rigidity for Coulomb gases at any temperature*, Ann. Probab. **49** (2021), no. 1, 46–121, DOI 10.1214/20-AOP1445. MR4203333

[AS22] S. Armstrong and S. Serfaty, *Thermal approximation of the equilibrium measure and obstacle problem* (English, with English and French summaries), Ann. Fac. Sci. Toulouse Math. (6) **31** (2022), no. 4, 1085–1110. MR4506138

[ASZ14] S. N. Armstrong, S. Serfaty, and O. Zeitouni, *Remarks on a constrained optimization problem for the Ginibre ensemble*, Potential Anal. **41** (2014), no. 3, 945–958, DOI 10.1007/s11118-014-9402-0. MR3264829

[BAB13] G. Ben Arous and P. Bourgade, *Extreme gaps between eigenvalues of random matrices*, Ann. Probab. **41** (2013), no. 4, 2648–2681, DOI 10.1214/11-AOP710. MR3112927

[BAG97] G. Ben Arous and A. Guionnet, *Large deviations for Wigner's law and Voiculescu's non-commutative entropy*, Probab. Theory Related Fields **108** (1997), no. 4, 517–542, DOI 10.1007/s004400050119. MR1465640

[Bax63] R. J. Baxter, *Statistical mechanics of a one-dimensional Coulomb system with a uniform charge background*, Proc. Cambridge Philos. Soc. **59** (1963), 779–787. MR155636

[BAZ98] G. Ben Arous and O. Zeitouni, *Large deviations from the circular law*, ESAIM Probab. Statist. **2** (1998), 123–134, DOI 10.1051/ps:1998104. MR1660943

[BBDR05] J. Barré, F. Bouchet, T. Dauxois, and S. Ruffo, *Large deviation techniques applied to systems with long-range interactions*, J. Stat. Phys. **119** (2005), no. 3-4, 677–713, DOI 10.1007/s10955-005-3768-8. MR2151219

[BBH94] F. Bethuel, H. Brezis, and F. Hélein, *Ginzburg-Landau vortices*, Progress in Nonlinear Differential Equations and their Applications, vol. 13, Birkhäuser Boston, Inc., Boston, MA, 1994, DOI 10.1007/978-1-4612-0287-5. MR1269538

[BBNY17] R. Bauerschmidt, P. Bourgade, M. Nikula, and H.-T. Yau, *Local density for two-dimensional one-component plasma*, Comm. Math. Phys. **356** (2017), no. 1, 189–230, DOI 10.1007/s00220-017-2932-8. MR3694026

[BBNY19] R. Bauerschmidt, P. Bourgade, M. Nikula, and H.-T. Yau, *The two-dimensional Coulomb plasma: quasi-free approximation and central limit theorem*, Adv. Theor. Math. Phys. **23** (2019), no. 4, 841–1002, DOI 10.4310/ATMP.2019.v23.n4.a1. MR4063572

[BBWN11] R. Berman, S. Boucksom, and D. Witt Nyström, *Fekete points and convergence towards equilibrium measures on complex manifolds*, Acta Math. **207** (2011), no. 1, 1–27, DOI 10.1007/s11511-011-0067-x. MR2863909

[BDJ24] D. Bresch, M. Duerinckx, and P. E. Jabin, *A duality method for mean-field limits with singular interactions*, `arXiv:2402.04695`, 2024.

[BdMPS95] A. Boutet de Monvel, L. Pastur, and M. Shcherbina, *On the statistical mechanics approach in the random matrix theory: integrated density of states*, J. Statist. Phys. **79** (1995), no. 3-4, 585–611, DOI 10.1007/BF02184872. MR1327898

[BDMS23] B. De Bruyne, P. Le Doussal, S. N. Majumdar, and G. Schehr, *Linear statistics for Coulomb gases: higher order cumulants*, J. Phys. A **57** (2024), no. 15, Paper No. 155002, 19, DOI 10.1088/1751-8121/ad329f. MR4733584

[BDS14] J. S. Brauchart, P. D. Dragnev, and E. B. Saff, *Riesz external field problems on the hypersphere and optimal point separation*, Potential Anal. **41** (2014), no. 3, 647–678, DOI 10.1007/s11118-014-9387-8. MR3264814

[Ber14] R. J. Berman, *Determinantal point processes and fermions on complex manifolds: large deviations and bosonization*, Comm. Math. Phys. **327** (2014), no. 1, 1–47, DOI 10.1007/s00220-014-1891-6. MR3177931

[Ber19] R. J. Berman, *Sharp deviation inequalities for the 2D Coulomb gas and Quantum hall states, I*, `arXiv:1906.08529`, 2019.

[BEY12] P. Bourgade, L. Erdős, and H.-T. Yau, *Bulk universality of general β-ensembles with non-convex potential*, J. Math. Phys. **53** (2012), no. 9, 095221, 19, DOI 10.1063/1.4751478. MR2905803

[BEY14] P. Bourgade, L. Erdős, and H.-T. Yau, *Universality of general β-ensembles*, Duke Math. J. **163** (2014), no. 6, 1127–1190, DOI 10.1215/00127094-2649752. MR3192527

[BLZ25] P. Bourgade, P. Lopatto, and O. Zeitouni, *Optimal rigidity and maximum of the characteristic polynomial of Wigner matrices*, Geom. Funct. Anal. **35** (2025), no. 1, 161–253, DOI 10.1007/s00039-025-00701-5. MR4865133

[BF80a] D. Brydges and P. Federbush, *Debye screening in classical statistical mechanics*, Mathematical problems in theoretical physics (Proc. Internat. Conf. Math. Phys., Lausanne, 1979), Lecture Notes in Phys., vol. 116, Springer, Berlin, 1980, pp. 151–155, DOI 10.1007/3-540-09964-6_313. MR582617

[BF80b] D. C. Brydges and P. Federbush, *Debye screening*, Comm. Math. Phys. **73** (1980), no. 3, 197–246. MR574172

[BF23] S.-S. Byun and P. J. Forrester, *Progress on the study of the Ginibre ensembles I: GinUE*, `arXiv:2211.16223`, 2023.

[BFG15] F. Bekerman, A. Figalli, and A. Guionnet, *Transport maps for β-matrix models and universality*, Comm. Math. Phys. **338** (2015), no. 2, 589–619, DOI 10.1007/s00220-015-2384-y. MR3351052

[BG13] G. Borot and A. Guionnet, *Asymptotic expansion of β matrix models in the one-cut regime*, Comm. Math. Phys. **317** (2013), no. 2, 447–483, DOI 10.1007/s00220-012-1619-4. MR3010191

[BG16a] W. Bietenholz and U. Gerber, *Berezinskii-Kosterlitz-Thouless transition and the Haldane conjecture: Highlights of the physics Nobel prize 2016*, `arXiv:1612.06132`, 2016.

[BG16b] A. Borodin and V. Gorin, *Lectures on integrable probability*, Probability and statistical physics in St. Petersburg, Proc. Sympos. Pure Math., vol. 91, Amer. Math. Soc., Providence, RI, 2016, pp. 155–214, DOI 10.1090/pspum/091/01533. MR3526828

[BG24] G. Borot and A. Guionnet, *Asymptotic expansion of β matrix models in the multi-cut regime*, Forum Math. Sigma **12** (2024), Paper No. e13, 93, DOI 10.1017/fms.2023.129. MR4695502

[BGG17] A. Borodin, V. Gorin, and A. Guionnet, *Gaussian asymptotics of discrete β-ensembles*, Publ. Math. Inst. Hautes Études Sci. **125** (2017), 1–78, DOI 10.1007/s10240-016-0085-5. MR3668648

[BGK15] G. Borot, A. Guionnet, and K. K. Kozlowski, *Large-N asymptotic expansion for mean field models with Coulomb gas interaction*, Int. Math. Res. Not. IMRN **20** (2015), 10451–10524, DOI 10.1093/imrn/rnu260. MR3455872

[BH19] C. Beltrán and A. Hardy, *Energy of the Coulomb gas on the sphere at low temperature*, Arch. Ration. Mech. Anal. **231** (2019), no. 3, 2007–2017, DOI 10.1007/s00205-018-1316-3. MR3902480

[BH20] R. Bardenet and A. Hardy, *Monte Carlo with determinantal point processes*, Ann. Appl. Probab. **30** (2020), no. 1, 368–417, DOI 10.1214/19-AAP1504. MR4068314

[BHS12] J. S. Brauchart, D. P. Hardin, and E. B. Saff, *The next-order term for optimal Riesz and logarithmic energy asymptotics on the sphere*, Recent advances in orthogonal polynomials, special functions, and their applications, Contemp. Math., vol. 578, Amer. Math. Soc., Providence, RI, 2012, pp. 31–61, DOI 10.1090/conm/578/11483. MR2964138

[BHS14] S. V. Borodachov, D. P. Hardin, and E. B. Saff, *Low complexity methods for discretizing manifolds via Riesz energy minimization*, Found. Comput. Math. **14** (2014), no. 6, 1173–1208, DOI 10.1007/s10208-014-9202-3. MR3273676

[BHS19] S. V. Borodachov, D. P. Hardin, and E. B. Saff, *Discrete energy on rectifiable sets*, Springer Monographs in Mathematics, Springer, New York, 2019, DOI 10.1007/978-0-387-84808-2. MR3970999

[BIK15] P. Biler, C. Imbert, and G. Karch, *The nonlocal porous medium equation: Barenblatt profiles and other weak solutions*, Arch. Ration. Mech. Anal. **215** (2015), no. 2, 497–529, DOI 10.1007/s00205-014-0786-1. MR3294409

[Bil20] D. Bilyk, *On some recent developments in uniform distribution and discrepancy theory*, Discrepancy theory, Radon Ser. Comput. Appl. Math., vol. 26, De Gruyter, Berlin, 2020, pp. 1–20, DOI 10.1515/9783110652581-001. MR4391423

[BJS08] F. Bethuel, R. L. Jerrard, and D. Smets, *On the NLS dynamics for infinite energy vortex configurations on the plane*, Rev. Mat. Iberoam. **24** (2008), no. 2, 671–702, DOI 10.4171/RMI/552. MR2459209

[BJS23] D. Bresch, P.-E. Jabin, and J. Soler, *A new approach to the mean-field limit of Vlasov-Fokker-Planck equations*, Anal. PDE **18** (2025), no. 4, 1037–1064, DOI 10.2140/apde.2025.18.1037. MR4885878

[BJW23] D. Bresch, P.-E. Jabin, and Z. Wang, *Mean field limit and quantitative estimates with singular attractive kernels*, Duke Math. J. **172** (2023), no. 13, 2591–2641, DOI 10.1215/00127094-2022-0088. MR4658923

[BKS23] S.-S. Byun, N.-G. Kang, and S.-M. Seo, *Partition functions of determinantal and Pfaffian Coulomb gases with radially symmetric potentials*, Comm. Math. Phys. **401** (2023), no. 2, 1627–1663, DOI 10.1007/s00220-023-04673-1. MR4610282

[BL75] H. J. Brascamp and E. H. Lieb, *Some inequalities for Gaussian measures and the long-range order of the one-dimensional plasma*, Functional Integration and its Applications. Clarendon Press, Oxford, 1975.

[BL15] X. Blanc and M. Lewin, *The crystallization conjecture: a review*, EMS Surv. Math. Sci. **2** (2015), no. 2, 225–306, DOI 10.4171/EMSS/13. MR3429164

[BL18] F. Bekerman and A. Lodhia, *Mesoscopic central limit theorem for general β-ensembles* (English, with English and French summaries), Ann. Inst. Henri Poincaré Probab. Stat. **54** (2018), no. 4, 1917–1938, DOI 10.1214/17-AIHP860. MR3865662

[BLJO12] F. Bach, S. Lacoste-Julien, and G. Obozinski, *On the equivalence between herding and conditional gradient algorithms*, Proceedings of the 29th International Conference on International Conference on Machine Learning, ICML'12, pages 1355–1362, Madison, WI, USA, 2012. Omnipress.

[BLS18] F. Bekerman, T. Leblé, and S. Serfaty, *CLT for fluctuations of β-ensembles with general potential*, Electron. J. Probab. **23** (2018), Paper no. 115, 31, DOI 10.1214/18-EJP209. MR3885548

[BMP22] P. Bourgade, K. Mody, and M. Pain, *Optimal local law and central limit theorem for β-ensembles*, Comm. Math. Phys. **390** (2022), no. 3, 1017–1079, DOI 10.1007/s00220-022-04311-2. MR4389077

[Bor11] A. Borodin, *Determinantal point processes*, The Oxford handbook of random matrix theory, Oxford Univ. Press, Oxford, 2011, pp. 231–249. MR2932631

[BOS05] F. Bethuel, G. Orlandi, and D. Smets, *Collisions and phase-vortex interactions in dissipative Ginzburg-Landau dynamics*, Duke Math. J. **130** (2005), no. 3, 523–614, DOI 10.1215/S0012-7094-05-13034-4. MR2184569

[BOS07] F. Bethuel, G. Orlandi, and D. Smets, *Dynamics of multiple degree Ginzburg-Landau vortices*, Comm. Math. Phys. **272** (2007), no. 1, 229–261, DOI 10.1007/s00220-007-0206-6. MR2291808

[Boua] J. Boursier, *Decay of correlations and thermodynamic limit for the circular Riesz gas*, `arXiv:2209.00396`, 2022.

[Boub] J. Boursier, *Optimal local laws and CLT for the circular Riesz gas*, `arXiv:2112.05881`, 2021.

[BP19] L. Bétermin and M. Petrache, *Optimal and non-optimal lattices for non-completely monotone interaction potentials*, Anal. Math. Phys. **9** (2019), no. 4, 2033–2073, DOI 10.1007/s13324-019-00299-6. MR4038122

[BP23] I. Ben Porat, *Derivation of Euler's equations of perfect fluids from von Neumann's equation with magnetic field*, J. Stat. Phys. **190** (2023), no. 7, Paper No. 121, 44, DOI 10.1007/s10955-023-03131-5. MR4614118

[BPT14] D. P. Bourne, M. A. Peletier, and F. Theil, *Optimality of the triangular lattice for a particle system with Wasserstein interaction*, Comm. Math. Phys. **329** (2014), no. 1, 117–140, DOI 10.1007/s00220-014-1965-5. MR3206999

[BR95] F. Bethuel and T. Rivière, *Vortices for a variational problem related to superconductivity* (English, with English and French summaries), Ann. Inst. H. Poincaré C Anal. Non Linéaire **12** (1995), no. 3, 243–303, DOI 10.1016/S0294-1449(16)30157-3. MR1340265

[Bra02] A. Braides, *Γ-convergence for beginners*, Oxford Lecture Series in Mathematics and its Applications, vol. 22, Oxford University Press, Oxford, 2002, DOI 10.1093/acprof:oso/9780198507840.001.0001. MR1968440

[Bra08] J. S. Brauchart, *Optimal logarithmic energy points on the unit sphere*, Math. Comp. **77** (2008), no. 263, 1599–1613, DOI 10.1090/S0025-5718-08-02085-1. MR2398782

[Bra14] A. Braides, *Local minimization, variational evolution and Γ-convergence*, Lecture Notes in Mathematics, vol. 2094, Springer, Cham, 2014, DOI 10.1007/978-3-319-01982-6. MR3137417

[Bre72] H. Brézis, *Problèmes unilatéraux*, J. Math. Pures Appl. (9) **51** (1972), 1–168. MR428137

[BS02] H. Brezis and S. Serfaty, *A variational formulation for the two-sided obstacle problem with measure data*, Commun. Contemp. Math. **4** (2002), no. 2, 357–374, DOI 10.1142/S0219199702000671. MR1901150

[BS13] A. Borodin and S. Serfaty, *Renormalized energy concentration in random matrices*, Comm. Math. Phys. **320** (2013), no. 1, 199–244, DOI 10.1007/s00220-013-1716-z. MR3046995

[BS18] L. Bétermin and E. Sandier, *Renormalized energy and asymptotic expansion of optimal logarithmic energy on the sphere*, Constr. Approx. **47** (2018), no. 1, 39–74, DOI 10.1007/s00365-016-9357-z. MR3742809

[BS24] J. Boursier and S. Serfaty, *Dipole formation in the two-component plasma*, `arXiv:2410.01025`, 2024.

[BS25] J. Boursier and S. Serfaty, *Multipole and Berezinskii-Kosterlitz-Thouless transitions in the two-component plasma*, `arXiv:2509.09449`, 2025.

[BST66] S. G. Brush, H. L. Sahlin, and E. Teller, *Monte-Carlo study of a one-component plasma*, J. Chem. Phys. **45** (1966), 2102–2118.

[BSY25] S.-S. Byun, S.-M. Seo, and M. Yang, *Free energy expansions of a conditional GinUE and large deviations of the smallest eigenvalue of the LUE*, Comm. Pure Appl. Math. **78** (2025), no. 12, 2247–2304, DOI 10.1002/cpa.70005. MR4969076

[Byu24] S.-S. Byun, *Planar equilibrium measure problem in the quadratic fields with a point charge*, Comput. Methods Funct. Theory **24** (2024), no. 2, 303–332, DOI 10.1007/s40315-023-00494-4. MR4750409

[Caf98] L. A. Caffarelli, *The obstacle problem revisited*, J. Fourier Anal. Appl. **4** (1998), no. 4-5, 383–402, DOI 10.1007/BF02498216. MR1658612

[Cas59] J. W. S. Cassels, *On a problem of Rankin about the Epstein zeta-function*, Proc. Glasgow Math. Assoc. **4** (1959), 73–80 (1959). MR117193

[CC83] P. Choquard and J. Clerouin, *Cooperative phenomena below melting of the one-component two-dimensional plasma*, Phys. Rev. Lett., **50** (1983), no. 26, 2086.

[CD22] L.-P. Chaintron and A. Diez, *Propagation of chaos: a review of models, methods and applications. I. Models and methods*, Kinet. Relat. Models **15** (2022), no. 6, 895–1015, DOI 10.3934/krm.2022017. MR4489768

[CdCRS25] A. Chodron de Courcel, M. Rosenzweig, and S. Serfaty, *Sharp uniform-in-time mean-field convergence for singular periodic Riesz flows*, Ann. Inst. H. Poincaré C Anal. Non Linéaire **42** (2025), no. 2, 391–472, DOI 10.4171/aihpc/105. MR4877368

[CDFR14] A. Campa, T. Dauxois, D. Fanelli, and S. Ruffo, *Physics of long-range interacting systems*, Oxford University Press, Oxford, 2014. With a foreword by David Mukamel, DOI 10.1093/acprof:oso/9780199581931.001.0001. MR3241433

[CDR09] A. Campa, T. Dauxois, and S. Ruffo, *Statistical mechanics and dynamics of solvable models with long-range interactions*, Phys. Rep. **480** (2009), no. 3-6, 57–159, DOI 10.1016/j.physrep.2009.07.001. MR2569625

[CF25] G. Colombo and A. Figalli, *Generic regularity of equilibrium measures for the logarithmic potential with external fields*, 2025.

[CFGW24] S. Cai, X. Feng, Y. Gong, and Z. Wang, *Propagation of chaos for 2D log gas on the whole space*, 2024.

[CFK13] C. Cotar, G. Friesecke, and C. Klüppelberg, *Density functional theory and optimal transportation with Coulomb cost*, Comm. Pure Appl. Math. **66** (2013), no. 4, 548–599, DOI 10.1002/cpa.21437. MR3020313

[CFTW15] T. Can, P. J. Forrester, G. Téllez, and P. Wiegmann, *Exact and asymptotic features of the edge density profile for the one component plasma in two dimensions*, J. Stat. Phys. **158** (2015), no. 5, 1147–1180, DOI 10.1007/s10955-014-1152-2. MR3313622

[CGZ14] D. Chafaï, N. Gozlan, and P.-A. Zitt, *First-order global asymptotics for confined particles with singular pair repulsion*, Ann. Appl. Probab. **24** (2014), no. 6, 2371–2413, DOI 10.1214/13-AAP980. MR3262506

[CH25] A. Cohen and F. Hernández, *A simple bound on fluctuations in the 3D Coulomb gas*, `arXiv:2510.27129`, 2025.

[Cha] D. Chafaï, `https://djalil.chafai.net/blog/2017/04/09/carre-du-champ/`.

[Cha19] S. Chatterjee, *Rigidity of the three-dimensional hierarchical Coulomb gas*, Probab. Theory Related Fields **175** (2019), no. 3-4, 1123–1176, DOI 10.1007/s00440-019-00912-6. MR4026615

[Cha23a] D. Chafaï, *Aspects of Coulomb gases*, Topics in statistical mechanics, Panor. Synthèses, vol. 59, Soc. Math. France, Paris, 2023, pp. 1–40. MR4619310

[Cha23b] C. Charlier, *Hole probabilities and balayage of measures for planar Coulomb gases*, `arXiv:2311.15285`, 2023.

[Cha24] C. Charlier, *Large gap asymptotics on annuli in the random normal matrix model*, Math. Ann. **388** (2024), no. 4, 3529–3587, DOI 10.1007/s00208-023-02603-z. MR4721776

[Cha25] C. Charlier, *Smallest gaps of the two-dimensional Coulomb gas*, 2025.

[CHM18] D. Chafaï, A. Hardy, and M. Maïda, *Concentration for Coulomb gases and Coulomb transport inequalities*, J. Funct. Anal. **275** (2018), no. 6, 1447–1483, DOI 10.1016/j.jfa.2018.06.004. MR3820329

[CJ98] J. E. Colliander and R. L. Jerrard, *Vortex dynamics for the Ginzburg-Landau-Schrödinger equation*, Internat. Math. Res. Notices **7** (1998), 333–358, DOI 10.1155/S1073792898000221. MR1623410

[CJ24] K. Courteaut and K. Johansson, *Partition function for the 2d Coulomb gas on a Jordan curve* (English, with English and Finnish summaries), Ann. Fenn. Math. **50** (2025), no. 1, 109–144, DOI 10.54330/afm.159822. MR4877952

[CK80] L. A. Caffarelli and D. Kinderlehrer, *Potential methods in variational inequalities*, J. Analyse Math. **37** (1980), 285–295, DOI 10.1007/BF02797689. MR583641

[CK07] H. Cohn and A. Kumar, *Universally optimal distribution of points on spheres*, J. Amer. Math. Soc. **20** (2007), no. 1, 99–148, DOI 10.1090/S0894-0347-06-00546-7. MR2257398

[CKM+17] H. Cohn, A. Kumar, S. D. Miller, D. Radchenko, and M. Viazovska, *The sphere packing problem in dimension 24*, Ann. of Math. (2) **185** (2017), no. 3, 1017–1033, DOI 10.4007/annals.2017.185.3.8. MR3664817

[CKM+22] H. Cohn, A. Kumar, S. D. Miller, D. Radchenko, and M. Viazovska, *Universal optimality of the E_8 and Leech lattices and interpolation formulas*, Ann. of Math. (2) **196** (2022), no. 3, 983–1082, DOI 10.4007/annals.2022.196.3.3. MR4502595

[CLMP92] E. Caglioti, P.-L. Lions, C. Marchioro, and M. Pulvirenti, *A special class of stationary flows for two-dimensional Euler equations: a statistical mechanics description*, Comm. Math. Phys. **143** (1992), no. 3, 501–525. MR1145596

[CLWH82] J. M. Caillol, D. Levesque, J.J Weis, and J. P. Hansen, *A Monte-Carlo study of the classical two-dimensional one-component plasma*, J. Stat. Phys. **28** (1982), 325–349.

[CMG+18] W. Y. Chen, L. Mackey, J. Gorham, F.-X. Briol, and C. Oates, *Stein points*, in Proceedings of the 35th International Conference on Machine Learning (J. Dy and A. Krause, editors), volume 80 of *Proceedings of Machine Learning Research*, pp. 844–853. PMLR, 10–15 Jul 2018.

[CMN18] R. Chhaibi, T. Madaule, and J. Najnudel, *On the maximum of the CβE field*, Duke Math. J. **167** (2018), no. 12, 2243–2345, DOI 10.1215/00127094-2018-0016. MR3848391

[CMT17] R. Choksi, C. B. Muratov, and I. Topaloglu, *An old problem resurfaces nonlocally: Gamow's liquid drops inspire today's research and applications*, Notices Amer. Math. Soc. **64** (2017), no. 11, 1275–1283. MR3727064

[CO07] X. Chen and Y. Oshita, *An application of the modular function in nonlocal variational problems*, Arch. Ration. Mech. Anal. **186** (2007), no. 1, 109–132, DOI 10.1007/s00205-007-0050-z. MR2338353

[Coh17] H. Cohn, *A conceptual breakthrough in sphere packing*, Notices Amer. Math. Soc. **64** (2017), no. 2, 102–115, DOI 10.1090/noti1474. MR3587715

[Cos] S. Coste, *Order, fluctuations, rigidities*, `https://scoste.fr/assets/survey_hyperuniformity.pdf`.

[CP19] C. Cotar and M. Petrache, *Next-order asymptotic expansion for N-marginal optimal transport with Coulomb and Riesz costs*, Adv. Math. **344** (2019), 137–233, DOI 10.1016/j.aim.2018.12.008. MR3895648

[CROS17] L. Caffarelli, X. Ros-Oton, and J. Serra, *Obstacle problems for integro-differential operators: regularity of solutions and free boundaries*, Invent. Math. **208** (2017), no. 3, 1155–1211, DOI 10.1007/s00222-016-0703-3. MR3648978

[CRS96] S. J. Chapman, J. Rubinstein, and M. Schatzman, *A mean-field model of superconducting vortices*, European J. Appl. Math. **7** (1996), no. 2, 97–111, DOI 10.1017/S0956792500002242. MR1388106

[CS99] J. H. Conway and N. J. A. Sloane, *Sphere packings, lattices and groups*, 3rd ed., Grundlehren der mathematischen Wissenschaften [Fundamental Principles of Mathematical Sciences], vol. 290, Springer-Verlag, New York, 1999. With additional contributions by E. Bannai, R. E. Borcherds, J. Leech, S. P. Norton, A. M. Odlyzko, R. A. Parker, L. Queen and B. B. Venkov, DOI 10.1007/978-1-4757-6568-7. MR1662447

[CS07] L. Caffarelli and L. Silvestre, *An extension problem related to the fractional Laplacian*, Comm. Partial Differential Equations **32** (2007), no. 7-9, 1245–1260, DOI 10.1080/03605300600987306. MR2354493

[CSA21] G. Cardoso, J.-M. Stéphan, and A. G. Abanov, *The boundary density profile of a Coulomb droplet. Freezing at the edge*, J. Phys. A **54** (2021), no. 1, Paper No. 015002, 24, DOI 10.1088/1751-8121/abcab9. MR4190120

[CSS08] L. A. Caffarelli, S. Salsa, and L. Silvestre, *Regularity estimates for the solution and the free boundary of the obstacle problem for the fractional Laplacian*, Invent. Math. **171** (2008), no. 2, 425–461, DOI 10.1007/s00222-007-0086-6. MR2367025

[CSW22] D. Chafaï, E. B. Saff, and R. S. Womersley, *On the solution of a Riesz equilibrium problem and integral identities for special functions*, J. Math. Anal. Appl. **515** (2022), no. 1, Paper No. 126367, 29, DOI 10.1016/j.jmaa.2022.126367. MR4435918

[CSW23] D. Chafaï, E. B. Saff, and R. S. Womersley, *Threshold condensation to singular support for a Riesz equilibrium problem*, Anal. Math. Phys. **13** (2023), no. 1, Paper No. 19, 30, DOI 10.1007/s13324-023-00779-w. MR4544160

[CV11] L. Caffarelli and J. L. Vazquez, *Nonlinear porous medium flow with fractional potential pressure*, Arch. Ration. Mech. Anal. **202** (2011), no. 2, 537–565, DOI 10.1007/s00205-011-0420-4. MR2847534

[CV12] P. Constantin and V. Vicol, *Nonlinear maximum principles for dissipative linear nonlocal operators and applications*, Geom. Funct. Anal. **22** (2012), no. 5, 1289–1321, DOI 10.1007/s00039-012-0172-9. MR2989434

[CWS10] Y. Chen, M. Welling, and A. Smola, *Super-samples from kernel herding*, in Proceedings of the Twenty-Sixth Conference on Uncertainty in Artificial Intelligence, UAI'10, pp. 109–116, Arlington, Virginia, USA, 2010. AUAI Press.

[Dah78] B. E. J. Dahlberg, *On the distribution of Fekete points*, Duke Math. J. **45** (1978), no. 3, 537–542. MR507457

[DE97] P. Dupuis and R. S. Ellis, *A weak convergence approach to the theory of large deviations*, Wiley Series in Probability and Statistics: Probability and Statistics, John Wiley & Sons, Inc., New York, 1997. A Wiley-Interscience Publication, DOI 10.1002/9781118165904. MR1431744

[DE02] I. Dumitriu and A. Edelman, *Matrix models for beta ensembles*, J. Math. Phys. **43** (2002), no. 11, 5830–5847, DOI 10.1063/1.1507823. MR1936554

[DF17] P. Diaconis and P. J. Forrester, *Hurwitz and the origins of random matrix theory in mathematics*, Random Matrices Theory Appl. **6** (2017), no. 1, 1730001, 26, DOI 10.1142/S2010326317300017. MR3612265

[DFGZ23] B. Dadoun, M. Fradelizi, O. Guédon, and P.-A. Zitt, *Asymptotics of the inertia moments and the variance conjecture in Schatten balls*, J. Funct. Anal. **284** (2023), no. 2, Paper No. 109741, 32, DOI 10.1016/j.jfa.2022.109741. MR4500732

[DG07] P. Deift and D. Gioev, *Universality at the edge of the spectrum for unitary, orthogonal, and symplectic ensembles of random matrices*, Comm. Pure Appl. Math. **60** (2007), no. 6, 867–910, DOI 10.1002/cpa.20164. MR2306224

[dH00] F. den Hollander, *Large deviations*, Fields Institute Monographs, vol. 14, American Mathematical Society, Providence, RI, 2000, DOI 10.1007/s00440-009-0235-5. MR1739680

[DHLM21] D. Dereudre, A. Hardy, T. Leblé, and M. Maïda, *DLR equations and rigidity for the sine-beta process*, Comm. Pure Appl. Math. **74** (2021), no. 1, 172–222, DOI 10.1002/cpa.21963. MR4178183

[Dia64] P. H. Diananda, *Notes on two lemmas concerning the Epstein zeta-function*, Proc. Glasgow Math. Assoc. **6** (1964), 202–204 (1964), DOI 10.1090/s0025-5718-52-99389-7. MR168537

[DKM+99] P. Deift, T. Kriecherbauer, K. T-R. McLaughlin, S. Venakides, and X. Zhou, *Strong asymptotics of orthogonal polynomials with respect to exponential weights*, Comm. Pure Appl. Math. **52** (1999), no. 12, 1491–1552, DOI 10.1002/(sici)1097-0312(199912)52:12<1491::aid-cpa2>3.0.co;2-#. MR1711036

[DL74] C. Deutsch and C. Lavaud, *Equilibrium properties of a two-dimensional Coulomb gas*, Phys. Rev. A **9** (1974), no. 6, 2598–2616.

[DLDMS19] D. S. Dean, P. Le Doussal, S. N. Majumdar, and G. Schehr, *Noninteracting fermions in a trap and random matrix theory*, J. Phys. A **52** (2019), no. 14, 144006, 32, DOI 10.1088/1751-8121/ab098d. MR3935533

[DM93] G. Dal Maso, *An introduction to Γ-convergence*, Progress in Nonlinear Differential Equations and their Applications, vol. 8, Birkhäuser Boston, Inc., Boston, MA, 1993, DOI 10.1007/978-1-4612-0327-8. MR1201152

[Doo01] J. L. Doob, *Classical potential theory and its probabilistic counterpart*, Classics in Mathematics, Springer-Verlag, Berlin, 2001. Reprint of the 1984 edition, DOI 10.1007/978-3-642-56573-1. MR1814344

[DS89] J.-D. Deuschel and D. W. Stroock, *Large deviations*, Pure and Applied Mathematics, vol. 137, Academic Press, Inc., Boston, MA, 1989. MR997938

[Due16] M. Duerinckx, *Mean-field limits for some Riesz interaction gradient flows*, SIAM J. Math. Anal. **48** (2016), no. 3, 2269–2300, DOI 10.1137/15M1042620. MR3514713

[DV23] D. Dereudre and T. Vasseur, *Number-rigidity and β-circular Riesz gas*, Ann. Probab. **51** (2023), no. 3, 1025–1065, DOI 10.1214/22-aop1606. MR4583062

[DVJ88] D. J. Daley and D. Vere-Jones, *An introduction to the theory of point processes*, Springer Series in Statistics, Springer-Verlag, New York, 1988. MR950166

[Dys62] F. J. Dyson, *Statistical theory of the energy levels of complex systems. I, II, III*, J. Mathematical Phys. **3** (1962), 140–175, DOI 10.1063/1.1703775. MR0143556

[DZ10] A. Dembo and O. Zeitouni, *Large deviations techniques and applications*, Stochastic Modelling and Applied Probability, vol. 38, Springer-Verlag, Berlin, 2010. Corrected reprint of the second (1998) edition, DOI 10.1007/978-3-642-03311-7. MR2571413

[E94] W. E, *Dynamics of vortices in Ginzburg-Landau theories with applications to superconductivity*, Phys. D **77** (1994), no. 4, 383–404, DOI 10.1016/0167-2789(94)90298-4. MR1297726

[EG15] L. C. Evans and R. F. Gariepy, *Measure theory and fine properties of functions*, Revised edition, Textbooks in Mathematics, CRC Press, Boca Raton, FL, 2015. MR3409135

[EHL21] M. Erbar, M. Huesmann, and T. Leblé, *The one-dimensional log-gas free energy has a unique minimizer*, Comm. Pure Appl. Math. **74** (2021), no. 3, 615–675, DOI 10.1002/cpa.21977. MR4201295

[EL81] S. Oxford and E. H. Lieb, *Improved lower bound on the indirect Coulomb energy*, Int. J. Quant. Chem. **19** (1981), 427–439.

[Enn64a] V. Ennola, *A lemma about the Epstein zeta-function*, Proc. Glasgow Math. Assoc. **6** (1964), 198–201 (1964). MR168536

[Enn64b] V. Ennola, *On a problem about the Epstein zeta-function*, Proc. Cambridge Philos. Soc. **60** (1964), 855–875. MR168535

[EPR$^+$10] L. Erdős, S. Péché, J. A. Ramírez, B. Schlein, and H.-T. Yau, *Bulk universality for Wigner matrices*, Comm. Pure Appl. Math. **63** (2010), no. 7, 895–925, DOI 10.1002/cpa.20317. MR2662426

[ES06] G. L. Eyink and K. R. Sreenivasan, *Onsager and the theory of hydrodynamic turbulence*, Rev. Modern Phys. **78** (2006), no. 1, 87–135, DOI 10.1103/RevModPhys.78.87. MR2214822

[F88] H. Föllmer, *Random fields and diffusion processes*, École d'Été de Probabilités de Saint-Flour XV–XVII, 1985–87, Lecture Notes in Math., vol. 1362, Springer, Berlin, 1988, pp. 101–203, DOI 10.1007/BFb0086180. MR983373

[FFGW03] P. J. Forrester, N. E. Frankel, T. M. Garoni, and N. S. Witte, *Finite one-dimensional impenetrable bose systems: Occupation numbers*, Phys. Rev. A, **67** (2003), 043607.

[FJ96] P. J. Forrester and B. Jancovici, *Two-dimensional one-component plasma in a quadrupolar field*, Internat. J. Modern Phys. A **11** (1996), no. 5, 941–949, DOI 10.1142/S0217751X96000432. MR1371262

[FJM01] P. J. Forrester, B. Jancovici, and D. S. McAnally, *Analytic properties of the structure function for the one-dimensional one-component log-gas*, Proceedings of the Baxter Revolution in Mathematical Physics (Canberra, 2000), J. Statist. Phys. **102** (2001), no. 3-4, 737–780, DOI 10.1023/A:1004846818738. MR1832069

[FKS82] E. B. Fabes, C. E. Kenig, and R. P. Serapioni, *The local regularity of solutions of degenerate elliptic equations*, Comm. Partial Differential Equations **7** (1982), no. 1, 77–116, DOI 10.1080/03605308208820218. MR643158

[FL22] M. Fenzl and G. Lambert, *Precise deviations for disk counting statistics of invariant determinantal processes*, Int. Math. Res. Not. IMRN **10** (2022), 7420–7494, DOI 10.1093/imrn/rnaa341. MR4418712

[FMS22] A. Flack, S. N. Majumdar, and G. Schehr, *Gap probability and full counting statistics in the one-dimensional one-component plasma*, J. Stat. Mech. Theory Exp. **5** (2022), Paper No. 053211, 35, DOI 10.1088/1742-5468/ac6a59. MR4473215

[FO88] H. Föllmer and S. Orey, *Large deviations for the empirical field of a Gibbs measure*, Ann. Probab. **16** (1988), no. 3, 961–977. MR942749

[For92] P. J. Forrester, *Some statistical properties of the eigenvalues of complex random matrices*, Phys. Lett. A **169** (1992), no. 1-2, 21–24, DOI 10.1016/0375-9601(92)90798-Q. MR1181356

[For06] P. J. Forrester, *Particles in a magnetic field and plasma analogies: doubly periodic boundary conditions*, J. Phys. A **39** (2006), no. 41, 13025–13036, DOI 10.1088/0305-4470/39/41/S14. MR2277465

[For10] P. J. Forrester, *Log-gases and random matrices*, London Mathematical Society Monographs Series, vol. 34, Princeton University Press, Princeton, NJ, 2010, DOI 10.1515/9781400835416. MR2641363

[For16] P. J. Forrester, *Analogies between random matrix ensembles and the one-component plasma in two-dimensions*, Nuclear Phys. B **904** (2016), 253–281, DOI 10.1016/j.nuclphysb.2016.01.014. MR3458536

[Fre72] J. Frehse, *On the regularity of the solution of a second order variational inequality* (English, with Italian summary), Boll. Un. Mat. Ital. (4) **6** (1972), 312–315. MR318650

[Fri88] A. Friedman, *Variational principles and free-boundary problems*, 2nd ed., Robert E. Krieger Publishing Co., Inc., Malabar, FL, 1988. MR1009785

[Fro35] O. Frostman, *Potentiel d'équilibre et capacité des ensembles avec quelques applications à la théorie des fonctions*, Meddelanden Mat. Sem. Univ. Lund 3, 115 s, 1935.

[Frö76] J. Fröhlich, *Classical and quantum statistical mechanics in one and two dimensions: two-component Yukawa- and Coulomb systems*, Comm. Math. Phys. **47** (1976), no. 3, 233–268. MR434278

[FROS20] A. Figalli, X. Ros-Oton, and J. Serra, *Generic regularity of free boundaries for the obstacle problem*, Publ. Math. Inst. Hautes Études Sci. **132** (2020), 181–292, DOI 10.1007/s10240-020-00119-9. MR4179834

[FS81a] J. Fröhlich and T. Spencer, *The Kosterlitz-Thouless transition in two-dimensional abelian spin systems and the Coulomb gas*, Comm. Math. Phys. **81** (1981), no. 4, 527–602. MR634447

[FS81b] J. Fröhlich and T. Spencer, *On the statistical mechanics of classical Coulomb and dipole gases*, J. Statist. Phys. **24** (1981), no. 4, 617–701, DOI 10.1007/BF01011379. MR610687

[FTW19] R. Feng, G. Tian, and D. Wei, *Small gaps of GOE*, Geom. Funct. Anal. **29** (2019), no. 6, 1794–1827, DOI 10.1007/s00039-019-00520-5. MR4034920

[FV18] S. Friedli and Y. Velenik, *Statistical mechanics of lattice systems: A concrete mathematical introduction*, Cambridge University Press, Cambridge, 2018. MR3752129

[FW21] R. Feng and D. Wei, *Small gaps of circular β-ensemble*, Ann. Probab. **49** (2021), no. 2, 997–1032, DOI 10.1214/20-aop1468. MR4255136

[GBR+12] A. Gretton, K. M. Borgwardt, M. J. Rasch, B. Schölkopf, and A. Smola, *A kernel two-sample test*, J. Mach. Learn. Res. **13** (2012), 723–773. MR2913716

[Geo93] H.-O. Georgii, *Large deviations and maximum entropy principle for interacting random fields on $\mathbf{Z}^d$*, Ann. Probab. **21** (1993), no. 4, 1845–1875. MR1245292

[Gin65] J. Ginibre, *Statistical ensembles of complex, quaternion, and real matrices*, J. Mathematical Phys. **6** (1965), 440–449, DOI 10.1063/1.1704292. MR173726

[Gir84] V. L. Girko, *The circle law*, Theory Probab. Appl. **29** (1984), 694–706. MR727271

[Gir05] S. M. Girvin, *Introduction to the fractional quantum Hall effect*, The quantum Hall effect, Prog. Math. Phys., vol. 45, Birkhäuser, Basel, 2005, pp. 133–162, DOI 10.1007/3-7643-7393-8_4. MR2180172

[GK24] V. Gorin and V. Kleptsyn, *Universal objects of the infinite beta random matrix theory*, J. Eur. Math. Soc. (JEMS) **26** (2024), no. 9, 3429–3496, DOI 10.4171/jems/1336. MR4767497

[GL65] V. L Ginzburg and L. D. Landau, Collected papers of L. D. Landau, 1965.

[GLM78] Ch. Gruber, Ch. Lugrin, and Ph. A. Martin, *Equilibrium equations for classical systems with long range forces and application to the one-dimensional Coulomb gas*, Helv. Phys. Acta **51** (1978), no. 5-6, 829–866 (1979). MR542803

[GLM80] Ch. Gruber, Ch. Lugrin, and Ph. A. Martin, *Equilibrium properties of classical systems with long-range forces. BBGKY equation, neutrality, screening, and sum rules*, J. Statist. Phys. **22** (1980), no. 2, 193–236, DOI 10.1007/BF01008049. MR560555

[GMS13] D. Goldman, C. B. Muratov, and S. Serfaty, *The Γ-limit of the two-dimensional Ohta-Kawasaki energy. I. Droplet density*, Arch. Ration. Mech. Anal. **210** (2013), no. 2, 581–613, DOI 10.1007/s00205-013-0657-1. MR3101793

[GMS14] D. Goldman, C. B. Muratov, and S. Serfaty, *The Γ-limit of the two-dimensional Ohta-Kawasaki energy. Droplet arrangement via the renormalized energy*, Arch. Ration. Mech. Anal. **212** (2014), no. 2, 445–501, DOI 10.1007/s00205-013-0711-z. MR3176350

[GN18] S. Ghosh and A. Nishry, *Point processes, hole events, and large deviations: random complex zeros and Coulomb gases*, Constr. Approx. **48** (2018), no. 1, 101–136, DOI 10.1007/s00365-018-9418-6. MR3825948

[Gol22] François Golse, *Mean-field limits in statistical dynamics* (English, with English and French summaries), Panor. Synthèses, vol. 65, Soc. Math. France, Paris, 2025, pp. 125–131. MR5008877

[Gor21] V. Gorin, *Lectures on random lozenge tilings*, Cambridge Studies in Advanced Mathematics, vol. 193, Cambridge University Press, Cambridge, 2021, DOI 10.1017/9781108921183. MR4299268

[GP77] J. Gunson and L. S. Panta, *Two-dimensional neutral Coulomb gas*, Comm. Math. Phys. **52** (1977), no. 3, 295–304. MR452403

[GP17] S. Ghosh and Y. Peres, *Rigidity and tolerance in point processes: Gaussian zeros and Ginibre eigenvalues*, Duke Math. J. **166** (2017), no. 10, 1789–1858, DOI 10.1215/00127094-2017-0002. MR3679882

[GP21] F. Golse and T. Paul, *Mean-field and classical limit for the N-body quantum dynamics with Coulomb interaction*, Comm. Pure Appl. Math. **75** (2022), no. 6, 1332–1376, DOI 10.1002/cpa.21986. MR4415778

[Gra14] L. Grafakos, *Classical Fourier analysis*, 3rd ed., Graduate Texts in Mathematics, vol. 249, Springer, New York, 2014, DOI 10.1007/978-1-4939-1194-3. MR3243734

[GS95] G. M. Graf and D. Schenker, *On the molecular limit of Coulomb gases*, Comm. Math. Phys. **174** (1995), no. 1, 215–227. MR1372807

[GS16] I. M. Gel'fand and G. E. Shilov, *Generalized functions. Vol. I: Properties and operations*, Academic Press, New York-London, 1964. Translated by Eugene Saletan. MR166596

[GS20] S. Ganguly and S. Sarkar, *Ground states and hyperuniformity of the hierarchical Coulomb gas in all dimensions*, Probab. Theory Related Fields **177** (2020), no. 3-4, 621–675, DOI 10.1007/s00440-019-00955-9. MR4126928

[GS21] Y. Ge and E. Sandier, *Lattices with finite renormalized Coulombian interaction energy in the plane*, Tunis. J. Math. **3** (2021), no. 1, 93–120, DOI 10.2140/tunis.2021.3.93. MR4149084

[GT01] D. Gilbarg and N. S. Trudinger, *Elliptic partial differential equations of second order*, Classics in Mathematics, Springer-Verlag, Berlin, 2001. Reprint of the 1998 edition. MR1814364

[GWT01] M. D. Girardeau, E. M. Wright, and J. M. Triscari, *Ground-state properties of a one-dimensional system of hard-core bosons in a harmonic trap*, Phys. Rev. A, **63** (2001), 033601.

[GZ93] H.-O. Georgii and H. Zessin, *Large deviations and the maximum entropy principle for marked point random fields*, Probab. Theory Related Fields **96** (1993), no. 2, 177–204, DOI 10.1007/BF01192132. MR1227031

[GZ19] D. García-Zelada, *A large deviation principle for empirical measures on Polish spaces: application to singular Gibbs measures on manifolds* (English, with English and French summaries), Ann. Inst. Henri Poincaré Probab. Stat. **55** (2019), no. 3, 1377–1401, DOI 10.1214/18-aihp922. MR4010939

[Har12] A. Hardy, *A note on large deviations for 2D Coulomb gas with weakly confining potential*, Electron. Commun. Probab. **17** (2012), no. 19, 12, DOI 10.1214/ECP.v17-1818. MR2926763

[HCR25] E. Hess-Childs and K. Rowan, *Higher-order propagation of chaos in L^2 for interacting diffusions*, Probab. Math. Phys. **6** (2025), no. 2, 581–646, DOI 10.2140/pmp.2025.6.581. MR4879390

[HCRS] E. Hess-Childs, M. Rosenzweig, and S. Serfaty, *Optimal quantization for Riesz Maximum Mean Discrepancies*, In preparation.

[HCRS25] M. Rosenzweig and S. Serfaty, *Sharp commutator estimates of all order for Coulomb and Riesz modulated energies*, Comm. Pure Appl. Math. **79** (2026), no. 2, 207–292, DOI 10.1002/cpa.70010. MR4974720

[Hir78] F. Hirsch, *Opérateurs carré du champ (d'après J. P. Roth)* (French), Séminaire Bourbaki, 29e année (1976/77), Lecture Notes in Math., vol. 677, Springer, Berlin, 1978, pp. Exp. No. 501, pp. 167–182. MR521767

[HKI21] D. Han-Kwan and M. Iacobelli, *From Newton's second law to Euler's equations of perfect fluids*, Proc. Amer. Math. Soc. **149** (2021), no. 7, 3045–3061, DOI 10.1090/proc/15349. MR4257814

[HKPV09] J. B. Hough, M. Krishnapur, Y. Peres, and B. Virág, *Zeros of Gaussian analytic functions and determinantal point processes*, University Lecture Series, vol. 51, American Mathematical Society, Providence, RI, 2009, DOI 10.1090/ulect/051. MR2552864

[HL21] A. Hardy and G. Lambert, *CLT for circular beta-ensembles at high temperature*, J. Funct. Anal. **280** (2021), no. 7, Paper No. 108869, 40, DOI 10.1016/j.jfa.2020.108869. MR4211032

[HL24] M. Huesmann and T. Leblé, *The link between hyperuniformity, Coulomb energy and Wasserstein distance to Lebesgue for two-dimensional point processes*, Probab. Math. Phys. **7** (2026), no. 1, 123–173, DOI 10.2140/pmp.2026.7.123. MR4998452

[HLS08] C. Hainzl, M. Lewin, and J. P. Solovej, *The thermodynamic limit of quantum Coulomb systems: a new approach*, Mathematical results in quantum mechanics, World Sci. Publ., Hackensack, NJ, 2008, pp. 97–116, DOI 10.1142/9789812832382_0008. MR2466681

[HLSS18] D. P. Hardin, T. Leblé, E. B. Saff, and S. Serfaty, *Large deviation principles for hypersingular Riesz gases*, Constr. Approx. **48** (2018), no. 1, 61–100, DOI 10.1007/s00365-018-9431-9. MR3825947

[HM13] H. Hedenmalm and N. Makarov, *Coulomb gas ensembles and Laplacian growth*, Proc. Lond. Math. Soc. (3) **106** (2013), no. 4, 859–907, DOI 10.1112/plms/pds032. MR3056295

[HP00] F. Hiai and D. Petz, *The semicircle law, free random variables and entropy*, Mathematical Surveys and Monographs, vol. 77, American Mathematical Society, Providence, RI, 2000, DOI 10.1090/surv/077. MR1746976

[HRSV19] D. P. Hardin, A. Reznikov, E. B. Saff, and A. Volberg, *Local properties of Riesz minimal energy configurations and equilibrium measures*, Int. Math. Res. Not. IMRN **16** (2019), 5066–5086, DOI 10.1093/imrn/rnx262. MR4009530

[Hua63] K. Huang, *Statistical mechanics*, John Wiley & Sons, Inc., New York-London, 1963. MR154659

[HW21] H. Hedenmalm and A. Wennman, *Planar orthogonal polynomials and boundary universality in the random normal matrix model*, Acta Math. **227** (2021), no. 2, 309–406, DOI 10.4310/acta.2021.v227.n2.a3. MR4366416

[Imb83] J. Z. Imbrie, *Debye screening for jellium and other Coulomb systems*, Comm. Math. Phys. **87** (1982/83), no. 4, 515–565. MR691043

[Jab14] P.-E. Jabin, *A review of the mean field limits for Vlasov equations*, Kinet. Relat. Models **7** (2014), no. 4, 661–711, DOI 10.3934/krm.2014.7.661. MR3317577

[Jan95] B. Jancovici, *Classical Coulomb systems: screening and correlations revisited*, J. Statist. Phys. **80** (1995), no. 1-2, 445–459, DOI 10.1007/BF02178367. MR1340566

[JC96] M. D. Jones and D. M. Ceperley, *Crystallization of the one-component plasma at finite temperature*, Phys. Rev. Lett. **76** (1996), 4572–4575.

[JGP94] B. Jancovici, Manificat G., and C. Pisani, *Coulomb systems seen as critical systems: Finite-size effects in two dimensions*, J. Statist. Phys. **76** (1994), 307–329.

[JLM93] B. Jancovici, J. L. Lebowitz, and G. Manificat, *Large charge fluctuations in classical Coulomb systems*, J. Statist. Phys. **72** (1993), no. 3-4, 773–787, DOI 10.1007/BF01048032. MR1239571

[Joh98] K. Johansson, *On fluctuations of eigenvalues of random Hermitian matrices*, Duke Math. J. **91** (1998), no. 1, 151–204, DOI 10.1215/S0012-7094-98-09108-6. MR1487983

[JS98] R. L. Jerrard and H. M. Soner, *Dynamics of Ginzburg-Landau vortices*, Arch. Rational Mech. Anal. **142** (1998), no. 2, 99–125, DOI 10.1007/s002050050085. MR1629646

[JS09] R. L. Jerrard and P. Sternberg, *Critical points via Γ-convergence: general theory and applications*, J. Eur. Math. Soc. (JEMS) **11** (2009), no. 4, 705–753, DOI 10.4171/JEMS/164. MR2538502

[JT96] B. Jancovici and G. Téllez, *Coulomb systems seen as critical systems: ideal conductor boundaries*, J. Statist. Phys. **82** (1996), no. 3-4, 609–632, DOI 10.1007/BF02179788. MR1372423

[JV23] K. Johansson and F. Viklund, *Coulomb gas and the Grunsky operator on a Jordan domain with corners*, arXiv:2309.00308, 2023.

[JW17] P.-E. Jabin and Z. Wang, *Mean field limit for stochastic particle systems*, Active particles. Vol. 1. Advances in theory, models, and applications, Model. Simul. Sci. Eng. Technol., Birkhäuser/Springer, Cham, 2017, pp. 379–402. MR3644596

[JW18] P.-E. Jabin and Z. Wang, *Quantitative estimates of propagation of chaos for stochastic systems with $W^{-1,\infty}$ kernels*, Invent. Math. **214** (2018), no. 1, 523–591, DOI 10.1007/s00222-018-0808-y. MR3858403

[Kö05] W. König, *Orthogonal polynomial ensembles in probability theory*, Probab. Surv. **2** (2005), 385–447, DOI 10.1214/154957805100000177. MR2203677

[KAFMA21] A. Korba, P.-C. Aubin-Frankowski, S. Majewski, and P. Ablin, *Kernel stein discrepancy descent*, in Proceedings of the 38th International Conference on Machine Learning (M. Meila and T. Zhang, editors), volume 139 of *Proceedings of Machine Learning Research*, pp. 5719–5730. PMLR, 18–24 Jul 2021.

[Kie93] M. K.-H. Kiessling, *Statistical mechanics of classical particles with logarithmic interactions*, Comm. Pure Appl. Math. **46** (1993), no. 1, 27–56, DOI 10.1002/cpa.3160460103. MR1193342

[KK16] S. A. Khrapak and A. G. Khrapak, *Internal energy of the classical two- and three-dimensional one- component-plasma*, Contrib. Plasma Phys. **56** (2016), no. 3–4, 270–280.

[KKK+21] J. Kethepalli, M. Kulkarni, A. Kundu, S. N. Majumdar, D. Mukamel, and G. Schehr, *Harmonically confined long-ranged interacting gas in the presence of a hard wall*, J. Stat. Mech. Theory Exp. **10** (2021), Paper No. 103209, 36, DOI 10.1088/1742-5468/ac2896. MR4406025

[KKK+22] J. Kethepalli, M. Kulkarni, A. Kundu, S. N. Majumdar, D. Mukamel, and G. Schehr, *Edge fluctuations and third-order phase transition in harmonically confined long-range systems*, J. Stat. Mech. Theory Exp. **3** (2022), Paper No. 033203, 44, DOI 10.1088/1742-5468/ac52b2. MR4455479

[KLDMS23] M. Kulkarni, P. Le Doussal, S. N. Majumdar, and G. Schehr, *Density profile of noninteracting fermions in a rotating two-dimensional trap at finite temperature*, Phys. Rev. A **107** (2023), no. 2, Paper No. 023302, 16, DOI 10.1103/physreva.107.023302. MR4558790

[Kle14] S. Klevtsov, *Random normal matrices, Bergman kernel and projective embeddings*, J. High Energy Phys. **1** (2014), 133, front matter+18, DOI 10.1007/JHEP01(2014)133. MR3599064

[Kle16] S. Klevtsov, *Geometry and large N limits in Laughlin states*, Travaux mathématiques. Vol. XXIV, Trav. Math., vol. 24, Fac. Sci. Technol. Commun. Univ. Luxemb., Luxembourg, 2016, pp. 63–127. MR3643934

[Kle19] S. Klevtsov, *Laughlin states on higher genus Riemann surfaces*, Comm. Math. Phys. **367** (2019), no. 3, 837–871, DOI 10.1007/s00220-019-03318-6. MR3943483

[KM13] N.-G. Kang and N. G. Makarov, *Gaussian free field and conformal field theory* (English, with English and French summaries), Astérisque **353** (2013), viii+136. MR3052311

[KMMW17] S. Klevtsov, X. Ma, G. Marinescu, and P. Wiegmann, *Quantum Hall effect and Quillen metric*, Comm. Math. Phys. **349** (2017), no. 3, 819–855, DOI 10.1007/s00220-016-2789-2. MR3602817

[KN04] R. Killip and I. Nenciu, *Matrix models for circular ensembles*, Int. Math. Res. Not. **50** (2004), 2665–2701, DOI 10.1155/S1073792804141597. MR2127367

[KS99] M. K.-H. Kiessling and H. Spohn, *A note on the eigenvalue density of random matrices*, Comm. Math. Phys. **199** (1999), no. 3, 683–695, DOI 10.1007/s002200050516. MR1669669

[KS00] D. Kinderlehrer and G. Stampacchia, *An introduction to variational inequalities and their applications*, Classics in Applied Mathematics, vol. 31, Society for Industrial and Applied Mathematics (SIAM), Philadelphia, PA, 2000. Reprint of the 1980 original, DOI 10.1137/1.9780898719451. MR1786735

[KS09] R. Killip and M. Stoiciu, *Eigenvalue statistics for CMV matrices: from Poisson to clock via random matrix ensembles*, Duke Math. J. **146** (2009), no. 3, 361–399, DOI 10.1215/00127094-2009-001. MR2484278

[Kui11] A. B. J. Kuijlaars, *Universality*, The Oxford handbook of random matrix theory, Oxford Univ. Press, Oxford, 2011, pp. 103–134. MR2932626

[Kun74] H. Kunz, *The one-dimensional classical electron gas*, Ann. Physics **85** (1974), 303–335, DOI 10.1016/0003-4916(74)90413-8. MR426742

[Kwa17] M. Kwaśnicki, *Ten equivalent definitions of the fractional Laplace operator*, Fract. Calc. Appl. Anal. **20** (2017), no. 1, 7–51, DOI 10.1515/fca-2017-0002. MR3613319

[Lac23] D. Lacker, *Hierarchies, entropy, and quantitative propagation of chaos for mean field diffusions*, Probab. Math. Phys. **4** (2023), no. 2, 377–432, DOI 10.2140/pmp.2023.4.377. MR4595391

[Lam21] G. Lambert, *Poisson statistics for Gibbs measures at high temperature* (English, with English and French summaries), Ann. Inst. Henri Poincaré Probab. Stat. **57** (2021), no. 1, 326–350, DOI 10.1214/20-aihp1080. MR4255177

[Lan72] N. S. Landkof, *Foundations of modern potential theory*, Die Grundlehren der mathematischen Wissenschaften, Band 180, Springer-Verlag, New York-Heidelberg, 1972. Translated from the Russian by A. P. Doohovskoy. MR350027

[Lan87] S. Lang, *Elliptic functions*, 2nd ed., Graduate Texts in Mathematics, vol. 112, Springer-Verlag, New York, 1987. With an appendix by J. Tate, DOI 10.1007/978-1-4612-4752-4. MR890960

[Lau] R. B. Laughlin, *Elementary theory: the incompressible quantum fluid*, Springer.

[Lau83] R. B. Laughlin, *Anomalous quantum Hall effect: an incompressible quantum fluid with fractionally charged excitations*, Phys. Rev. Lett. **50** (1983), no. 18, 1395.

[Lau99] R. B. Laughlin, *Nobel lecture: fractional quantization*, Rev. Modern Phys. **71** (1999), no. 4, 863–874, DOI 10.1103/RevModPhys.71.863. MR1712324

[Lau21] A. B. Lauritsen, *Floating Wigner crystal and periodic jellium configurations*, J. Math. Phys. **62** (2021), no. 8, Paper No. 083305, 17, DOI 10.1063/5.0053494. MR4298023

[Leb] T. Leblé, *The two-dimensional one-component plasma is hyperuniform*, `arXiv:2104.05109`, 2021.

[Leb83] J. Lebowitz, *Charge fluctuations in Coulomb systems*, Phys. Rev. A, **27** (1983), no. 3, 1491.

[Leb15] T. Leblé, *A uniqueness result for minimizers of the 1D log-gas renormalized energy*, J. Funct. Anal. **268** (2015), no. 7, 1649–1677, DOI 10.1016/j.jfa.2014.11.023. MR3315574

[Leb16] T. Leblé, *Logarithmic, Coulomb and Riesz energy of point processes*, J. Stat. Phys. **162** (2016), no. 4, 887–923, DOI 10.1007/s10955-015-1425-4. MR3456982

[Leb17] T. Leblé, *Local microscopic behavior for 2D Coulomb gases*, Probab. Theory Related Fields **169** (2017), no. 3-4, 931–976, DOI 10.1007/s00440-016-0744-y. MR3719060

[Leb21] T. Leblé, *CLT for fluctuations of linear statistics in the sine-beta process*, Int. Math. Res. Not. IMRN **8** (2021), 5676–5756, DOI 10.1093/imrn/rnz020. MR4251261

[Leb24] T. Leblé, *DLR equations, number-rigidity and translation-invariance for infinite-volume limit points of the 2DOCP*, `arXiv:2410.04958`, 2024.

[Leb25] T. Leblé, *The hexagonal lattice is universally locally optimal*, `arXiv:2511.03353`, 2025.

[Len61] A. Lenard, *Exact statistical mechanics of a one-dimensional system with Coulomb forces*, J. Mathematical Phys. **2** (1961), 682–693, DOI 10.1063/1.1703757. MR129874

[Len63] A. Lenard, *Exact statistical mechanics of a one-dimensional system with Coulomb forces. III. Statistics of the electric field*, J. Mathematical Phys. **4** (1963), 533–543, DOI 10.1063/1.1703988. MR148411

[Lew22] M. Lewin, *Coulomb and Riesz gases: the known and the unknown*, J. Math. Phys. **63** (2022), no. 6, Paper No. 061101, 77, DOI 10.1063/5.0086835. MR4431907

[Lie09] E. Lieb, Personal communication, 2009.

[Lin96] F. H. Lin, *Some dynamical properties of Ginzburg-Landau vortices*, Comm. Pure Appl. Math. **49** (1996), no. 4, 323–359, DOI 10.1002/(SICI)1097-0312(199604)49:4<323::AID-CPA1>3.0.CO;2-E. MR1376654

[Lin23] Z. Lin, *Nonlocal energy functionals and determinantal point processes on non-smooth domains*, Math. Z. **307** (2024), no. 3, Paper No. 56, 22, DOI 10.1007/s00209-024-03540-6. MR4761438

[LL69] E. H. Lieb and J. L. Lebowitz, *Existence of thermodynamics for real matter with Coulomb forces*, Phys. Rev. Lett. **22** (1969), 631–634.

[LL01] E. H. Lieb and M. Loss, *Analysis*, 2nd ed., Graduate Studies in Mathematics, vol. 14, American Mathematical Society, Providence, RI, 2001, DOI 10.1090/gsm/014. MR1817225

[LLK$^+$23] Lingxiao L., Q. Liu, A. Korba, M. Yurochkin, and J. Solomon, *Sampling with Mollified Interaction Energy Descent*, in The Eleventh International Conference on Learning Representations, 2023.

[LLS18] M. Lewin, E. H. Lieb, and R. Seiringer, *Statistical mechanics of the uniform electron gas* (English, with English and French summaries), J. Éc. polytech. Math. **5** (2018), 79–116, DOI 10.5802/jep.64. MR3732693

[LLS22] M. Lewin, E. H. Lieb, and R. Seiringer, *Improved Lieb-Oxford bound on the indirect and exchange energies*, Lett. Math. Phys. **112** (2022), no. 5, Paper No. 92, 36, DOI 10.1007/s11005-022-01584-5. MR4483896

[LLW19] G. Lambert, M. Ledoux, and C. Webb, *Quantitative normal approximation of linear statistics of β-ensembles*, Ann. Probab. **47** (2019), no. 5, 2619–2685, DOI 10.1214/18-AOP1314. MR4021234

[LLZ24] G. Lambert, T. Leblé, and O. Zeitouni, *Law of large numbers for the maximum of the two-dimensional Coulomb gas potential*, Electron. J. Probab. **29** (2024), Paper No. 42, 36, DOI 10.1214/24-ejp1102. MR4718451

[LM16] S.-Y. Lee and N. G. Makarov, *Topology of quadrature domains*, J. Amer. Math. Soc. **29** (2016), no. 2, 333–369, DOI 10.1090/jams828. MR3454377

[LN75] E. H. Lieb and H. Narnhofer, *The thermodynamic limit for jellium*, J. Statist. Phys. **12** (1975), 291–310, DOI 10.1007/BF01012066. MR401029

[LP23] G. Lambert and E. Paquette, *Strong approximation of Gaussian β ensemble characteristic polynomials: the hyperbolic regime*, Ann. Appl. Probab. **33** (2023), no. 1, 549–612, DOI 10.1214/22-aap1823. MR4551558

[LR10] M. Ledoux and B. Rider, *Small deviations for beta ensembles*, Electron. J. Probab. **15** (2010), no. 41, 1319–1343, DOI 10.1214/EJP.v15-798. MR2678393

[LRY18] E. H. Lieb, N. Rougerie, and J. Yngvason, *Rigidity of the Laughlin liquid*, J. Stat. Phys. **172** (2018), no. 2, 544–554, DOI 10.1007/s10955-018-2082-1. MR3824950

[LS10] E. H. Lieb and R. Seiringer, *The stability of matter in quantum mechanics*, Cambridge University Press, Cambridge, 2010. MR2583992

[LS17] T. Leblé and S. Serfaty, *Large deviation principle for empirical fields of log and Riesz gases*, Invent. Math. **210** (2017), no. 3, 645–757, DOI 10.1007/s00222-017-0738-0. MR3735628

[LS18] T. Leblé and S. Serfaty, *Fluctuations of two dimensional Coulomb gases*, Geom. Funct. Anal. **28** (2018), no. 2, 443–508, DOI 10.1007/s00039-018-0443-1. MR3788208

[LSZ17] T. Leblé, S. Serfaty, and O. Zeitouni, *Large deviations for the two-dimensional two-component plasma*, Comm. Math. Phys. **350** (2017), no. 1, 301–360, DOI 10.1007/s00220-016-2735-3. MR3606477

[LW16] Q. Liu and D. Wang, *Stein variational gradient descent: a general purpose bayesian inference algorithm*, in Proceedings of the 30th International Conference on Neural Information Processing Systems, NIPS'16, pp. 2378–2386, Red Hook, NY, USA, 2016. Curran Associates Inc.

[LW20] W. Liu and L. Wu, *Large deviations for empirical measures of mean-field Gibbs measures*, Stochastic Process. Appl. **130** (2020), no. 2, 503–520, DOI 10.1016/j.spa.2019.01.008. MR4046507

[LWL00] D. Levesque, J.-J. Weis, and J. L. Lebowitz, *Charge fluctuations in the two-dimensional one-component plasma*, J. Statist. Phys. **100** (2000), no. 1-2, 209–222, DOI 10.1023/A:1018643829340. Special issue dedicated to George Stell. MR1791555

[LX99a] F.-H. Lin and J. X. Xin, *On the dynamical law of the Ginzburg-Landau vortices on the plane*, Comm. Pure Appl. Math. **52** (1999), no. 10, 1189–1212, DOI 10.1002/(SICI)1097-0312(199910)52:10<1189::AID-CPA1>3.0.CO;2-T. MR1699965

[LX99b] F.-H. Lin and J. X. Xin, *On the incompressible fluid limit and the vortex motion law of the nonlinear Schrödinger equation*, Comm. Math. Phys. **200** (1999), no. 2, 249–274, DOI 10.1007/s002200050529. MR1674000

[LZ21] T. Leblé and O. Zeitouni, *A local CLT for linear statistics of 2D Coulomb gases*, Markov Process. Related Fields **27** (2021), no. 4, 607–630. MR4396198

[M2́4] M. Ménard, *Mean-field limit of point vortices for the lake equations*, Commun. Math. Sci. **22** (2024), no. 8, 2167–2228, DOI 10.4310/cms.240918050058. MR4836955

[Mag12] F. Maggi, *Sets of finite perimeter and geometric variational problems: An introduction to geometric measure theory*, Cambridge Studies in Advanced Mathematics, vol. 135, Cambridge University Press, Cambridge, 2012, DOI 10.1017/CBO9781139108133. MR2976521

[Mar88] Ph. A. Martin, *Sum rules in charged fluids*, Rev. Modern Phys. **60** (1988), no. 4, 1075–1127, DOI 10.1103/RevModPhys.60.1075. MR969999

[Mar18] M. Mariani, *A Γ-convergence approach to large deviations*, Ann. Sc. Norm. Super. Pisa Cl. Sci. (5) **18** (2018), no. 3, 951–976. MR3807592

[Maz11] M. Mazars, *Long ranged interactions in computer simulations and for quasi-2D systems*, Phys. Reports **500** (2011), 43–116.

[McK67] H. P. McKean Jr., *Propagation of chaos for a class of non-linear parabolic equations*, Stochastic Differential Equations (Lecture Series in Differential Equations, Session 7, Catholic Univ., 1967), Lecture Series in Differential Equations, Session 7, Air Force Office of Scientific Research, Office of Aerospace Research, United States Air Force, Arlington, VA, 1967, pp. 41–57. MR233437

[Meh04] M. L. Mehta, *Random matrices*, 3rd ed., Pure and Applied Mathematics (Amsterdam), vol. 142, Elsevier/Academic Press, Amsterdam, 2004. MR2129906

[MO69] S. A. Molvcanov and E. Ostrovskiĭ, *Symmetric stable processes as traces of degenerate diffusion processes* (Russian, with English summary), Teor. Verojatnost. i Primenen. **14** (1969), 127–130. MR247668

[Mon88] H. L. Montgomery, *Minimal theta functions*, Glasgow Math. J. **30** (1988), no. 1, 75–85, DOI 10.1017/S0017089500007047. MR925561

[MP84] C. Marchioro and M. Pulvirenti, *Vortex methods in two-dimensional fluid dynamics*, Lecture Notes in Physics, vol. 203, Springer-Verlag, Berlin, 1984. MR750980

[MR23] F. Marceca and J. L. Romero, *Improved discrepancy for the planar Coulomb gas at low temperatures*, J. Anal. Math. **157** (2025), no. 1, 113–153, DOI 10.1007/s11854-025-0388-x. MR4999811

[MY80] Ph. A. Martin and T. Yalcin, *The charge fluctuations in classical Coulomb systems*, J. Statist. Phys. **22** (1980), no. 4, 435–463, DOI 10.1007/BF01012866. MR574007

[NRS22] Q.-H. Nguyen, M. Rosenzweig, and S. Serfaty, *Mean-field limits of Riesz-type singular flows*, Ars Inven. Anal. (2022), Paper No. 4, 45. MR4462479

[NSV08] F. Nazarov, M. Sodin, and A. Volberg, *The Jancovici-Lebowitz-Manificat law for large fluctuations of random complex zeroes*, Comm. Math. Phys. **284** (2008), no. 3, 833–865, DOI 10.1007/s00220-008-0646-7. MR2452596

[NT18] F. Nakano and K. D. Trinh, *Gaussian beta ensembles at high temperature: eigenvalue fluctuations and bulk statistics*, J. Stat. Phys. **173** (2018), no. 2, 295–321, DOI 10.1007/s10955-018-2131-9. MR3860215

[NT20] F. Nakano and K. D. Trinh, *Poisson statistics for beta ensembles on the real line at high temperature*, J. Stat. Phys. **179** (2020), no. 2, 632–649, DOI 10.1007/s10955-020-02542-y. MR4091568

[NV21] J. Najnudel and B. Virág, *Uniform point variance bounds in classical beta ensembles*, Random Matrices Theory Appl. **10** (2021), no. 4, Paper No. 2150033, 52, DOI 10.1142/S2010326321500337. MR4379538

[NY24] A. Nishry and O. Yakir, *Large charge fluctuations in the hierarchical Coulomb gas*, Probab. Math. Phys. **6** (2025), no. 4, 1245–1290, DOI 10.2140/pmp.2025.6.1245. MR4975380

[Oll88] S. Olla, *Large deviations for Gibbs random fields*, Probab. Theory Related Fields **77** (1988), no. 3, 343–357, DOI 10.1007/BF00319293. MR931502

[OPS88] B. Osgood, R. Phillips, and P. Sarnak, *Extremals of determinants of Laplacians*, J. Funct. Anal. **80** (1988), no. 1, 148–211, DOI 10.1016/0022-1236(88)90070-5. MR960228

[OS08] H. Osada and T. Shirai, *Variance of the linear statistics of the Ginibre random point field*, Proceedings of RIMS Workshop on Stochastic Analysis and Applications, RIMS

Kôkyûroku Bessatsu, B6, Res. Inst. Math. Sci. (RIMS), Kyoto, 2008, pp. 193–200. MR2407564

[Pei24] L. Peilen, *Local laws and a mesoscopic CLT for β-ensembles*, Comm. Pure Appl. Math. **77** (2024), no. 4, 2452–2567, DOI 10.1002/cpa.22175. MR4705298

[Pei25] L. Peilen, *On the maximum of the potential of a general two-dimensional Coulomb gas*, Electron. Commun. Probab. **30** (2025), Paper No. 41, 17, DOI 10.1214/25-ecp692. MR4904092

[PG] D. Padilla-Garza, *Large deviations principle for the tagged empirical field of a general interacting gas*, J. Math. Phys. **67** (2026), no. 1, Paper No. 013301, 30, DOI 10.1063/5.0283483. MR5012337

[PG23] D. Padilla-Garza, *Concentration inequality around the thermal equilibrium measure of Coulomb gases*, J. Funct. Anal. **284** (2023), no. 1, Paper No. 109733, 44, DOI 10.1016/j.jfa.2022.109733. MR4494564

[PG24] D. Padilla-Garza, *Large deviations principles for Coulomb gases at intermediate temperature regime*, Electron. J. Probab. **29** (2024), Paper No. 186, 42, DOI 10.1214/24-ejp1255. MR4840490

[PR93] L. Peres and J. Rubinstein, *Vortex dynamics in* U(1) *Ginzburg-Landau models*, Phys. D **64** (1993), no. 1-3, 299–309, DOI 10.1016/0167-2789(93)90261-X. MR1214555

[PRN18] M. Petrache and S. Rota Nodari, *Equidistribution of jellium energy for Coulomb and Riesz interactions*, Constr. Approx. **47** (2018), no. 1, 163–210, DOI 10.1007/s00365-017-9395-1. MR3742814

[PS] L. Peilen and S. Serfaty, *Local laws and fluctuations for super-coulombic Riesz gases.*

[PS72] O. Penrose and E. R. Smith, *Thermodynamic limit for classical systems with Coulomb interactions in a constant external field*, Comm. Math. Phys. **26** (1972), 53–77. MR303866

[PS17] M. Petrache and S. Serfaty, *Next order asymptotics and renormalized energy for Riesz interactions*, J. Inst. Math. Jussieu **16** (2017), no. 3, 501–569, DOI 10.1017/S1474748015000201. MR3646281

[PS20] M. Petrache and S. Serfaty, *Crystallization for Coulomb and Riesz interactions as a consequence of the Cohn-Kumar conjecture*, Proc. Amer. Math. Soc. **148** (2020), no. 7, 3047–3057, DOI 10.1090/proc/15003. MR4099791

[PZ20] L. Pronzato and A. Zhigljavsky, *Bayesian quadrature, energy minimization, and space-filling design*, SIAM/ASA J. Uncertain. Quantif. **8** (2020), no. 3, 959–1011, DOI 10.1137/18M1210332. MR4133484

[Rad81] C. Radin, *The ground state for soft disks*, J. Statist. Phys. **26** (1981), no. 2, 365–373, DOI 10.1007/BF01013177. MR643714

[Ran53] R. A. Rankin, *A minimum problem for the Epstein zeta-function*, Proc. Glasgow Math. Assoc. **1** (1953), 149–158. MR59300

[RAS09] F. Rassoul-Agha and T. Seppäläinen, *A course on large deviations with an introduction to Gibbs measures*, Graduate Studies in Mathematics, vol. 162, American Mathematical Society, Providence, RI, 2015, DOI 10.1090/gsm/162. MR3309619

[RBM25] M. Rouault, R. Bardenet, and M. Maïda, *Monte Carlo with kernel-based Gibbs measures: Guarantees for probabilistic herding*, 2025.

[RNS15] S. R. Nodari and S. Serfaty, *Renormalized energy equidistribution and local charge balance in 2D Coulomb systems*, Int. Math. Res. Not. IMRN **11** (2015), 3035–3093, DOI 10.1093/imrn/rnu031. MR3373044

[RO] X. Ros-Otón, *Some remarks about the fractional obstacle problem, appendix of [PS].*

[Ros20] M. Rosenzweig, *The Mean-Field Limit of Stochastic Point vortex systems with multiplicative noise*, `arXiv:2011.12180`, 2020.

[Ros22] M. Rosenzweig, *Mean-field convergence of point vortices to the incompressible Euler equation with vorticity in L^∞*, Arch. Ration. Mech. Anal. **243** (2022), no. 3, 1361–1431, DOI 10.1007/s00205-021-01735-3. MR4381143

[Ros23] M. Rosenzweig, *On the rigorous derivation of the incompressible Euler equation from Newton's second law*, Lett. Math. Phys. **113** (2023), no. 1, Paper No. 13, 32, DOI 10.1007/s11005-023-01630-w. MR4540782

[Rou22a] N. Rougerie, *On the stability of Laughlin's fractional quantum Hall phase*, In Encyclopedia of condensed matter physics, 2nd edition. Elsevier, 2022.

[Rou22b] N. Rougerie, *The classical jellium and the Laughlin phase*, The physics and mathematics of Elliott Lieb—the 90th anniversary. Vol. II, EMS Press, Berlin, 2022, pp. 179–217. MR4531361

[RS15] N. Rougerie and S. Serfaty, *Higher-dimensional Coulomb gases and renormalized energy functionals*, Comm. Pure Appl. Math. **69** (2016), no. 3, 519–605, DOI 10.1002/cpa.21570. MR3455593

[RS23] M. Rosenzweig and S. Serfaty, *Global-in-time mean-field convergence for singular Riesz-type diffusive flows*, Ann. Appl. Probab. **33** (2023), no. 2, 754–798, DOI 10.1214/22-aap1833. MR4564418

[RS24a] M. Rosenzweig and S. Serfaty, *Relative entropy and modulated free energy without confinement via self-similar transformation*, arXiv:2402.13977, 2024.

[RS24b] M. Rosenzweig and S. Serfaty, *Sharp commutator estimates of all order for Coulomb and Riesz modulated energies*, Comm. Pure Appl. Math. **79** (2026), no. 2, 207–292, DOI 10.1002/cpa.70010. MR4974720

[RS25a] M. Rosenzweig and S. Serfaty, *The lake equation as a supercritical mean-field limit* (English, with English and French summaries), J. Éc. polytech. Math. **12** (2025), 1019–1068, DOI 10.5802/jep.305. MR4926319

[RS25b] M. Rosenzweig and S. Serfaty, *Modulated logarithmic Sobolev inequalities and generation of chaos* (English, with English and French summaries), Ann. Fac. Sci. Toulouse Math. (6) **34** (2025), no. 1, 107–134. MR4927284

[RSY14] N. Rougerie, S. Serfaty, and J. Yngvason, *Quantum Hall phases and plasma analogy in rotating trapped Bose gases*, J. Stat. Phys. **154** (2014), no. 1-2, 2–50, DOI 10.1007/s10955-013-0766-0. MR3162531

[RSZ94] E. A. Rakhmanov, E. B. Saff, and Y. M. Zhou, *Minimal discrete energy on the sphere*, Math. Res. Lett. **1** (1994), no. 6, 647–662, DOI 10.4310/MRL.1994.v1.n6.a3. MR1306011

[Rue99] D. Ruelle, *Statistical mechanics: Rigorous results*, World Scientific Publishing Co., Inc., River Edge, NJ; Imperial College Press, London, 1999. Reprint of the 1989 edition, DOI 10.1142/4090. MR1747792

[RV07] B. Rider and B. Virág, *The noise in the circular law and the Gaussian free field*, Int. Math. Res. Not. IMRN **2** (2007), Art. ID rnm006, 33, DOI 10.1093/imrn/rnm006. MR2361453

[RY15] N. Rougerie and J. Yngvason, *Incompressibility estimates for the Laughlin phase*, Comm. Math. Phys. **336** (2015), no. 3, 1109–1140, DOI 10.1007/s00220-014-2232-5. MR3324139

[Sak91] M. Sakai, *Regularity of a boundary having a Schwarz function*, Acta Math. **166** (1991), no. 3-4, 263–297, DOI 10.1007/BF02398888. MR1097025

[SBO07] D. Smets, F. Bethuel, and G. Orlandi, *Quantization and motion law for Ginzburg-Landau vortices*, Arch. Ration. Mech. Anal. **183** (2007), no. 2, 315–370, DOI 10.1007/s00205-006-0018-4. MR2278409

[Ser07] S. Serfaty, *Vortex collisions and energy-dissipation rates in the Ginzburg-Landau heat flow. II. The dynamics*, J. Eur. Math. Soc. (JEMS) **9** (2007), no. 3, 383–426, DOI 10.4171/JEMS/84. MR2314103

[Ser15] S. Serfaty, *Coulomb gases and Ginzburg-Landau vortices*, Zurich Lectures in Advanced Mathematics, European Mathematical Society (EMS), Zürich, 2015, DOI 10.4171/152. MR3309890

[Ser17] S. Serfaty, *Mean field limits of the Gross-Pitaevskii and parabolic Ginzburg-Landau equations*, J. Amer. Math. Soc. **30** (2017), no. 3, 713–768, DOI 10.1090/jams/872. MR3630086

[Ser20] S. Serfaty, *Mean field limit for Coulomb-type flows*, Duke Math. J. **169** (2020), no. 15, 2887–2935, DOI 10.1215/00127094-2020-0019. With an appendix by Mitia Duerinckx and Serfaty. MR4158670

[Ser23] S. Serfaty, *Gaussian fluctuations and free energy expansion for Coulomb gases at any temperature* (English, with English and French summaries), Ann. Inst. Henri Poincaré Probab. Stat. **59** (2023), no. 2, 1074–1142, DOI 10.1214/22-aihp1285. MR4575026

[Sha11] Sh. Shakirov, *Exact solution for mean energy of 2d Dyson gas at $\beta = 1$*, Phys. Lett. A **375** (2011), no. 6, 984–989, DOI 10.1016/j.physleta.2011.01.004. MR2756341

[Shc13] M. Shcherbina, *Fluctuations of linear eigenvalue statistics of β matrix models in the multi-cut regime*, J. Stat. Phys. **151** (2013), no. 6, 1004–1034, DOI 10.1007/s10955-013-0740-x. MR3063494

[Shi06] T. Shirai, *Large deviations for the fermion point process associated with the exponential kernel*, J. Stat. Phys. **123** (2006), no. 3, 615–629, DOI 10.1007/s10955-006-9026-x. MR2252160

[Sil07] L. Silvestre, *Regularity of the obstacle problem for a fractional power of the Laplace operator*, Comm. Pure Appl. Math. **60** (2007), no. 1, 67–112, DOI 10.1002/cpa.20153. MR2270163

[Sim08] B. Simon, *The Christoffel-Darboux kernel*, Perspectives in partial differential equations, harmonic analysis and applications, Proc. Sympos. Pure Math., vol. 79, Amer. Math. Soc., Providence, RI, 2008, pp. 295–335, DOI 10.1090/pspum/079/2500498. MR2500498

[SK97] E. B. Saff and A. B. J. Kuijlaars, *Distributing many points on a sphere*, Math. Intelligencer **19** (1997), no. 1, 5–11, DOI 10.1007/BF03024331. MR1439152

[SM76] R. R. Sari and D. Merlini, *On the ν-dimensional one-component classical plasma: the thermodynamic limit problem revisited*, J. Statist. Phys. **14** (1976), no. 2, 91–100, DOI 10.1007/BF01011761. MR449401

[Spe97] T. Spencer, *Scaling, the free field and statistical mechanics*, The Legacy of Norbert Wiener: A Centennial Symposium (Cambridge, MA, 1994), Proc. Sympos. Pure Math., vol. 60, Amer. Math. Soc., Providence, RI, 1997, pp. 373–389, DOI 10.1090/pspum/060/1460292. MR1460292

[SS04] E. Sandier and S. Serfaty, *Gamma-convergence of gradient flows with applications to Ginzburg-Landau*, Comm. Pure Appl. Math. **57** (2004), no. 12, 1627–1672, DOI 10.1002/cpa.20046. MR2082242

[SS06] P. Sarnak and A. Strömbergsson, *Minima of Epstein's zeta function and heights of flat tori*, Invent. Math. **165** (2006), no. 1, 115–151, DOI 10.1007/s00222-005-0488-2. MR2221138

[SS07] E. Sandier and S. Serfaty, *Vortices in the magnetic Ginzburg-Landau model*, Progress in Nonlinear Differential Equations and their Applications, vol. 70, Birkhäuser Boston, Inc., Boston, MA, 2007. MR2279839

[SS12] E. Sandier and S. Serfaty, *From the Ginzburg-Landau model to vortex lattice problems*, Comm. Math. Phys. **313** (2012), no. 3, 635–743, DOI 10.1007/s00220-012-1508-x. MR2945619

[SS15a] E. Sandier and S. Serfaty, *1D log gases and the renormalized energy: crystallization at vanishing temperature*, Probab. Theory Related Fields **162** (2015), no. 3-4, 795–846, DOI 10.1007/s00440-014-0585-5. MR3383343

[SS15b] E. Sandier and S. Serfaty, *2D Coulomb gases and the renormalized energy*, Ann. Probab. **43** (2015), no. 4, 2026–2083, DOI 10.1214/14-AOP927. MR3353821

[SS18] S. Serfaty and J. Serra, *Quantitative stability of the free boundary in the obstacle problem*, Anal. PDE **11** (2018), no. 7, 1803–1839, DOI 10.2140/apde.2018.11.1803. MR3810473

[SSS19] A. Seeger, C. K. Smart, and B. Street, *Multilinear singular integral forms of Christ-Journé type*, Mem. Amer. Math. Soc. **257** (2019), no. 1231, v+134, DOI 10.1090/memo/1231. MR3897981

[ST97] E. B. Saff and V. Totik, *Logarithmic potentials with external fields*, Grundlehren der mathematischen Wissenschaften [Fundamental Principles of Mathematical Sciences], vol. 316, Springer-Verlag, Berlin, 1997. Appendix B by Thomas Bloom, DOI 10.1007/978-3-662-03329-6. MR1485778

[STG99] H. Stormer, D. Tsui, and A. Gossard. *The fractional quantum hall effect*, Rev. Modern Phys. **71** (1999), no. 2, S298.

[Szn91] A.-S. Sznitman, *Topics in propagation of chaos*, École d'Été de Probabilités de Saint-Flour XIX—1989, Lecture Notes in Math., vol. 1464, Springer, Berlin, 1991, pp. 165–251, DOI 10.1007/BFb0085169. MR1108185

[TF99] G. Téllez and P. J. Forrester, *Exact finite-size study of the 2D OCP at $\Gamma = 4$ and $\Gamma = 6$*, J. Statist. Phys. **97** (1999), no. 3-4, 489–521, DOI 10.1023/A:1004654923170. MR1736107

[The06] F. Theil, *A proof of crystallization in two dimensions*, Comm. Math. Phys. **262** (2006), no. 1, 209–236, DOI 10.1007/s00220-005-1458-7. MR2200888

[Tho23] E. Thoma, *Non-rigidity Properties of the Coulomb Gas*, arXiv:2303.11486, 2023.

[Tho24] E. Thoma, *Overcrowding and separation estimates for the Coulomb gas*, Comm. Pure Appl. Math. **77** (2024), no. 7, 3227–3276, DOI 10.1002/cpa.22188. MR4764743

[Tor18] S. Torquato, *Hyperuniform states of matter*, Phys. Rep. **745** (2018), 1–95, DOI 10.1016/j.physrep.2018.03.001. MR3815253

[TS03] S. Torquato and F. H. Stillinger, *Local density fluctuations, hyperuniformity, and order metrics*, Phys. Rev. E (3) **68** (2003), no. 4, 041113, 25, DOI 10.1103/PhysRevE.68.041113. MR2060843

[TV11] T. Tao and V. Vu, *Random matrices: universality of local eigenvalue statistics*, Acta Math. **206** (2011), no. 1, 127–204, DOI 10.1007/s11511-011-0061-3. MR2784665

[Var88] S. R. S. Varadhan, *Large deviations and applications*, École d'Été de Probabilités de Saint-Flour XV–XVII, 1985–87, Lecture Notes in Math., vol. 1362, Springer, Berlin, 1988, pp. 1–49, DOI 10.1007/BFb0086178. MR983371

[Ven13] M. Venker, *Particle systems with repulsion exponent β and random matrices*, Electron. Commun. Probab. **18** (2013), no. 83, 12, DOI 10.1214/ECP.v18-2864. MR3141792

[Via17] M. S. Viazovska, *The sphere packing problem in dimension 8*, Ann. of Math. (2) **185** (2017), no. 3, 991–1015, DOI 10.4007/annals.2017.185.3.7. MR3664816

[VV09] B. Valkó and B. Virág, *Continuum limits of random matrices and the Brownian carousel*, Invent. Math. **177** (2009), no. 3, 463–508, DOI 10.1007/s00222-009-0180-z. MR2534097

[Wan24] Z. Wang, Personal communication, 2024.

[Wig55] E. P. Wigner, *Characteristic vectors of bordered matrices with infinite dimensions*, Ann. of Math. (2) **62** (1955), 548–564, DOI 10.2307/1970079. MR77805

[ZW06] A. Zabrodin and P. Wiegmann, *Large-N expansion for the 2D Dyson gas*, J. Phys. A **39** (2006), no. 28, 8933–8963, DOI 10.1088/0305-4470/39/28/S10. MR2240466

Table of Notation

- δ_{ij} denotes the Kronecker symbol, 1 if $i = j$ and 0 otherwise
- For any real number x, x_+ denotes $\max(x, 0)$ and x_- denotes $\min(x, 0)$
- For any two real numbers a and b, $a \wedge b$ denotes the minimum of a and b
- $B(x, R)$ denotes the open ball centered at x and of radius R, B_R denotes a ball of radius R, not necessarily centered at 0
- $\square_R$ denotes a closed cube of sidelength R, not necessarily centered at 0, $\square_{tR}$ arising in conjunction then denotes the cube of same center and lengthscale tR
- Given a set $A \subset \mathbb{R}^\mathsf{d}$, $|A|$ denotes the Lebesgue measure of A. Given a hypersurface Γ, $|\Gamma|$ denotes its measure
- For a set A, $\bar{A}$ denotes its closure and $\overset{\circ}{A}$ its interior
- For any set or event A, $\mathbf{1}_A$ denotes the indicator function of that set or event
- $a \ll b$ means that a/b tends to 0, in the asymptotic regime considered
- For a set S, $\#S$ denotes the cardinality of the set. For a measure μ and a map Φ, $\Phi\#\mu$ denotes the push-forward of μ by Φ (see Definition 9.1)
- $\triangle$ denotes the diagonal of $\mathbb{R}^\mathsf{d} \times \mathbb{R}^\mathsf{d}$
- supp denotes the support of a function or of a measure
- $\mu|_Q$ denotes the restriction of a measure density μ to a set Q
- $\fint_A f$ denotes the average of f over the set A
- $\mathcal{P}(X)$ is the space of probability measures on X
- $\mathbb{E}_P$ denotes the expectation with respect to the probability P, Var denotes the variance and Cov the covariance
- $L^p(\mathbb{R}^\mathsf{d})$, $1 \le p \le \infty$ denote the standard Lebesgue spaces, L^p_w the weighted Lebesgue spaces with weight w
- For $0 < \alpha \le 1$, $|\varphi|_{C^\alpha}$ denotes the Hölder seminorm
$$|\varphi|_{C^\alpha} = \sup_{x,y} \frac{|\varphi(x) - \varphi(y)|}{|x - y|^\alpha}$$
and for $0 < \alpha \le 1$ and k integer, $|\varphi|_{C^{k+\alpha}}$ denotes the C^α seminorm of $\nabla^{\otimes k}\varphi$.
- C^0_c denotes the space of compactly supported continuous functions, C^1_c denotes the space of compactly supported functions in C^1
- $C^{k,\alpha}$ denotes the standard Hölder space of functions with k-th derivative in C^α
- $\dot{H}^s(\mathbb{R}^\mathsf{d})$ denotes the homogeneous fractional Sobolev space, defined via its seminorm in (2.2.9), $H^s(\mathbb{R}^\mathsf{d})$ is the corresponding inhomogeneous Sobolev space.
- H^1 is the Sobolev space with norm $\|f\|_{H^1} = \|f\|_{L^2} + \|\nabla f\|_{L^2}$.

- Given a set $\Omega \subset \mathbb{R}^{\mathsf{d}}$, $H^1_0(\Omega)$ denotes the completion of $C^1_c(\Omega)$ for the H^1 Sobolev norm.
- the outer unit normal to a set is generally denoted ν
- $\mathcal{H}_N$ is the main energy, defined in (1.1.1)
- g is the interaction kernel, defined in (1.1.2)
- s is the power in the interaction kernel (1.1.2)
- V is the external (confinement) potential in (1.1.1)
- $\mathbb{P}_{N,\beta}$ is the Gibbs measure, defined in (1.1.5)
- $Z_{N,\beta}$ is the partition function, defined in (1.1.6).
- $\mathcal{E}(\mu)$ is the mean-field energy, defined in (2.0.3)
- $\theta = \beta N^{1-\frac{\mathsf{s}}{\mathsf{d}}}$ is the effective temperature at the macroscopic scale
- $\mathcal{E}_\theta$ is the thermal mean-field energy, defined in (2.0.4)
- μ_V is the equilibrium measure, see Theorem 2.8
- Σ is the support of μ_V or of a generic measure μ, $\Sigma' = N^{1/\mathsf{d}}\Sigma$ its blow-up
- ζ is the effective confinement potential, defined in (2.1.12)
- μ_θ is the thermal equilibrium measure, see Proposition 2.18
- $\alpha = \frac{\mathsf{d}-\mathsf{s}}{2}$
- $\delta_{\mathbb{R}^{\mathsf{d}}}$ is defined in (2.2.2)
- k is the dimension-extension, equal to 0 in the Coulomb case, and 1 in the (non-Coulomb) Riesz cases $\mathsf{d}-2<\mathsf{s}<\mathsf{d}$
- $\gamma = \mathsf{s}+2-\mathsf{d}-\mathsf{k}$, see (2.2.5)
- h^μ is the potential generated by μ, defined by $h^\mu = \mathsf{g} * \mu$
- For a point configuration X_N in $(\mathbb{R}^{\mathsf{d}})^N$, $\widehat{\mu_N}$ denotes the empirical measure $\frac{1}{N}\sum_{i=1}^N \delta_{x_i}$, see (3.1.1)
- For a point configuration X_N in $(\mathbb{R}^{\mathsf{d}})^N$, I_Ω denotes the set of indices i such that $x_i \in \Omega$
- $\mathsf{F}_N(X_N,\mu)$ is the modulated or next-order electric energy, defined in (4.1.2)
- $\mathsf{F}(X_N,\mu)$ is the blown-up version of the next-order electric energy, defined in (5.2.7)
- $\mathsf{F}_N^\Omega(X_N,\mu)$ is the localized version of F_N, defined in (4.5.2)
- $\mathsf{F}(X_N,\mu,U)$ is the Neumann energy in blown-up scale, relative to a domain U, defined in (7.1.22). $\mathsf{F}_N(X_N,\mu,U)$ is its blown-down version, see (9.2.1)
- $\mathsf{F}^\Omega(X_N,\mu,U)$ is the version of $\mathsf{F}(X_N,\mu,U)$ localized in Ω, as defined in (7.1.23)
- For a point configuration X_N in $(\mathbb{R}^{\mathsf{d}})^N$ and a reference measure μ, $h_N^\mu[X_N]$, often abbreviated h_N, denotes $\mathsf{g} * (\sum_{i=1}^N \delta_{x_i} - N\mu)$, or in blown-up scale $\mathsf{g} * (\sum_{i=1}^N \delta_{x_i} - \mu)$
- $h_{\vec\eta}$ denotes the truncation of a potential h with radii η_i, as in (4.1.9), in the same way $h_{N,\vec\eta}$ denotes the truncation of h_N and $E_{\vec\eta}$ denotes the truncation of a vector field, as in (7.1.26).
- $\delta_x^{(\eta)}$ denotes the η-smearing of the Dirac mass at x, see (4.1.10) and (4.1.21)
- g_η is the truncation of g at scale η, equal to $\min(\mathsf{g},\mathsf{g}(\eta))$ and defined in (4.1.18)
- $\mathsf{f}_\eta = \mathsf{g} - \mathsf{g}_\eta$
- λ is the microscopic lengthscale $(N\|\mu\|_{L^\infty})^{-1/\mathsf{d}}$, defined in (4.2.11)

- r_i denote the nearest neighbor distance (4.2.14) or (5.2.18) in blown-up scale
- $\tilde{\mathsf{r}}_i$ is the variant defined in (4.5.1) or in (7.1.25) depending on the context
- $\hat{\mathsf{r}}_i$ is the variant defined in (7.1.21) or (7.1.24) depending on the context
- r_p and $\tilde{\mathsf{r}}_p$ are the variants for infinite point configurations, defined in (11.2.3) and (11.2.4)
- $D(\Omega)$ denotes the discrepancy in the set Ω, defined in (4.4.1)
- $\mathsf{K}_{N,\beta}(\mu)$ is the reduced partition function, defined in (5.1.12)
- $\mathsf{K}_{N,\beta}(\mu, U)$ is the Neumann reduced partition function, defined in (9.2.3), and $\mathsf{K}_{N,\beta}(\mu) = \mathsf{K}_{N,\beta}(\mu, \mathbb{R}^{\mathsf{d}})$
- $\mathsf{L}_{N,\beta}(\mu, U)$ is the Dirichlet partition function, defined in (7.1.15)
- $\mathsf{K}_\beta(\mu)$ is the blown-up version of $\mathsf{K}_{N,\beta}(\mu)$, defined in (5.2.10)
- $\mathbb{Q}_{N,\beta}(\mu)$ is the modulated Gibbs measure, defined in (5.1.13)
- $\mathbb{Q}_\beta(\mu)$ is its blown-up version, defined in (5.2.11)
- $\mathbb{Q}_{N,\beta}(\mu, U)$ is the Neumann modulated Gibbs measure at usual scale, defined in (9.2.2)
- $\mathbb{Q}_\beta(\mu, U)$ is the Neumann modulated Gibbs measure in blown-up scale, defined in (7.1.14)
- $\tilde{\mathsf{K}}_{N,\beta}(\mu, \zeta)$ is defined in (5.1.5)
- $\tilde{\mathbb{Q}}_{N,\beta}(\mu, \zeta)$ is defined in (5.1.6)
- H_N is the normalized relative entropy defined in (5.1.16)
- $\chi(\beta)$ is defined in (5.2.27)
- ρ_β denotes the minimal lengthscale defined in (8.0.2)
- $\mathsf{A}_n(X_N, \mu, v)$ is defined in (6.1.3), $\mathsf{A}_n(X_{\bar{n}}, \mu, v, U)$ is the Neumann version introduced in (9.2.5)
- For $\vec{\alpha}$ an N-tuple of nonnegative numbers, $\mathcal{F}^{\vec{\alpha}}$ is defined in (4.5.7)
- $[\cdot, \cdot]$ denotes the stress-tensor quadratic form, defined in (6.1.5)
- $\mathbb{M}$ is a symmetric nonnegative matrix, appearing in the dynamical law (6.2.2)
- f_N^t is the joint law of the point locations in the dynamics
- $\mathcal{F}_N^\theta$ is the modulated free energy, defined in (6.2.16)
- $\mathcal{I}(f|g)$ is the Fisher information, defined in (6.2.22)
- μ' denotes the rescaled density given by $\mu'(x) = \mu(xN^{-1/\mathsf{d}})$
- X'_N or x'_i denote the rescaled configuration/points, given by $x'_i = N^{1/\mathsf{d}} x_i$
- G_U is defined in (7.1.43)
- H_U is defined in (7.1.44)
- $\mathsf{E}_\infty(\mu, U)$ is the minimum of the Neumann energy, defined in (7.1.57)
- $S(X_n, w)$ and $S'(X_n, w)$ are defined in (7.2.5)
- $\bar{S}(X_n)$ and $\bar{S}'(X_n)$ are defined in (7.2.11)–(7.2.12)
- $\mathsf{G}_U^{\text{inn/out}}$ are defined in (7.2.9), resp. (7.2.10)
- Λ denotes a set where the reference measure μ is bounded below by some $m > 0$
- $\mathcal{R}$ is the rate in (9.4.2)
- $f_{\mathsf{d}}(\infty)$ and $\mathsf{f}_{\mathsf{d}}(\beta)$ denote the pressure at inverse temperature β, defined in (8.2.1) and Theorem 8.14
- d_0 is a distance defined in (8.3.2), $\bar{d}_0$ in (9.3.1)
- $\mathrm{Fluct}_\mu(\varphi)$ is the fluctuations of linear statistics, defined in (9.0.1)

- $\mathcal{X}(A)$ is the set of locally finite point configurations on the set A, see Section 11.1
- $\mathcal{C} \in \mathcal{X}(A)$ is a locally finite point configuration
- $\mathrm{Elec}(\mathcal{C}, m)$ is the set of electric fields compatible with the configuration $\mathcal{C}$ and the density m, see Definition 11.1. The sets Elec_m and Elec are also defined in Definition 11.1
- $\mathrm{ent}(\cdot|\cdot)$ denotes the usual relative entropy defined in (3.2.4) and $\mathsf{ent}(\cdot|\cdot)$ the specific relative entropy, defined in (13.1.1)
- Π^m denotes the Poisson point process of intensity m, see Section 13.1
- $\mathcal{W}$ is the jellium renormalized energy for electric fields, defined in (11.2.13)
- $\mathbb{W}$ is the jellium renormalized energy for point configurations, defined in (11.2.15)
- $\overline{\mathbb{W}}$ is the averaged jellium renormalized energy for a tagged empirical point process, defined in (12.2.1)
- $\mathcal{F}^{\square_R}(E, m)$ is defined in (11.2.7)
- $\mathcal{P}_s(\mathcal{X})$ denotes the set of stationary (translation-invariant) point processes
- $\mathcal{P}_s(\Sigma \times \mathcal{X})$ denotes the set of stationary tagged point processes, and $\mathcal{P}_{s,\mu}(\Sigma \times \mathcal{X})$ denotes the set of such point processes such that for almost every $x \in \Sigma$, the disintegration measure $\bar{P}^x$ has intensity $\mu(x)$, see Section 12.1
- $\bar{P}_N[X_N]$ is the tagged empirical field associated to the configuration X_N, defined in (12.1.1)
- $P_N^{x_0,R}[X_N]$ is the local empirical field averaged in $\square_R(N^{1/\mathsf{d}}x_0)$, defined in (12.1.2)
- $\bar{P}^{\mathrm{e}}$ or $P_N^{\mathrm{e}}[X_N]$ denote tagged electric field processes, defined in Section 12.2.1 and (12.2.6)
- $\mathcal{I}_\beta^m(P)$ denotes the rate function of (13.2.1)

Index

SELECTED PUBLISHED TITLES IN THIS SERIES

70 **Sylvia Serfaty,** Lectures on Coulomb and Riesz Gases, 2026
69 **Jean-Luc Chabert,** Integer-Valued Polynomials, 2025
68 **Gerald B. Folland,** The Heisenberg Group, 2025
67 **Fritz Gesztesy, Roger Nichols, and Maxim Zinchenko,** Sturm–Liouville Operators, Their Spectral Theory, and Some Applications, 2024
66 **Eli Aljadeff, Antonio Giambruno, Claudio Procesi, and Amitai Regev,** Rings with Polynomial Identities and Finite Dimensional Representations of Algebras, 2020
65 **László Lovász,** Graphs and Geometry, 2019
64 **Kathrin Bringmann, Amanda Folsom, Ken Ono, and Larry Rolen,** Harmonic Maass Forms and Mock Modular Forms: Theory and Applications, 2017
63 **Cornelia Druţu and Michael Kapovich,** Geometric Group Theory, 2018
62 **Martin Olsson,** Algebraic Spaces and Stacks, 2016
61 **James Arthur,** The Endoscopic Classification of Representations, 2013
60 **László Lovász,** Large Networks and Graph Limits, 2012
59 **Kai Cieliebak and Yakov Eliashberg,** From Stein to Weinstein and Back, 2012
58 **Freydoon Shahidi,** Eisenstein Series and Automorphic L-Functions, 2010
57 **John Friedlander and Henryk Iwaniec,** Opera de Cribro, 2010
56 **Richard Elman, Nikita Karpenko, and Alexander Merkurjev,** The Algebraic and Geometric Theory of Quadratic Forms, 2008
55 **Alain Connes and Matilde Marcolli,** Noncommutative Geometry, Quantum Fields and Motives, 2008
54 **Barry Simon,** Orthogonal Polynomials on the Unit Circle: Part 1: Classical Theory; Part 2: Spectral Theory, 2005
53 **Henryk Iwaniec and Emmanuel Kowalski,** Analytic Number Theory, 2004
52 **Dusa McDuff and Dietmar Salamon,** J-holomorphic Curves and Symplectic Topology, Second Edition, 2012
51 **Alexander Beilinson and Vladimir Drinfeld,** Chiral Algebras, 2004
50 **E. B. Dynkin,** Diffusions, Superdiffusions and Partial Differential Equations, 2002
49 **Vladimir V. Chepyzhov and Mark I. Vishik,** Attractors for Equations of Mathematical Physics, 2002
48 **Yoav Benyamini and Joram Lindenstrauss,** Geometric Nonlinear Functional Analysis, 2000
47 **Yuri I. Manin,** Frobenius Manifolds, Quantum Cohomology, and Moduli Spaces, 1999
46 **J. Bourgain,** Global Solutions of Nonlinear Schrödinger Equations, 1999
45 **Nicholas M. Katz and Peter Sarnak,** Random Matrices, Frobenius Eigenvalues, and Monodromy, 1999
44 **Max-Albert Knus, Alexander Merkurjev, Markus Rost, and Jean-Pierre Tignol,** The Book of Involutions, 1998
43 **Luis A. Caffarelli and Xavier Cabré,** Fully Nonlinear Elliptic Equations, 1995
42 **Victor W. Guillemin and Shlomo Sternberg,** Variations on a Theme by Kepler, 1991
40 **R. H. Bing,** The Geometric Topology of 3-Manifolds, 1983
38 **O. Ore,** Theory of Graphs, 1962
37 **N. Jacobson,** Structure of Rings, 1956
36 **Walter Helbig Gottschalk and Gustav Arnold Hedlund,** Topological Dynamics, 1955
32 **R. L. Wilder,** Topology of Manifolds, 1949
31 **E. Hille and R. S. Phillips,** Functional Analysis and Semi-groups, 1957

For a complete list of titles in this series, visit the
AMS Bookstore at **www.ams.org/bookstore/collseries/**.

Kolibri 360 GmbH
Ferdinand-Jühlke-Straße 7
99095 Erfurt, Deutschland
produktsicherheit@kolibri360.de